DVD Studio Pro 3
In the Studio

DVD Studio Pro 3

In the Studio

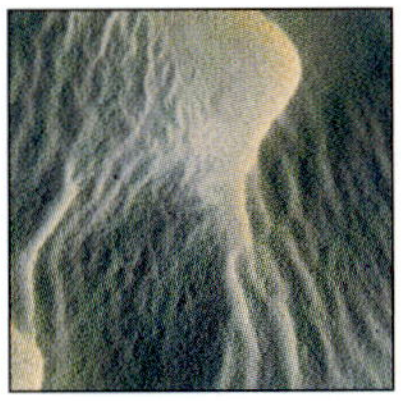
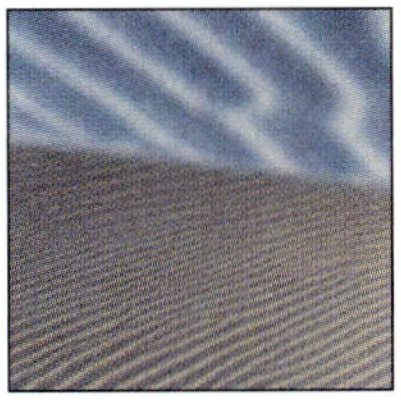
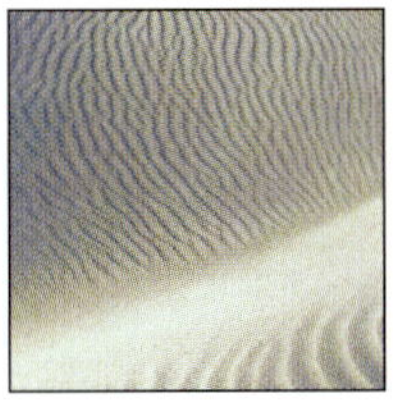

Marc Loy

O'REILLY®

BEIJING • CAMBRIDGE • FARNHAM • KÖLN • PARIS • SEBASTOPOL • TAIPEI • TOKYO

DVD Studio Pro 3: In the Studio

by Marc Loy

Printed in the United States of America.

Published by O'Reilly Media, Inc., 1005 Gravenstein Highway North, Sebastopol, CA 95472.

O'Reilly books may be purchased for educational, business, or sales promotional use. Online editions are also available for most titles (*safari.oreilly.com*). For more information, contact our corporate/institutional sales department: 800-998-9938 or *corporate@oreilly.com*.

Print History:

October 2004: First Edition.

Editor: Robert Eckstein

Production Editor: Genevieve d'Entremont

Cover Designer: Emma Colby

Interior Designer: Melanie Wang

RepKover™ This book uses RepKover™, a durable and flexible lay-flat binding.

0-596-00588-1
[C]

Contents

Preface

The software behind this book, DVD Studio Pro, is an amazing tool for building complex and exciting discs. With its great-looking templates and its powerful scripting language, DVD Studio Pro 3 (often abbreviated DVDSP 3 or just DVDSP) can make quick work of simple projects, yet still enable you to build some of the most complex discs available today.

What if you aren't using Version 3? Well, that's a valid question. At the time this book went to press, DVDSP 2 was less than a year old, and several folks didn't have a pressing need to upgrade. If you use DVDSP 2, you can work with 90% of this book; although you can complete the projects relying on transitions (in Chapters 4 and 11) only if you're using DVDSP 3, you can build and burn every other project with Version 2. If you're still using Version 1.x, however, this is not the book for you. The chasm between Versions 1.5 and 2 is just about as vast as a software chasm can be. With only so much space before us, we decided to concentrate on the future of DVDSP.

This book is part of O'Reilly's *Digital Studio* series, a new set of books that are geared toward turning novices into subject-matter experts, regardless of their area of digital expertise. At O'Reilly, we forever strive to raise the bar with this series—and with our other books as well—providing you a volume that not only offers a tremendous amount of value for your money, but also relieves your "information pain." In other words, we want to address your simple questions, as well as those arcane and annoying problems that you'll face as your DVD authoring capabilities mature. These are the problems that other authors tend to think of as being "outside the scope" of their books.

In this chapter

Organization of This Book

So, what lies in your immediate future? The chapters in this book cover several dozen projects. Each project focuses on a particular aspect of DVD production. Depending on the types of DVDs you produce, you might not need to complete all the projects. That's OK. The chapters collect related projects and stand on their own. Earlier chapters spend a bit more time outlining the precise steps required to complete the projects, but even the most advanced chapters try to delineate the key parts of any project.

This book is divided into 11 chapters, each of which is described briefly here.

Chapter 1, Getting Started
: If you're new to DVD Studio Pro or just recently made the leap from Version 1.x, this chapter is a good place to start. It covers two simple projects: one using the spiffy built-in templates and a second that builds a disc from scratch. If you've used DVDSP 2 or 3 at all, you can safely skip this chapter.

Chapter 2, Media Formats
: This chapter covers several DVD basics. It provides great background information on how DVDs are encoded and what formats are supported. The projects in this chapter illustrate the use of the various formats for video, audio, and graphics. They aren't critical to any future projects, but they provide a good overview for readers new to this technology.

Chapter 3, Great Menus: Layers
: DVDs are all about the menus. We have two chapters devoted to them, in fact, and this is the first one. Here we look at still menus and how to create graphically rich menus that enable easy navigation. We also take a small detour to build some home-movie slideshows.

Chapter 4, Great Menus: Motion
: We continue the menu projects in this chapter. Here we look specifically at menus with video and audio elements. The buttons on these menus are more limited than the buttons available for still menus, but we cover the ways to make the most interesting buttons possible.

Chapter 5, Le DVD Entwurf Internacionale (An International DVD Project)
: DVDs are built with multilingual distribution in mind. As such, the DVD format supports multiple audio and subtitle tracks. In this chapter we look at adding such tracks to a simple project. We also look at a simple subtitle format that you can use outside DVDSP for more efficient project management.

Chapter 6, The Director's Cut
: Another feature that has propelled the popularity of DVDs is the addition of "bonus material" to discs. We look at some tips for including your own bonus material, as well as the details of handling multiple video streams for an individual clip.

Chapter 7, DVD Scripting

Scripting provides you with fine-grained control over DVD navigation. In this chapter we look at the basics of scripting to create projects with conditional viewing of clips.

Chapter 8, Games and Easter Eggs

Probably the single most enjoyable feature of video distribution via DVD is the ability to build hidden bonus material into your discs. In this chapter, we augment the preceding scripting discussion by creating several games and treasure-hunt projects. We also cover the use of passwords.

Chapter 9, DVDs and Computers

More and more computers are shipping with DVD players (and even burners!) built right in. DVDs have such a large capacity that adding computer content to video discs was a natural step. This chapter covers the DVD@ccess feature of DVDSP and includes information on using DVD@ccess with Windows-based computers.

Chapter 10, Templates and Styles

DVDSP 2 introduced the template feature for creating quick, stylish projects. You can save buttons, text, and even complete menus as templates for use on future projects. DVDSP 3 adds a few new templates to the already impressive list. We look at getting the most out of the built-in templates as well as building your own templates.

Chapter 11, Transitions

Many discs include smooth transitions between individual menus. This chapter covers a project for building transitions in Adobe After Effects that you can use in both DVDSP 2 and 3. We also cover building your own transitions for the very extensible transition architecture available in DVDSP 3.

Appendix A, Script Command Reference

This appendix takes an in-depth look at the scripting commands available to you in DVD Studio Pro 3. In addition, it enumerates the types of arguments the commands can use, as well as various conditional arguments that can be used.

Appendix B, DVD Style Guide

This short appendix provides some useful tips on creating effective, professional-looking DVDs.

Appendix C, DVD-9 Projects

Just as we went to press, dual-layer media and burners became available. This appendix looks at the basics of burning a dual-layer project to disc with the help of Toast 6.

The Companion DVD

The DVD included with this book contains both the completed projects and the raw materials to build the projects yourself. You can pop it into a consumer player and check out any project you want. You also can pop it into a computer and grab the project disk image to tackle hands-on work. Just double-click the disk image (*dvdsp3its.dmg*) and then drag the *projects* folder to your hard disk. That should leave you with all the resources you need to build and burn every project in the book.

The templates and transitions we build in Chapters 10 and 11 also are included separately to make it easy to install these elements.

Comments and Questions

Please address comments and questions concerning this book to the publisher:

O'Reilly Media, Inc.
1005 Gravenstein Highway North
Sebastopol, CA 95472
(800) 998-9938 (in the United States or Canada)
(707) 829-0515 (international or local)
(707) 829-0104 (fax)

We have a web page for this book, where we list errata, examples, and any additional information. You can access this page at:

http://www.oreilly.com/catalog/dvdstudio3/

To comment or ask technical questions about this book, send email to:

bookquestions@oreilly.com

For more information about our books, conferences, Resource Centers, and the O'Reilly Network, see our web site at:

http://www.oreilly.com

Conventions Used in This Book

Because DVD Studio Pro provides multiple methods for performing the same action (such as menu items or keyboard shortcuts), this book provides these alternatives when introducing new operations. You can use the main menus, or a right-click context menu, or a fly-out palette menu, or a dedicated button or tool in a window.

Menu commands are given in hierarchical order, with a → preceding each new appearance of a cascading menu, like this: Image → Adjustments → Levels. If a menu appears from a palette or dialog menu, the name of the menu or dialog will precede the name of the command hierarchy.

The following typographical conventions are used in this book:

Italic
: Indicates new terms, URLs, email addresses, filenames, file extensions, pathnames, directories, and Unix utilities.

`Constant width`
: Indicates commands, options, switches, variables, attributes, keys, functions, types, classes, namespaces, methods, modules, properties, parameters, values, objects, events, event handlers, HTML tags, macros, the contents of files, or the output from commands.

`Constant width bold`
: Shows summaries, or indicates text that should be typed literally by the user.

Acknowledgments

I'd like to start off by thanking my editor, Bob Eckstein, for his patience and input throughout the long process of getting this book on the shelves. Several people supplied both time and content to the book and have my sincerest gratitude: Damian Moshak, Amy Hammond, Brooks Graham, and Jason Weaver. The good folks at Studio Anti-Thesis, co-dependent productions, and ThemeWorks Music also contributed material that added to the overall professionalism of the book. I would like to thank Paul Wesselmann at The Ripples Project and Aaron Sherer at the Paine Art Center and Museum for their time and generous permission to use their video content. Christian Hessler, Eric Bolden, Jim Elliott, Tom Berry, and Michael Faurot all provided invaluable reviews and early critiques of both the book and the companion DVD. The book is better for their participation.

I also owe a great deal of thanks to my partner, Ron Becker. His unwavering support and encouragement helped me to overcome several hurdles. His continued reminders to relax and have fun also translated to a more enjoyable book—both to write and (hopefully!) to read.

Getting Started

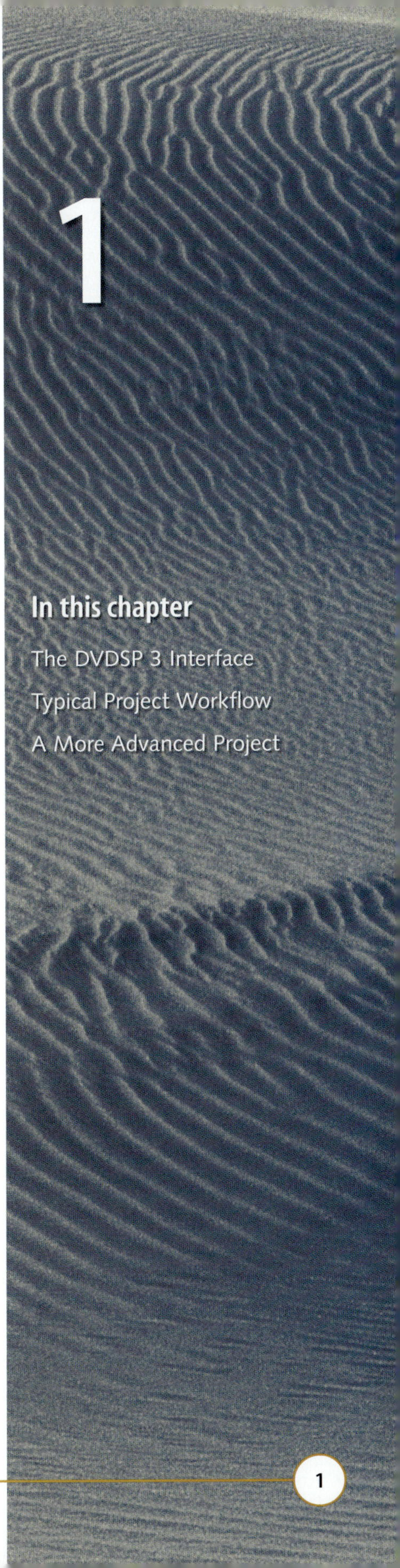

1

In this chapter

Welcome to DVD Studio Pro 3! This chapter provides a detailed overview of Apple's DVD Studio Pro 3 (DVDSP 3) software to make sure everyone is on the same page.

Frankly speaking, DVDSP 3 gives you access to all the best features of DVDs: great video, multiple audio tracks, interactive activities, and fantastic menus. If you're new to DVDSP, you should read this chapter and try building the simple projects we cover. They give you a chance to play with the key parts of DVDSP 3 and leave you in good shape for the remaining chapters. Those of you upgrading from Version 2 might want to skim the section on using templates, which Apple introduced in Version 2 of the software. If you're upgrading from Version 2 or you're already comfortable with the basics, feel free to skip this chapter and get into the fun stuff. Chapter 2 covers the supported media types and various conversion notes. Chapter 3 is where we begin the real projects. Pick your starting point and get burning!

As with anything, the more you practice and experiment, the more skilled you'll become. We heartily recommend you pick up a spool of DVD-R media and give yourself permission to make a few (dozen!) coasters. You'll learn a lot along the way, and future projects will only benefit.

With that said, let's get started!

The DVDSP 3 Interface

DVDSP 3 gives you an abundant workspace for building your DVD projects. You can arrange your projects in a way that is visually intuitive and easy to follow. You also have easy access to all the various properties of the DVD and your media. In short, you'll find everything you need to lay out great projects. That's not to say that there aren't a few quirks, but we'll warn you about these as we come across them.

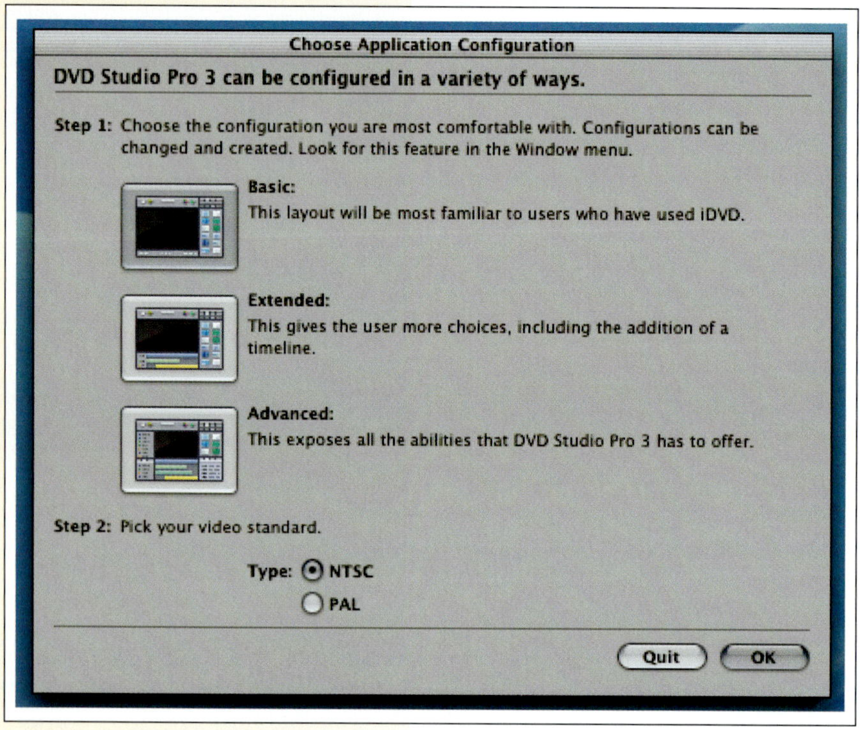

Figure 1-1. The Application Configuration dialog

After you install DVDSP 3, and you run it for the first time, you'll be presented with the dialog shown in Figure 1-1. From this dialog you can choose an interface configuration that you're comfortable with. Your choices are as follows.

Basic
: This layout is somewhat similar to the layout in iDVD, Apple's consumer-level DVD authoring tool that ships as part of iLife. If you're used to using iDVD, you might want to start off with this configuration for simplicity.

Extended
: This layout expands the Basic layout, adding two additional subwindows and a Property Inspector window below the Palette window. This is a good place to start if you enjoy using DVDSP for simple projects.

Advanced
: This layout brings out all the options and tabs that you can configure with DVDSP. This interface can be a little overwhelming at first, so I recommend this to users who are either intensely curious or already familiar with using previous versions of DVDSP for creating advanced projects.

The Advanced dialog is where you can choose the default video standard that you will be using for your DVDs. You probably already know which one you want, but here's a brief rundown of each.

NTSC (National Television System Committee)
: The NTSC standard has a fixed vertical resolution of 525 horizontal lines. It has varying amounts of horizontal resolution, depending on the electronics and formats involved. A total of 59.94 *fields* (sets of even or odd lines) are displayed per second. The odd and even fields are displayed sequentially, thus *interlacing* the full frame. Consequently, one full *frame* is composed of two interlaced fields, and is displayed about every 1/30 of a second. You should choose NTSC if you are creating DVDs for North America, including the United States and its territories, Canada, and Mexico.

PAL (Phase Alternation by Line)

PAL has a vertical resolution of 625 lines. Instead of 60 fields, however, PAL interlaces 50 fields per second, resulting in a 25-frame-per-second display system. PAL is used throughout Europe and in parts of Africa and Asia.

Key windows and controls

Let's start off with the Advanced interface configuration so that we can explain all the windows and their options. Figure 1-2 shows DVDSP 3's Main window, which has a very similar interface to Version 2. For the simple projects in this chapter, the only real difference you'll notice between Versions 2 and 3 of the software is the number of templates available in the Palette. In this chapter we tried to include projects you can accomplish in either version of the software, but there are a few new features that exist only in DVDSP 3. We'll be sure to point those out where appropriate.

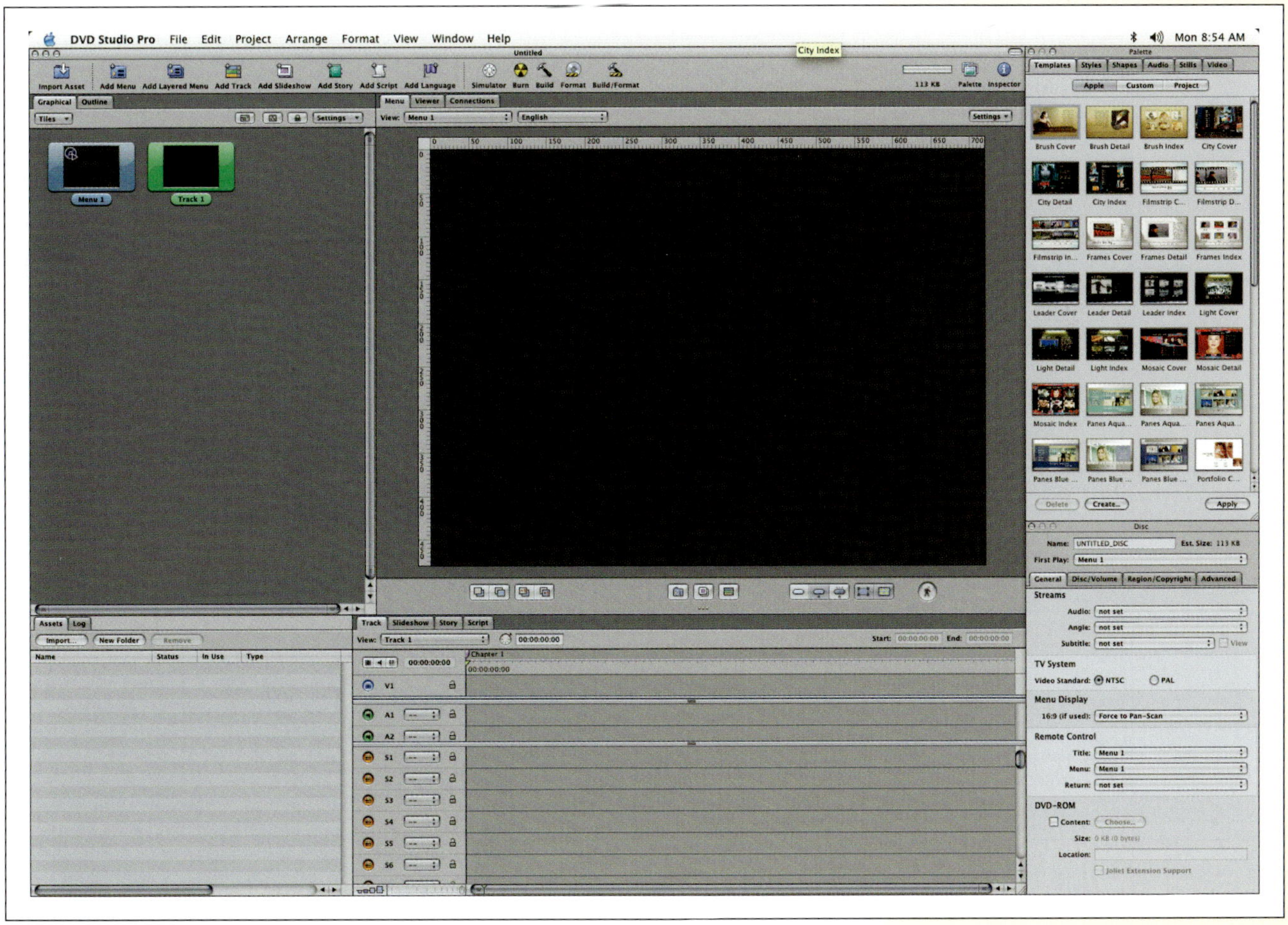

Figure 1-2. The main DVDSP 3 interface

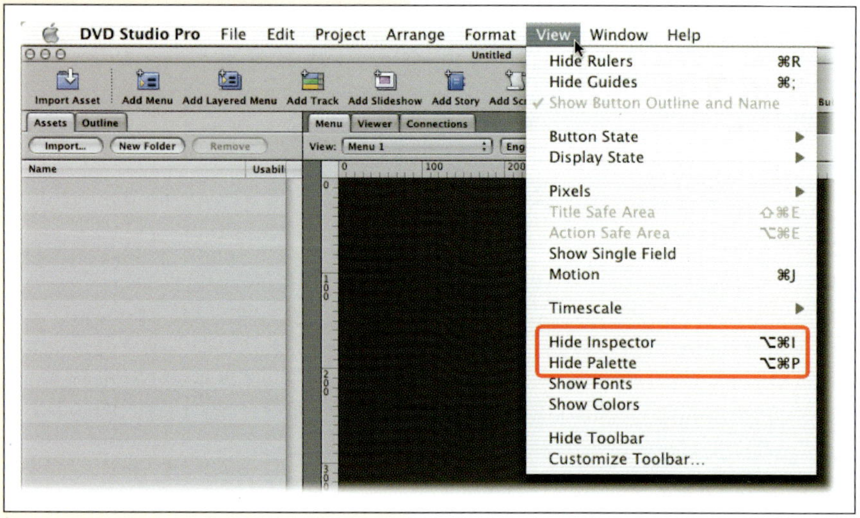

Figure 1-3. The View menu

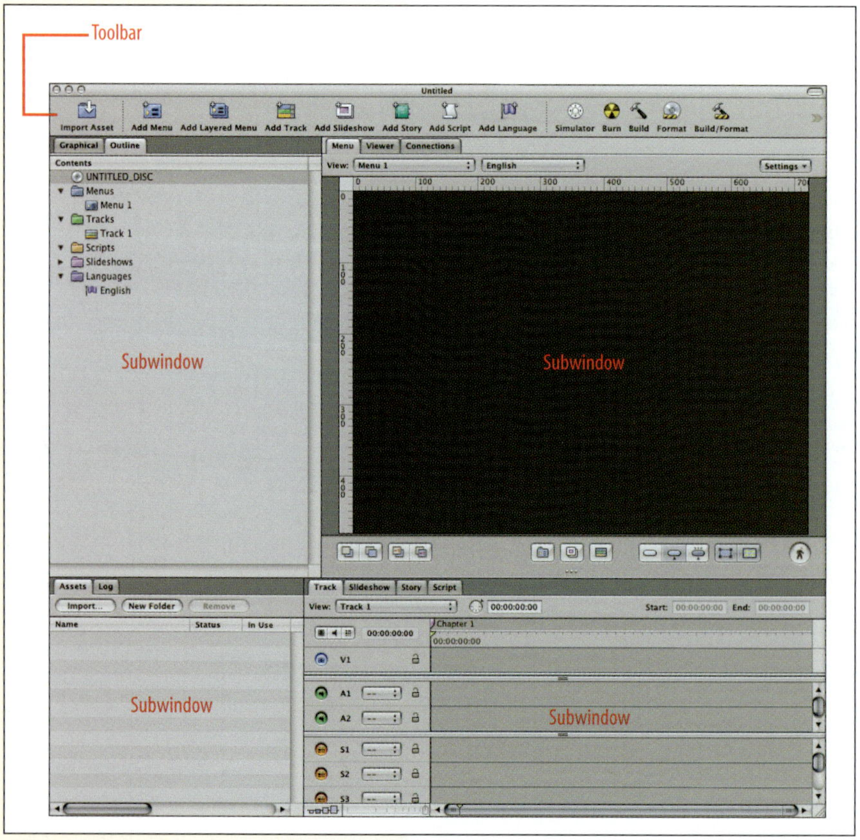

Figure 1-4. DVDSP 3's Main window

In Figure 1-2 you can see three windows: the Main window, the Property Inspector (labeled Disc in the screenshot), and the Palette. If you ever misplace these windows, you can toggle them through the View menu (see Figure 1-3) or via the icons on the toolbar.

At first glance, this looks like a mind-numbing number of components and options. So, let's quickly go over the purpose of these windows, and then we'll pare things down to the few pieces you really need to get started.

The Main window

The Main window, shown in Figure 1-4, is the largest of the three windows; this is where you create your DVD projects. Note that the Main window defaults to having four subwindows, and each subwindow consists of one or more tabs.In a typical workflow, first you import various resources (e.g., menus, buttons, etc.) into DVDSP 3 as *assets*, which then are stored under the Assets tab (as shown in the lower-left subwindow in Figure 1-4). Alternatively, you can use the various Add buttons (e.g., Add Menu, Add Track, Add Story, etc.) along the toolbar at the top to do the same thing. Then you can build a DVD by dragging and dropping these items from the Assets tab to an appropriate tab somewhere else. At this point, the assets become *objects* in your project, and you can "edit" them. Note that I enclosed *edit* in quotation marks, as it can mean different things based on context.

For example:

- Editing a *menu* means creating the buttons and setting their actions. This means dragging assets onto the Menu tab in the upper-right subwindow.
- Editing a *movie clip* means setting the chapter stops, audio tracks, and subtitles. You can edit clips by dragging them onto the Track tab on the lower right.
- Editing a *script* doesn't mean editing a movie script. Instead, it refers to editing programmatic instructions that you want the DVD player to follow when it reaches a certain point. You can edit scripts using the Script tab, also on the lower right.

You also can edit other things such as stories and slideshows, but we'll get to those later.

Finally, right-clicking (or holding down the Control key and clicking, for you one-button mouse folks) in any subwindow tab area enables you to choose the tabs for that subwindow from a pop-up menu, as shown in Figure 1-5.

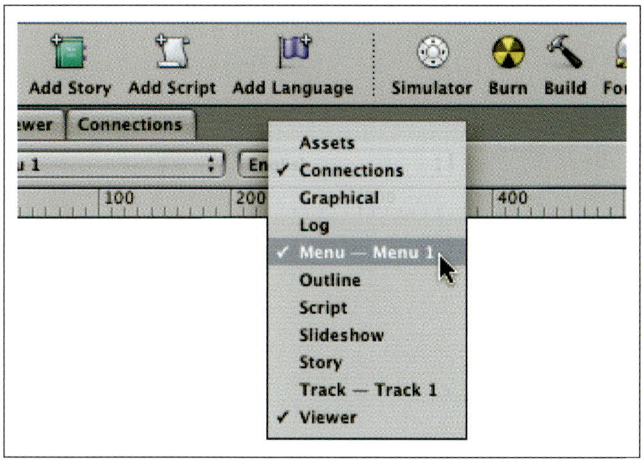

Figure 1-5. The right-click tab configuration pop-up window

The Property Inspector window

The Property Inspector tracks the object currently selected in the Main window and enables you to configure all the relevant options for that object. For example, in Figure 1-6, you can see the properties available for the DVD disc itself. Other property sheets appear when

TIP

What if I Don't Like My Interface Configurations?

If you find that the current interface configuration just isn't what you hoped for, you can switch to a different one by selecting Window → Configurations. Note that you have four choices there, not three: the Advanced configuration has both a regular and a "cinema" mode, depending on which type of video you're creating. If you prefer using hotkeys, you also can cycle through the interface configurations by pressing F1 through F4.

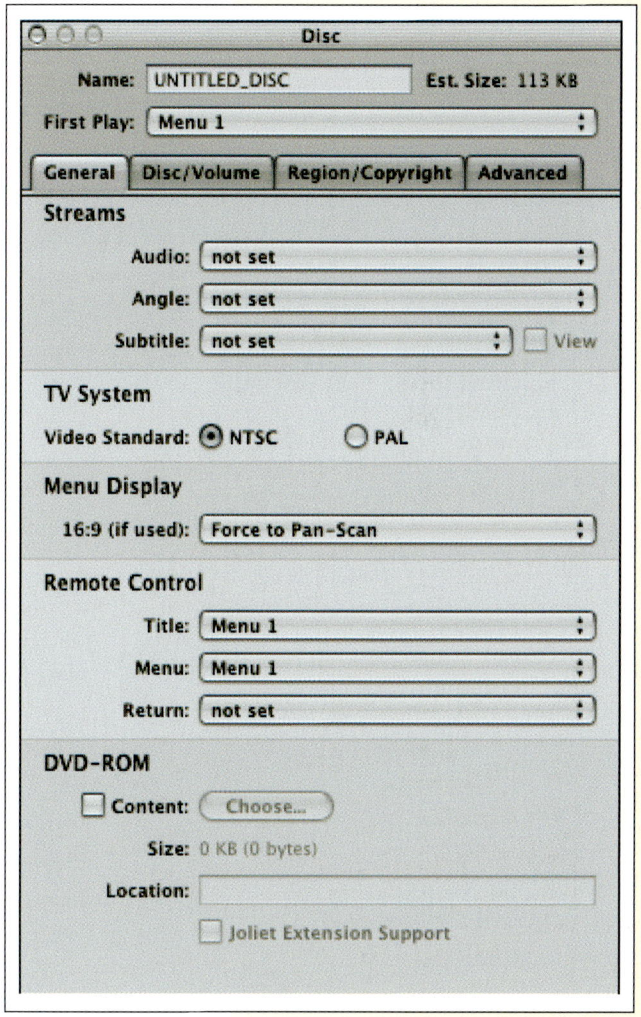

Figure 1-6. The Property Inspector window (DVD view)

TIP

Don't Like the Layout in the Main Window Either?

You can tailor each subwindow to your liking by dragging a tab from one subwindow to another, or even outside the application to create a separate window. This means you can have one huge area with lots of tabs, or several areas, each with a few tabs. Note that each tab can live in only one subwindow, but you are free to move any of them to any position you want. If you find a configuration that works well for you, you can save it through the Window → Configurations menu.

you work on menus, movie clips, or buttons. Note that they all have a slightly different format. For example, compare the option sheets for both menus and buttons, shown in Figure 1-7.

If you want to know which object the Property Inspector is currently "inspecting," the easiest way is to check out the listing of objects in the Outline tab of the Main window. Whatever object is selected there is the object that shows up in the Inspector.

In my experience, it's easy to lose the Property Inspector window when you're working on a DVD project. However, as we mentioned earlier, you can get your Inspector back by selecting View → Show Inspector, pressing Option-⌘I, or clicking the Inspector icon in the toolbar.

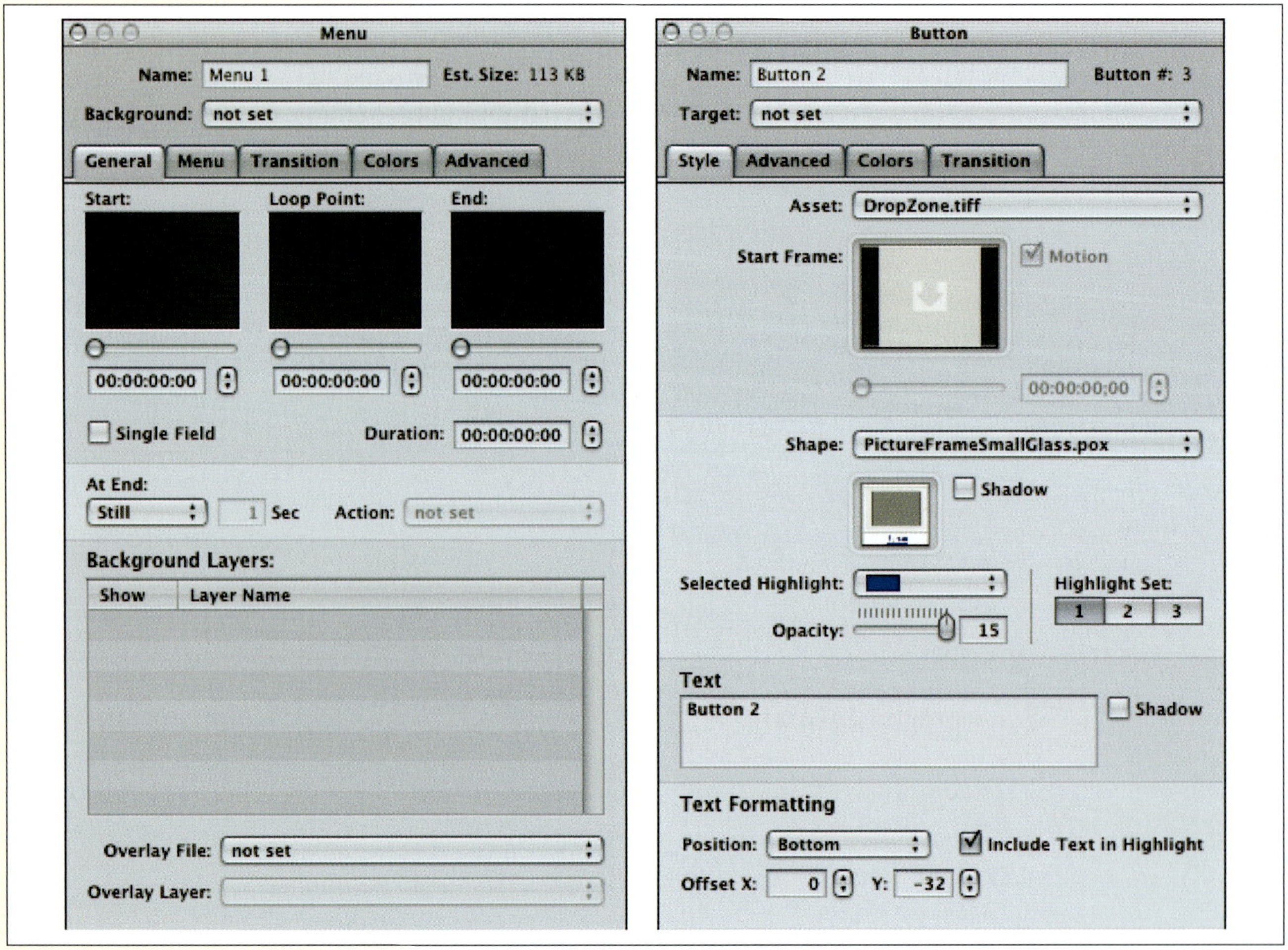

Figure 1-7. The Inspector showing menu and button properties

The Palette window

The view shown in Figure 1-8 is a complete list of your raw materials. All your assets—your clips, audio files, and images—appear here. The Palette window organizes assets available to your project into quick-glance lists. Through the tabs you can access lists for things such as the currently defined templates, styles of text and buttons available, music you have in iTunes, etc. You can quickly grab items from the Palette and drop them into your project without importing them explicitly. (But any asset you grab from the Palette will be added to your regular asset list as well.)

The Palette is essential for playing with templates and building menus. Or rather, we should say that the Palette makes child's play of working with templates and building simple menus. You still can roll your own if you need to.

As with the Property Inspector, the Palette can get lost. You can get it back by selecting View → Show Palette, by pressing Command-Option-P, or by clicking the Palette icon on the toolbar.

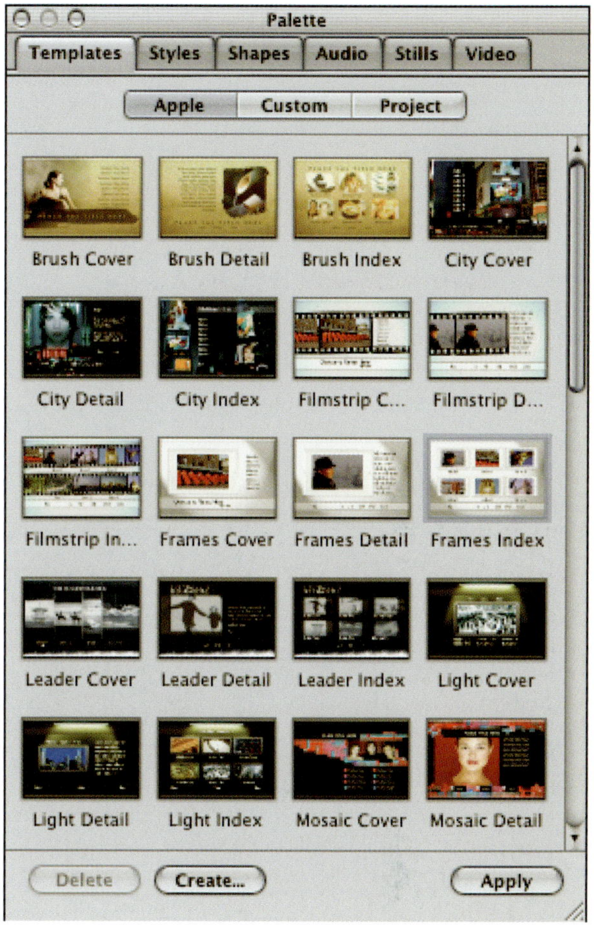

Figure 1-8. DVDSP 3's Palette window

The toolbar and other useful things

The toolbar sits at the top of your Main window and gives you quick access to a number of common tasks. Figure 1-9 shows a sample toolbar.

We say *sample* because the toolbar is customizable and enables you to keep your favorite actions close by. You can configure the toolbar by Control-clicking or right-clicking anywhere along the toolbar. If you want to remove an item from the toolbar, click above that item

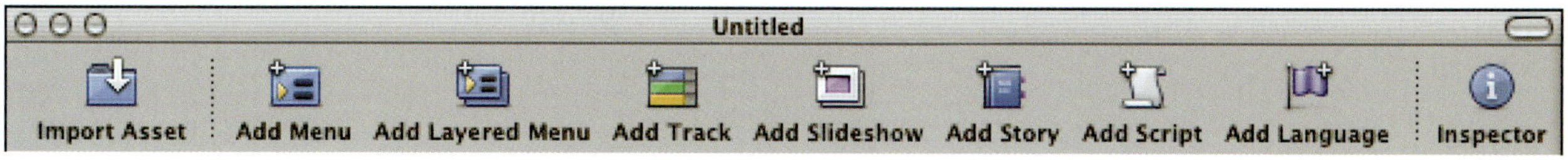

Figure 1-9. A sample toolbar in DVDSP

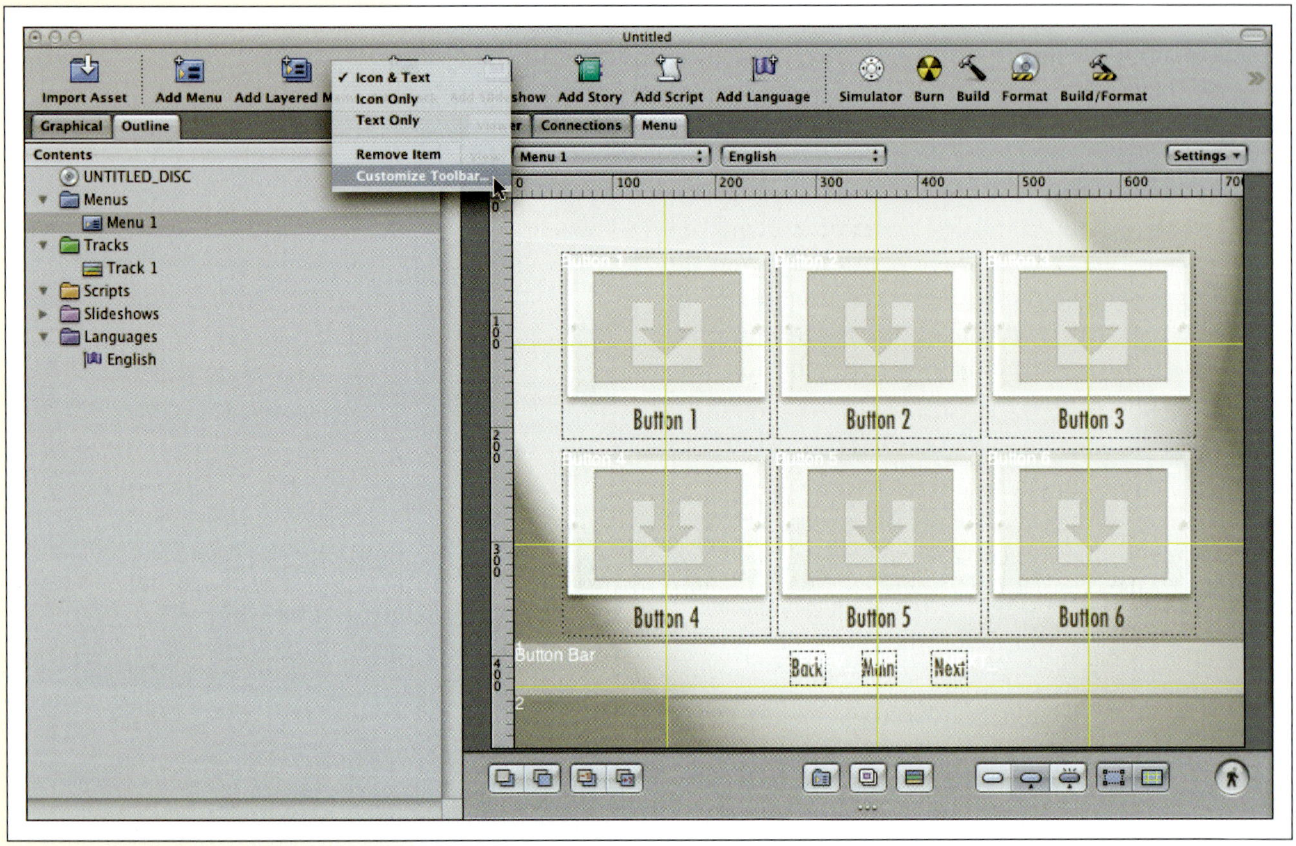

Figure 1-10. You can customize the toolbar to include functionality as you see fit.

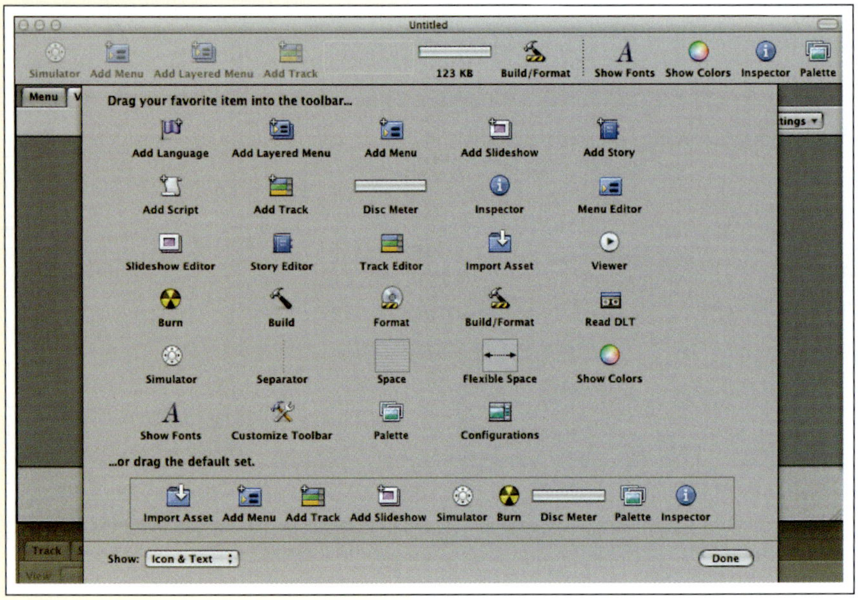

Figure 1-11. The Toolbar Customizer dialog

and select Remove Item. If you want to add one or more items, select the Customize Toolbar option from the menu that pops up, as shown in Figure 1-10.

This brings up the Toolbar Customizer dialog, shown in Figure 1-11. Here, you can drag items to and from the toolbar to suit your needs. Or you can drag to the toolbar the entire default setup at the bottom and go with that.

Simply click the Done button in the lower-right corner when you're satisfied with the toolbar's configuration.

Interesting tabs

A few of the tabs available are worth mentioning here. Look at Figure 1-12. We've chopped up the interface a bit to zoom in on the interesting areas.

We'll be using the following tabs on every project:

Assets
: This tab shows you the various audio, video, and still resources (files) you have associated with your project. To use a video clip, you can import it so that it is listed here in the Assets tab. You also can just drag a clip from the Finder (or another application) and drop it into DVDSP. It will be imported automatically and listed here as well.

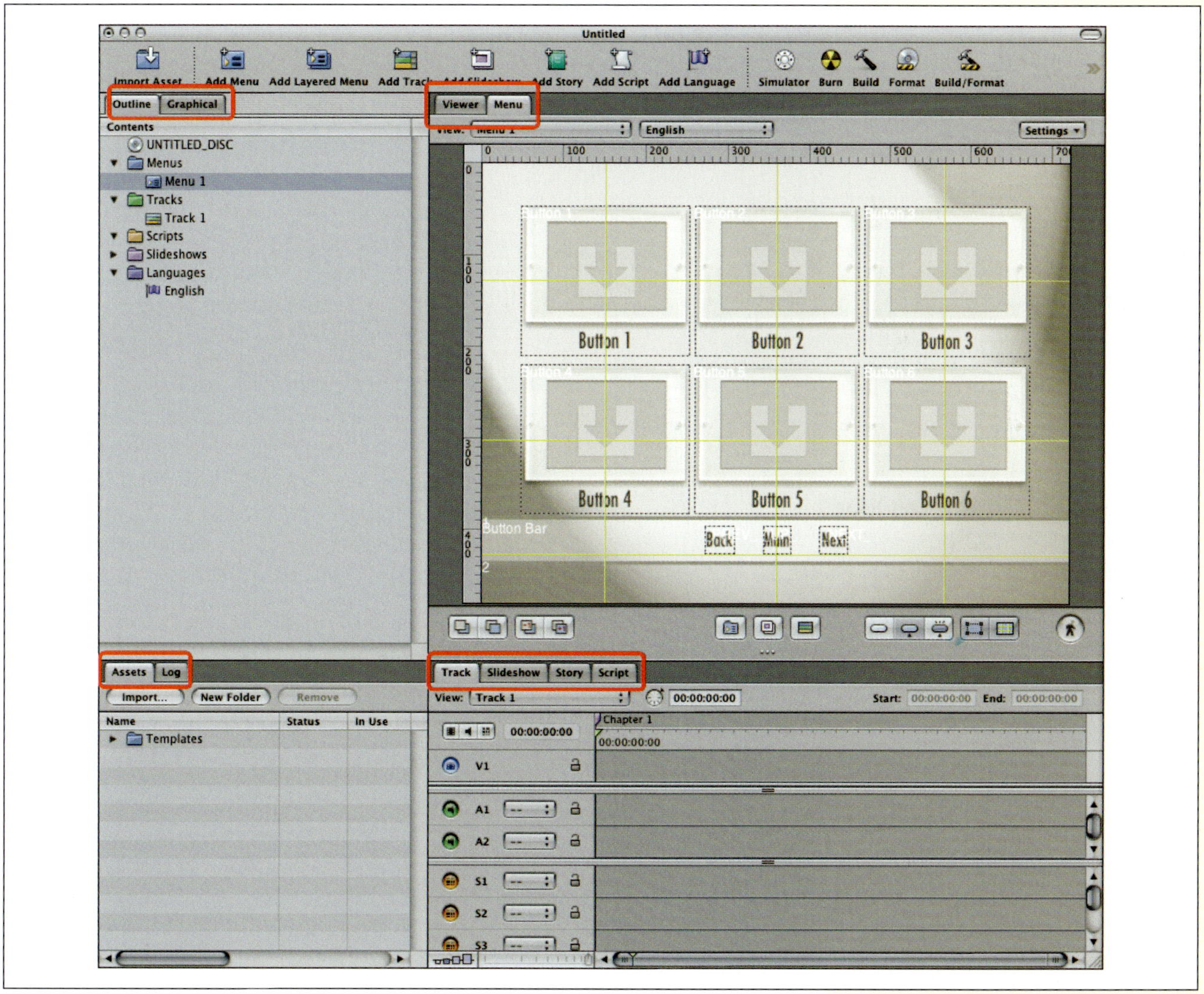

Figure 1-12. Several commonly used tabs

Outline

This tab shows you the contents of your DVD in a hierarchical view. You can use it to get a quick overview of what's going on in the DVD. You'll also use this tab quite often to select a particular item so that you can edit its properties.

Track

This tab details the video, audio, and subtitle streams you have active. You can use the pop-up window to select the track to view. You can organize multiple clips into one track, and define multiple audio, video, and subtitle streams in this tab.

Menu

In this tab you arrange your DVD menu items and hook up their *jump targets* (where the user is taken when he selects a menu item). You also can configure any background audio or video you want for the menu. Most of the work we'll be doing in the first four chapters of this book will happen in this tab. The Menu tab shows one menu at a time, but you quickly can switch between them using the pop-up button in the upper-left corner.

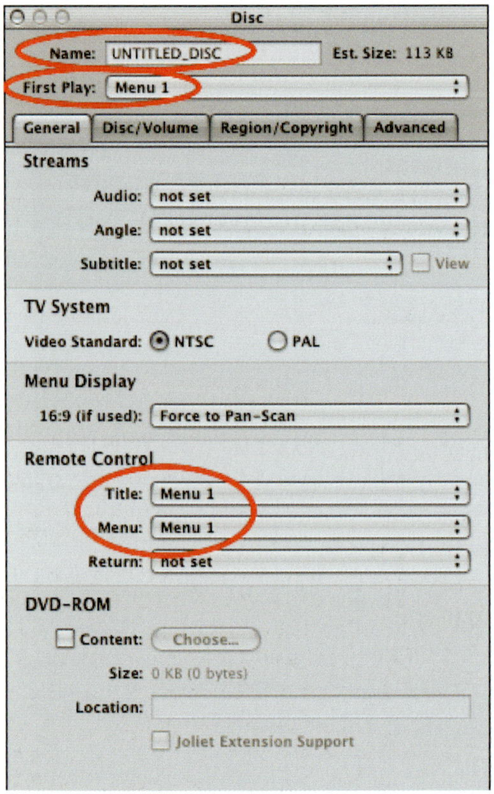

Figure 1-13. The must-set properties for any project

Viewer (contents not shown in Figure 1-12*)*

This tab enables you to preview an asset without starting up the Simulator. This can be handy if you haven't yet set up a menu for the project.

We'll explore the other tabs, such as the Log and Graphical views, later in specific projects. One quick note for DVDSP 2 users: the Graphical tab is new to DVDSP 3. It provides you with a quick, visual view of your project, which can be great for dealing with complex projects, but it does not provide any new functionality. You can do nothing in the Graphical view that you can't accomplish elsewhere in DVDSP.

Things you can't forget

The Property Inspector for the Disc object contains a few options that you should set on every project. They are highlighted in Figure 1-13.

1. **The Disc Name.** You should always name your disc. Otherwise, inserting the DVD into a computer will bring up a disc with the ever-popular default name, Untitled Disc.
2. **The First Play.** This is critical. Be sure to pick an initial menu or movie to show when you put the disc

in a player. DVDSP 3 always defaults to the first menu that you create. If you want your movie to start immediately when the disc is inserted into a player, this is the place to specify it.

3. **The Remote Control Settings.** These properties determine how the remote control will work with the disc. At a minimum, you'll want the Title and Menu controls to take you to your main standard menu. The default on these settings is also the first menu.

You should get into the habit of specifying these values as soon as possible. That usually happens just before you save the project for the first time. That being said, I burned a few coasters before learning these little tidbits, so don't feel bad if you forget.

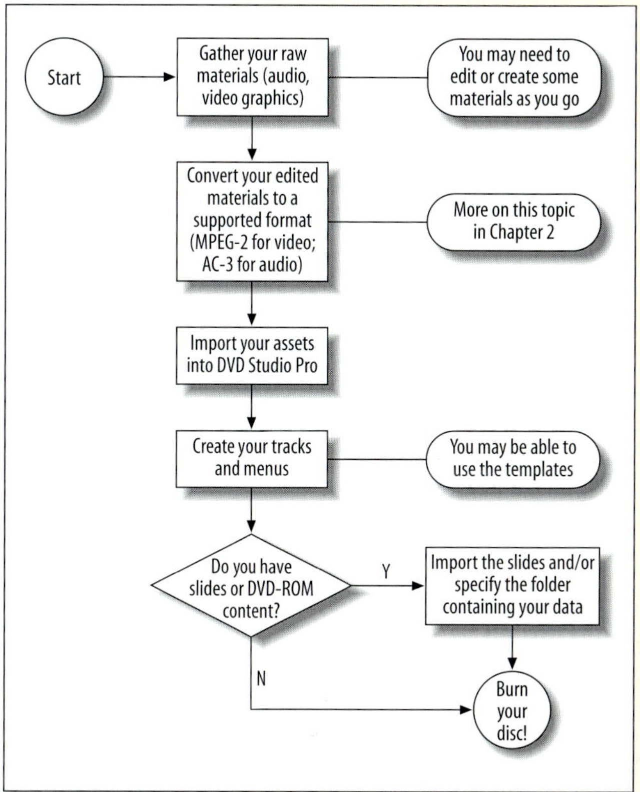

Figure 1-14. Project flow in DVDSP 3

Typical Project Workflow

Every DVD project you work on will flow in a fairly standard way. Although you'll certainly encounter projects in which you must add a new clip at the last minute or delete a bonus item that wasn't ready to be included, most of your projects will follow the five basic steps shown in Figure 1-14.

We'll work through these five steps in the following sections, using the sample media from the *projects/ch01* folder on the accompanying DVD. Again, you should feel free to skip this project if you're already familiar with DVDSP. If you're just starting out or want to play with the new version, dive on in!

Step 1: collecting and formatting the raw media

The first step is, of course, gathering your raw materials. You need to collect your movie clips, audio to accompany the movies, and images (or more movies) to use for the menus. Chapter 2 details the specific formats that DVDSP supports. However, the big thing to realize/remember here is that you usually do all of this work *outside of DVDSP*. DVDSP is the last application along the path to getting video onto a DVD. For the earlier steps, Apple's Final Cut Pro and Adobe's Photoshop are particularly well suited to producing the right type of media.

Make Sure You Have Enough Disk Space

Something else to keep in mind while you're collecting your media: make sure you leave enough room on your hard drive to build your final DVD project. If you're running low on disk space, you'll have to be careful. DVDSP creates a full copy of the DVD on your hard drive before burning it. If you don't leave enough space, the burn will fail. It's best to have a separate drive handy for the burn process anyway.

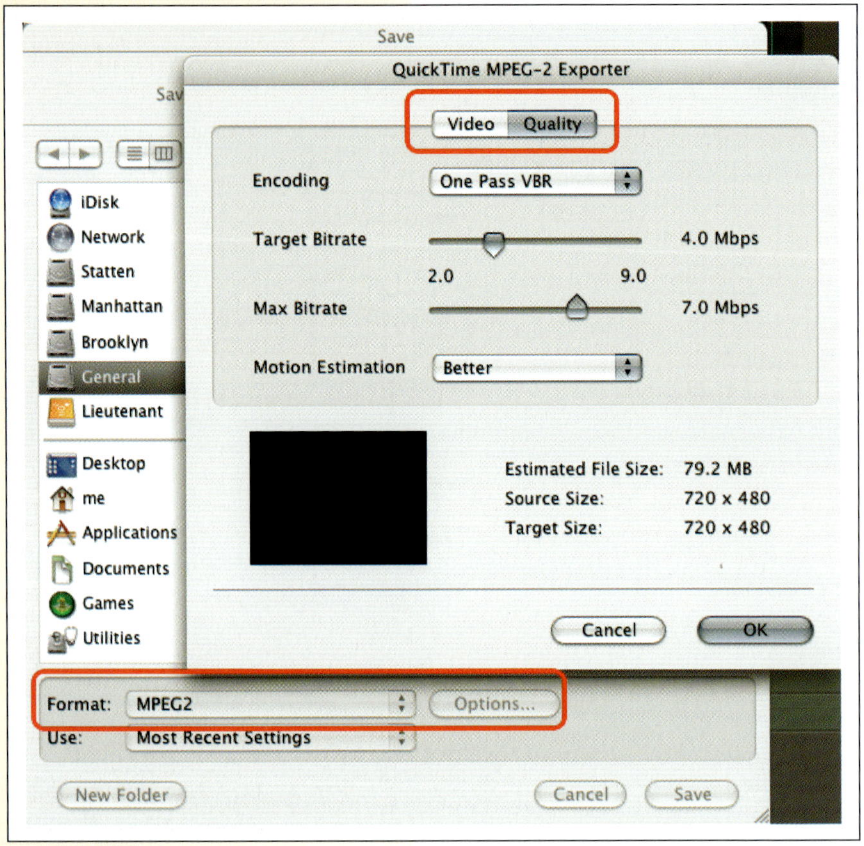

Figure 1-15. The Final Cut Pro export options for MPEG-2

Apple Final Cut Pro

Final Cut Pro enables you to edit your movie and add fancy things such as titles and special effects. Once you have your movie in good shape, you can export it to MPEG-2. Figure 1-15 shows the export dialog for Final Cut Pro.

The export will create two files: a movie file (*.m2v*) and an audio file (*.aif*). You'll need both files once you're inside DVDSP. We actually converted the audio file to the AC3 format to save a bit of space, but more on that in Chapter 2.

> **NOTE**
>
> *DVDSP 3 does enable you to import other types of audio/video files that QuickTime understands, such as Digital Video streams. But before you burn your disc, you'll have to encode those files to MPEG-2. In other words, DVDSP 3 will transcode it for you. The downside to this approach is that it takes a lot longer to build the project.*

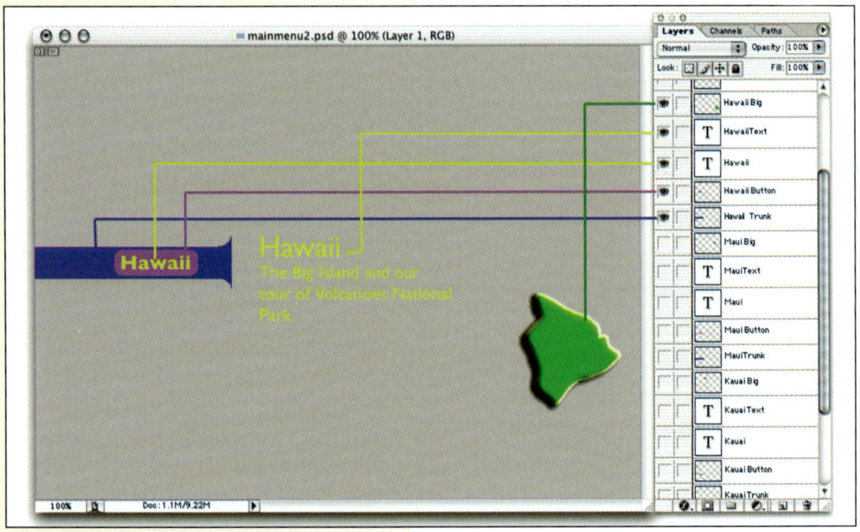

Figure 1-16. One menu associated with the Hawaiian Vacation DVD project (from Chapter 3)

Adobe Photoshop

Adobe Photoshop enables you to create extremely intricate menus using its Layers feature. The layers form the various buttons and screens available in your menus, as shown in Figure 1-16. Layered Photoshop images are best saved in Photoshop's native PSD format, so you don't need to export them into a non-native format. You also can use your favorite editing software to create still images for simple menus. Chapters 2 and 3 provide more details on this.

You can see the various files for another volcano-related project in Figure 1-17.

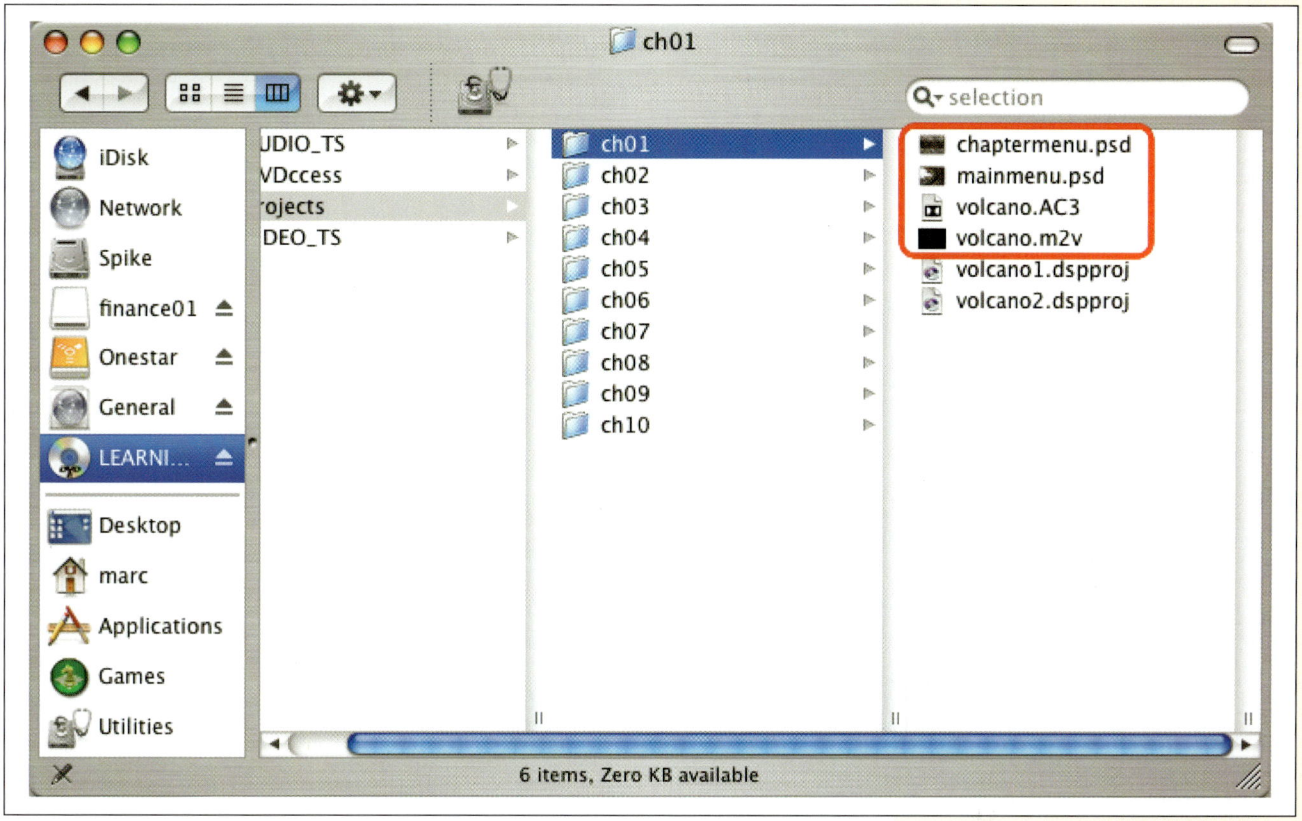

Figure 1-17. The files associated with the volcano DVD project in this chapter

Step 2: importing media into DVDSP 3

From DVDSP's File menu, you can import files by selecting Import → Asset... (or just pressing Shift-⌘I). Select all the files you want to load and click the Import button. Your imported files should show up in the Assets tab.

For our first project, go ahead and import the following files from the *Chapter 1* folder on the companion DVD:

- *volcano.m2v*
- *volcano.aif*

For the moment, we'll rely on the graphics built into the templates for all the menu images. Templates make quick work of simple projects! (But more on templates in Chapter 10.)

Checking References

Imported media is *referenced*, not copied. You still can edit the individual files outside of DVDSP, but the changes might not be immediately visible inside of DVDSP. Often, restarting DVDSP or deleting and reloading the asset will make the changes show up.

Step 3: creating a menu

First things first: let's switch to a simpler interface configuration for this project. Press the F1 key or select the Basic configuration in the Window → Configurations menu to set up the two key windows you need to get started. You should have the Main layout window and the Palette open, as shown in Figure 1-18.

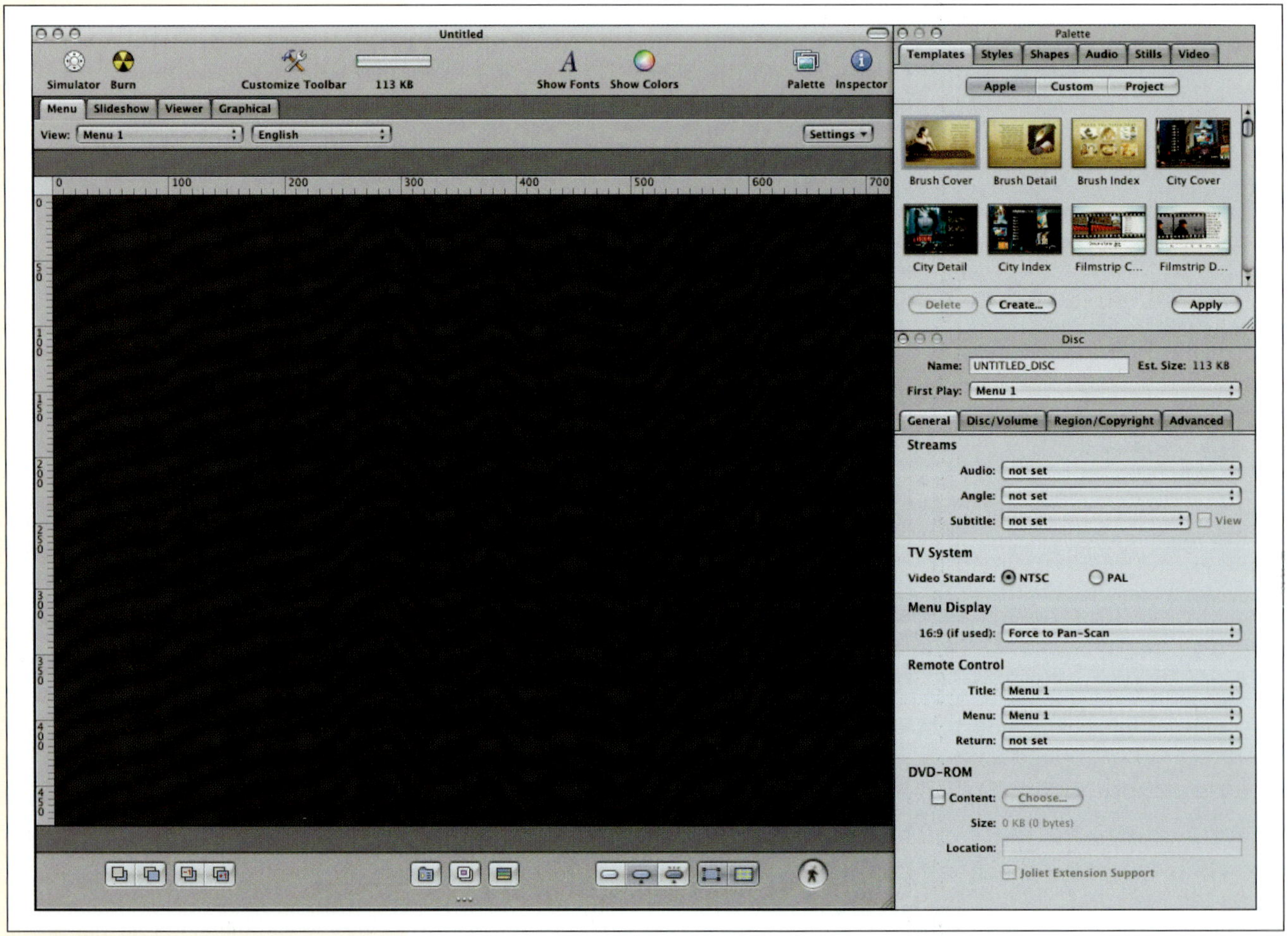

Figure 1-18. The Basic layout option for your windows

In the Main window on the left, you should have four tabs: Menu, Slideshow, Viewer, and Graphical. Click the Menu tab so that it is raised and highlighted (if it isn't already), as shown in Figure 1-19. You'll be rewarded with a drab, blank gray canvas! Don't worry, we're about to change that.

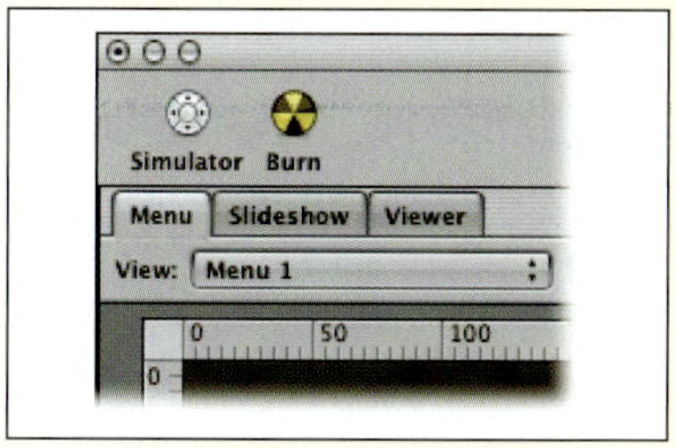

Figure 1-19. The Menu tab shows an editable view of the menus in your project

In the Palette, select the Templates tab. Scroll down to find the Light Cover template. Double-click that icon. It should transform that blank gray canvas into a spiffy, nearly completed menu. Check out Figure 1-20 to see the results.

We say *nearly* completed because you've still got some work to do. First you want to fill in that big box in the middle with something a bit more interesting. For that, you need your assets, so grab the Assets tab by pressing ⌘1

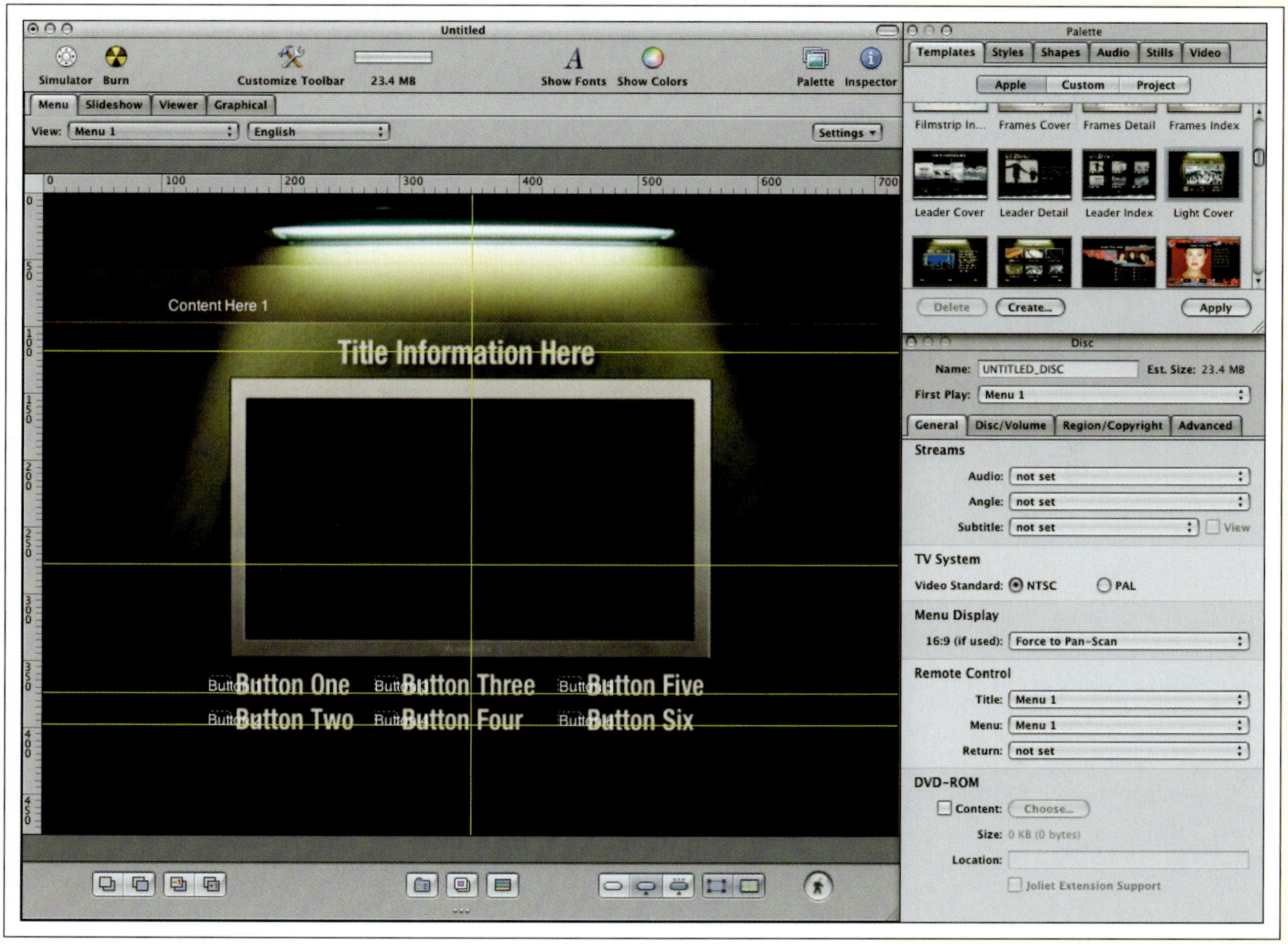

Figure 1-20. The amazing results of applying a template

or selecting Window → Assets. Drag the *volcano.m2v* asset over the box, as in Figure 1-21, and pause for a moment. You should see a set of options magically appear.

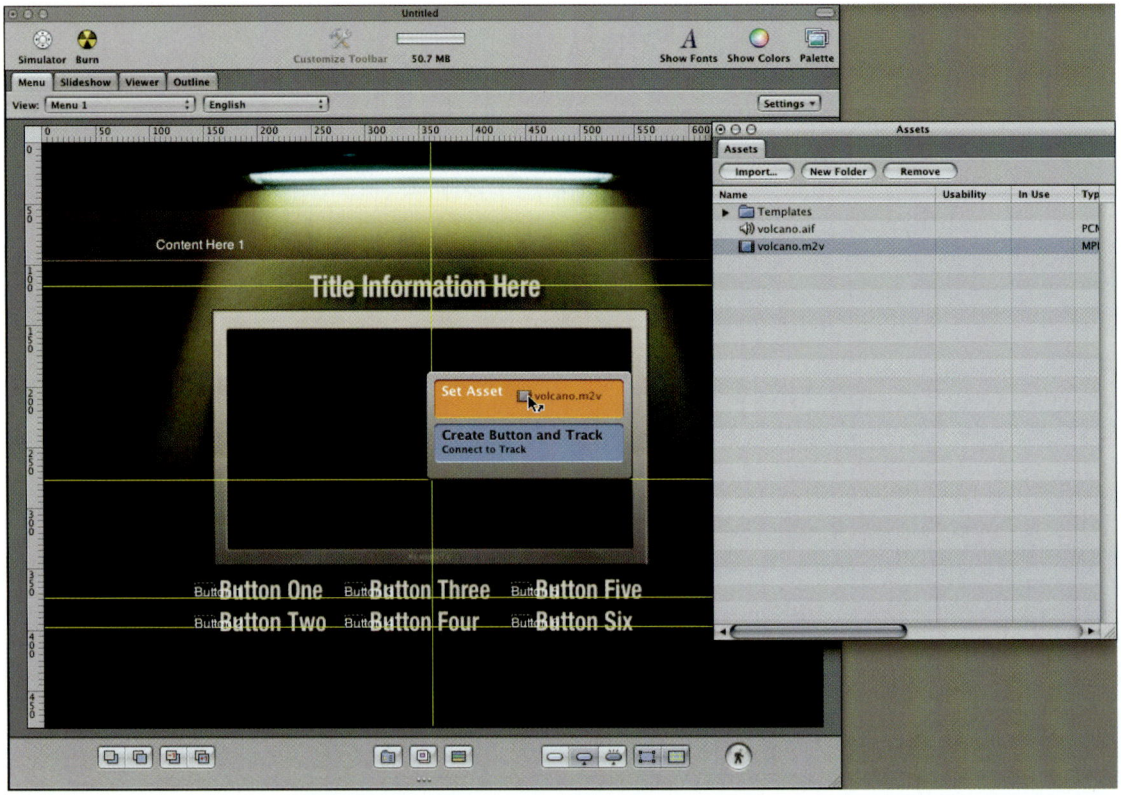

Figure 1-21. Setting the asset for a template drop zone

You'll want to choose the Set Asset option. It shows you the first frame of the video—which in this case is all black. You can check that the drop really worked by clicking the Toggle Motion button, which is represented by the little walking person at the bottom of the Main window shown in Figure 1-22.

You'll also want to change the default text; "Title Information Here" just doesn't seem very interesting. Give your project a nice title (we used "Volcano Flight") by double-clicking the default text and then typing your new title. Be careful, as DVDSP 3 can be a bit twitchy when you're editing

> **NOTE**
>
> *Using motion backgrounds can be a great way to create eye-popping menus. You'll get to see more details on using video for menus in Chapter 4. We just want to warn you that turning on motion with that Toggle Motion button can tax your system. It's great for quick previews, but it is not meant to be left on indefinitely. Of course, with each evolution of Mac hardware making leaps and bounds, this warning might not apply for long!*

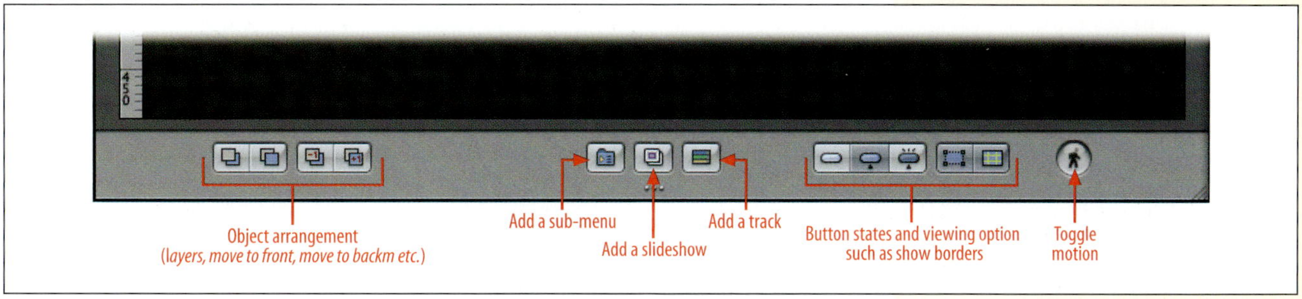

Figure 1-22. The volcano project in progress

text. Words to the wise: make sure you have a blinking cursor before editing the text.

Next, delete the lower three buttons (Button Two, Button Four, and Button Six). You select a button by clicking inside its dashed border. To delete it, press the Delete key. We'll deal with the upper three buttons in a few minutes, but for now, your project should look like Figure 1-23.

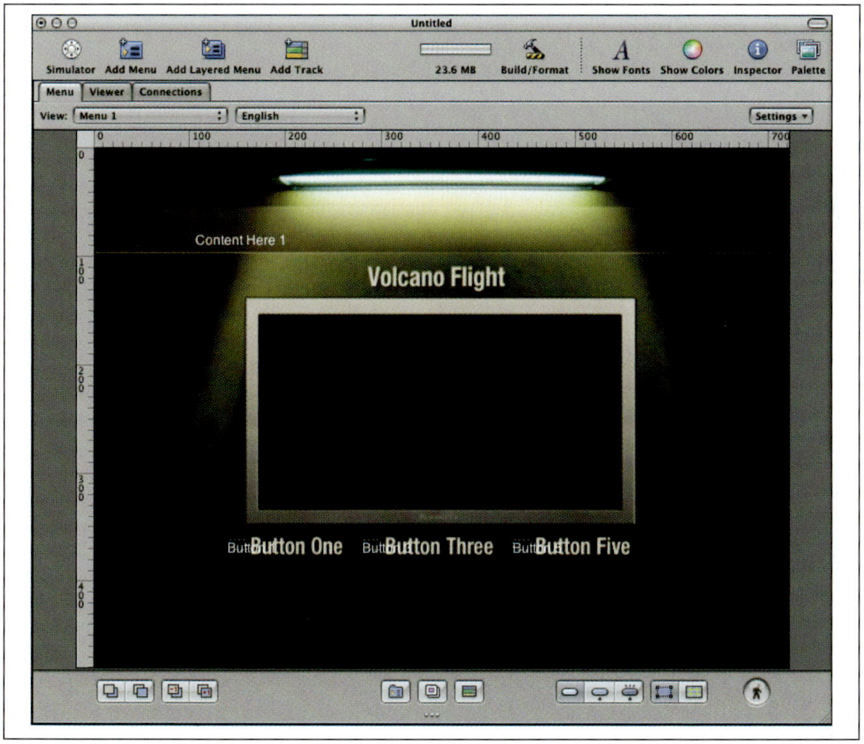

Figure 1-23. The volcano project in progress

Step 4: adding the video

Next, you need to add a new track for the video itself. Click the Add a Track button at the bottom of the Main window. You should see the Track tab pop up. You need that Assets tab again, so go ahead and raise it (⌘1). Then,

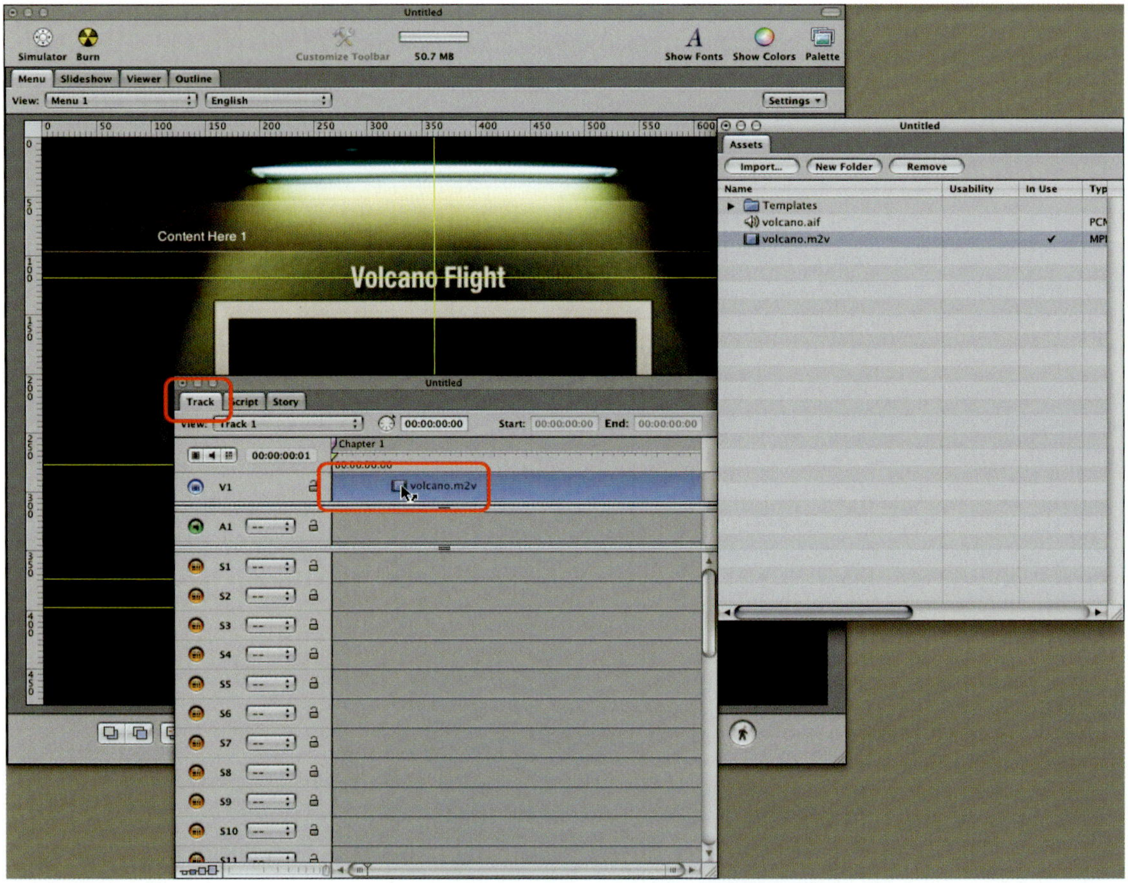

Figure 1-24. Adding a video track to the volcano project

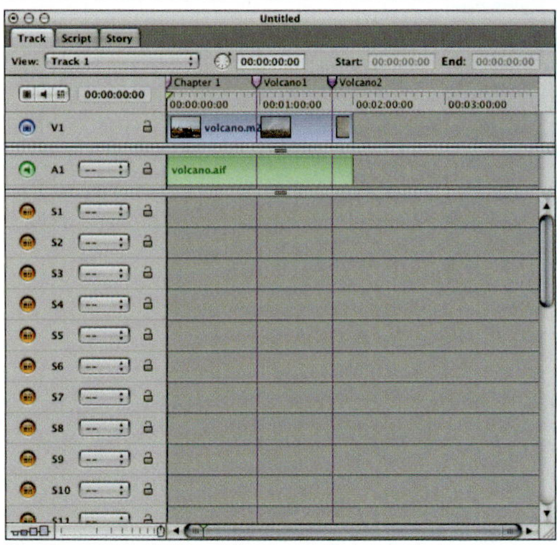

Figure 1-25. The Track window after adding the video

drag the *volcano.m2v* file from the Assets window to the Track window, as shown in Figure 1-24. Be sure you use the Track tab and not the Menu tab; otherwise you'll end up with an extra button on your menu!

Notice in Figure 1-25 that the audio file was added automatically, along with the video. You won't always want that, but it's perfect for what we're trying to do this time around. You might be wondering about those purple lines running down the track listing. Those are actually the chapter markers we set inside Final Cut Pro. You also can set new markers here if you need to, but we'll talk about that later.

That's actually all there is to the video track. You'll have to go back and link your menu to the video, but for now you can go ahead and close the Track window.

Step 5: linking the buttons

Setting up your buttons is easy with templates. For each button, you should Control-click (or right-click if you have a three-button mouse) the button and set the target. Figure 1-26 shows the pop-up menus you should see.

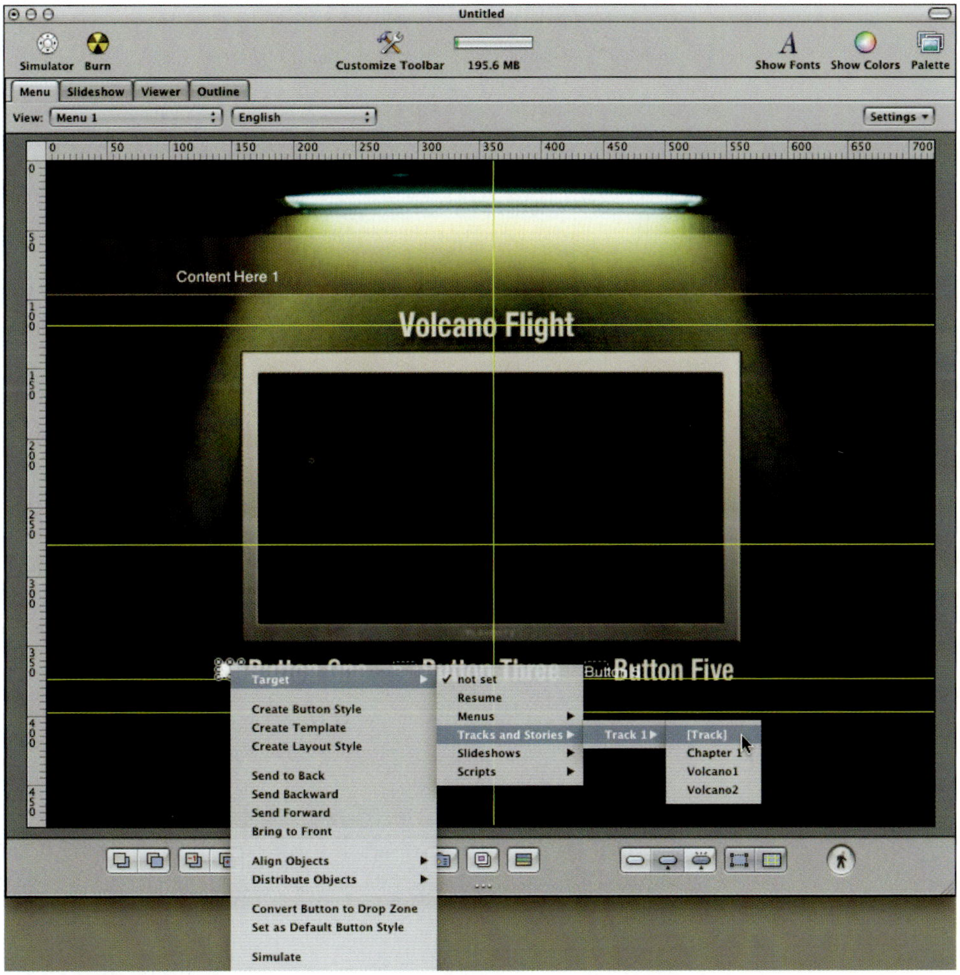

Figure 1-26. Setting a button's target

TIP

Rodentia Tributtonus

If you have an original Apple mouse, you are confined to the world of Control-clicking to bring up the context menus similar to the one shown in Figure 1-26. However, if you have upgraded to any of the three-button USB mice available out there, you have the great advantage of that third button. *Right-clicking* (clicking the button furthest from your index finger—the right button on a right-handed mouse or the left button on a left-handed mouse) brings up the same menu as holding down the Control key and pressing the "left" button.

Today you can find three-button USB mice in the $10–$20 range. We highly recommend picking one up.

You'll also want to double-click the text to the right of the button and change it to match the actual video content. Table 1-1 shows the targets and text we used for the buttons.

Table 1-1. Our template project button targets

Button	Target	Text
One	Tracks and Stories → Track 1 → [Track]	Taking off
Three	Tracks and Stories → Track 1 → volcano1	Approaching
Five	Tracks and Stories → Track 1 → volcano2	Over the Volcano

Testing your project

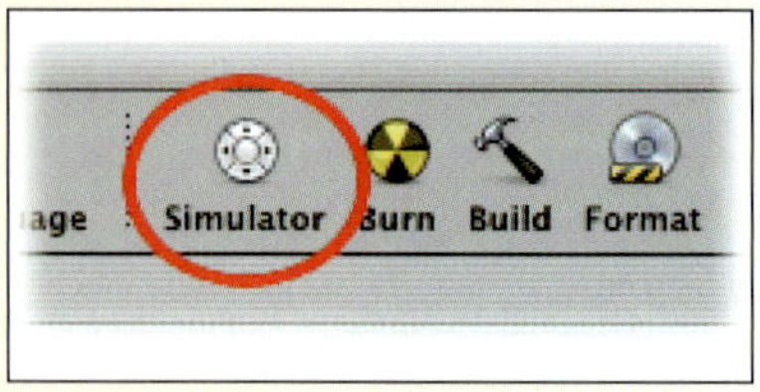

Figure 1-27. The Simulator button in the toolbar

That should be it! Make sure you save your project before moving on. Once you do, you should be able to launch the Simulator and check out your project. The Simulator is easy to launch from the toolbar, as shown in Figure 1-27.

Figure 1-28 shows our current project in the Simulator. If you're happy with how the project works in the Simulator, you're set. All you have left to do is to burn the disc. But we'll save that for the grand finale of the chapter. (If you can't wait, skip to the section "Burning the disc," later in this chapter.)

In our next project we'll re-create much of what we've done here, but this time from scratch. Although templates work great in many situations, you'll find some projects demand more. You can deliver on those needs with DVDSP; you just need to do a bit more of the legwork yourself.

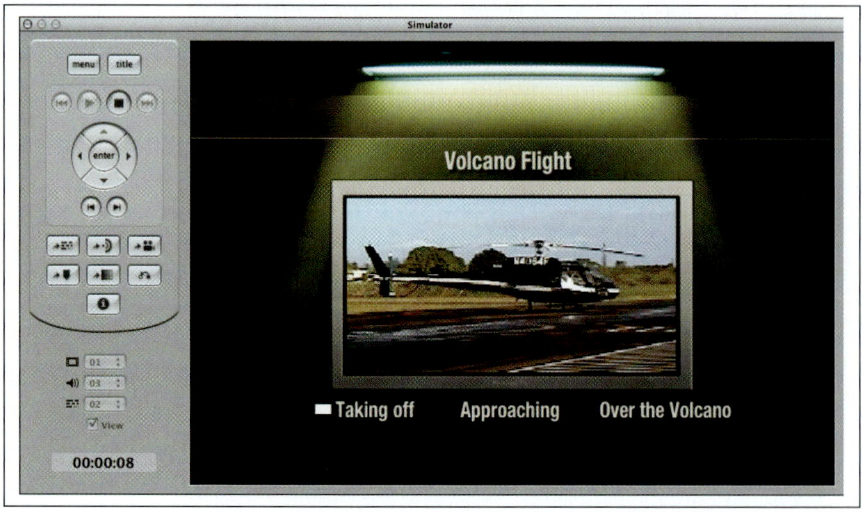

Figure 1-28. Our first project in the DVDSP Simulator

A More Advanced Project

The second project in this chapter covers the same basic process as the first project, but this time we'll create our own menu in Photoshop instead of using one of the template menus.

For the second project, you'll need to import the following files from the *Chapter 1* folder:

- *volcano.m2v*
- *volcano.AC3*
- *mainmenu.psd*
- *chaptermenu.psd*

You'll probably want to switch to the Extended layout for this project. Press F2 or use the Window → Configurations menu. You should end up with a screen similar to Figure 1-29.

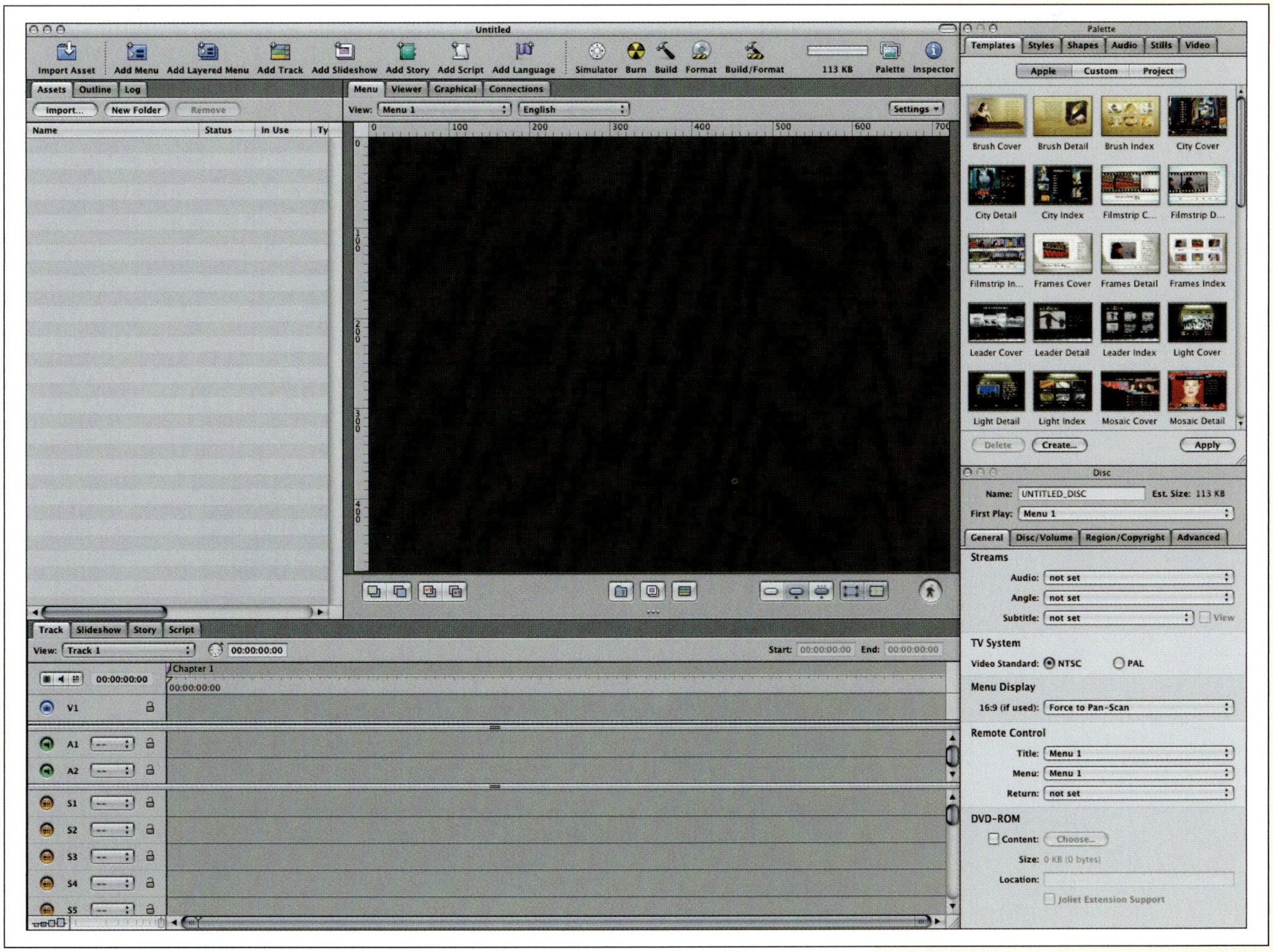

Figure 1-29. The Extended layout you'll use for the next project

Audio and Video

One audio file is associated with the movie by default—provided it shares the same name as the video clip. You still can add more audio clips (or replace the default one) if you need to. If it *doesn't* show up for some reason, just drag the audio file down and drop it in the A1 track.

Step 1: adding a movie clip

Usually you don't have to do much work to add a movie clip to your DVD projects. In most cases, you can just drag the movie clip from the Assets tab to one of the editors and drop it. Occasionally you'll have to set the chapter markers manually. You'll probably also want to specify where to go after finishing the clip—back to the main menu, for example.

For this project, we set the chapters while editing the video in Final Cut Pro (which, incidentally, was very easy to do), so we're OK there. After the movie is done, we want to go back to the main menu. We'll set that property in a few minutes after we add the menu.

So, here's the next task: drag the *volcano.m2v* file from the Assets tab to the Track tab. Your Track editor now should have one movie item, as shown in Figure 1-30.

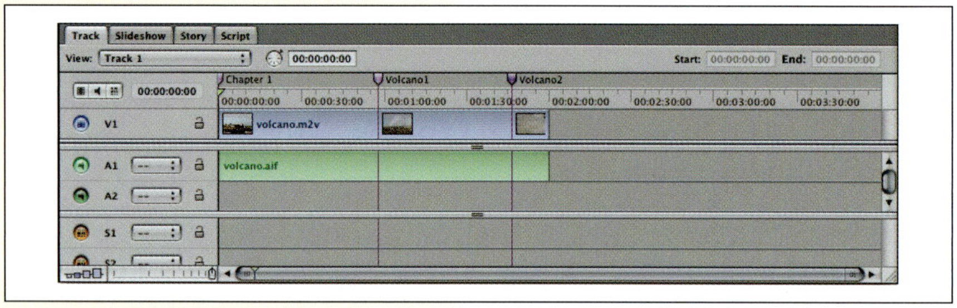

Figure 1-30. The first step in our second project: adding the movie

Step 2: adding a menu and buttons

Adding menus is just as simple as adding movie clips. First make the Menu tab active in your Main window. Now drag the *mainmenu.psd* file from the Assets view to the menu area. You should see a pop-up list similar to Figure 1-31. Choose the first option to set your image as the background for the menu.

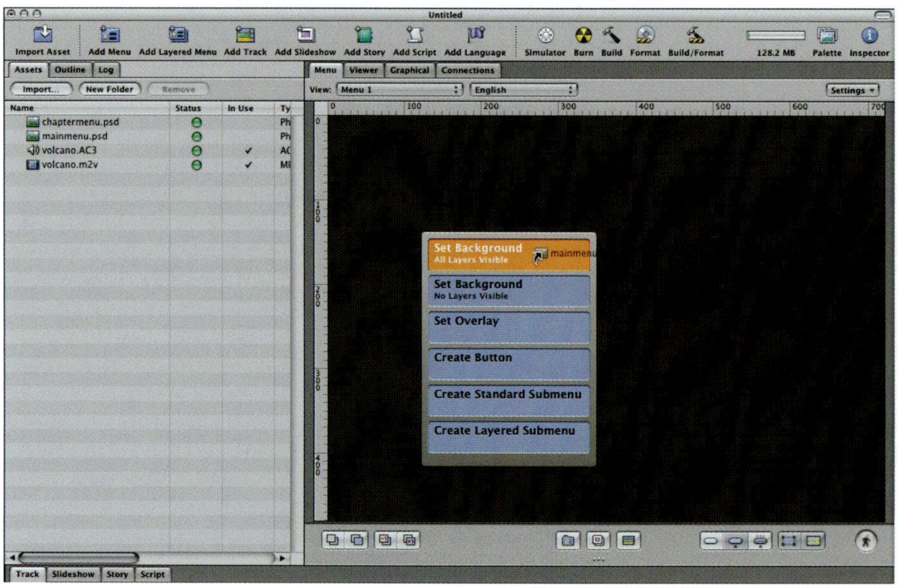

Figure 1-31. The next step in our second project: setting up the main menu asset

So, at this point, the menu itself exists. However, you still have to create the buttons for the menu. If you're making a simple menu with the standard box highlights over the buttons, be sure to pick your highlight colors in the Menu Inspector. See Figure 1-32 for an example.

If your Photoshop document consists of multiple layers, you'll also need to specify which layers of the Photoshop document you want to use. The Menu Inspector can help you there as well. Choose the options highlighted in Figure 1-33.

You'll see some examples of really complex layered buttons and menus later. For now we'll just use the one layer we have. Figure 1-34 shows that menu in the Menu tab.

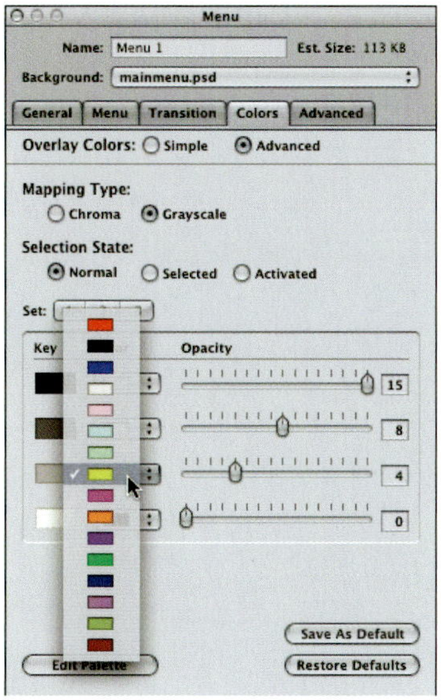

Figure 1-32. Setting the highlight colors for your menu buttons

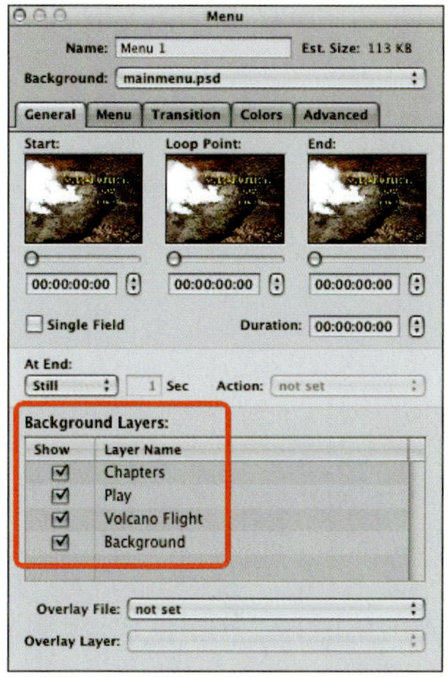

Figure 1-33. Setting the active layers for your menu background

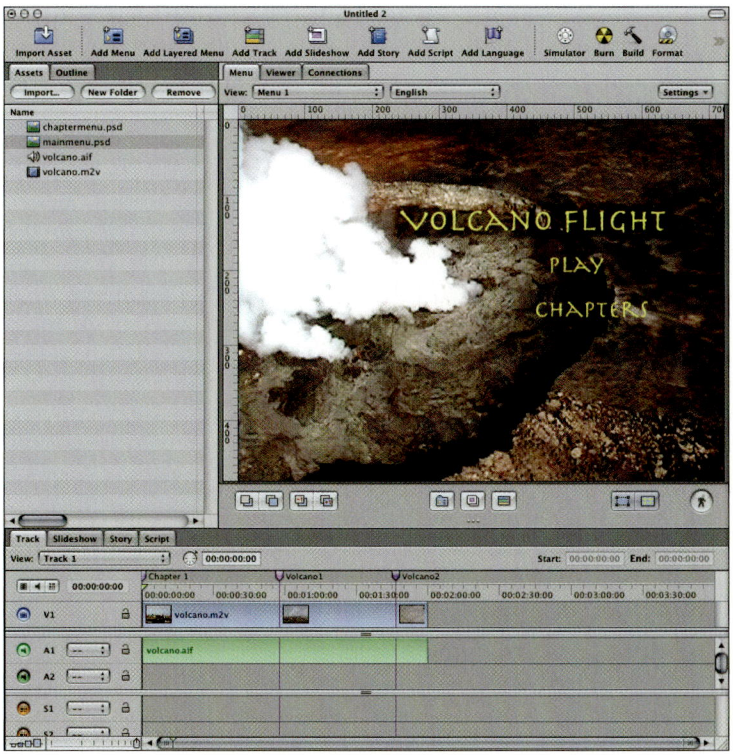

Figure 1-34. The Menu editor window showing our main menu

Here's a nifty thing to know: you can create new buttons by simply dragging an outline over part of the background. Then you can select an existing button by clicking inside its border. Figure 1-35 shows the action of creating our first button.

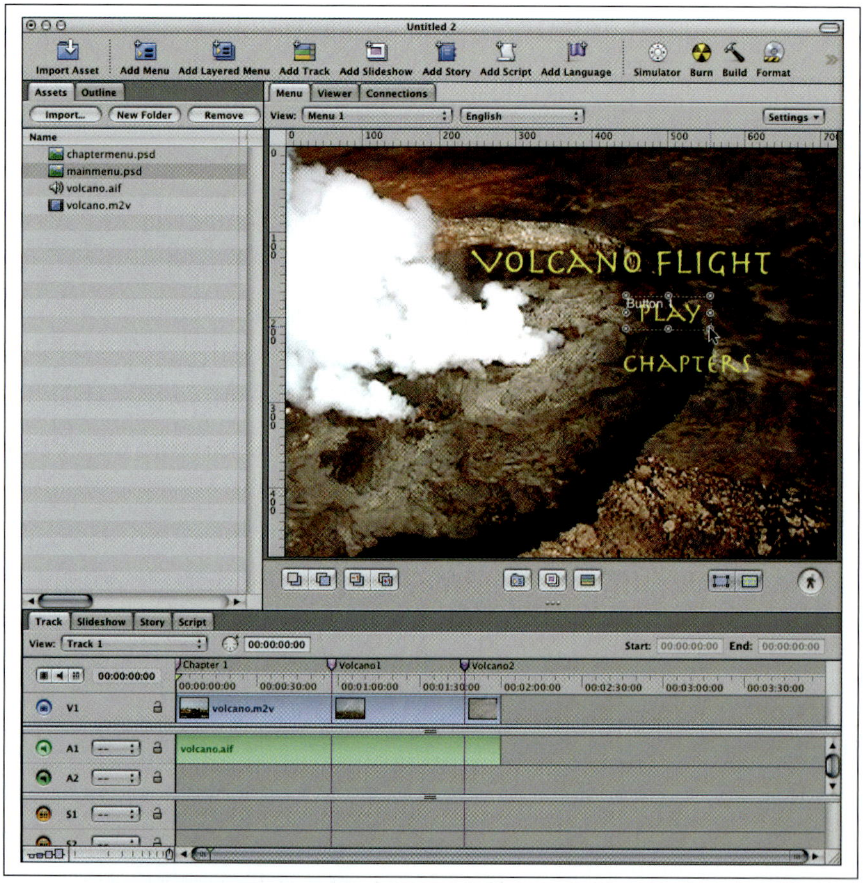

Figure 1-35. Adding a button by hand to our main menu

Once you have the boundary set to the right size and position, right-click the button to bring up a menu of quick options. You can set the target for the button here or through the Inspector. (I recommend always relying on the Inspector, as shown in Figure 1-36.) You can use the Inspector to set a few other key properties: specifically, the button's name and its navigation properties (e.g., which button comes next when you press the left arrow on your remote). You might not be able to specify the proper navigation entries until you have created all the buttons in your menu. That's fine. Just click the interior of a button's outline in the menu window to make that the "active" button in the Property Inspector.

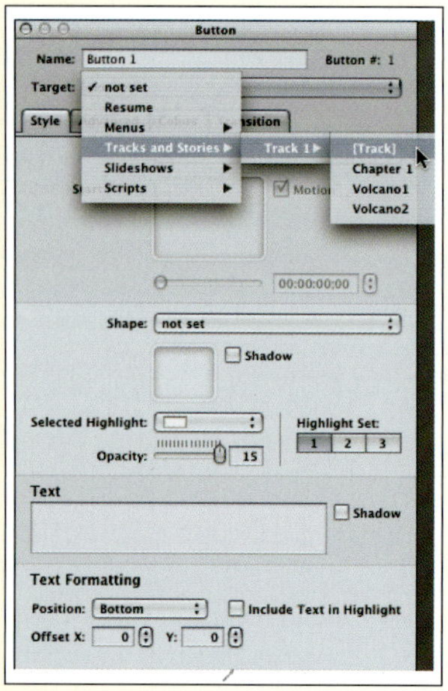

Figure 1-36. Setting a target with the Button Inspector

For this project, you'll need to make menus from the *mainmenu.psd* and *chaptermenu.psd* assets. You can rename the menus via their property sheets if you don't like using the filename to tag them. Double-click the *mainmenu.psd* icon to build the buttons shown in Figure 1-37.

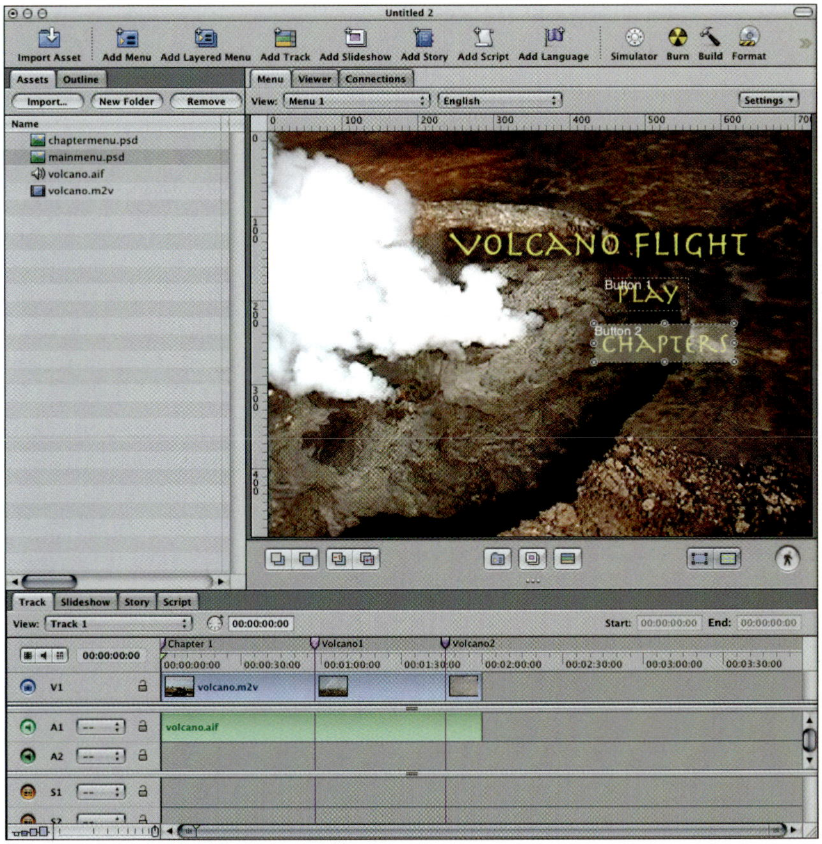

Figure 1-37. The Menu tab showing both buttons for our main menu

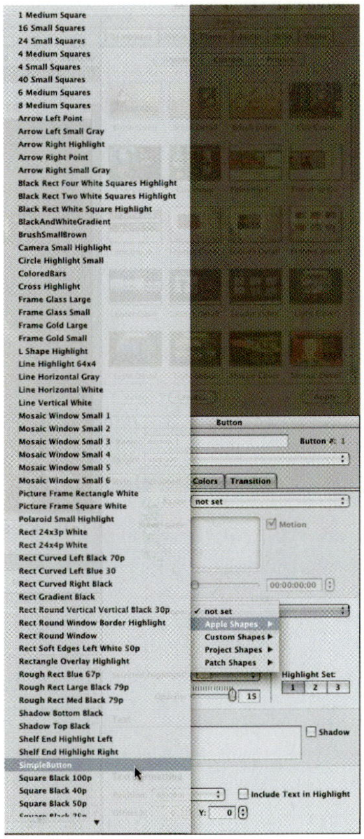

Figure 1-38. The pop-up dialog for choosing a button shape

You'll also need to pick a shape for your buttons. We just went with the Simple button option, but you can play around a bit if you want. As you can see in Figure 1-38, there are plenty of options.

Once you've got the shape, you also might need to set the level of transparency, or *opacity*. If the shape is off to the side of your text, the default of 100% opaque might be fine. Our buttons cover the text we built in Photoshop, so we want translucent buttons. You can set that opacity value in the Button Inspector. (See Figure 1-39.)

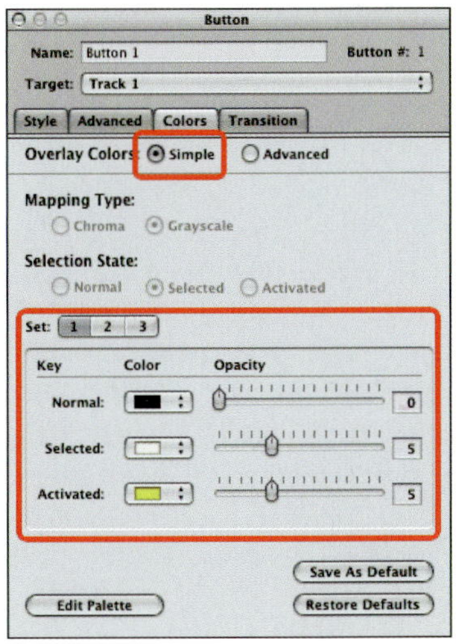

Figure 1-39. Setting the buttons' opacity in the Colors tab of the Button Inspector

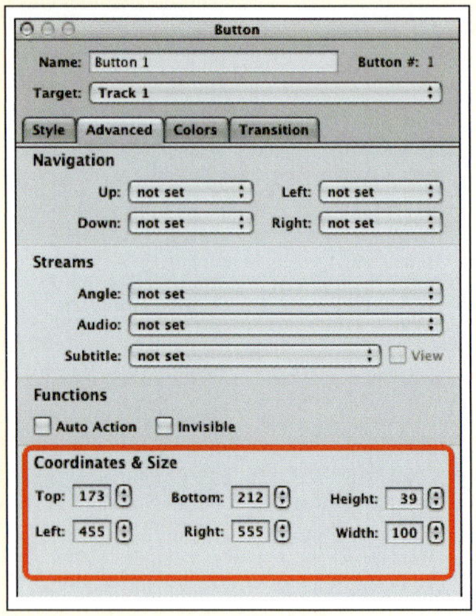

Figure 1-40. Controlling a button's coordinates using the Button Inspector

If you want to set the precise values for the buttons' coordinates, use the Advanced tab in the button property sheets, and enter the numbers shown in Table 1-2.

Table 1-2. Button coordinates for the main menu

Button	Top	Left	Bottom	Right
PlayButton	173	445	212	555
ChapterButton	228	431	274	616

Figure 1-40 shows you the Advanced tab of the Button Inspector where you can enter these values.

Click Add Menu from the toolbar. You should now have a Menu 2 option in the pull-down on the Menu tab. Select Menu 2 and set the *chaptermenu.psd* file as the background, as you did for the first menu. Go ahead and set up the four buttons shown in Figure 1-41.

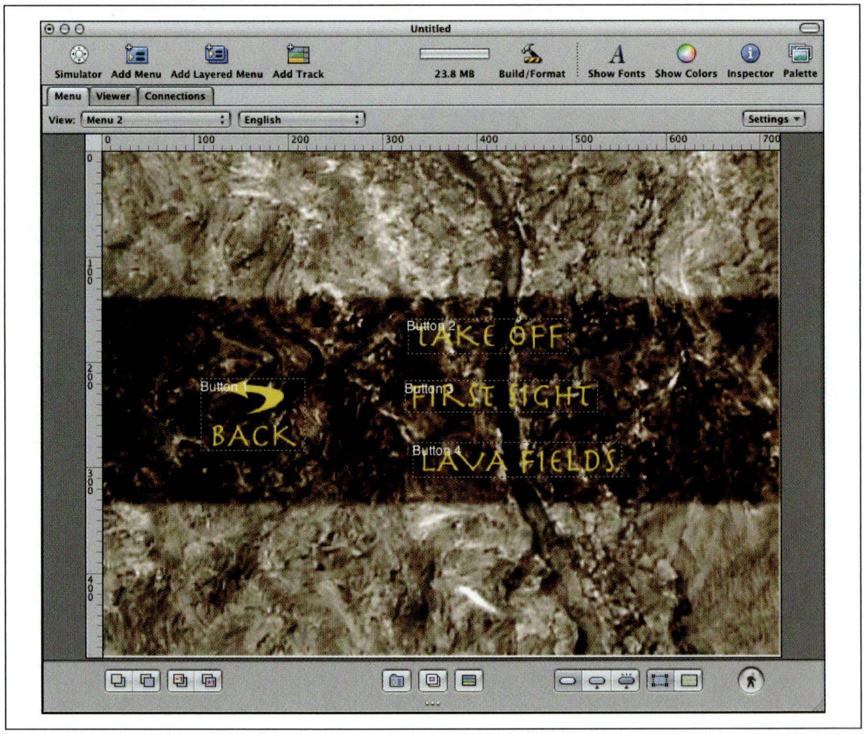

Figure 1-41. The Chapters menu button layouts

> **NOTE**
>
> *The Height values for the button coordinates in Table 1-2 and Table 1-3 are dependent on the Top and Bottom values. Changing either the Top or Bottom value will automatically change the Height value. Changing the Height value will, in turn, update the Bottom value.*
>
> *Similarly, the Width value is dependent on the Left and Right values. Changing either the Left or Right value will automatically change the Width value. Changing the Width value will, in turn, update the Right value.*

Table 1-3 lists the values for the buttons in this menu.

Table 1-3. Button coordinates for the Chapters menu

Button	Top	Left	Bottom	Right	Target
1. BackButton	160	323	193	504	Menu 1
2. TakeoffButton	210	323	210	528	Track 1 → [Track]
3. FirstSightButton	273	330	311	552	Track 1 → Volcano 1
4. LavaButton	209	101	288	223	Track 1 → Volcano 2

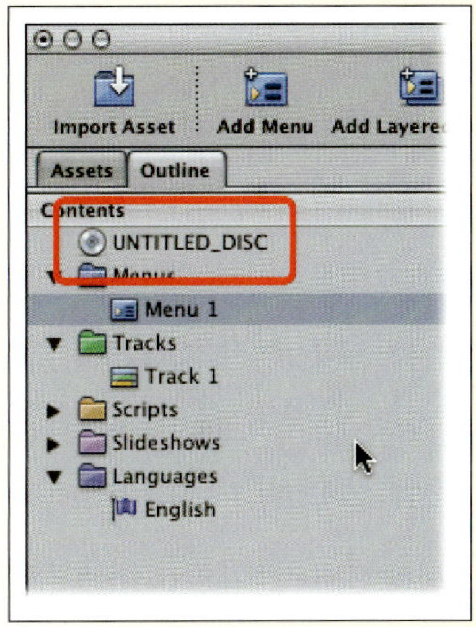

Figure 1-42. Using the Outline tab to select the disc

You'll need to go back to Menu 1 and set the target for the Chapters button now that you actually have a chapter menu. And you should definitely save your project at this point if you haven't already.

Step 3: setting the mandatory disc options

Now is a good time to make sure you've specified those "Things You Can't Forget" that we covered earlier. In the Outline tab (see Figure 1-42), click the DVD icon at the top and bring up the Inspector.

Be sure to set the following values in the Inspector and then save the project:

- Name: VOLCANO_FLIGHT
- First Play: Menu 1
- Remote-Control → Title: Menu 1
- Remote-Control → Menu: Menu 1

Previewing the disc

At this point, your DVD project should be in really good shape. You can test that theory by previewing the project. Launch the Simulator just as you did before. (You can call it up by selecting File → Simulate... as well.) Figure 1-43 highlights the Simulator button. You also can just press Option-⌘0.

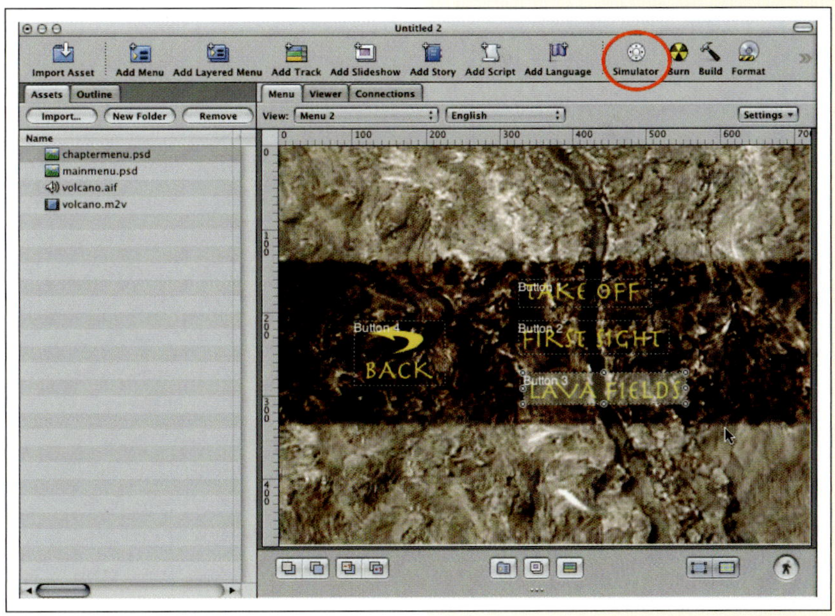

Figure 1-43. Launching the DVD Simulator

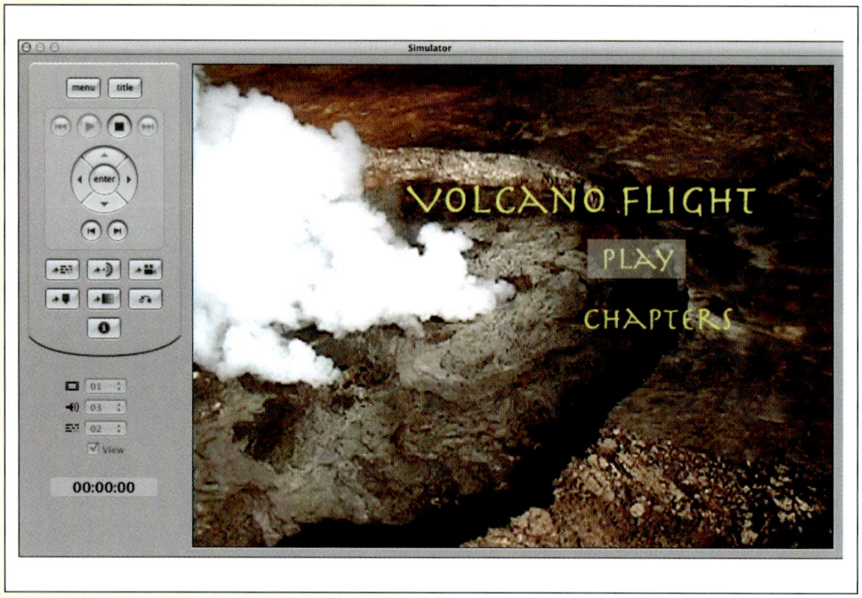

Figure 1-44. Our volcano project in Preview mode

That should launch you into a fully functional preview of your disc. You can navigate your menus using the arrow keys and the Return key. See Figure 1-44 for an example of our disc running in Preview mode.

Simulating is a great way of debugging things such as the menu buttons. It's easy to forget to specify the navigation information for every button. When you actually try to use the disc, though, things such as broken buttons pop out almost immediately. Simulating can save you the hassle of finding problems *after* you've burned the disc. Although you shouldn't worry about creating a few bad discs, that shouldn't be your goal!

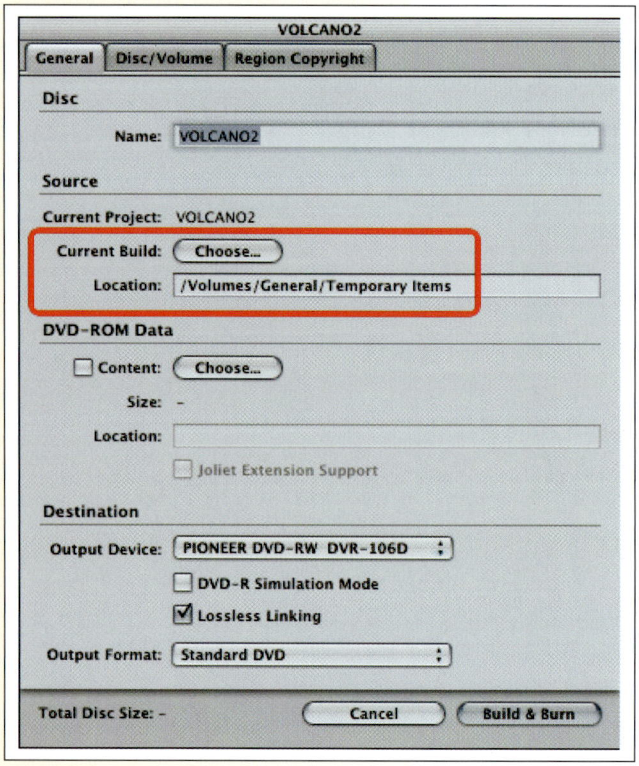

Figure 1-45. Specifying a build location

Burning the disc

OK, you're almost there. All you have to do now is build and burn the project. Although the assets already should be encoded in the proper format, they need to be assembled properly for a DVD. From the File menu, choose the Burn option. That should launch a build and format session that picks reasonable defaults for the burn.

If you need to, you can select File → Advanced Burn → Build & Format. This option enables you to customize the burn session. If you have separate drives for building the disc or multiple DVD burners, you'll need to take this route. Figure 1-45 shows you where to set your build location.

Note that you created a *Volcano* folder on a separate hard disk. You certainly can use the same disk if you have space, but separate disks make the process go a little bit faster. That way, you can read continuously from one disk and write to the other at the same time rather than reading, stopping, writing, stopping, reading, stopping, etc., on the same disk.

Next, you have to tell DVDSP to write the project to an actual DVD. If you're lucky enough to have a variety of DVD writers in your machine, pick the one you want to use. As you can see in Figure 1-46, we have the built-in SuperDrive that shipped with our G4 and an external Pioneer 106 drive in an external FireWire enclosure. We'll use the built-in burner for this example.

Now you should be prompted to insert a blank piece of DVD media. Once DVDSP recognizes the media, it'll start burning. First it builds the project and encodes your menus. Then it actually writes the project to the disc. Watch for any errors in the Log window, which is shown in Figure 1-47.

When it's done writing, you should have an honest-to-goodness, ready-to-play DVD. Most players can play these discs—but not all. You can find a number of compatibility charts online. The folks at DVDRHelp.com (*http://www.dvdrhelp.com/dvdplayers.php*) even maintain an interactive search tool that you can use to determine which players will work and which ones will not.

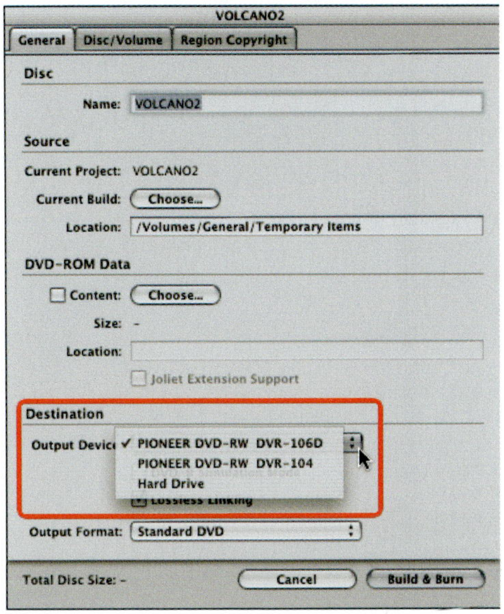

Figure 1-46. Picking a burner

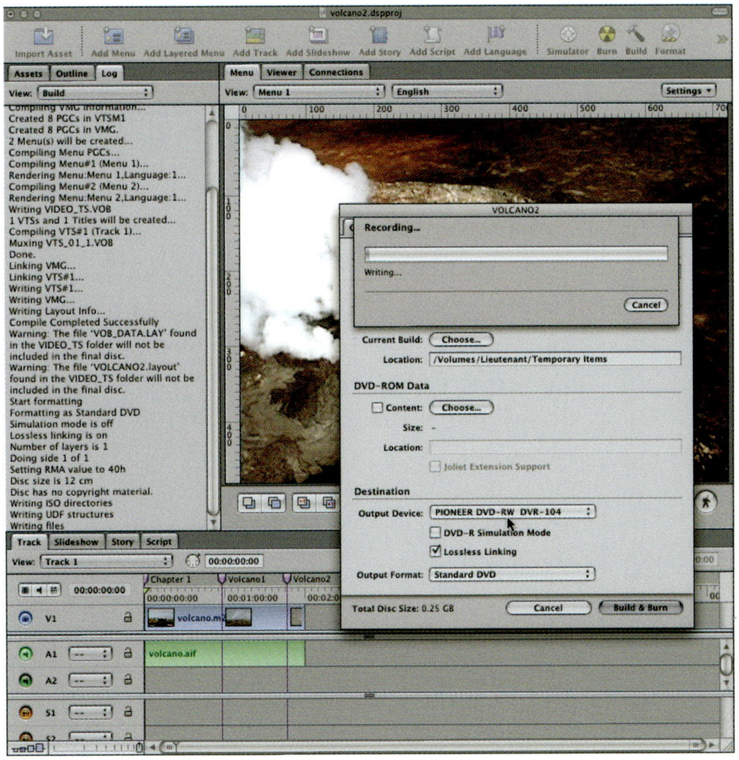

Figure 1-47. The actual burn in progress

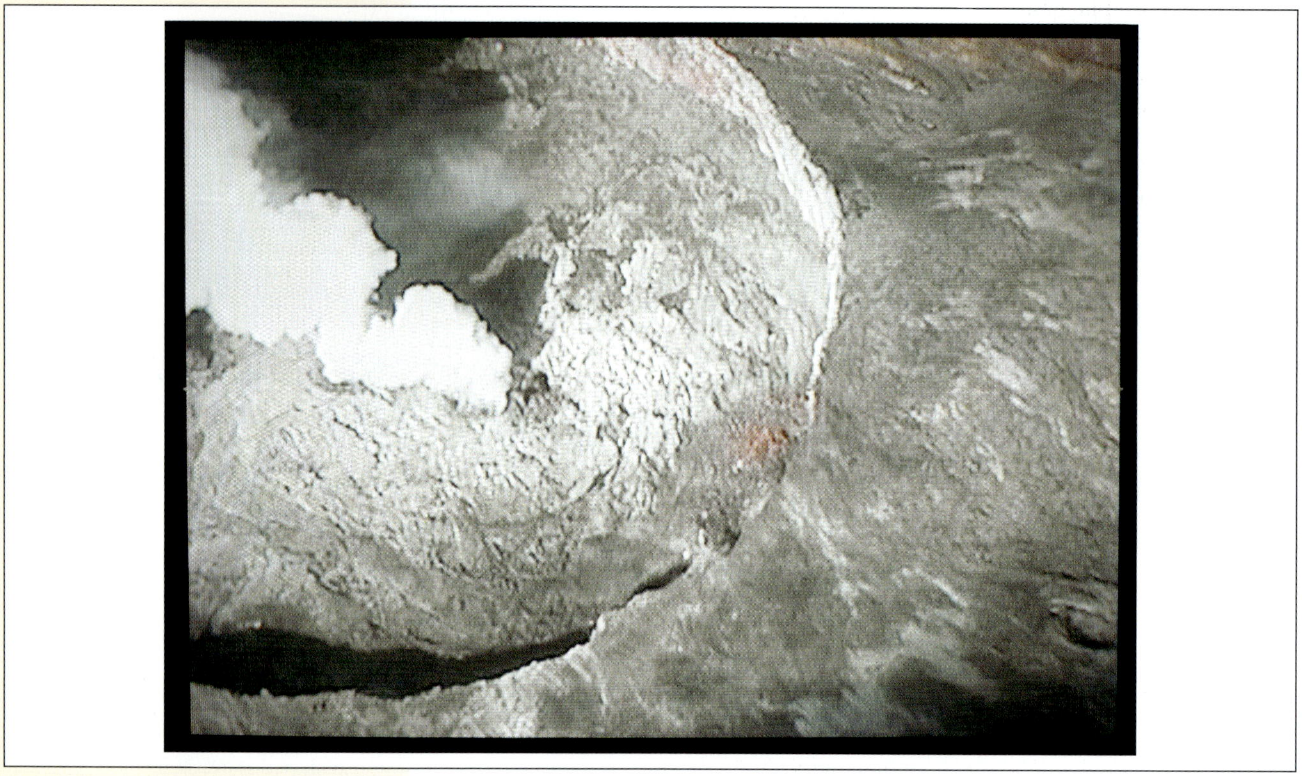

Figure 1-48. Our volcano project playing on a consumer television set

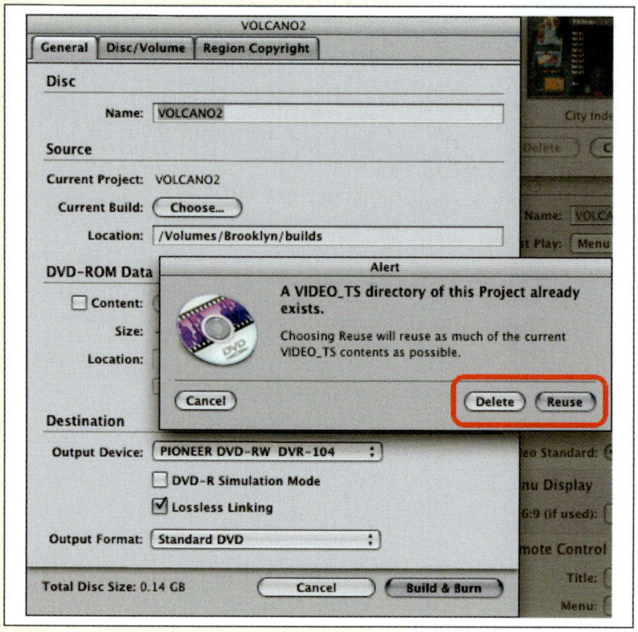

Figure 1-49. Reusing information from a previous burn when making another copy

Figure 1-48 shows a screen capture from a TV playing our spiffy volcano DVD.

Making multiple copies

If you want multiple copies of this disc, you can just go back and choose the Format option from the toolbar. This time around you'll be prompted to reuse the existing project's data, as shown in Figure 1-49. Click the Reuse button, and you can skip over the menu encoding step.

Playing back your DVD on consumer DVD players

For those of you who are new to the DVD world, let's take a quick detour and talk about playing your DVD in a consumer device, which is usually the goal for these projects. Let's start off by making sure you know how to navigate your menus with the remote controls that come with such players. Figures 1-50 through 1-54 show a consumer remote (from a Yamaha DVD-S796), the Simulator remote control, and the DVD Player remote control with various function categories highlighted.

The VCR functions (Figure 1-50), which are available on both consumer remotes and the software DVD players, behave exactly as you would expect them to behave. You can play, stop, pause, and fast-forward/fast-rewind through the video. Most fast-forward/fast-rewind buttons double as chapter-forward/chapter-back buttons as well.

The arrow and Enter buttons highlighted in Figure 1-51 enable you to move the "cursor" around the onscreen menus and select a highlighted option. The Enter button is sometimes labeled Select or OK.

The more interesting buttons to note are the stream controls (audio, video, and subtitles) and the various menu buttons, shown in Figure 1-52.

DVDs support a fancy variety of audio, video, and text streams. Most DVDs have menus that enable you to configure which streams you want to see (a Spanish audio track with French subtitles, for example). In addition,

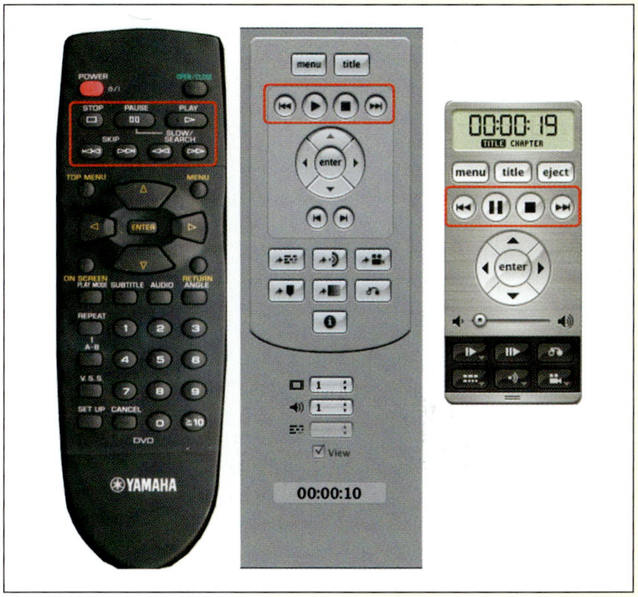

Figure 1-50. VCR functions on consumer player and software remotes

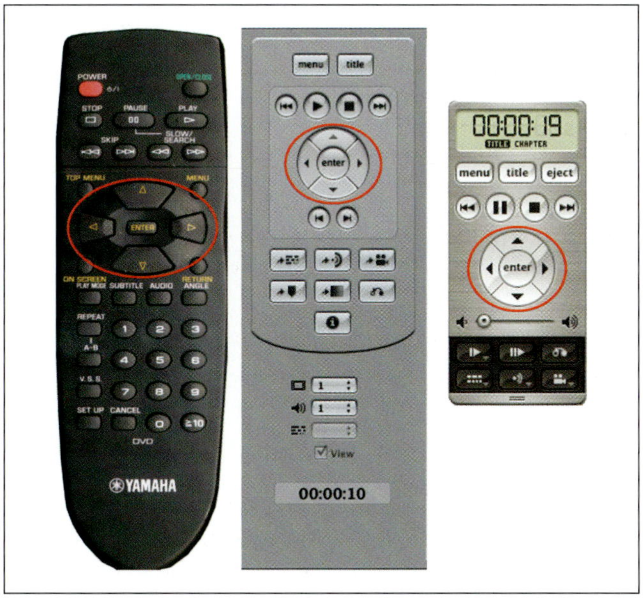

Figure 1-51. Onscreen navigation functions on consumer player and software remotes

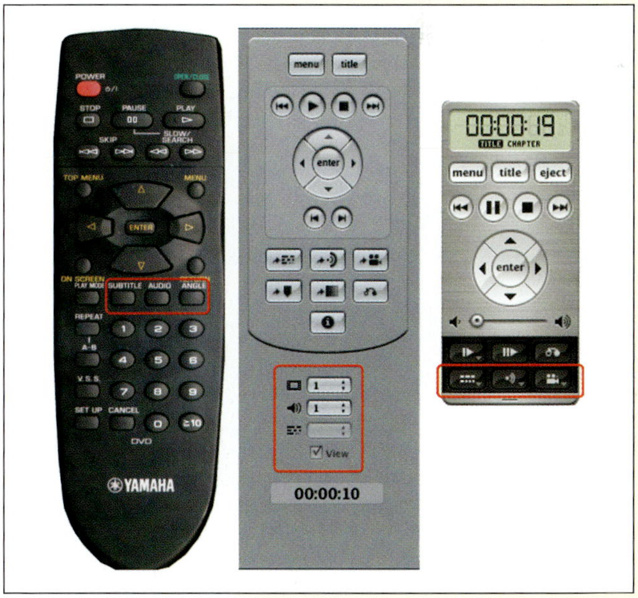

Figure 1-52. Stream control functions on consumer player and software remotes

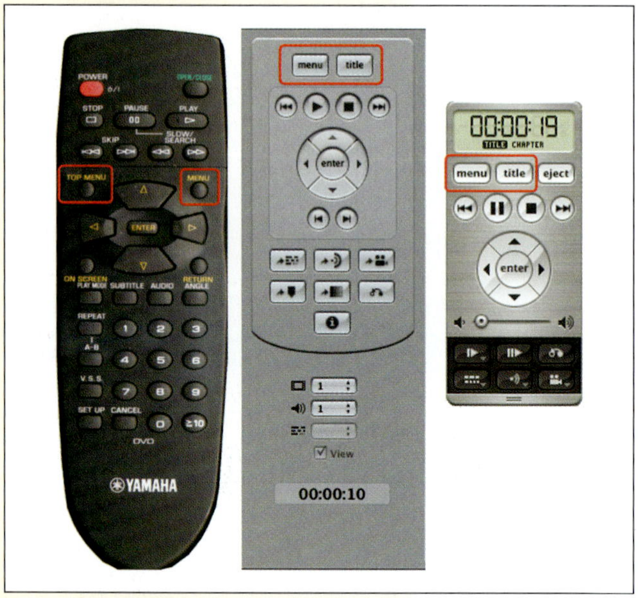

Figure 1-53. Menu functions on consumer player and software remotes

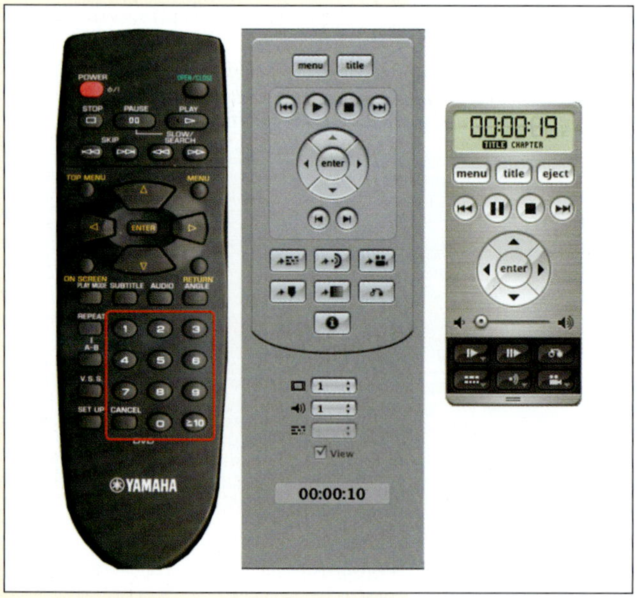

Figure 1-54. Numeric input functions on the consumer remote

most DVD players also enable you to switch streams on the fly. The stream control buttons perform that service. You can cycle through all the available streams. Not every DVD will have multiple streams, of course. We'll be discussing streams and how to use them extensively in Chapter 5, where we'll cover our international project.

Most DVD remotes also have two separate menu buttons: a button labeled Top Menu (sometimes called the Title, Root, or Disc menu) and one labeled Menu, highlighted on the far left in Figure 1-53. The Top Menu button is supposed to take you back to the first menu of the disc. The Menu button should take you back to the most recently viewed menu. (Or, if you just used the Menu button to get to a menu, pressing it again often resumes playback where you left off.) Chapters 3 and 4 dive into menus and show you how to configure your remote control menu buttons if you don't like the default behaviors.

Last but not least, you probably noticed that the consumer remote has numeric input keys (shown in Figure 1-54). You can use these buttons for various things. Some discs allow time searching so that you can enter a time—usually in minutes and seconds—and jump to that point in the video. While viewing a menu, you often can use the numeric buttons to immediately activate a given onscreen item. Chapter 8 covers using numeric input for fun and games.

So, what's next, you ask? Chapter 2 covers the various media types (audio, video, and still images) you can use in DVDSP. If you're already up to speed on that topic, skip Chapter 2 and get to the fun stuff in Chapter 3.

Media Formats

2

As with any computer application, DVDSP 3 has a well-defined set of document types it understands. Thankfully, that set is large. Although you'll probably still format your media as MPEG-2 before getting into DVDSP, now you can import almost any file format that QuickTime understands. (DVDSP can transcode most QuickTime files to MPEG-2 for you.)

As we mentioned in the Preface, you can't use DVDSP by itself. You need the MPEG-2 codec that is bundled with QuickTime Pro. In addition, you need A.Pack for your audio encoding. Technically, you could get by with those three applications, but that's not a realistic setup. You definitely want Photoshop, and probably Final Cut Pro and After Effects as well. With those applications in hand, you easily can manage any type of DVD project.

In this chapter

DVD Explained

Before we dive into DVD media formats, let's look a bit closer at the media itself. Here's an interesting bit of trivia: what do the letters D-V-D stand for? Those of us who remember the original advertising for DVDs in the mid-1990s recall that the letters stood for Digital Video Disc. Later, however, the acronym somehow revised itself to stand for Digital Versatile Disc, which truthfully implied that the disc format could hold much more than video (e.g., data and audio). Depending on which corporate patent holder you ask, you're likely to get one of those two responses. However, the DVD Forum, the current governing consortium of more than 200 corporate members, has not come to an official decision as to what the letters stand for. And neither of the preceding definitions is considered valid today. So, if someone asks what DVD stands for ... well, it just stands for "DVD."

As with most standards, DVD started off with two fiercely competing video standards. Fortunately for us, it didn't stay that way. In 1993, companies such as Philips, Sony, and Nokia released the Multimedia Compact Disc (MMCD). Soon afterward, the competitive Super Density (SD) format was proposed, backed by companies such as Matsushita, Thomson, and Toshiba. Because both sides had extensive patents in the field and were unwilling to give up their share of potential royalties, an agreement on a common video format stalled out. In 1995, however, IBM and various Hollywood studios pressured all the companies involved to bypass the mistakes of the VHS/Betamax era, and in September 1996, a unified standard was agreed upon: DVD.

Here are some of the features of the DVD format that make it superior to older video formats:

- DVDs have interactive, animated menus supporting simple list selection and modest amounts of memory.
- DVDs can hold up to eight hours of high-quality (MPEG-2) digital video, although most commercial DVDs can hold up to four and most burnable DVDs hold up to two hours.
- DVDs can contain up to eight digital audio tracks, which you can use for alternate languages, directors' or actors' commentaries, or music-only soundtracks, as well as 32 subtitle tracks.
- You can encode video to represent common aspect ratios of television (4:3) and anamorphic ("widescreen") film (1.85:1 and 2.35:1).
- DVDs can randomly access program content. This means you can jump to any point in a disc instantaneously. This is useful for chapter searches and video branching features, which can offer everything from alternative storylines to animated "choose-your-path" adventure games.

In short, DVDs had so much going for them over the previous formats of choice, VHS and Laserdisc, that their adoption was practically guaranteed.

How do DVDs work?

The short answer is: much like normal CDs. In fact, if you get out your ruler, you'll see that they're identical in size: 120mm in diameter, 1.2mm deep, with a 15mm diameter hole in the center. The beauty of DVDs is, in fact, a little deeper.

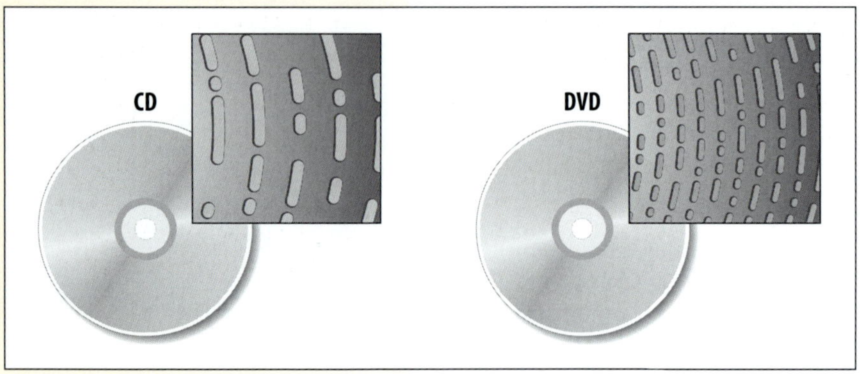

Figure 2-1. Physical construction of a CD and DVD

With both CDs and DVDs, an extremely accurate laser is used to burn depressions, called *pits*, in a substrate material sandwiched between two clear, protective plastic layers. CDs consist of only a single layer on the bottom side; DVDs, on the other hand, can have up to two layers on each side, for a total of four layers. A similar laser then can be used to "scan" the pits, reconstructing the data (see Figure 2-1).

Pits on both CDs and DVDs are organized in a running track that continuously spirals around the disc. In the first layer of a DVD, the spiral starts at the center and expands outward to the edge. In the second layer, however, the path can start at either the inside of the disc and spiral outward like the first, or start at the outside of the disc and spiral inward. The latter is called Opposite Track Path (OTP), and is frequently used to keep video playing continuously; resetting the heads back to the interior of the disc can cause older players to pause anywhere from 1 to 3 seconds.

The *1*s and *0*s on CDs and DVDs happen based on the basis of the pits and lands. A 1 occurs when you switch from a pit to a land or back. A 0 occurs wherever there is no change (inside a pit, or along a land). The length of the pit along the spiral determines the value of the data, as shown in Figure 2-2.

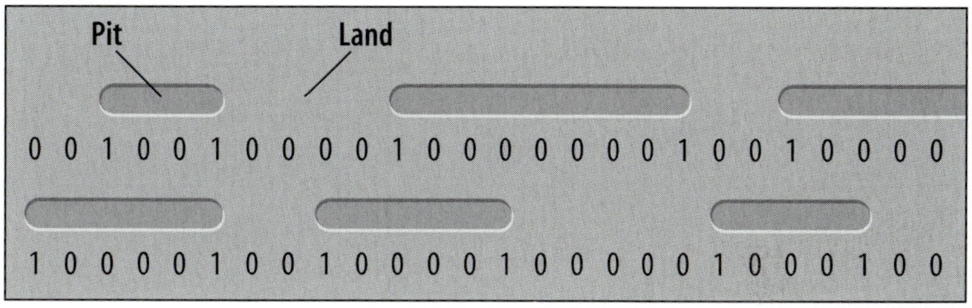

Figure 2-2. Data storage on CDs and DVDs

With DVDs, the spirals are denser (i.e., more tightly packed along the radius) than on CDs. In addition, the data-pit length value is interpreted as half that of a CD. For example, the minimum pit length of a CD is .83 micrometers. The minimum pit length of a DVD, on the other hand, is around .4 micrometers—that's half the size!

Because of the density of the information, DVD players also differ from CD players in that they use a narrower laser to read the pits, either 635 or 650 nanometers depending on the purpose of the DVD. A CD player, on the other hand, uses a 780-nanometer laser. And although it's difficult to tell, a DVD spins twice as fast as a CD inside the player. What all this boils down to is that DVDs have a much larger data-transfer rate: up to 10.08 megabits per second. That's more than seven times the transfer rate of CDs, at 1.4112 Mb/s.

BACKGROUND INFO

Sequential Data?

Just for the record, the *1*s and *0*s that occur along the spiral don't correspond to sequential data bits. The data bits are re-encoded to channel bits in a way that aids players in performing error correction and laser tracking. The exact process that a DVD player follows is extremely complex, and you'll probably never need to know it. However, if you're curious about the details, visit the specification web site for Pioneer Corporation in Japan, *http://www.pioneer.co.jp/crdl/tech/dvd/2-3-e.html*.

DVD types

Earlier, we mentioned that DVDs can have up to two layers on either side, for a total of four layers. Of course, if you've rented a DVD recently, you've likely discovered that the top of many commercial DVDs contain a "poster" print of the program they carry. Why is this? Because two layers are usually good enough to hold four hours of movies—most movies do not go beyond this.

Actually, you can mix and match layers and sides into four different combinations, the designations of which are shown in Table 2-1. Looking at the table, you'll quickly note that the second layer of a DVD-9 contains slightly less data than the first layer. This is because the data pits are spaced slightly further apart, which allows for more accurate reading by the laser through the semi-transparent first layer. Also, you might have noticed that dual-layer, dual-sided DVDs are not common. This is because DVD-18 discs are more difficult to create than DVD-5, DVD-9, or DVD-10 discs, so they are not in widespread use yet.

Table 2-1. DVD types and their capacities

DVD Type	Sides	Layers Per Side	Storage Capacity
DVD-5	1	1	4.7GB
DVD-9	1	2	8.54GB
DVD-10	2	1	9.4GB
DVD-18	2	2	17.08GB

Figure 2-3 shows a cross section of the four types of DVDs listed in Table 2-1.

In the next two sections, we'll talk about the physical and application formats of a DVD, and the differences between the two.

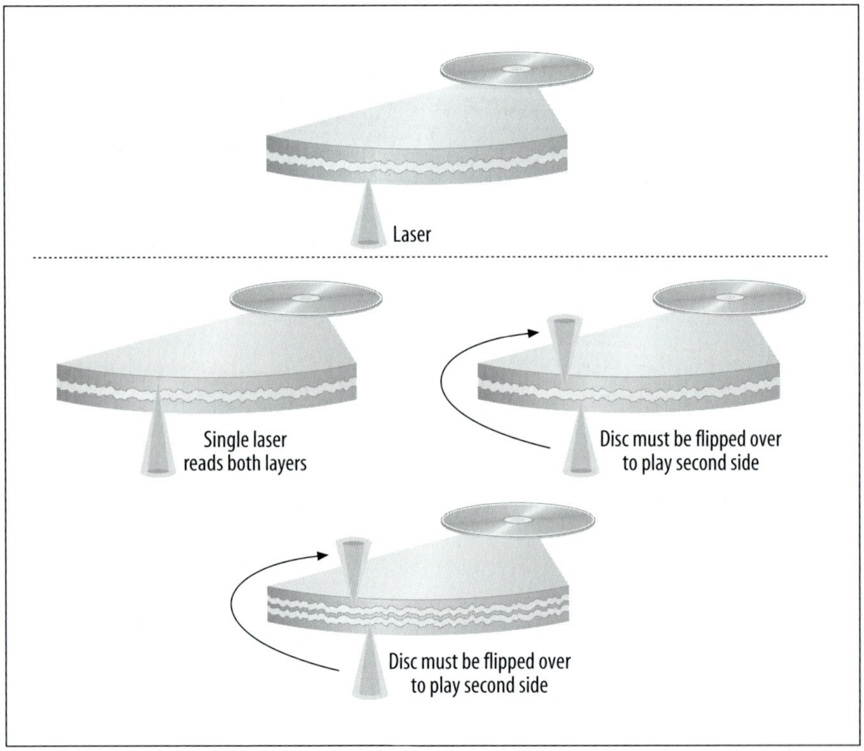

Figure 2-3. Cross section of four DVD types

DVD physical formats

There are nine types of physical formats, and they are shown in Table 2-2.

Table 2-2. DVD physical formats

DVD Format	Read/Write	Comments
DVD-ROM	Read-only	Base-level format. Works in all players.
DVD-R (General)	Write-once	Uses a 650nm laser.
DVD-R DL	Write-once	Dual-layer DVD-R.
DVD-R (Authoring)	Write-once	Can be used to generate Cutting Master Format (CMF) information, which is used for DVD replication (although it currently is not recommended). Uses a 635nm laser instead of a 650nm laser. Most replication houses now accept DLTs or normal DVD-R (General) discs.
DVD-RW	Rewritable	Can be rewritten about 1,000 times, but the entire disk must be written in one session (i.e., all at once). Not very compatible with consumer players.
DVD-RAM	Multiple session rewritable	Provides random access for writing, much like a floppy disk. Can be rewritten up to 100,000 times. Not compatible with many DVD players.
DVD+R	Write-once	The +R format is a media standard that competes with –R. DVDSP 2 supports only the –R format. Most Apple systems ship with –R writers, but with a +R or +/-R drive, DVDSP 3 can burn to +R media.
DVD+R DL	Write-once	Dual-layer DVD+R.
DVD+RW	Rewritable	The +R equivalent of the –RW format. Also can be rewritten about 1000 times.

Here are a few additional notes to keep in mind:

- DVD-R has two specific formats, *general* and *authoring*. The general format is what most developers will use for data or video that they do not intend to replicate. You can, of course, still duplicate these discs—the big difference between duplication and replication boils down to quantity. Duplication is cheap and good for short runs. Replication, on the other hand, can be expensive. It involves creating a glass master and is good for large runs.

- Authoring DVDs can contain more information, such as CMF data, which is designed to be used during the replication process. If you're using an Apple Superdrive inside your Macintosh, you can burn only DVD-R General discs. If you need to burn DVD-R Authoring discs, you'll need the Panasonic DVR-S201 or a similar DVD-R burner attached to your Macintosh.
- DVD-RAM is a random-access format that you can use to write information much like a floppy disk (in fact, these discs often come in a cartridge). In other words, the entire disk does not have to be written in a single session, unlike DVD-RW.
- DVD+R and DVD+RW are newer formats. DVD+RW promises the ability to have random access much like DVD-RAM discs. DVDSP 3 supports these formats.
- At the time of this writing, several dual-layer burners were available. The media is still fairly expensive, but that price will certainly drop. Appendix C has more details on this technology and how to use it with DVDSP.

As you likely guessed, the multitude of physical formats is due primarily to corporate competition. Table 2-3 outlines the major players and their backing.

Table 2-3. Physical formats and corporate backers

DVD Format	Corporate Backers
DVD-R/RW	Pioneer, Apple
DVD-RAM	Panasonic, Toshiba
DVD+R/RW	Sony, Philips, Hewlett-Packard

As noted in Table 2-3, Apple currently ships systems with DVD-R drives, so you'll likely spend most of your time creating DVD-Rs with DVDSP. In DVDSP 2, that was your only option. With the advent of dual-format drives, it won't be long before systems ship with the ability to burn to both formats. Currently, DVDSP 3 will happily burn to +R media if you have attached your own +R (or dual-format) DVD burner. Drives such as the Pioneer A06 and A07 work as advertised now with all single-layer media. It remains to be seen whether DVDSP 3 will support the consumer dual-layer burners for DVD-9 projects.

DVD application formats

DVD application formats are built on top of DVD physical formats. In effect, they define how the data is organized on the disc, and what device is the target of the data. There are many types of DVD application formats, including:

- DVD-Video
- DVD-Audio
- DVD-Video Recording (DVD-VR)
- DVD+RW Video Recording (DVD+VR)
- Super Audio CD (SACD)

In addition, there are special application formats for game consoles such as the Sony PlayStation 2 and Microsoft Xbox. For this book, however, we are interested primarily in the DVD-Video application format, but we'll take a short look at some of the other common formats as well.

DVD-Video

DVD-Video is the format that all DVD movies are created with. Video content is created from master footage and compressed using the MPEG-2 format. This is what we're here to deal with. As you probably know, you can have multiple video angles, multiple audio tracks, and multiple subtitle tracks for any given piece of video. DVD-Video also handles simple slideshows and has some reasonably fancy interactive capabilities.

DVD-Audio

DVD-Audio is not just the audio part of a video DVD. It's a separate high-fidelity audio format that improves on the world of CD audio. Instead of stereo-bit samples taken at a frequency of 44.1kHz, most DVD-Audio recordings are 24-bit samples at 96kHz over six channels. A proprietary packing scheme known as Meridian Lossless Packing allows such large quantities of information to fit in the throughput limits which are similar to DVD-Video (9.6Mbps).

DVD-VR and DVD+VR

The DVD-Video Recording format essentially enables you to create "editable" DVD movies. You can change menus or add new content to an existing disc. This format isn't terribly common at the moment, but it is gathering adherents. You can find high-end home recorders that use the VR format.

As with DVD-Video, the "+" and "–" here mean competing versions of the format. In the VR format, however, the "+" variation appears to have a distinct advantage in that most +VR discs should play in home players that normally handle +R media.

SACD

Super Audio CD (SACD) is a complete redesign of digital audio. Sony and Philips developed a new process known as Direct Stream Digital. The recording and playback happens at an astonishing 2.8Mbps. That's right—M—bits per second. This rate allows DSD to avoid chopping up the recording (and conversely interpolating the playback). The result is clean, analog-like sound.

SACD discs are not compatible with most CD or DVD players, but many discs have dual layers: one layer of standard CD music, and one SACD layer that SACD players can use.

DVD video recorders

Although clearly you're here because your computer has a DVD burner, you also might have one of the new home DVD video recorders. These devices hook up like good old-fashioned VCRs, but they record the video to DVD media. You can find both –R and +R format recorders. You even can find TiVo™ DVD recorder combinations out there. That opens up some really fun archiving options.

As the media and devices become cheaper and cheaper (and that trend has begun already), these types of recorder will become more and more popular. Currently, you can't use them directly with a computer, but that might change as computers and home entertainment become more intertwined. (The few recorders with FireWire inputs expect DV streams from your camcorder, not data streams from a computer.) In the meantime, back to DVDSP.

Can I copy a commercial DVD?

Theoretically, no. DVDs have two common forms of piracy protection: Macrovision and the Content Scrambling System (CSS). DVDSP supports both of them. That is, DVDSP enables you to add them to your own projects—not strip them from existing discs.

Macrovision is designed to prevent you from cabling the output of your DVD player into the input of your VCR and making a VHS duplication of a movie. It works by scrambling a synchronization signal that the VCR needs to accurately decode the analog picture. If you record a Macrovision-protected DVD onto a VHS tape, you'll get a signal with banding and a warped picture, similar to what you would see if you tuned into a scrambled cable channel.

With CSS, the data on a commercial DVD is protected from unauthorized applications by a 40-bit key (i.e., a number somewhere between 0 and 2^{40}). Data in each sector of the DVD is scrambled using this secret key, and the decoding key for the data is embedded in an unreadable section of the DVD, as well as in the hardware of some of the players.

Note that we say theoretically. The CSS copy protection scheme didn't work out quite as well as the DVD creators had planned. A few years after DVDs became mainstream, a group of programmers in Europe (the now well-known Jon Johansen among them) broke the CSS key using a "brute force attack." In other words, this team programmed a number of computers to divide and conquer deciphering a DVD, using each and every possible key. The hope was that they eventually would discover the one key that worked. And in September 1999, they did indeed find one. A short while later, a program known as DeCSS was released on the Internet.

The resulting uproar throughout the entertainment industry was swift. Police tracked down and raided Johansen's home in Norway. The Motion Picture Association of America (MPAA) quickly sent threatening legal letters to any web site that allowed the download of the DeCSS tool, as well as any web site that even linked to it. At present, a number of legal battles continue to work their way through both U.S. and international courts.

Spilling the Beans

In an unusual bit of irony, the MPAA actually disclosed much of the "confidential" CSS information in court as a piece of evidence. As a result, this information is part of the public court records of Los Angeles County and is available on the Internet.

In the meantime, several programs out there will offload the contents of a DVD and decrypt them. However, before you decide that you're going to be the next illegal copyright infringer, consider the following practical facts:

- The cost of buying a DVD is pretty low—typically \$15–\$20, sometimes more for extra features. The cost of renting one is even lower.
- If you've ever obtained any programs that attempt to copy DVDs by attacking the CSS algorithm (e.g., DVDXCopy), the first thing you probably noticed—and the last thing the program's creators tell you—is that you have to either compress the movie further, or use more than one DVD to store a perfect copy of the offloaded data. Why is that? Because commercial DVDs are single-side, dual-layer DVDs (DVD-9s) and the discs that you're copying them to are probably single-side, single-layer discs (DVD-5s). Further compression reduces the quality of the image, and splitting across discs often has the added nuisance of losing some settings when you eject a disc. For example, watching bonus options such as a director's commentary on the second disc can be difficult or impossible. (And even the pending DL media will be only moderately cheaper than just buying another original disc.)

Region codes

If you've ever considered purchasing a DVD from another country, you've probably run into region codes. Discs and players can be branded with a particular region (or set of regions). These region codes restrict the use of discs and players. Discs from region 1, for example, will play only in region 1 players. Many discs are "region-free" and will play in any player.

In addition, you can (legally) purchase region-free players that will play any DVD. You should be aware, however, that DVDs from different countries might not use NTSC for their encoding standard. Many region-free players output whatever signal the disc was encoded with. So, a PAL DVD from region 2 will play fine in a region-free player, but you have to have a PAL television set to watch it.

Often you can tell a software DVD player to use a particular region code. The DVD Player application that ships with DVD-equipped Mac computers, for example, can switch region codes five times before settling on a final region. (This makes it easier to ship the software with machines sold around the world.)

You can use DVDSP to set a region code if you need it. The following list identifies the countries associated with each region code shown in Figure 2-4:

1. U.S., Canada, and U.S. territories
2. Japan, Europe, South Africa, and Middle East (including Egypt)
3. Southeast Asia and East Asia (including Hong Kong)
4. Australia, New Zealand, Pacific Islands, Central America, Mexico, South America, and the Caribbean
5. Eastern Europe (former Soviet Union), Indian subcontinent, Africa, and North Korea
6. China and Tibet
7. Reserved
8. Special international venues (airplanes, cruise ships, etc.)

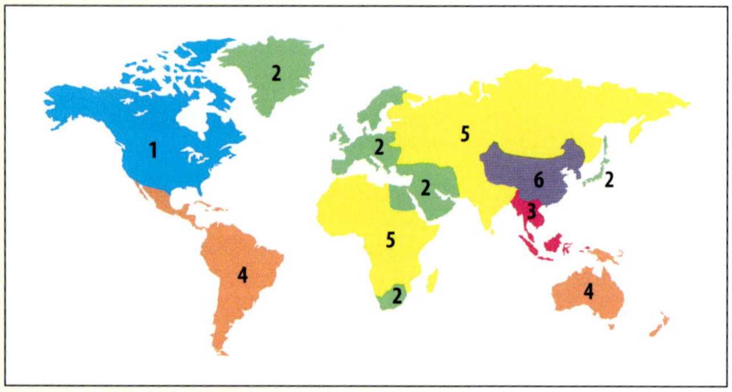

Figure 2-4. DVD region codes

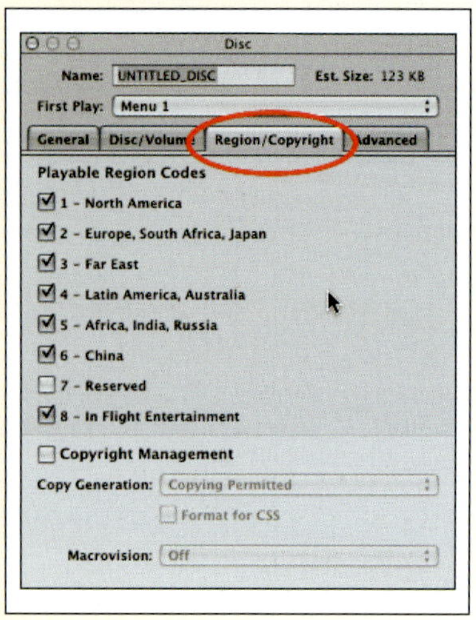

Figure 2-5. The Inspector region options for a disc

Figure 2-5 shows DVDSP 3's Disc Inspector and the region code options in the Advanced tab. Notice that you can select multiple regions for any given disc. After all, most homebrew projects don't require a region code restriction.

Supported Formats

The DVDSP User's Manual recommends MPEG-2 encoded video with AC3 encoded audio. For impatient readers, that's the primary format you need to be concerned with. To be honest, that recommendation came from the 1.5 release of DVDSP, but it still applies. If you create your own MPEG-2 (*.m2v* is the common file extension) and AC3 (*.AC3*) files, that leaves you with maximum control over your audio and video. DVDSP can accept many other formats, but it will convert them to MPEG-2 and AIFF (audio) before burning the disc. You'll get good results, but you probably will not get the best possible results. For the more curious and detail-oriented, read on!

Video

The technical details on what video formats are supported are fairly straightforward, as shown in Table 2-4. Owing to the vast improvement in quality provided by MPEG-2, we'll work with that encoding for all the projects in this book.

Table 2-4. Video standards in DVDSP

Format	Maximum Bit Rate (Mbps)	NTSC Resolution/ Frame Rate (fps)	PAL Resolution/ Frame Rate (fps)
MPEG-1	1.8	352×240/29.97	352×288/25
MPEG-2	9.8		
Full D1		720×480/29.97	720×576/25
Cropped D1		704×480/29.97	704×576/25
1/2 D1		352×240/29.97	352×288/25

You might have some legacy MPEG-1 video lying around, and if that's the case, try importing it into DVDSP. If the video is not in a format that DVDSP can understand, the software will generate an error when you try to import it. That's the easiest way to verify the file format. To use MPEG-1 with DVDSP, you need to make sure the resolution and the frame rate are correct. Also note that DVDSP expects elemental streams—one stream for audio, one for video. Many MPEG-1 movies out there have multiplexed (or *muxed*) audio and video. Tools are available to "demux" and extract tracks, but they aren't part of QuickTime Pro or DVDSP.

Compression rate project

Let's take a quick detour and build a test for these compression discussions. We'll use a fairly simple clip of time-lapse cars at an intersection compressed at different rates to compare them during playback. We'll also take this opportunity to make "video buttons" in our menu.

1. **Create a New Project and Import the Required Assets.** In the *projects/ch02* folder, you should find five video files:
    - *cars_high_cbr.m2v*
    - *cars_low_cbr.m2v*
    - *cars_medium_cbr.m2v*
    - *cars_ntsc_1pass.m2v*
    - *cars_ntsc_2pass.m2v*

LEARNING MORE

More Detail Than You Want

If you're looking for some of the really low-level stuff, there are plenty of resources on the web. One of the best overall starting points is the DVD FAQ maintained by Jim Taylor:

http://www.dvddemystified.com/dvdfaq.html

His book, *DVD Demystified, 2nd Ed.* (McGraw-Hill, 2000) is an unparalleled source of technical, readable details on DVDs and surrounding technology.

On the AC-3 front, you should take a look at 5dot1.com (*http://www.5dot1.com/*).

BACKGROUND INFO

What Is a Group of Pictures?

While we're on the subject of video encoding, it's probably worth mentioning the term GOP. Far from a political statement, an MPEG GOP stands for *group of pictures*, and represents a fundamental compression unit. A GOP can be closed (think self-contained) or open (relies on a neighboring GOP). The DVDSP User's Manual and the DVD Demystified site (*http://www.dvddemystified.com*) have more of the gory details, but here are a few quick notes of interest:

- Multi-angle video requires closed GOPs.
- QuickTime will encode video content to MPEG-2 with closed GOPs.

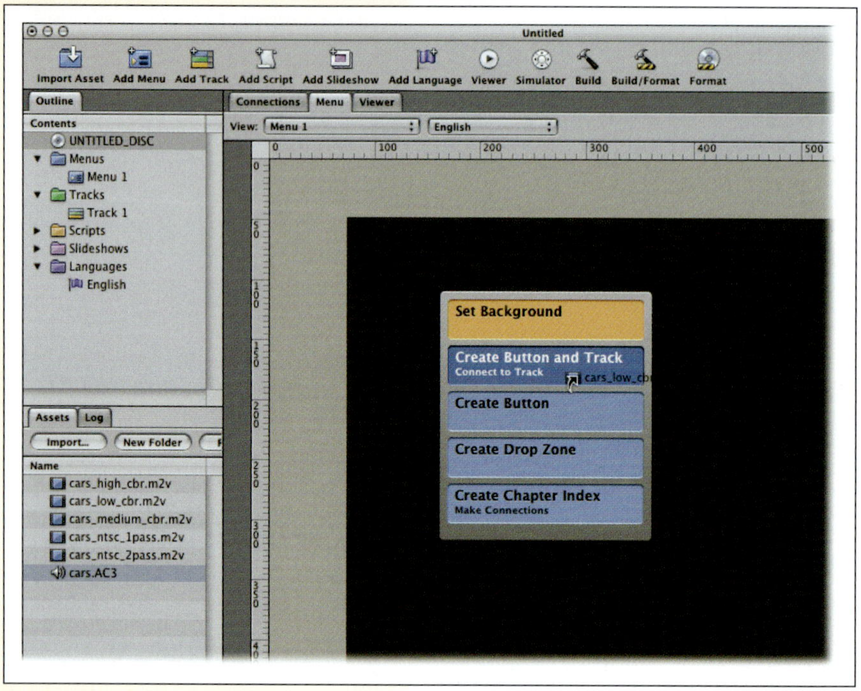

Figure 2-6. Creating a video button and an associated track for a menu

Start a new project in DVDSP and import these assets along with the *cars.AC3* file. If you don't want to play with all five files, you can pick just the ones you want by holding down the ⌘ key and clicking the desired files.

Because we're creating a test project, we won't worry about creating a fancy menu background. We can use the video assets to create buttons and tracks all at the same time. This works perfectly for this project because we just want separate clips that return to the menu when they finish playing. Bigger projects are not usually this easy, but DVDSP tries to make as many of these simple projects downright...simple.

2. **Create a Button.** Drag the *cars_low_cbr.m2v* file over to the menu and select the Create Button and Track option, as you see in Figure 2-6.

   That should leave you with a "video button" that you can arrange on the screen in any way you want. The default size is just that—a default. You can resize the button to fit your needs as well. If you want to see the button's motion, remember to press the walking-person icon in the lower-right corner of the Menu tab to get it started. Shut off the motion once you're satisfied the video plays.

3. **Adjust the Loop Time.** By default, you'll get the first 30 seconds of the clip in a loop. You can adjust the video-loop duration using the Property Inspector for the menu, as shown in Figure 2-7.

   We went with 20 seconds just to give our video buttons a little more leeway on their starting points. The clips themselves are only 30 seconds long. We want to make sure each button has a full 20 seconds of footage to show. This enables us to start the loop for any given button up to 10 seconds into the clip.

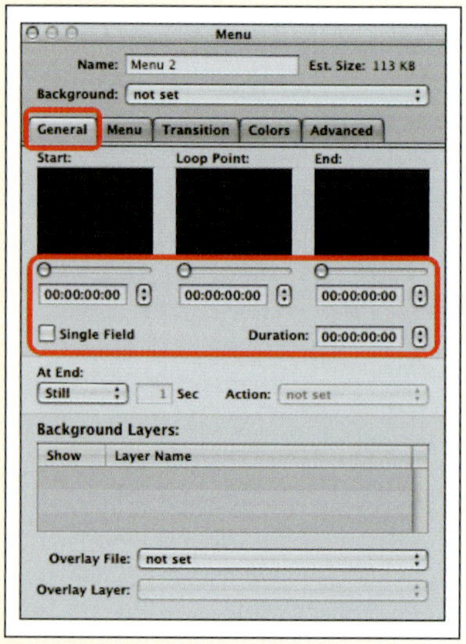

Figure 2-7. Adjusting the loop duration for all video buttons in the menu properties

4. **Adjust the Starting Time.** The particular starting point for your video button's loop is set on the button's Property Inspector, as shown in Figure 2-8.

Go ahead and label this button as the "Low Quality CBR" button. You can type the text and set its position in the same Inspector using the Style tab. Figure 2-9 highlights the text categories on the Style tab.

That should leave you with a screen similar to that shown in Figure 2-10, with the video button in place.

5. **Repeat the Process for the Other Clips.** Go ahead and add as many of the other test clips as you want. We ended up with the choices shown in Figure 2-11.

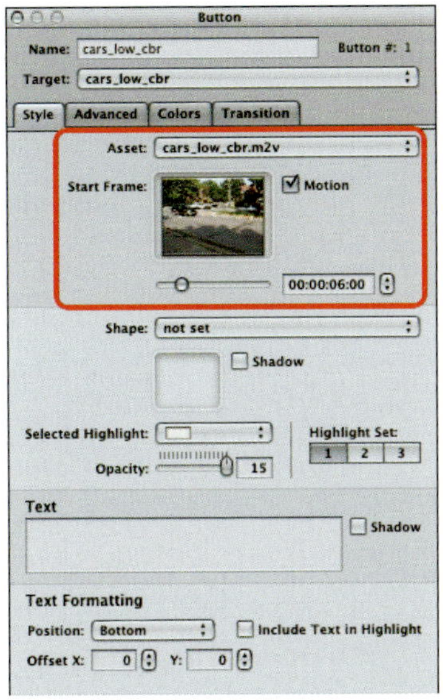

Figure 2-8. Setting the starting point for a video button loop

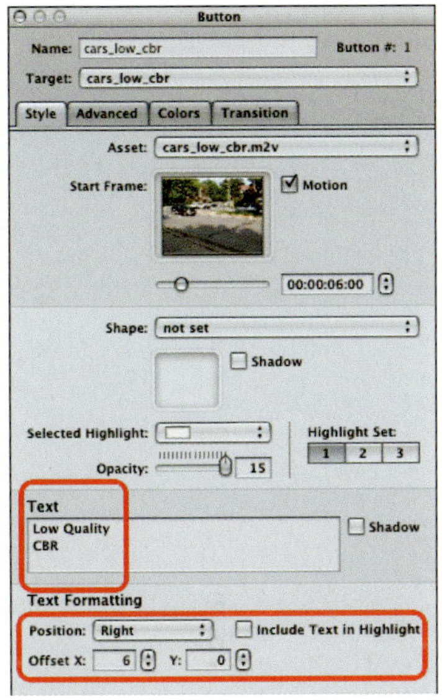

Figure 2-9. Setting the text label for a video button

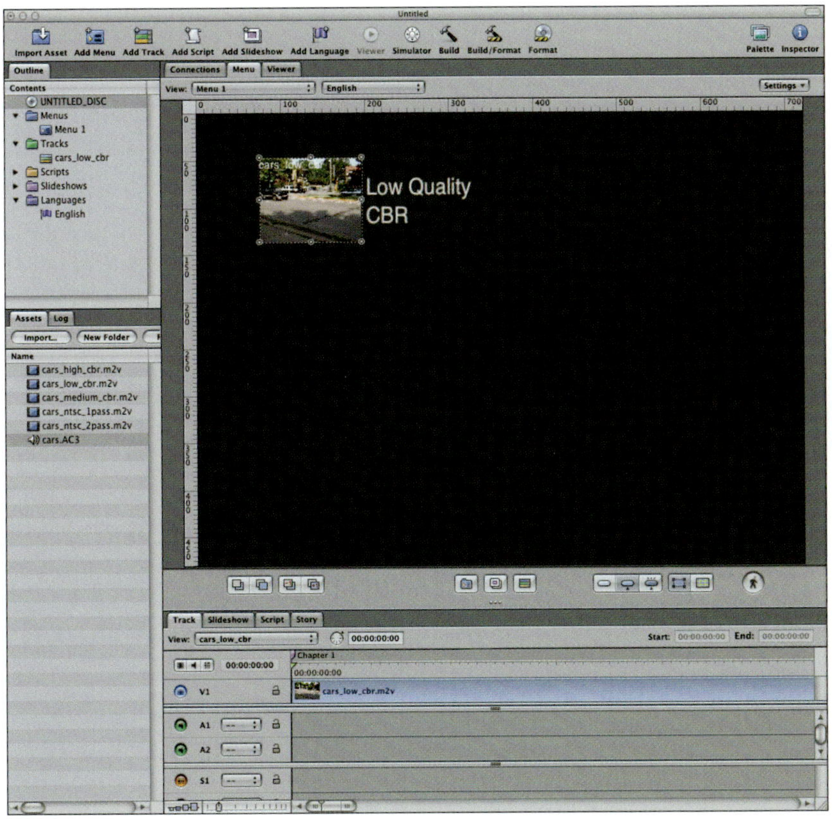

Figure 2-10. Our first video button (shown in the Menu editor)

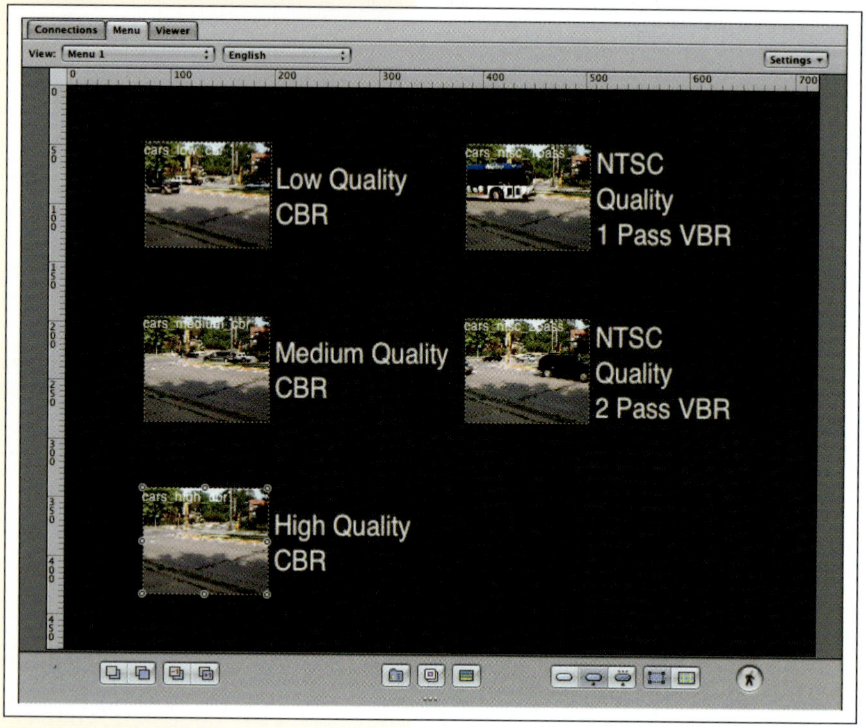

Figure 2-11. Completed bit-rate demonstration menu

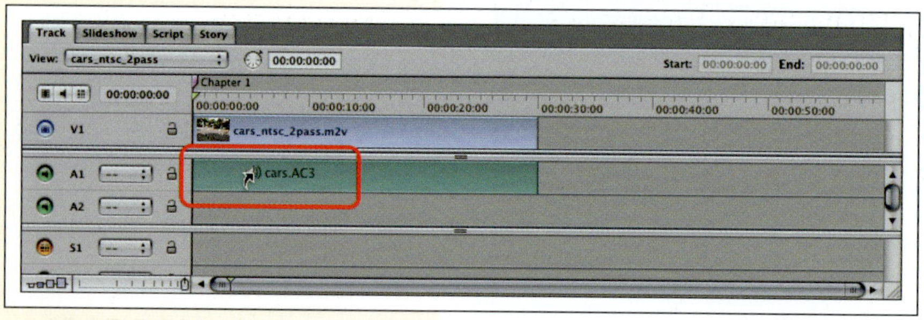

Figure 2-12. Adding an audio track to the video track created automatically by the video button

6. **Add Audio.** If you want the audio, you'll need to open each track in the Track window (⌘9) and simply drag the *cars.AC3* file down to the audio line for each track. Figure 2-12 shows how to add the audio to the ultimate version of the video.

   The clips are all set to jump back to this menu when they complete. That's fine for this demonstration, but you can change that behavior in the Track property sheet if you prefer to have all the clips play one after the other. The End Jump property of a track can point to any location. You just need to set each clip's End Jump property to point to the next clip. The last clip's End Jump property would lead back to the menu. We'll see more examples of this property in Chapters 3 and 4. We'll also use it heavily when tackling DVD scripting in Chapter 7.

That's it! Simulate it to make sure you can start each clip. If you haven't played with throughput rates before, it might be worth burning this disc and trying it out in your various players to see the difference in quality during playback.

You might find that on your television set, you can't really tell the difference between images that are of "OK" quality and "really high" quality. In the QuickTime player, you certainly can see differences (see Figure 2-13), but they might not be as stark as you expect. Of course, really fast action or sporting-event sequences can suffer more artifacts, but it is worth testing before you decide to use a particular compression setting.

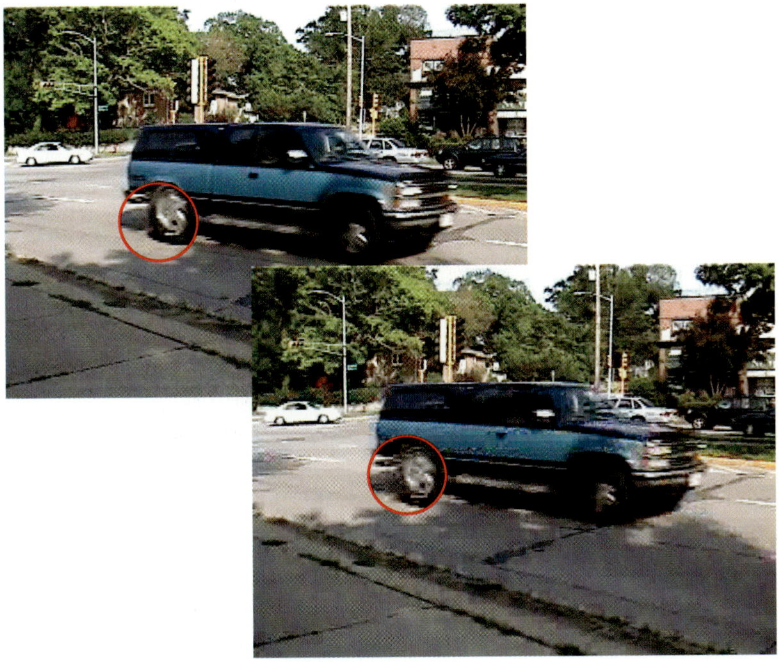

Figure 2-13. Quality comparison between the high- (left) and low- (right) quality MPEG settings

The project for this chapter on the disc actually bundles up the other small demonstration projects presented later into one big project. You might want to build your own demo disc that you can carry with you to test out new systems.

MPEG-2 aspect ratios

With MPEG-2 streams, you have the choice of a 4:3 aspect ratio (that is, 4 units wide and 3 units high, which is the size of most older TVs) or 16:9 (16 units wide by 9 units high, which commonly is referred to as *widescreen*). If your camera can capture at 16:9—and many of today's DV cameras can—you can preserve that aspect ratio all the way to the DVD. You can store the video in anamorphic format (horizontally squeezed to match the 4:3 aspect ratio) so that it plays back properly on widescreen TVs. Many DVD players will support auto letterboxing for anamorphic DVDs that really are being played on typical 4:3 television sets.

STUDIO ADVICE

Quantization and Quality

While it may seem obvious that higher bit rates provide better quality pictures, where do those extra bits actually come from? MPEG uses a process known as quantization to help compress the video. Blocks of the image are analyzed for their color and brightness. The results of that analysis are then quantized. Quantizing groups similar values. Instead of tracking five separate values such as 7, 9, 10, 11, and 12, you keep track of the fact that there are five values close to 10. In this example, 10 is a quantization step.

How many "steps" you use in your possible range of values determines the quality of the picture. Smaller, more frequent quantization steps mean better pictures, but require more storage and throughput. Larger, sparser quantization steps take less space, but lose more quality.

When you select the bit rate for your MPEG encoding, you're really determining what quantization information the encoder will use. That's important because different encoders can apply this information differently (read: some software encoders do a better job than others—even at the same bit rate). You'll want to experiment with your encoder to make sure you know what rates work well.

STUDIO ADVICE

Anamorphic Format

As the film industry fought to compete with the increasingly popular TV medium, one trick they used was presenting images far too wide for television screens. (Cinemascope!)

The problem was they needed to keep using the 35mm film that was available. They managed this by using an anamorphic lens with the camera. The lens optically compresses the incoming image to fit in the 4:3 aspect ratio of 35mm film. Then at playback time, the image would be projected through a complementary lens that would uncompress the image and display it at its original aspect ratio.

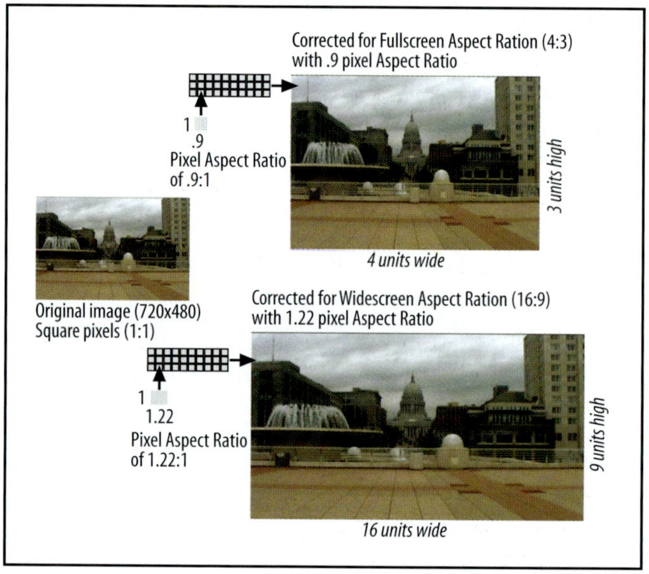

Starting with Version 1.2.5, Final Cut Pro can capture and eport 16:9 video. DVDSP handles such clips without problems. You can verify the video uses a 16:9 aspect ratio in the property sheet for the clip asset, which is shown in Figure 2-14.

If you are planning a 16:9 project, take care when designing the aspect ratio of your menus. The property sheet (shown in Figure 2-15) for the disc enables you to specify the menu settings for both the video standard (NTSC or PAL) and the aspect ratio (4:3 or 16:9).

A widescreen project

If you want to take a quick look at widescreen media and menus, you can build up this next (brief!) demonstration project. You'll need the following files from the *projects/ch02* directory:

- *wide_menu.psd*
- *MononaTer1.m2v*
- *MononaTer1.AC3*

1. **Start a New Menu.** You can include this 16:9 project with the other media demos from this chapter. Because we keep this menu and the video clip fairly encapsulated, DVDSP has no problem mixing 16:9 with 4:3 parts. (We couldn't do this for things such as multiple angle clips, though.) You are, of course, free to start a completely new disc as well.
2. **Set the Menu as Widescreen.** You'll need to tell the new menu that it is a widescreen menu. You do that through the property sheet shown in Figure 2-16.
3. **Set the Background for the Widescreen Menu.** Set the *wide_menu.psd* file as the background for the menu. Turn on all layers if you didn't choose that when dropping the file into the menu. You have only one button, so go ahead and define that now. We

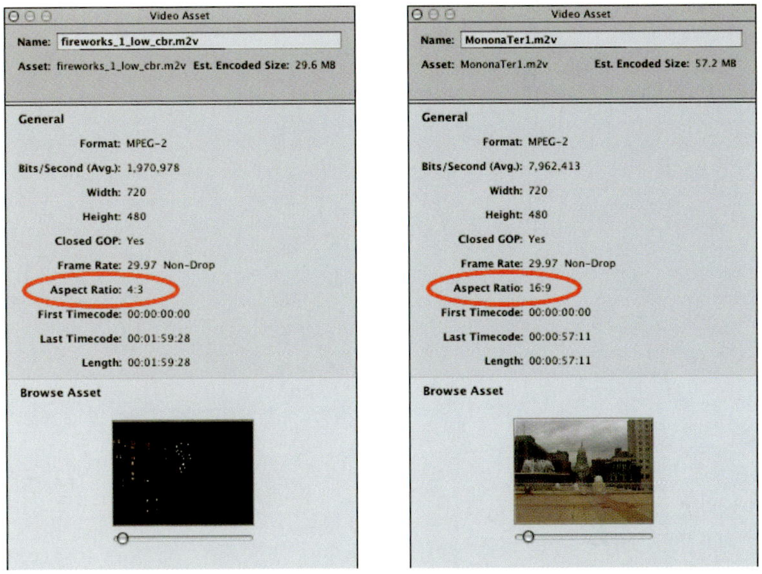

Figure 2-14. Video clips in the Inspector: a 4:3 clip (left) and a 16:9 clip (right)

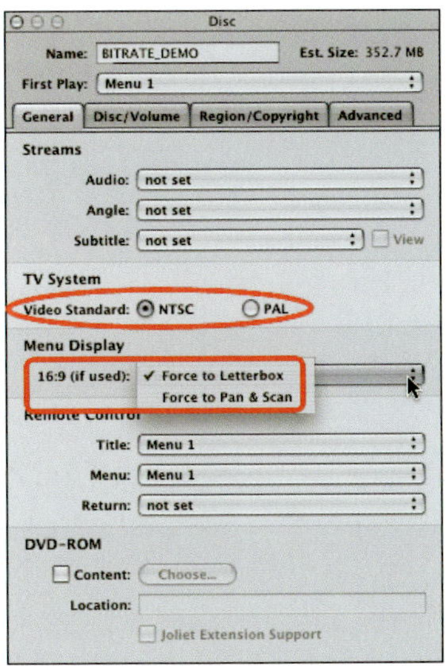

Figure 2-15. A 16:9 disc project

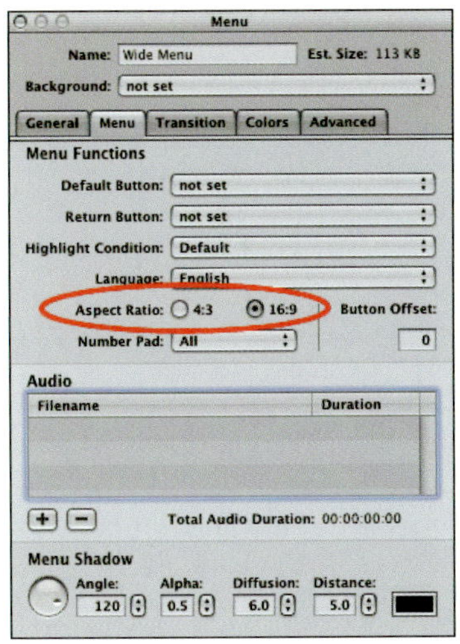

Figure 2-16. Setting the aspect ratio for a 16:9 menu

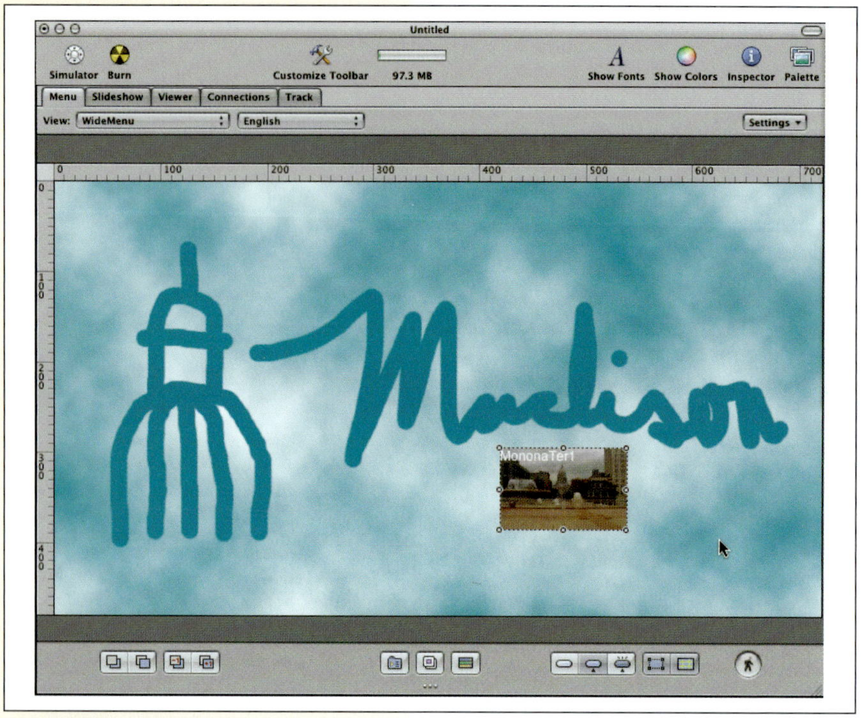

Figure 2-17. Our first widescreen menu with a lone button

just used the *MononaTer1.m2v* file to create a button and track at the same time, similar to how we set up the fireworks clips in the bit-rate demos. You should end up with a screen similar to that shown in Figure 2-17.

The track should be recognized as 16:9 media (check the Property Inspector). If it's not, you can set the mode to handle the 16:9 ratio however you want. Figure 2-18 shows the entry on the Track Property Inspector. We chose to go with letterboxing. That's it!

Figure 2-19 shows how the Simulator (in 4:3 Letterbox mode) accommodates our new widescreen media.

If you want to see how your project would look on a widescreen television set, you can change the

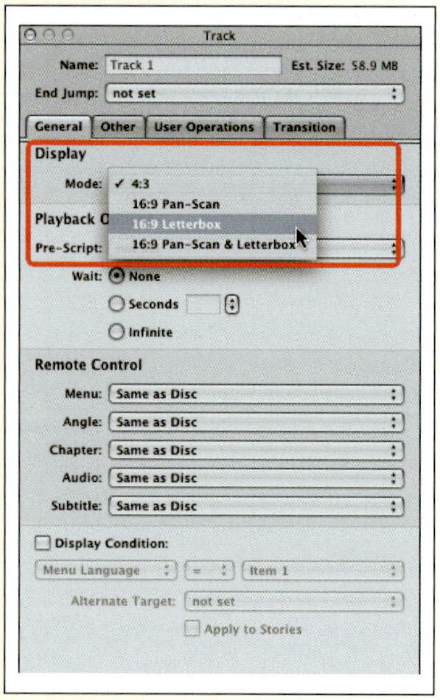

Figure 2-18. Setting the track properties to force a particular presentation

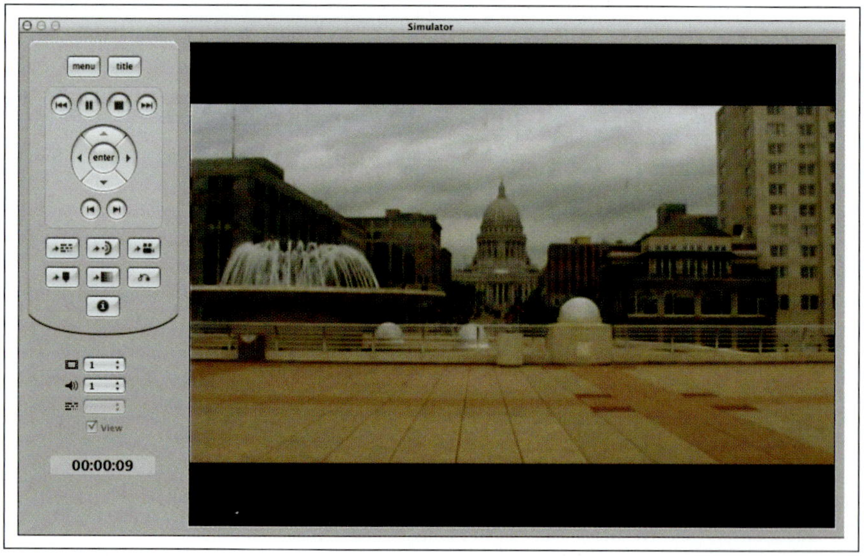

Figure 2-19. Our widescreen project on a simulated 4:3 screen

Simulator's settings in the preferences for DVDSP. Select Apple Menu → Preferences... or ⌘, (yep, that's Command-comma) to get the window you see in Figure 2-20.

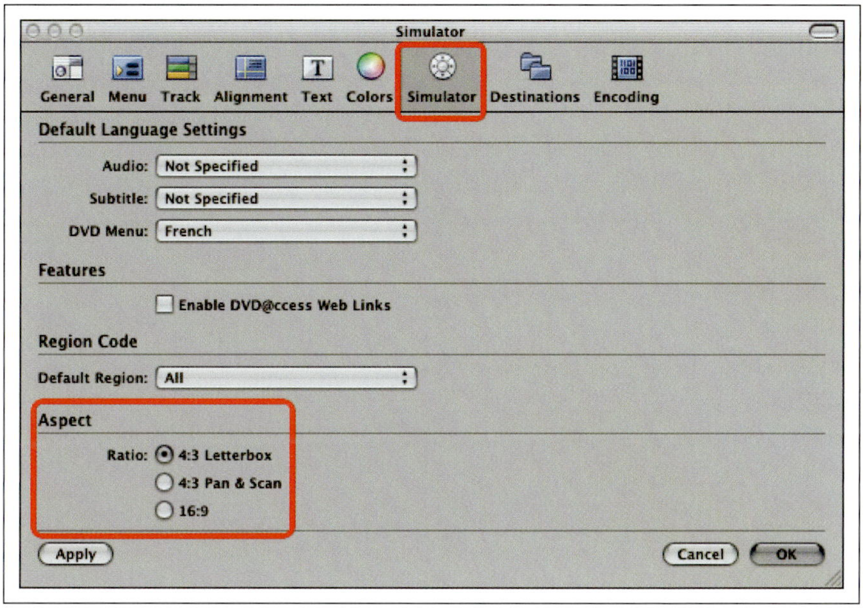

Figure 2-20. Setting the Simulator to present widescreen media

Figure 2-21 shows that same project being played back on our television test set. Here you can see the letterbox choices in effect.

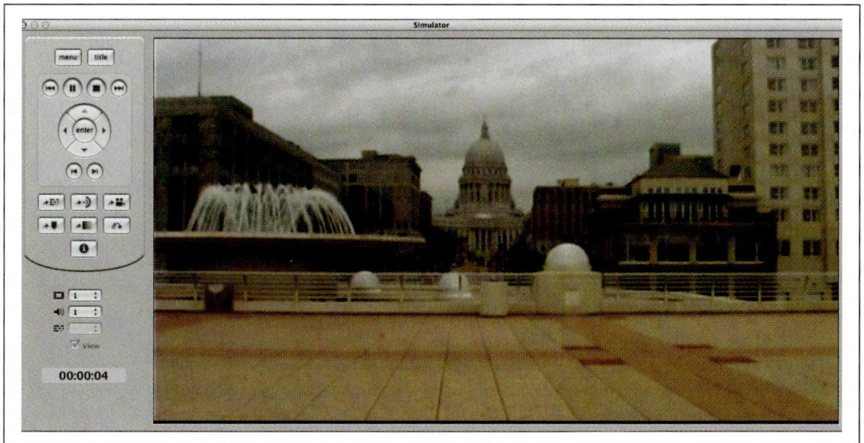

Figure 2-21. The Simulator in 16:9 mode showing our widescreen project

What About DTS?

DVDSP 3 includes support for Digital Theater Systems (DTS) surround streams. But *support* can mean so many things....

DVDSP 3 does not ship with any software to encode a DTS audio stream. If you have other software that can produce DTS audio (such as Steinberg's Nuendo with its DTS Encoder), you can use that stream in your project. But again, *support* is a nebulous term. You cannot use the Simulator (or even DVD Player) to listen to the DTS track. DTS requires separate software or even hardware to be played on the Mac. In a nice change from all the "go buy this multimillion-dollar package" statements, the free VLC Media Player from VideoLAN can decode DTS. (Check it out at *http://www.videolan.org*.)

Audio

DVDs support a variety of audio formats, as shown in Table 2-5.

Table 2-5. Supported audio formats

Format	Supported Sample Rates (kHz)	Supported Bit Depths	Maximum Supported Audio Channels	Maximum Bit Rates (bps)
PCM	48/96	16/20/24	8	6144
AC3 (Dolby Digital)	32/44.1/48	16/20/24	5.1	448
MPEG-1 Layer 2	48	16	2	384
MPEG-2 Layer 2	48	16	7.1	912
Digital Theater Systems (DTS)*	48	20	5.1	1536

Note that the new Advanced Audio Coding (AAC) format found in iTunes and MPEG-4 is not supported. Also, MP3 (which is technically MPEG-1 Layer 3, but often mistakenly denoted as MPEG-2 Layer 3) is not available.

With DVDSP, you can import these formats for use in your project. DVDSP will convert them to 48/16 PCM and then use the converted versions during the final build. Note that it might take longer for DVDSP to build and burn your project if it has to transcode any of the assets. (Remember, you can set a preference for doing this at build time or in the background.)

Using A.Pack

DVDSP includes a separate standalone audio encoding/decoding application called A.Pack. A.Pack can encode PCM audio streams (both AIFF and WAV) into Dolby Digital (previously known as AC3). It also can decode AC3 files into their separate channels. Note, however, that you can't import just any old audio file. It needs to be AIFF or WAV. (QuickTime Pro can help you get other formats converted to one of these.) You can have up to five regular audio channels and a low-frequency effects channel (commonly known as 5.1: left, center, right, surround left, surround right, and the subwoofer).

* These days, DTS tracks can be encoded with higher specs than we list here, but you need special decoding hardware to take full advantage of things such as 6.1 audio channels or their 96/24 encoding. Check out *http://www.dtstech.com* for the latest details.

It's also important to mention that A.Pack is an encoder, not an editor. For example, if you want to pan a sound from the left speaker to the right, you need to do that in an editing program such as Final Cut Pro or Soundtrack. A.Pack wants to see only a completed channel, ready for encoding. Your editing software should export each speaker as one track. Chapter 5 provides an example of using Soundtrack to do just this sort of thing.

Launching A.Pack lands you on a fairly straightforward screen, shown in Figure 2-22.

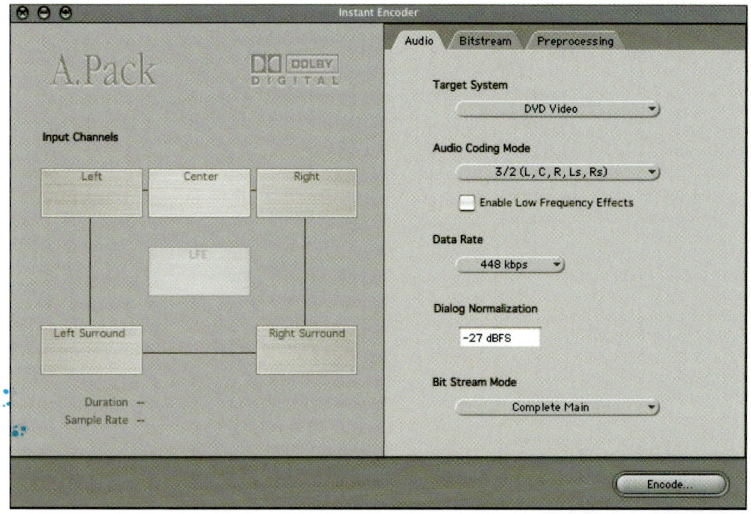

Figure 2-22. The A.Pack application

You can choose your target format (DVD-Video or DVD-Audio), the layout of the audio (mono, stereo, surround, LFE or not, etc.), and the bit rate on the Audio tab. Figure 2-23 shows the options available for layout.

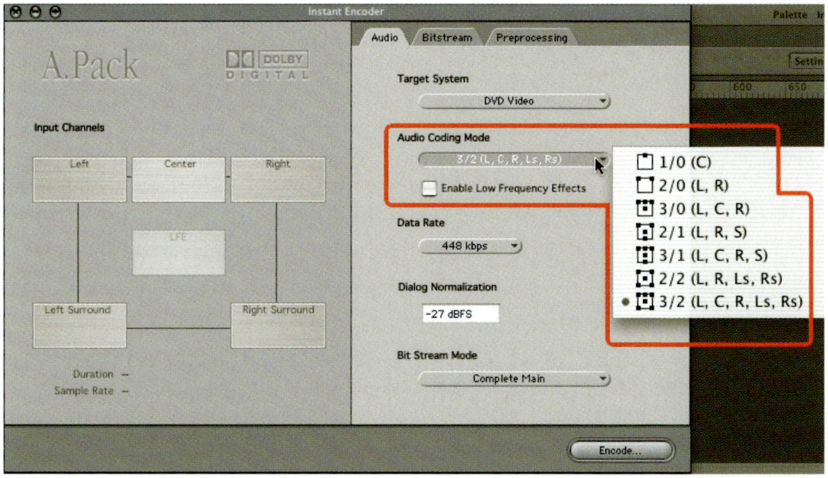

Figure 2-23. The audio layout options in A.Pack

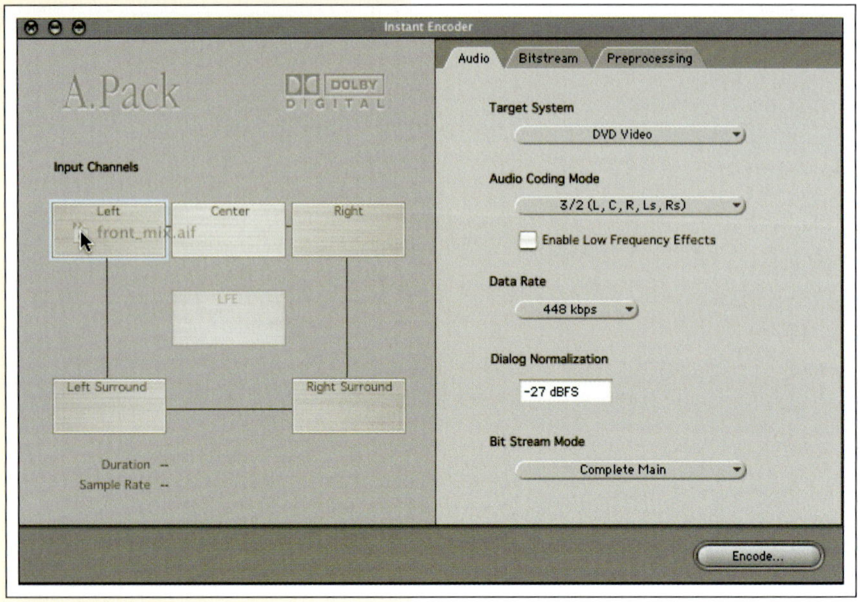

Figure 2-24. Adding a stereo AIFF file to A.Pack for use as the front left channel

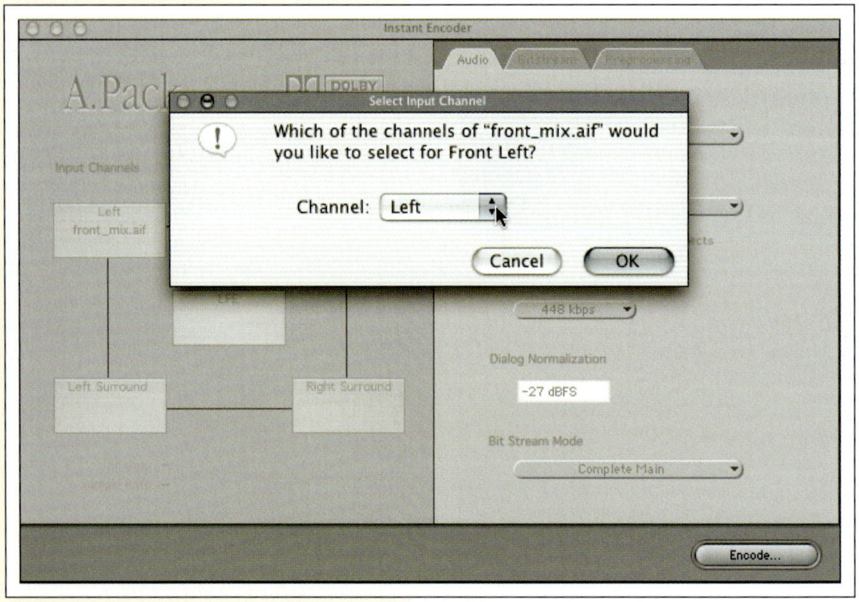

Figure 2-25. Choosing which channel to use during encoding

Once you have the mix set up the way you want, you can drag audio files into the channels on the left. If the audio file is a stereo file, A.Pack will enable you to select the left or right channel from the audio file. Note that you can use a file multiple times in one mix. For example, many people like to convert normal stereo AIFF files to Dolby Digital stereo files. Just drag the AIFF file to the left channel, as shown in Figure 2-24.

Once you drop the file, you should be prompted to select a channel, as in Figure 2-25. Choose the Left option.

Now drag that same AIFF file to the right channel, but this time select the Right option. You could duplicate these steps if you wanted to put the surround channels to use—even without fancy separate surround effects. With the audio files for all your channels specified, you can click the Encode button and an AC3 file will pop out a few minutes later.

If you need to encode several clips, you can set up a batch list. You just set up the various parameters for each clip, add them to the batch list, and then click Encode when your list is ready. Each clip will be processed serially.

To start, create a new batch list (from the File menu, or by pressing ⌘N). You should see the screen shown in Figure 2-26.

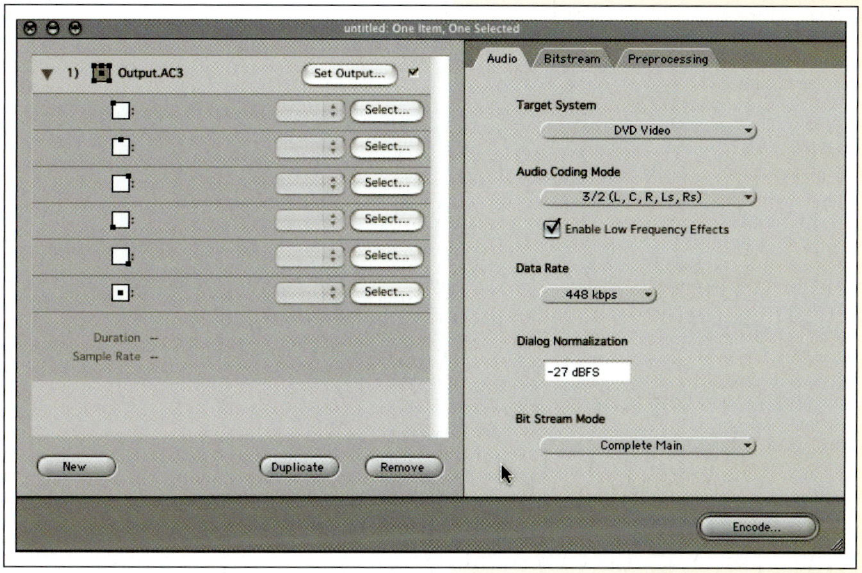

Figure 2-26. The batch list for processing multiple audio tracks in A.Pack

For each audio track you need to process, do the following:

1. Set the output by clicking the Set Output... button.
2. Set the Audio Coding Mode to your desired layout.
3. Specify the input files (and channels) for each channel in your selected layout.
4. For each remaining job, add a new job (⌘K) or duplicate the selected job (⌘D) and repeat steps 1–3.
5. Press the Encode button.

A.Pack will process all your tracks, leaving you free to go have a desperately needed cup of coffee. Trust us, after having to manually set each channel on a dozen 5.1 tracks, you'll want to at least stand up!

Audio stereo versus 5.1 demonstration

Now we'll build a simple project to demonstrate the differences between a stereo clip and a 5.1 clip just like we did for the compression discussion earlier in this chapter. Start a new project (or just a new menu in the current project if you want) and import the following files from the *projects/ch02* folder:

- *blank_card.pct*
- *51_card.pct*
- *stereo_card.pct*
- *51_demo.AC3*
- *stereo_demo.AC3*

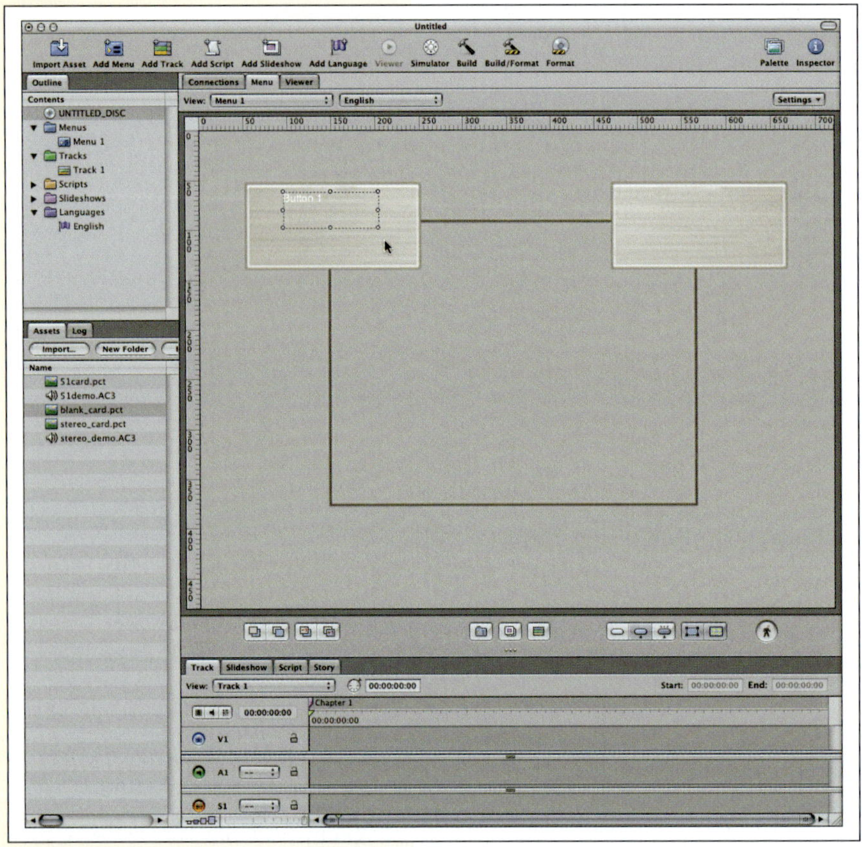

Figure 2-27. The start of our Audio Format demo project

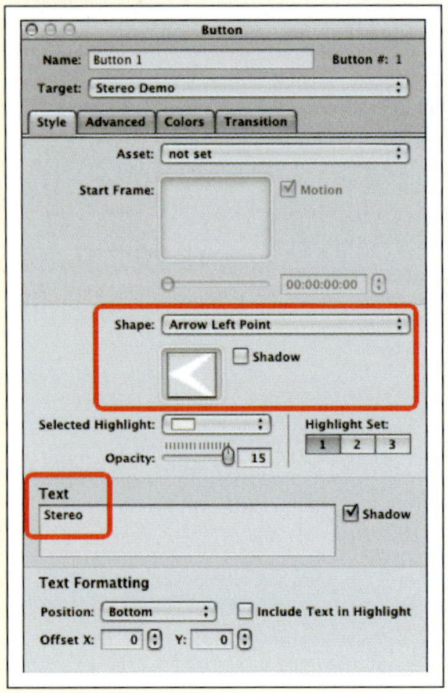

Figure 2-28. Setting the properties on our button

Now, work your way through the following steps:

1. **Create a Simple Menu.** You can do this just as you did earlier. This time, however, grab the *blank_card.pct* file for the background of the menu.

2. **Create the First Button.** Now click and drag to create a button over the "speaker" in the upper-left corner, as you see in Figure 2-27.

   We'll use the advanced features of the DVDSP menu buttons to create the actual shape and text of the button. Choose an arrow shape in the Property Inspector. Figure 2-28 shows the one we chose. Label this button Stereo in the text area. We haven't made any tracks yet, so we'll set the button's target later.

3. **Create the Second Button.** Perform the same steps for a second button on the right-hand speaker. Label this one 5.1. We used a right-pointing arrow for this button. We also tweaked the width and height in the Advanced tab to match our first button.

4. **Create a Slideshow for the Audio.** For the tracks, we really just want the audio. We can get around having to make a full video clip with our audio track by using the Slideshow feature built into DVDSP. In the toolbar click Add Slideshow (or select Project → Add to Project → Slideshow, or just press ⌘K). Add the *stereo_card.pct* file to the slideshow, as you see in Figure 2-29.

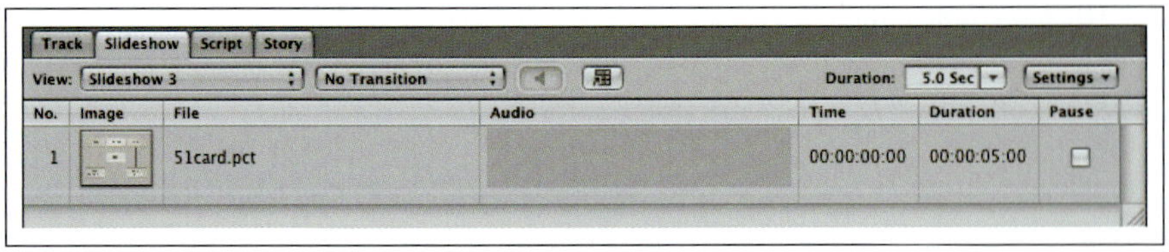

Figure 2-29. Creating a simple slideshow for our audio test

Notice the drop area for audio in the lower-right corner of the Slideshow window. Grab the *stereo.AC3* file and drop it on that area. Your slideshow should now look like Figure 2-30.

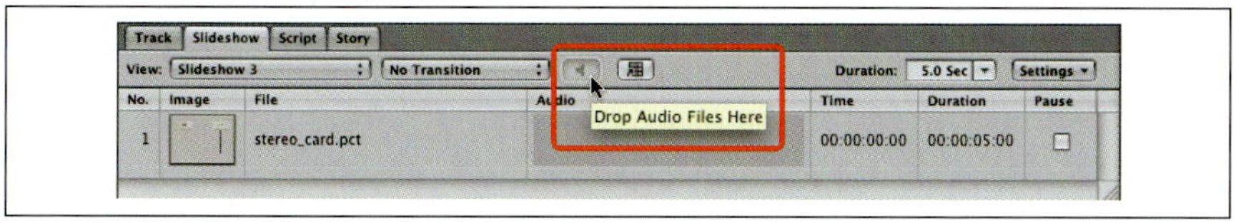

Figure 2-30. Making sure the display matches the audio test

5. **Set the End Jump for the Slideshow.** While this slideshow is on the Property Inspector, set the End Jump value to return to the menu we created, as shown in Figure 2-31. Make another slideshow following the same steps, but use the *51_card.pct* and *51_demo.AC3* files for the slides and the audio.

   Notice that you also can control the duration of the slideshow from the Property Inspector. You can load several audio clips here and then change the duration to Fit To Audio rather than the default. This can be useful if your audio track is more interesting than the slides—exactly the case for this demonstration.

6. **Set the Button Targets.** Now you can set the target on each button in the menu. The easiest way to do this is to right-click the button and pick the target, as you can see in Figure 2-32.

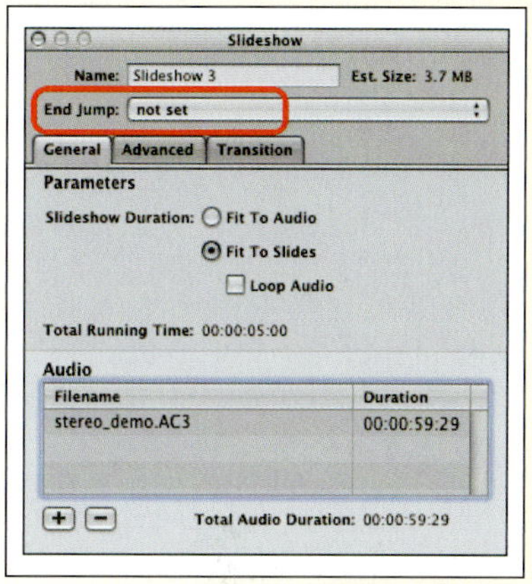

Figure 2-31. Setting the slideshow to jump back to the menu when it completes

DVD Project Minimums

You might also notice the "Convert to Track" button in the slideshow window. That can come in very handy. An odd place you might encounter this need happens when dealing with quick projects. We've done a number of quick-and-dirty PowerPoint conversions. Pretty easy, really, just one caveat: your DVD must have at least one video clip. It can't be all slideshows. We just made each slideshow its own track and everyone was happy.

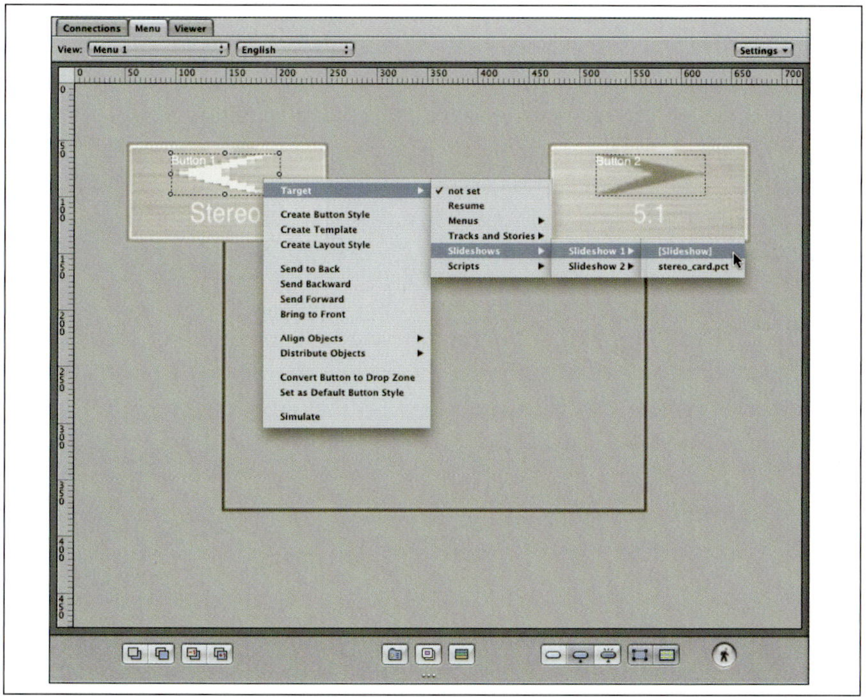

Figure 2-32. Setting the Stereo button's target

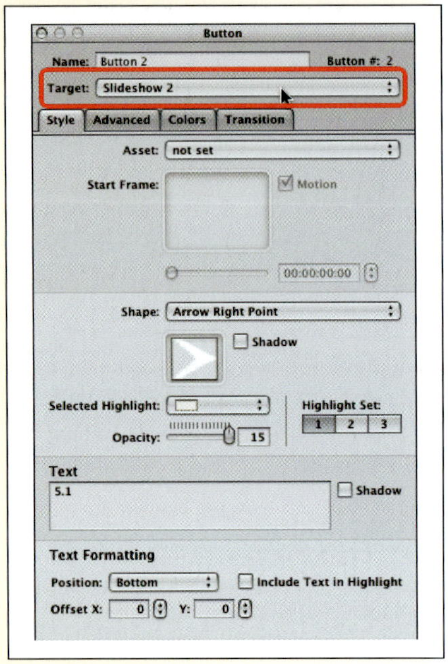

Figure 2-33. Setting the 5.1 button's target via the Property Inspector

You also can set the target in the button's Inspector, of course. Figure 2-33 illustrates this approach.

At this point, save the project, simulate it to test the button links, and then burn it. Most Macs aren't set up with Dolby Digital surround systems,[*] so you'll probably want to play this disc in a home system that does have all the right parts. The stereo version will have a nice mix and everything should be audible, but in the 5.1 clip you should be able to hear certain parts of the audio come from distinct areas, such as the center and rear channels.

Still Images

You can use still images in two distinct areas of DVD production: menus and slideshows. Menus often use multilayer Photoshop images. (Chapters 3 and 4 contain several menu projects that detail the specifics of creating menus.) Slideshows also are a fun way to present scanned photos in a video environment. If you've been building all the projects in this chapter, you've already encountered both scenarios. However, when you're preparing these still images, you should be aware of a few things, starting with *screen aspect ratios* and *pixel aspect ratios*.

Screen and pixel aspect ratios

Let's take a look at screen aspect ratios—sometimes called *display aspect ratios*—first. Regular television has a screen aspect ratio of 4:3. That is, the entire display screen is rectangular: 4 units (just pretend those units are squares of some size) wide and 3 units high. Most televisions available today, as well as those that have been sold over the past 50 years or so, use this 4:3 aspect ratio, which is often called a *full-screen aspect ratio*. Newer, widescreen televisions, on the other hand, have gained favor with consumers over the past few years, as they simulate the theatrical aspect ratios of feature films. These televisions have a 16:9 aspect ratio, which is often known as a *widescreen aspect ratio*.

DVD video frames are stored at the same image resolution (720×480 for NTSC or 720×576 for PAL), whether the video is intended for standard or widescreen display. But that doesn't mean they should be displayed at that resolution. In fact, if you view these raw DVD images on a computer, they will look distorted—a raw DVD frame intended for a widescreen presentation will look horizontally squished, and a raw DVD frame intended for a full-screen display will look slightly stretched.

* Of course, you can get the hardware if you want. There are a number of USB and PCI solutions (check out M-Audio at *www.m-audio.com*) that allow you to play all six channels. You still need the six speakers attached to hear everything, though.

Correcting these problems is where the pixel aspect ratio comes in. To make these images work in both full-screen and widescreen television formats, you need to adjust the aspect ratio of each pixel. So, for example, if you want to create a 16:9 screen aspect ratio out of a 720×480 image, you'll need to set the pixel aspect ratio to favor a pixel that is wider than it is tall. And if you want to create a 4:3 screen aspect ratio out of a 720×480 image, you'll need to use a pixel aspect ratio that is slightly taller than it is wide. See Figure 2-34.

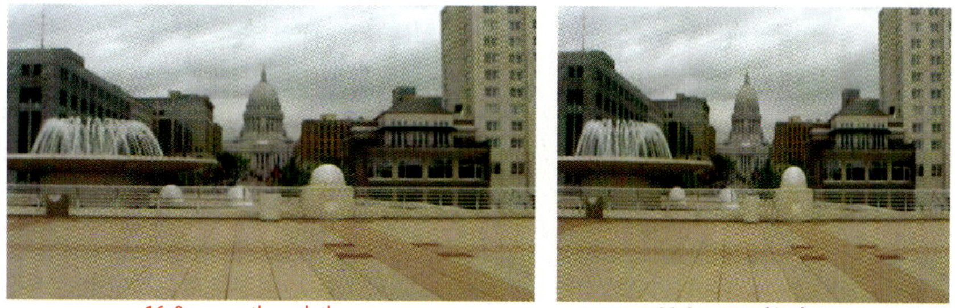

Figure 2-34. Creating full-screen and widescreen video by varying the pixel aspect ratio (NTSC)

Table 2-6 shows the approximate pixel aspect ratios used to convert standard NTSC and PAL resolutions to full-screen and widescreen television screen aspect ratios.

Table 2-6. Pixel aspect ratios for NTSC and PAL

	Full-screen Pixel Aspect Ratio (x:1)	Widescreen Pixel Aspect Ratio (x:1)
NTSC	0.9	1.21
PAL	1.09	1.45

A DVD can contain material intended for both full-screen and widescreen aspect ratios. DVD players are designed to correct the output video to display properly, whatever the output display device might be. For example, with a standard television, a DVD player can letterbox the video by placing black bars at the top or bottom. In the same manner, a DVD player can show standard-resolution video on a widescreen display with black areas on each side. Of course, this requires that the DVD player is set up to provide the correct type of output for the actual display device. Today's display monitors also provide additional options for specifying how different input video signals can be reformatted on the display.

Working with Photoshop 7.0 and below

If you start out building a menu for NTSC in Photoshop 7.0 or below, you might have noticed two possible screen sizes: 720×540 (NTSC) and 720×534 (NTSC DV). "But wait!" you cry. "My camera outputs DV at 720×480. Why don't I create menus at 720×480?" Well, actual NTSC is defined as 720×486. DV has its own variation coming in at 720×480.

720×540 is what Photoshop defines as NTSC. It is, indeed, the correct 4:3 aspect ratio for a television. And, as you might have guessed, 720×534 is the DV variation.

Remember that full-screen NTSC uses nonsquare pixels that are taller than they are wide. Computer monitors, though, use square pixels. So, on a computer, planning your graphics on a 720×534 template will give you a better sense of how things will look when they're actually played on a TV. Before the menu gets encoded into an MPEG-2 stream for DVD, though, you should scale it to 720×480 so that its contents come out in the proper 4:3 aspect ratio when displayed with the tall (nonsquare) pixels. Figures 2-35 and 2-36 show these various ratios in action.

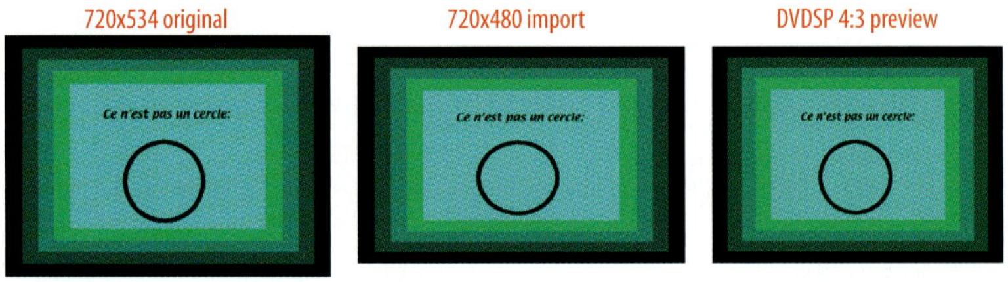

Figure 2-35. Properly scaled images for nonsquare pixels

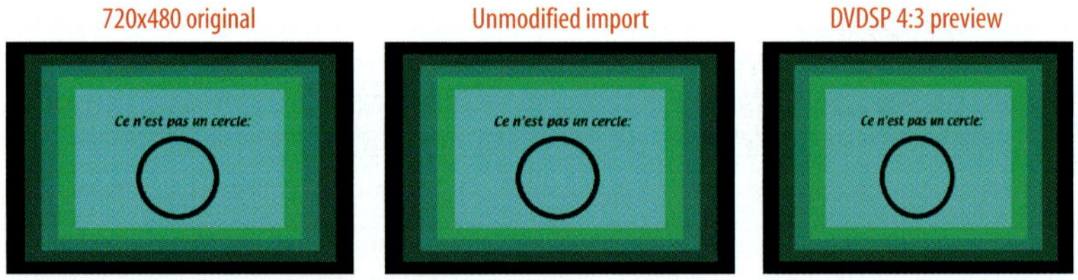

Figure 2-36. Improperly scaled images for nonsquare pixels

DVDSP shows you the television version of your menus, with the pixel aspect ratios adjusted, so you can catch these mistakes fairly early in the process. And to be honest, DVDSP is smart enough to squish any images built at 720×534 to 720×480 for you. Other applications might not be as smart, so there is some utility to squishing your own images. But clearly you can be a bit cavalier when dealing with quick and dirty projects and demos. (For example, our slides in the previous demo took advantage of DVDSP's intelligent use of 720×534 graphics.)

Table 2-7 shows a listing of the various formats common on DVDs with the square-pixel dimensions and their nonsquare counterparts. To demonstrate these various ratios, we also can bundle up a quick menu project with some graphics built in Photoshop. We'll take advantage of this project to demonstrate one other important note about video images: television sets don't show the whole image.

Table 2-7. Video formats and their dimensions

Format	Square Pixel Dimensions	Nonsquare (Scaled) Pixel Dimensions
NTSC DV	720×534	720×480
NTSC DV Widescreen	864×480	
PAL	768×576	720×576
PAL Widescreen	1024×576	

Figure 2-37 shows you how to resize the image in Photoshop. Select Image → Image size... and then change the height from 534 to 480.

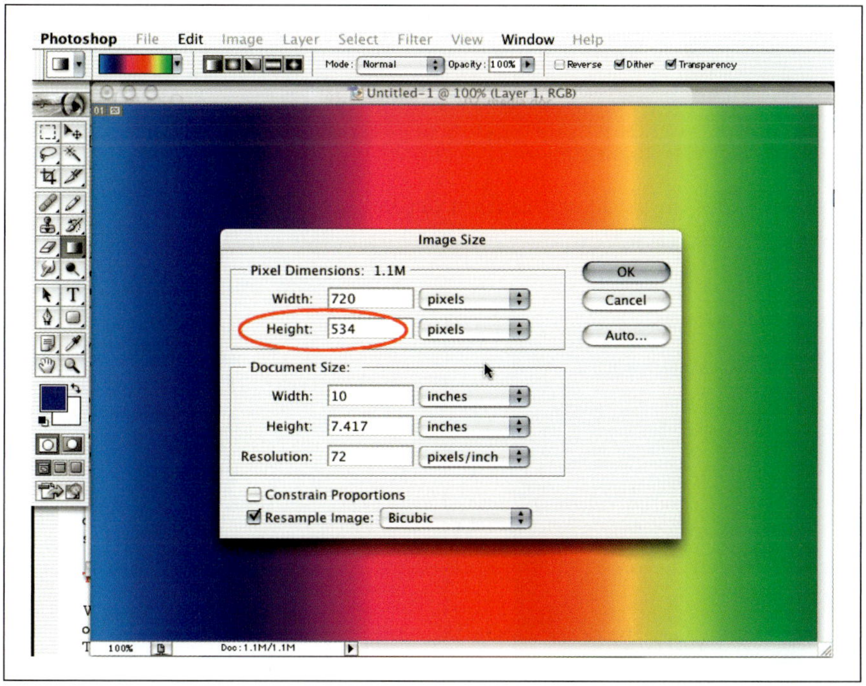

Figure 2-37. Resizing an image with square pixels

Working with Photoshop CS

If you have Photoshop CS (the professional Creative Suite edition of the software), you don't need to worry. Simply create your document at the target resolution (e.g., 720×480), then adjust the pixel aspect ratio (Image → Pixel Aspect Ratio) to the output format you're using. The resulting image will preview itself on the screen accordingly, without adjusting the resolution of the original image underneath. Whenever you want to view the image with square pixels again, you can readjust the pixel aspect ratio in the same manner.

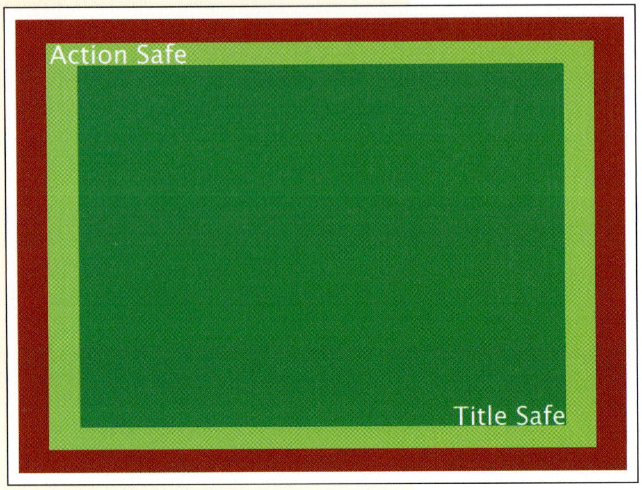

Figure 2-38. A menu with explicit title- and action-safe borders

Safe zones

When you watch normal, full-screen video on a TV, you can see that the image goes all the way to the edge. Indeed, television sets go *beyond* the edge to make sure the screen is filled. Add in the physical casing of your set and you can lose up to 10% of your screen. In building menus, you'll often hear about *action-safe* and *title-safe* boundaries. Action-safe is a 5% border; title-safe is 10%. Figure 2-38 shows a simple menu background with the boundaries clearly called out. It's worth burning this project to disc and trying it out in a few players hooked up to normal television sets. This should prove to you that keeping menus inside the title-safe area is a good idea.

Title-safe demonstration

Here's another quick demonstration detour. Grab the following assets for a new menu and track from the *projects/ch02* folder:

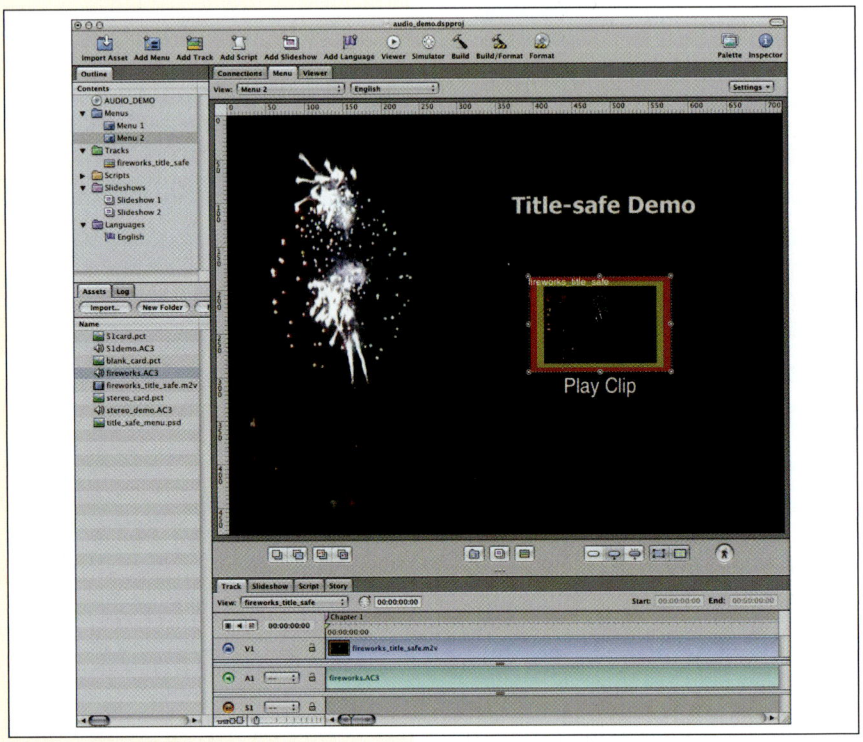

Figure 2-39. Our simple title-safe demo menu

- *title_safe_menu.psd*
- *fireworks_title_safe.m2v*
- *fireworks.AC3* (if you want audio)

1. **Create a New Menu.** Create a new menu—or a new project if you prefer—and set its background to the Photoshop file, *title_safe_menu.psd*.
2. **Add the Button and Audio.** Now drag the *fireworks_title_safe.m2v* clip onto the menu and make a video button plus track, as shown in Figure 2-39. Feel free to add the audio at this time as well.

If you thought that was too easy, remember that this is only a demonstration, so it wasn't meant to be difficult. Watch the clip in the Simulator. You should see the entire screen, including the masked edges we created. After you burn this project, though, try it out on a regular television set. You should notice some areas are cut off. We use an inexpensive consumer set to preview all our menus to make sure they're readable. Figure 2-40 shows a snapshot of the demo clip playing on that television.

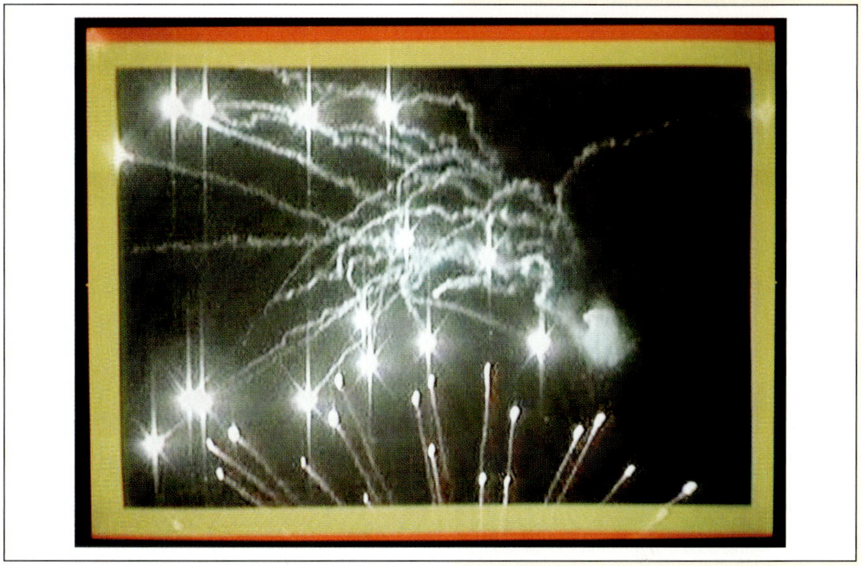

Figure 2-40. Our title-safe demo playing on a consumer TV

NTSC colors

Here's something else to keep in mind when creating menus or any media images for a DVD: if the DVD will be played in a consumer player hooked up to a regular TV in the United States, you can't use some colors. Photoshop has a nifty NTSC color filter that you can apply to any image (see Figure 2-41). It will bring your image into line with the acceptable colors that NTSC television sets can produce.

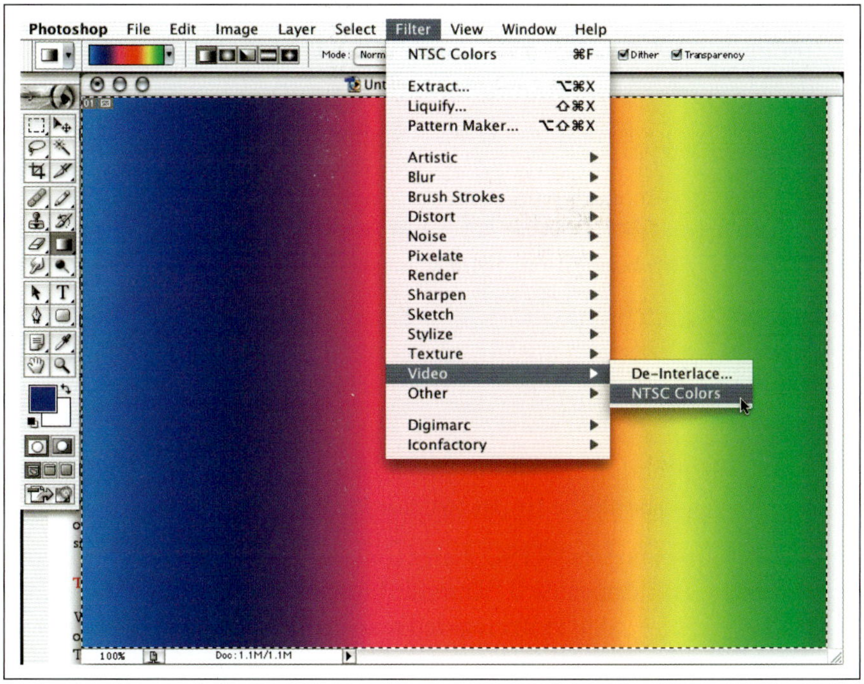

Figure 2-41. The NTSC filter in Photoshop

Computers vs. Televisions

The note about NTSC colors only applies to projects you expect to play on regular television sets. If you know your project is headed for playback on computers, you can use any color available to either Photoshop or the Overlay choices in DVD Studio Pro.

If you accidentally use a color from outside the set of NTSC colors, you'll still be able to play the DVD on a television. But you'll probably hear that telltale buzz as your TV tries to produce colors it simply was not designed to show. If at all possible, your video content and your menus should be done in NTSC-safe palettes.

You can find several websites detailing NTSC. One hot spot to check out is *http://www.ntsc-tv.com*.

Other applications such as Adobe's After Effects and Apple's Final Cut Pro also have such a filter. You should make sure that all incoming content is NTSC-safe. For video, you often handle this during the color correction process.

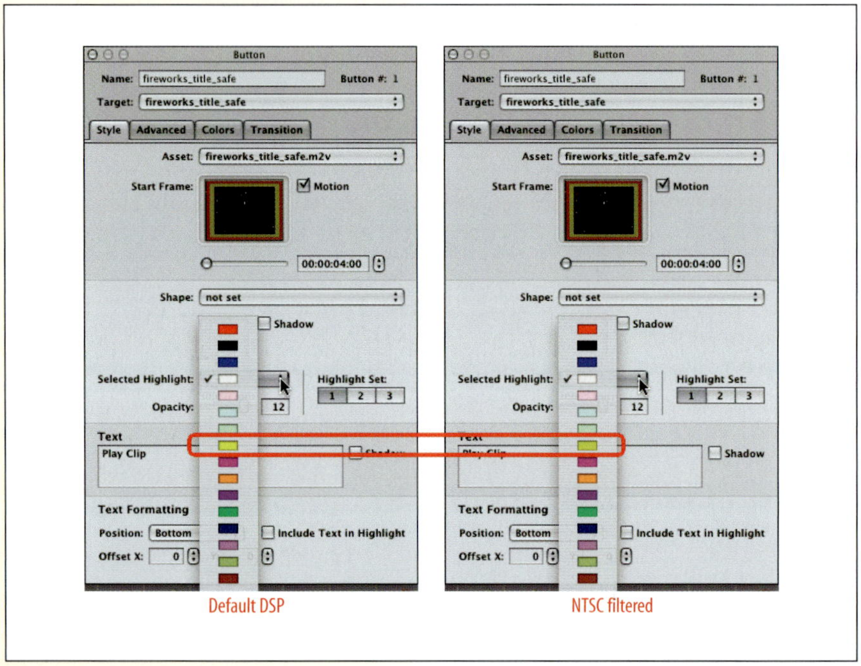

Figure 2-42. Default overlay colors in DVDSP (left) and NTSC-safe (right)

DVDSP contains one feature that you must use with caution. When you create a menu overlay (as we will in the second project in Chapter 3), you have to be careful what colors you use. Some of the selections available in DVDSP are not NTSC-safe. Figure 2-42 shows some possible overlay colors and a Photoshop-filtered variation that is more suitable for NTSC. We highlighted colors that are not NTSC-safe at 100% opaque in DVDSP. DVDSP 2 isn't off by much. Earlier versions were much more offensive.

You also can edit the palette of colors to make sure your projects never enter the danger zone. The Colors tab on any Button or Menu Inspector can help you with that task.

Just to make sure you don't play with colors you shouldn't, here are a few rules to keep in mind while creating your content. You don't have to follow these rules if you have a handy NTSC filter in your software, but doing so can keep you from being unpleasantly surprised:

- Stay away from bright colors. White and red, in particular, cause problems. NTSC white is actually closer to the RGB values of (234, 234, 234) than it is to the (255, 255, 255) values that computers display. In fact, that 234 number is a good cap to use for all your colors.
- At the other end of the spectrum, too much black is also a bad thing. The low end should be (16,16,16), not (0, 0, 0).
- Avoid sharp contrasts. NTSC shares some information between horizontal pixels. This can cause a "shimmer" around sharp edges of high-contrast objects (i.e., don't put crisp red text against a black background).

NTSC color demonstration

Let's look at one last demonstration project. We'll set up two menus: one with NTSC-safe colors and one without. You can use your remote control to jump back and forth between them. Once this project is on a DVD, you can try it out on your own system to see how the colors play back.

You'll need only one file for this project:

- *NTSC_Colors.psd*

1. **Create Two Menus with the Same Background.** You heard right: make that file the background for two new menus. It's only one file, but it's one of those fancy multilayered Photoshop documents we'll be detailing in Chapter 3. Choose which layers you want to show in the Property Inspector (see Figure 2-43).

2. **Select the Background Layers for Each Menu.** For this project, turn on the NTSC set, NTSC safe, and Gray card layers for the NTSC menu. For the non-NTSC menu, use the Non-NTSC set, Non-NTSC but ok, and Gray card layers. On the NTSC menu, make a button labeled Non-NTSC in the center and have it link to the non-NTSC menu. It should look something like what you see in Figure 2-44.

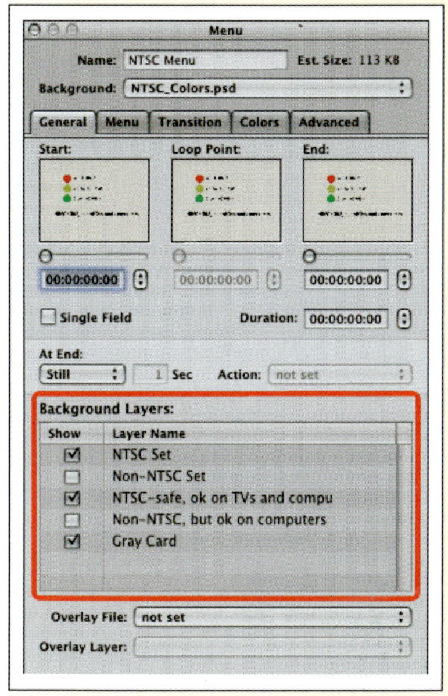

Figure 2-43. Selecting background layers from a multilayered Photoshop document

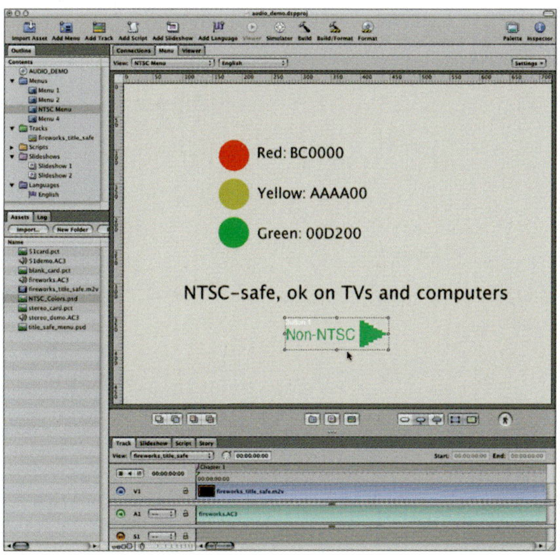

Figure 2-44. The NTSC-safe menu with its only button

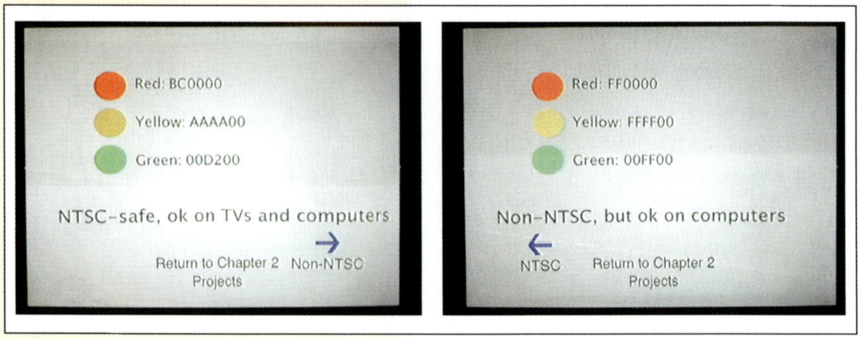

Figure 2-45. Safe and not-safe colors on a TV

Do the same for the Non-NTSC menu (but have it point at the NTSC menu, of course). Although this is not terribly exciting, it is interesting to see how different systems handle the colors. Figure 2-45 shows a comparison on a consumer television set. You might be able to hear that buzz when viewing the "illegal" colors. Our colors occupy such a small portion of the screen that it probably will only translate to some bleeding of the illegal colors.

Storage requirements

Table 2-8 lists the space requirements for common formats. Remember that your combined audio/video/menus can't exceed about 4.2GB after formatting.

Table 2-8. Storage sizes for DVD projects

Media Format	Raw Media Space (DV)	Formatted Media Space	DVD-5 Space
8Mbps MPEG-2	~15GB	~4GB/hour	1 hour
4Mbps MPEG-2	~30GB	~2GB/hour	2 hours
PCM audio (AIFF)	~3GB	~700MB/hour	6 hours (no video)
AC3 audio	~3GB	~200MB/hour	20 hours (no video)
Still image (PICT)		~250KB	~10,000 images (limited by DVD spec, not size)
Still image (multilayer Photoshop document)		~200KB/layer	~10,000 images (limited by DVD spec, not size)

Adobe Photoshop

Adobe Photoshop is absolutely essential for your DVDSP projects. You really, really need this application. Although PICT files certainly are supported for still images, menus are just a whole lot easier with Photoshop. This is primarily because DVDSP supports Photoshop layers. In fact, it not only supports layers, but also it expects them! Several parts of DVDSP enable you to choose which layers of a particular Photoshop document you want to use. We'll see more of that in Chapters 3 and 4.

Layer effects

Most Photoshop layer effects are not supported in DVDSP. But never fear; you can render the effects layer to its own standard layer as the last step in your menu preparation. That will make the effect available inside DVDSP. Note that having effects on a particular layer does not cause problems for DVDSP; it just can't reproduce the effects. Things such as drop shadows and outer glows are common effects that do not translate. See the example project in Chapter 3 for some examples and tips on effects rendering in Photoshop.

Menu overlays

You also can create 1- and 2-bit overlay masks for use with menus and special features on your DVD. DVDSP will accept a black-and-white image as a 1-bit overlay mask. You pick a color and transparency level for your mask. Anything that's black shows up in your selected color. Anything that's white is completely transparent.

For 2-bit-depth documents, your Photoshop file should contain only black, dark gray, light gray, and white. (See the template for overlays in Chapter 3.) You can use a simple overlay color (in which black represents 100% of the color-plus-transparency option you picked, dark gray is 66%, and light gray is 33%), or you can use a complex overlay color scheme, which enables you to pick separate color/transparency options for each of black, dark gray, and light gray. Technically, you also can pick a color for white, but you usually leave white as fully transparent. Figure 2-46 shows an example of such a complex overlay.

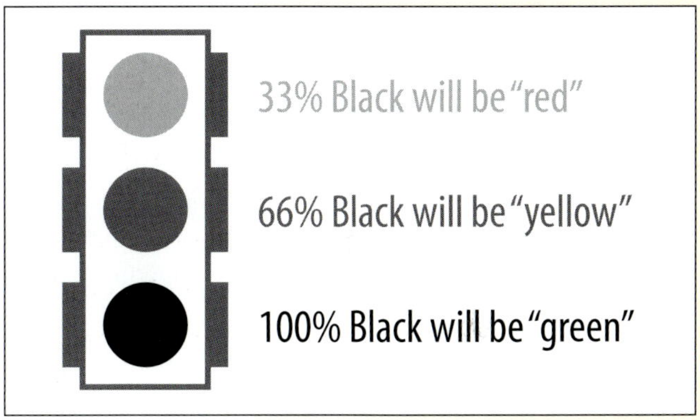

Figure 2-46. Shades of gray mapped to specific colors in DVDSP

Don't worry if this doesn't make much sense yet; we'll be talking about menu overlays later in the book.

Converting to a Supported Format

You can use a number of applications on the Mac platform to convert to and from a specific format.

Photoshop

Photoshop can open almost any graphics format. Once the file is open, you can save it as a Photoshop document with layers turned on. See Figure 1-15 in Chapter 1 for an example of verifying layers. Before you save the image, though, you should remember to adjust your image for use in DVDSP. Make sure the image (and this means every layer) is NTSC-safe by applying the NTSC Colors filter from the Video category (refer back to Figure 2-41).

Hopefully, Photoshop is in your arsenal already. If it's not, you really, really need to get it. The custom menus in Chapters 3 and 4 are very difficult to build without Photoshop. DVDSP does understand JPEG, TIFF, and PICT files, so any editing software that can save documents in those formats will work with basic menus. Building layered menus (Chapter 3) or custom shapes (Chapter 10) requires Photoshop documents. No ifs, ands, or buts.

A.Pack

You saw examples of encoding AIFF files with A.Pack earlier in this chapter. Although A.Pack is great at encoding AC3, it doesn't handle a lot of input types. You can use AIFF or WAV formatted files. Beyond that you'll have to convert the audio using one of the other tools mentioned in this section before bringing it into A.Pack.

QuickTime Pro

QuickTime Pro is a $30 upgrade to the QuickTime Player which you can download from Apple. (Actually, you should have received an upgrade key with your DVDSP purchase.) It has a fairly wide range of input media and can export to a good number of formats. When you install DVDSP, it adds a new video plug-in to the QuickTime library: MPEG-2. (Surprised?) Now you can write and read *m2v* files. We even have worked on projects using exported Flash presentations and they turned out flawless.

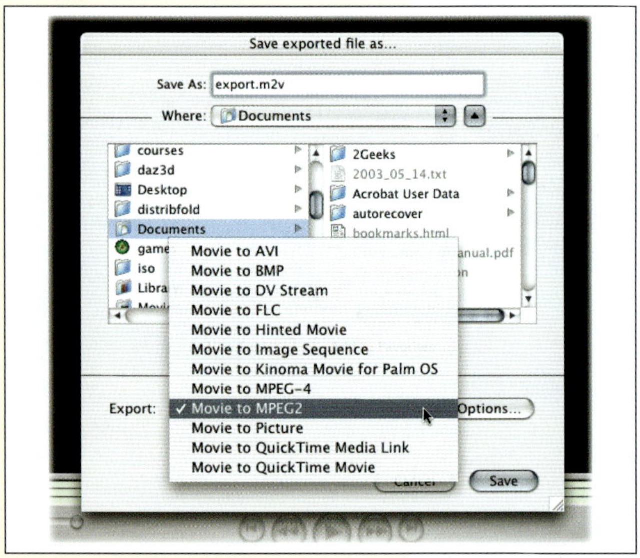

Figure 2-47. Different types of movies QuickTime can convert to MPEG-2

After you open a file, you should be able to select the Export... option from the File menu. The Export dialog enables you to choose your destination, filename, and format. Figure 2-47 shows a typical export list. (If a particular file cannot be exported to a given format for whatever reason, that format won't be in the list.)

If you select the MPEG-2 option, you'll have the opportunity to make adjustments such as the bit rate and aspect ratio. Recall the bit rate demo we put together earlier in this chapter.

With the MPEG codec included in DVDSP, you have more options for MPEG output than before (see Figure 2-48). Now you can choose between constant and variable bit rates. If you go the variable route, you even can choose a two-pass option. The two-pass VBR process yields the best video, but also takes the longest to process. Not every project requires the quality of two-pass VBR.

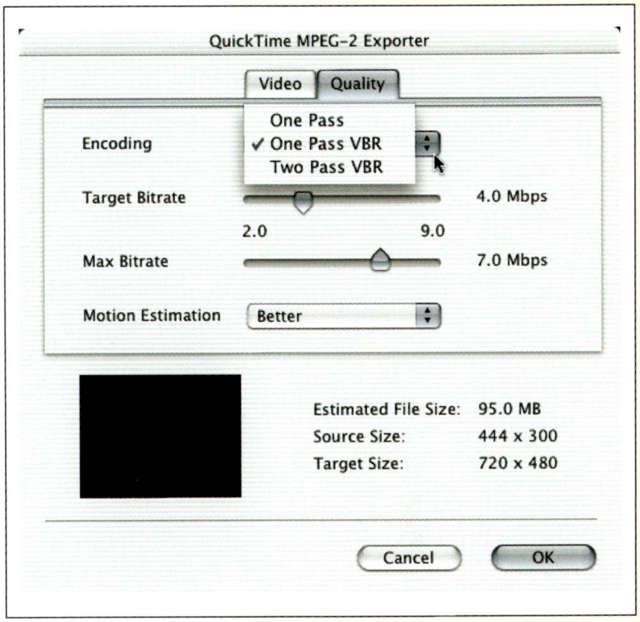

Figure 2-48. The MPEG-2 export options with the codec from DVDSP

You can do elementary editing (such as trimming and cut-and-paste) with QuickTime Pro, but you'll be much happier editing video with Final Cut Pro or Adobe Premiere (covered a bit later). But if you've got some edited video you just want to pop onto a DVD, QuickTime Pro can get you there with a minimum of fuss.

Final Cut Pro and Final Cut Express

Final Cut Pro is Apple's entry into the nonlinear digital video editing world. It's amazingly complete. You can build great-looking video and menus right in this single application. It plays nice with Photoshop and has good export options. Final Cut Pro is a great application to have (and consider After Effects as well!). These days Final Cut Pro comes bundled with some very nifty extras, such as Soundtrack and LiveType.

Final Cut Express is a great starting point for new DV/DVD shops and for hobbyists. It's got all the right parts of Final Cut Pro, but without the hefty price tag (and without the film options). If your video source is always video from DV cameras, Final Cut Express is definitely an option worth looking into. Many folks don't need the extra bits in Final Cut Pro.

Adobe Premiere

Premiere is Adobe's long-standing video editing software. It sports great integration with other Adobe products such as Photoshop, Illustrator, and After Effects. For DVD purposes, it has several export formats including MPEG-2. You even can export small chunks of video to MPEG-2 directly from the timeline. That can be quite useful for playing around with motion menus (see Chapter 4).

Adobe After Effects

After Effects is an amazing program. You can use it to create some kick-ass menus and special-feature sections. Technically it's not required for DVD production, but it's definitely one of those applications you should be thinking about purchasing.

Unfortunately, for all that splendiferousness, After Effects does not export to MPEG-2—even from the normal QuickTime export. However, you can export a full DV stream and then convert that to MPEG-2 in Premiere 6.5 or Final Cut Pro 4, or just let DVDSP handle it when you go to burn the disc. We built several of the menu projects you'll work through in Chapter 4 with After Effects 5.5 and Final Cut Pro.

Odds and Ends

As you probably are aware, DVDs allow for some fancy features, including watching multiple angles of the same video and listening to multiple audio tracks (the Director's Commentary feature). We'll go through some projects that create such tracks in later chapters, but there are a few rules you should keep in mind from here on out.

Multiple audio streams

If you plan to have multiple audio streams for a given video stream, you need to make sure you meet the following limitations:

- A maximum of eight audio streams is available.
- Each stream is the same length.

Note that the audio streams do not have to have the same encoding. In fact, it's common to have a stereo track and a 5.1 track for the same video clip.

Multiple and mixed video streams

People often confuse the terms multiple- and mixed-angle video streams. *Multiple* angles represent one clip that has two (or more) video tracks. They don't even have to be different angles of the same images, but they do have to be exactly the same length. *Mixed* angles represent alternate "paths" for small portions of a clip. The mixed-angle approach can save a lot of room on the DVD for the right video. Figure 2-49 illustrates the differences on a standard nonlinear editing timeline.

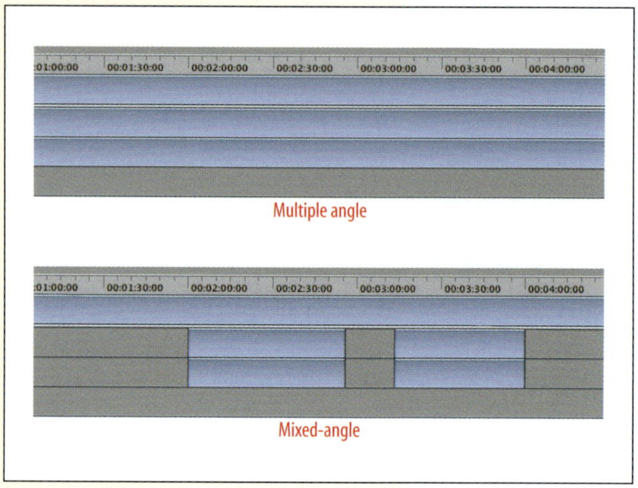

Figure 2-49. Mixed- versus multiple-angle tracks

In either case, the overlapping video has a few limitations you need to remember when preparing the streams. The following comes from the DVDSP User's Manual:

- There are different maximum bit rates for multiple- and mixed-angle video depending on the number of angles in the track:
  - Up to 5 angles: 8 Mbps
  - 6–8 angles: 7.5 Mbps
  - 9 angles: 7 Mbps
- Streams must have the same resolution and frame rate.
- The length and pattern of each angle's GOPs must be the same. (The GOPs should also be closed. You can't switch cleanly between angles with open GOPs.)

OK, now all the formalities are out of the way. You know what formats are supported and how to take maximum advantage of them with your content. This is stuff you'll want to keep experimenting with. Eventually you will get a sense of what bit rates and tools will do the job.

In the meantime, don't shy away from test projects. If you want to see how something looks or sounds on a real system, there's nothing like trying it out. Remember that stack of blank discs we told you to buy? Use it! To help you get started, Chapter 3 will dive into one of the most satisfying parts of DVDSP: custom menu creation.

Great Menus: Layers

3

Now that all the basics are out of the way, let's get on with the projects. The first set of projects we're going to tackle revolves around DVD menus. Menus are the first thing you see when you pop in a DVD. And we all know how important first impressions are! It stands to reason that creating menus consumes most of the time you spend when working on DVD projects. We just want to make sure it's time well spent.

In addition, our projects are going to get a bit more complex—and interesting!

In this chapter, we'll start with the most important part of your DVD: the menu. Everything else is definitely secondary to a fantastic menu. Don't get us wrong—some of the extra features such as audio commentaries and scripting can produce some fun, impressive projects. But nothing will get used as much as menus—every project will have them. We'll look at all the possibilities for making your menus as creative and engaging as possible.

Menus are so integral, in fact, that we can't fit a discussion of the topic into just one chapter. Therefore, this chapter will focus on the templates that are available in DVDSP and will teach you how to create your own layered menus, with highly customizable buttons and graphics. Chapter 4 will focus on using video assets to produce dynamic backgrounds and buttons for your menus.

In this chapter

Safety Reminder

Before we dive in and start creating DVD project menus, we've got to pause for a public service announcement. Well, OK, no one really is going to get hurt if you ignore this section, but your project definitely will suffer! So, you should follow these general guidelines:

- Build your images at 720×534 for NTSC DV or 768×576 for PAL and then resize them to 720×480 once you're done. (If you're lucky enough to have Photoshop CS, this is easy to do, thanks to the software's built-in templates and support for nonsquare pixels.)

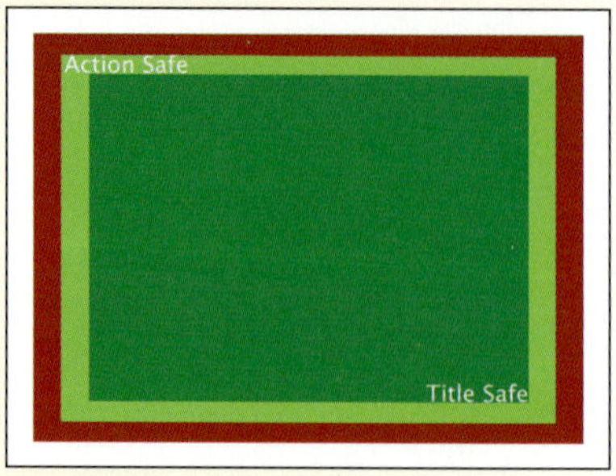

Figure 3-1. The action- and title-safe regions of a video screen

- Be sure your images use only NTSC-safe colors.
- You also need to keep your menu text inside the title-safe boundaries. Recall the garishly colored example from Chapter 2, shown again in Figure 3-1.

When building still menus, feel free to start with the *menu_template.psd* file in the *templates* folder on the included DVD. It has guides set up for both action-safe and title-safe boundaries. It also has an action to resize the document to 720×480.

Using DVDSP Templates

In Chapter 1, we looked at a quick example of the templates in DVDSP. These are extremely powerful components in your "quick and dirty" library. If you look at some of these built-in templates from DVDSP 3 shown in Figure 3-2, you'll see they are anything but dirty. (DVDSP 3 added several new templates. If you're still using DVDSP 2, it has an impressive set as well, and they function in the same way.)

We'll be building some of our own templates in Chapter 10. In the meantime, here's how to get the most out of the templates you already have.

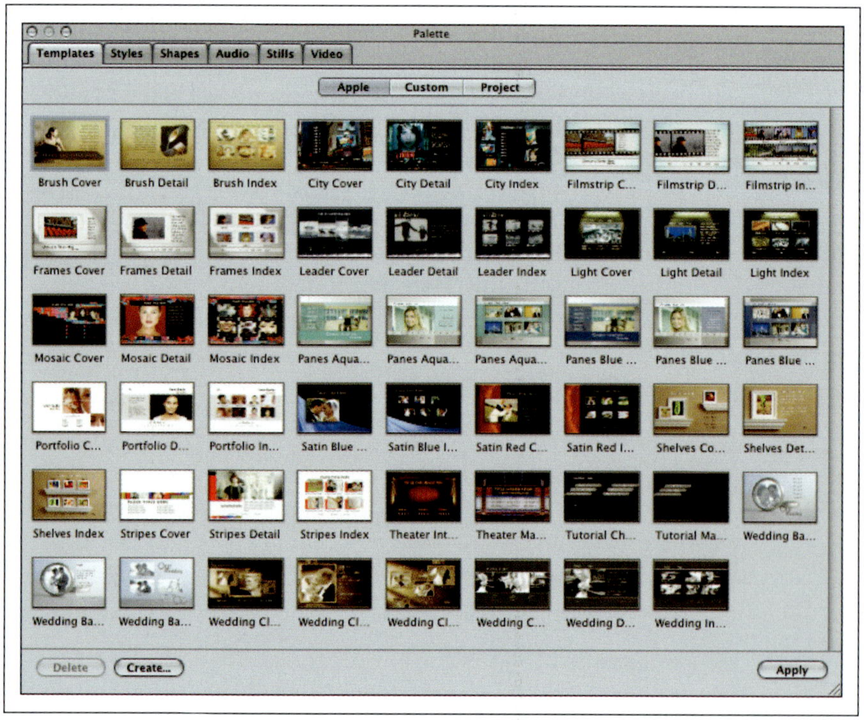

Figure 3-2. The menu templates built into DVDSP

Final Cut Pro and Inaccurate Borders

Several folks preparing motion menus and other elements for DVDSP with Final Cut Pro have noticed a small discrepancy between the two programs. If you recall, in DVDSP the border for content that should be action-safe is 5%. For the width of an NTSC screen, that should be 36 pixels on each side. But it would be only 27 pixels on the top and bottom. Final Cut Pro sets a border of 36 pixels on all sides. The same problem occurs with the title-safe (10%) region.

The miscalculation isn't terrible; it's just a bit more conservative than it needs to be. But at least it errs on the right side! If you have a template from somewhere else (such as our Photoshop menu template on the DVD) that doesn't quite line up in Final Cut Pro, don't be too alarmed. On the other hand, if you do all your title work in Final Cut Pro, you'll never go wrong staying inside its title-safe area.

Assets, buttons, and drop zones

Templates provide mechanisms for building both the background of the menu and the navigation components (buttons). *Drop zones* are areas on the menu where you can add image or video resources—but they are not buttons. Drop zones don't have targets or overlays. They're really just spots for your own images to sit and look pretty. Figure 3-3 shows an example of drop zones and buttons in a menu. The assets that fill them are labeled.

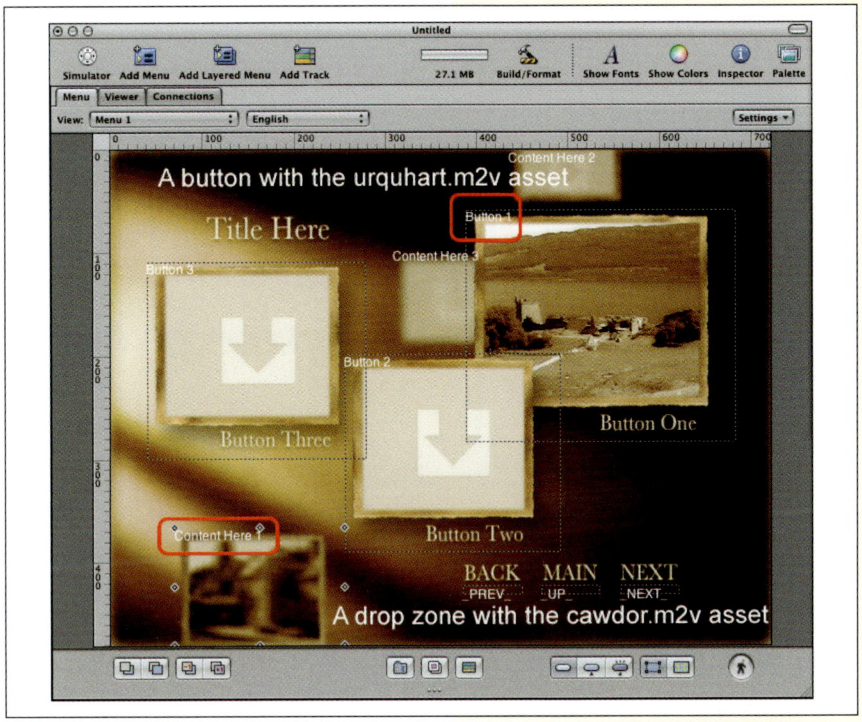

Figure 3-3. Drop zones and buttons in a menu; their assets are highlighted

Creating a button

All the templates provided with DVDSP come with several button areas for you to drop your own images and video. Figure 3-4 highlights a button area in an example template.

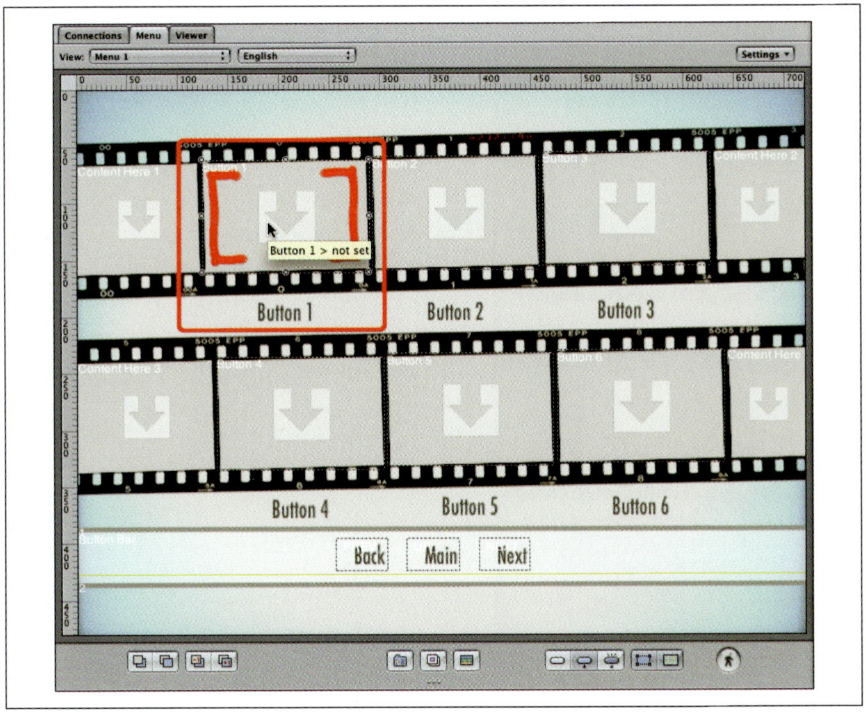

Figure 3-4. An example template with a button area highlighted

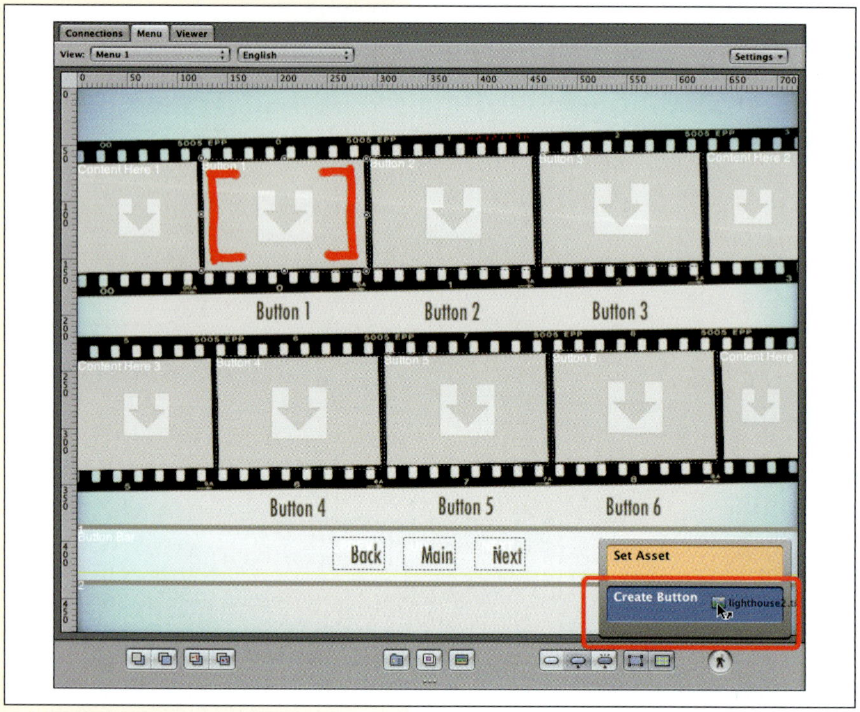

Figure 3-5. Creating a new button on a template by dropping an asset

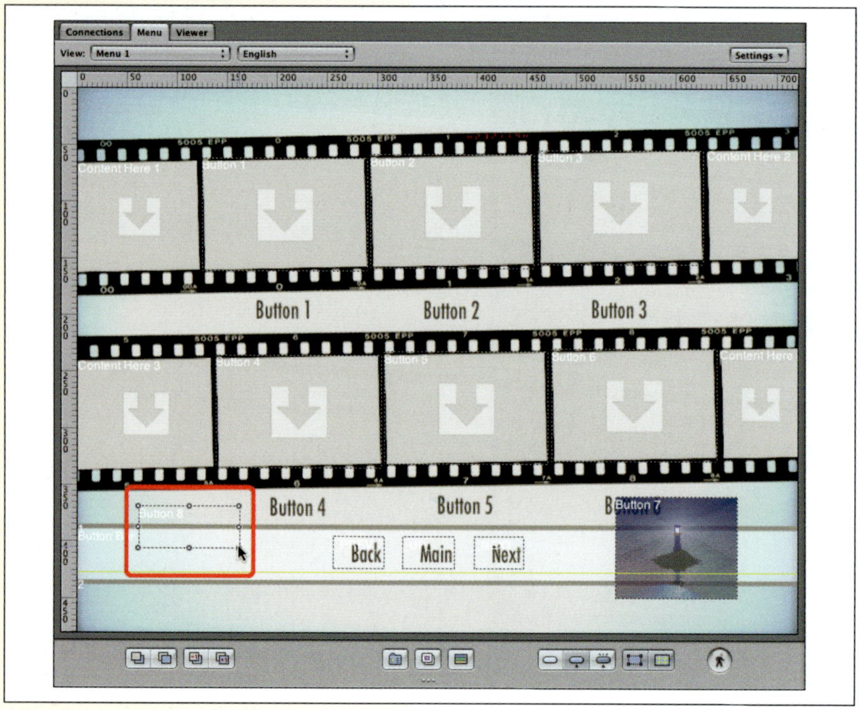

Figure 3-6. Creating a new button on a template by dragging an outline

It's very easy to add more buttons if you need them. You have two simple options:

- Drag an asset on the menu and select the Create Button option (see Figure 3-5). This creates a simple 120×90-pixel button roughly where you dropped the asset.
- Click and drag anywhere on the menu to create a button outline (see Figure 3-6). This provides you with more control over the button's dimensions and proportions.

In either case, you're left with a normal button area and you can use the right-click menu or the Property Inspector to set the action.

Creating a drop zone

The easiest way to create a drop zone on your menu is to start with a button. Drag a button outline on your menu, as you saw in Figure 3-6. Right-click that button and choose the Convert Button to Drop Zone option, as shown in Figure 3-7. *Voilà*! Now you have a drop zone. You also can convert buttons to drop zones (and vice versa) by selecting Project → Convert Button to Drop Zone.

Using drop zones

Drop zones are great for adding your own content to a generic background. We'll use these much more when creating our own templates in Chapter 10, but you'll see them in several of the templates DVDSP provides. Again, the big distinction between drop zones and buttons is that drop zones are for decoration *only*; they do not activate submenus or tracks.

When you drag an asset over a drop zone, you're offered two choices, as shown in Figure 3-8.

Be wary of the Create Button option. It has nothing to do with drop zones. It is simply an option for what you can do with that asset. You can either put it in the drop zone, or create a brand-new button. Fortunately, it's easy to undo things in DVDSP if you goof up. Just select the drop zone or button you don't want and press the Delete key.

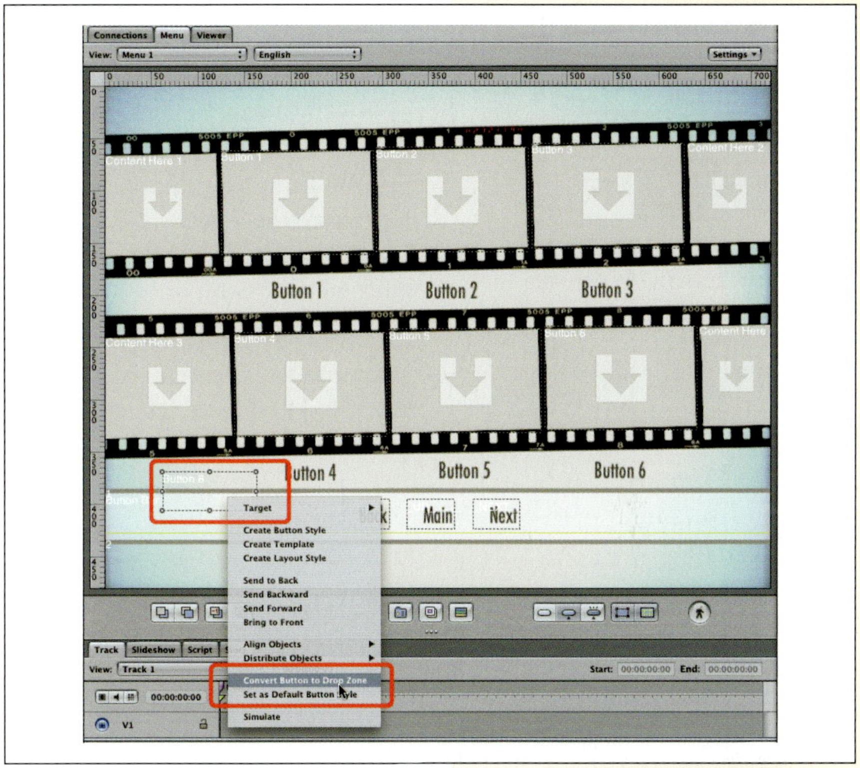

Figure 3-7. Converting a button to a drop zone

Drop zones and asset sizes

You might notice some cropping of your asset inside the drop zone. Drop zones always try to fill their content area. That usually means scaling your asset (with a fixed proportion) until it completely fills both the width and height of the drop zone and then chopping off any excess.

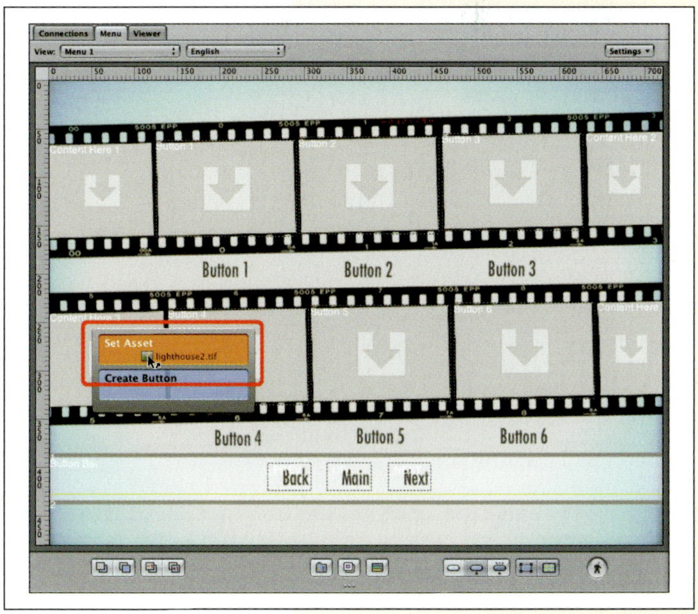

Figure 3-8. Setting the asset for a drop zone

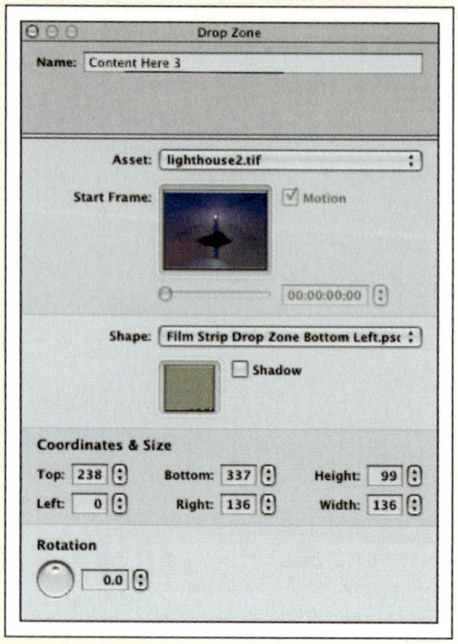

Figure 3-9. Controlling the size of a drop zone

The closer your drop zone is to the proportion of your asset, the more of your asset you'll see. You can play around with the exact size of the drop zone using the Property Inspector for your zone, as shown in Figure 3-9.

Fun with drop zones

One fun thing you can do with drop zones that you can't do with buttons is to rotate the asset. Figure 3-10 shows a stark rotation of a small graphic. Rotating video can make for eye-catching menus that don't take a lot of effort on your part.

You also should note that unlike buttons, drop zones can overlap. Although you want to be careful about cluttering your menu with crowded drop zones, you can create effects such as "stacks," as well as depth with proper overlapping.

Both drop zones and buttons should be able to use video and still image resources. Which you use depends on what you're highlighting. Video can be appealing and gives your menus flare, with very little effort. Still images enable you to emphasize important information and can be quicker to produce.

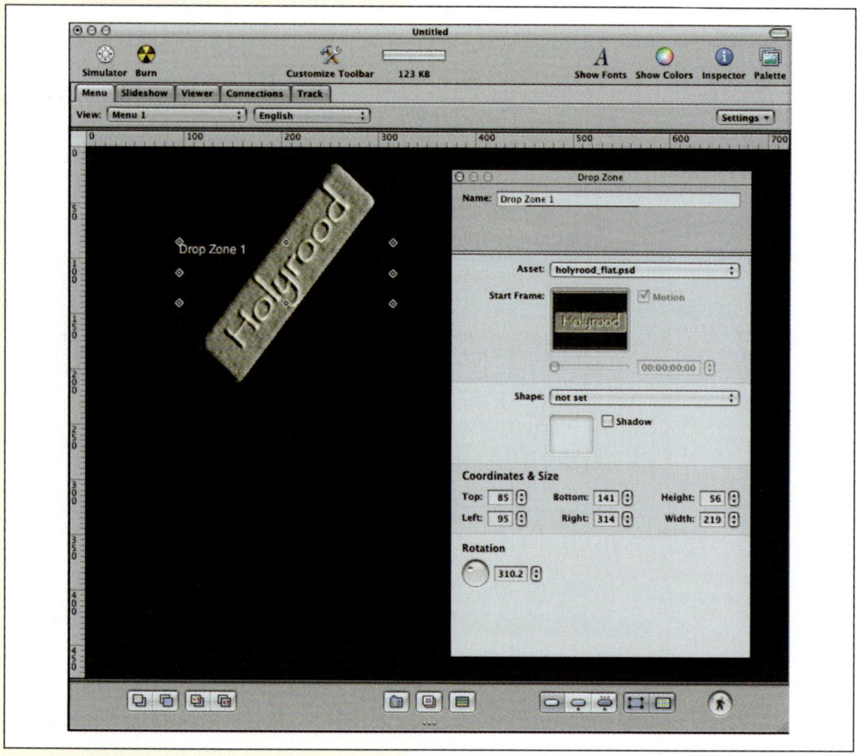

Figure 3-10. A rotated drop-zone asset

Playing around

One of the most powerful features of DVDSP's templates is the ability to switch them on the fly. As long as you stick with similar templates (with the same drop zones and such), you simply can click a new template and all your hard work will remain intact. Figure 3-11 shows a template we started filling out and customizing. Here, we've set a few assets and edited the button text.

Figure 3-12 shows those same resources after simply double-clicking a new template entry from the Palette.

Again, this works only on templates with similar resources, but it can be a great tool for making quick changes in front of clients or for even more demanding family members.

There's a lot more to the templates in DVDSP. You'll probably run into some fun features as you play around with them. But we're really here to show you what you can do when you're willing to get your hands dirty and build some fantastic menus from scratch. With that said, let's roll up those sleeves!

A Quick Castle Tour

Templates are great for putting out quick projects that still look professional. Many times, though, you really want custom menus that show off your project's content. You can create some fun menus with video backgrounds and video buttons, but we'll save that for the next chapter. Here we're going to tackle some intricate menus built with the all-powerful Photoshop layers. You need to make sure you learn to love Photoshop layers. They get used all over the place for all manner of clever tricks.

You might find it easier to keep both a "raw" and a "flat" copy of your document. You'll find that no project is ever really finished. Being able to go back to your raw document to quickly fix a typo might save your sanity one day.

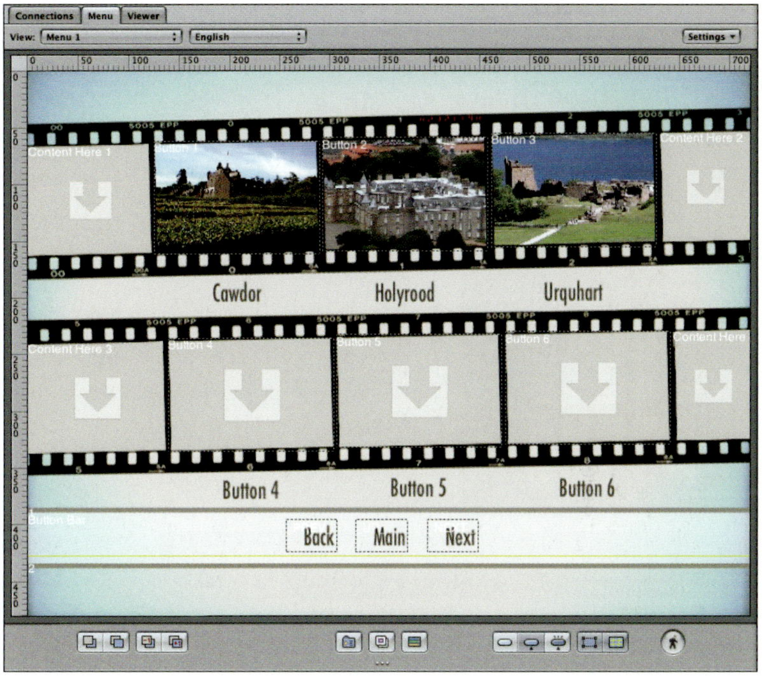

Figure 3-11. A template with our assets in place

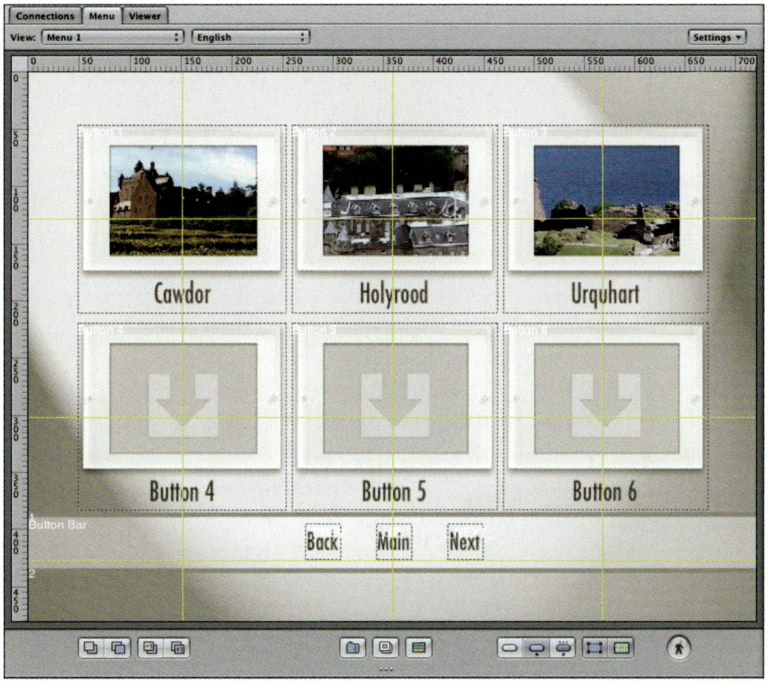

Figure 3-12. The same assets from Figure 3-11, now in a new template

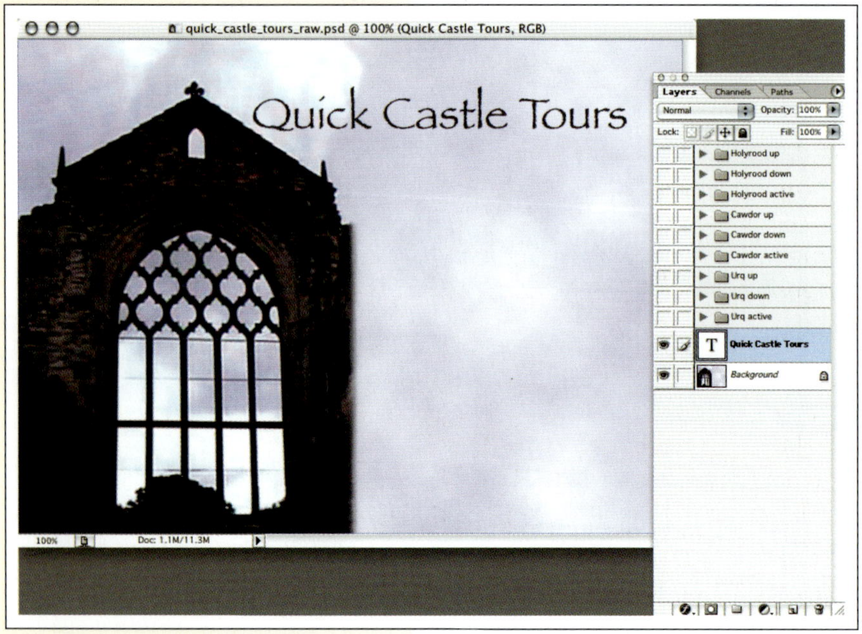

Figure 3-13. The background and title layers of a layered menu in Photoshop

Our first layered menu

We'll start this project in Photoshop to create the layered document we'll use for our castle tour menu. We need a background, a title, and a few buttons to help users navigate to the video clips. The background and title are pretty straightforward, as shown in Figure 3-13.

The buttons, however, are the interesting part here. We need to create three layers for each button: a normal layer, a selected layer, and an activated layer. DVDSP will turn those layers into the actual states of the button in your menu. One button on the screen will show its selected state, the others will show their normal states. When the user presses Enter or Play on his remote, the currently selected button will switch to its activated state to provide a brief confirmation that the user's request was received. Figure 3-14 shows the three required states for our first button.

WARNING

Layer Effects

One fantastic feature of the newest layers in Photoshop is *layer effects*. With these effects you can add spiffy things such as drop shadows and embossing without touching the source content. That means you can play around with different effects without harming your original material. Unfortunately, none of those effects shows up in DVDSP.

Basically this means that you still have to flatten any fancy layer effects you create. Remember that DVDSP understands all the layers in a Photoshop document, but does not import properties such as a layer's opacity or any of the drop shadow/embossing effects.

You can rasterize many effects in Photoshop into their own layers. Those layers then show up in DVDSP and work just as you'd expect—you just have more layers than usual. It's not hard to rasterize an effect. Just turn on only the layers you want flattened (i.e., turn off things you don't want—such as the background), use a marquee tool to select the area you need to flatten, copy it by selecting Edit → Copy Merged, and paste that back as a new layer. You might want to make that an Action.

Of course, in the real document these three layers stack right on top of each other. If they don't, the buttons will appear to magically jump around the screen as you try to navigate the menu!

The names we chose reflect the way we designed the buttons. DVDSP really doesn't care about names such as "up" or "down" or "active". As long as the button states stand out so that a user easily can determine which button currently is selected, you should be in good shape.

We have three buttons, so we'll have nine total layers (three layers for each button) beyond whatever we use for our background. In this particular project, we have the background image, with the title layer adding another layer, for a total of 10 layers. If you're running a recent version of Photoshop, you should consider using *Photoshop sets* to organize your button layers.

Figure 3-15 shows our raw version of the Photoshop document with all the buttons in their sets. Figure 3-16

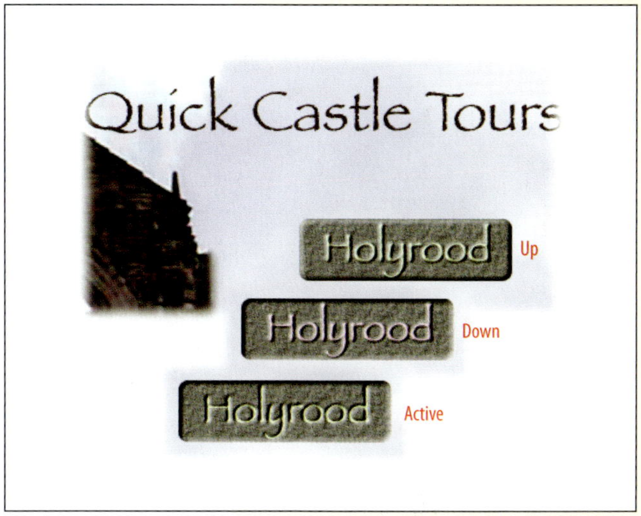

Figure 3-14. The normal (up), selected (down), and activated (active) states of a layered button

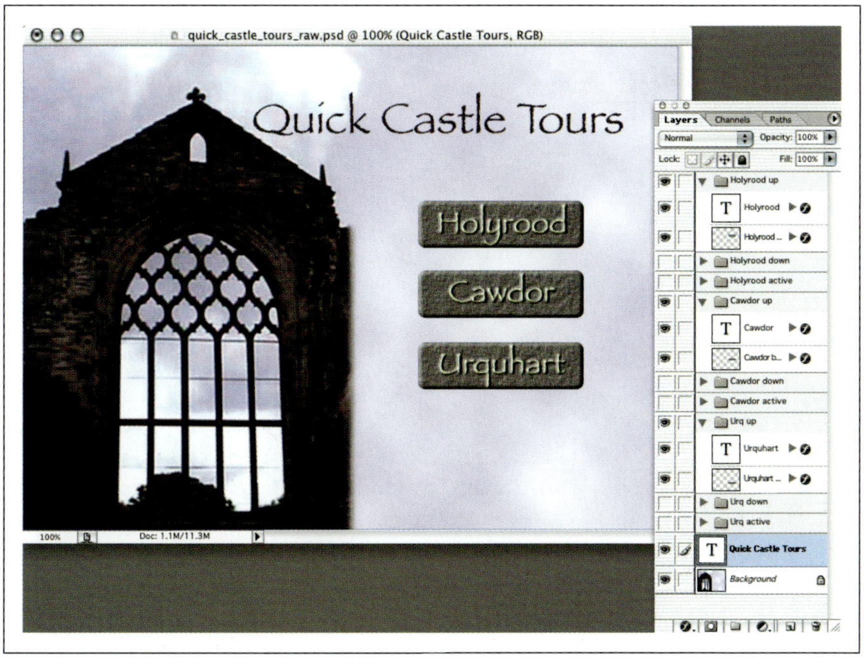

Figure 3-15. The raw "up" layers of our castle buttons (other layers are in their appropriate sets)

DEFINITION

Photoshop Sets

Photoshop sets function like "folders for layers." You can use them in Photoshop for organizing complex images. They are absolutely vital while building your DVD layered menus.

When you're done making your menu, though, you should flatten up your sets and scale the image before bringing it into DVDSP. Sets don't harm anything, but they aren't useful as such in DVDSP. (If you leave sets in a layered document and import that document into DVDSP, the sets show up as empty, unnamed layers.

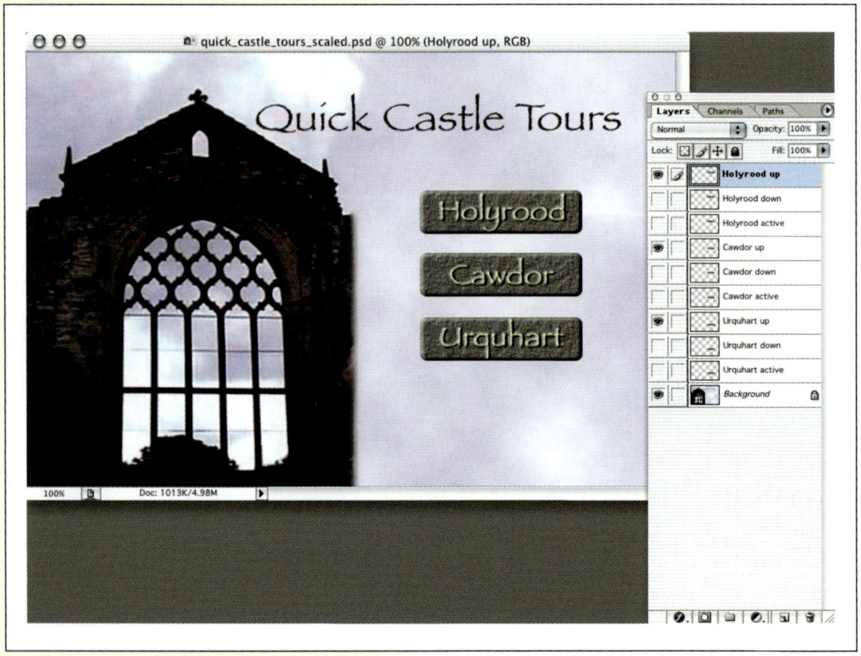

Figure 3-16. The same menu after flattening layer effects, collapsing sets, and scaling

shows the flat version scaled to the proper dimensions for use in the project.

Make sure you name your layers something useful. It'll make your life easier when you're working in DVDSP and when you're relying largely on those names to figure out what to turn on and off.

Importing layered documents

One of the odd things about layered menus is that one Photoshop document contains all our buttons. We need to import that document only once. We'll use it over and over, but we don't need to import lots of copies. Indeed, you can't. DVDSP is smart enough to make you reuse assets you already loaded.

Go ahead and import the menu and the video clips you need for this project from the *projects/ch03* folder:

- *cawdor.m2v*
- *cawdor.AC3*
- *holyrood.m2v*
- *holyrood.AC3*
- *urquhart.m2v*
- *urquhart.AC3*
- *quick_castle_tours_scaled.psd*

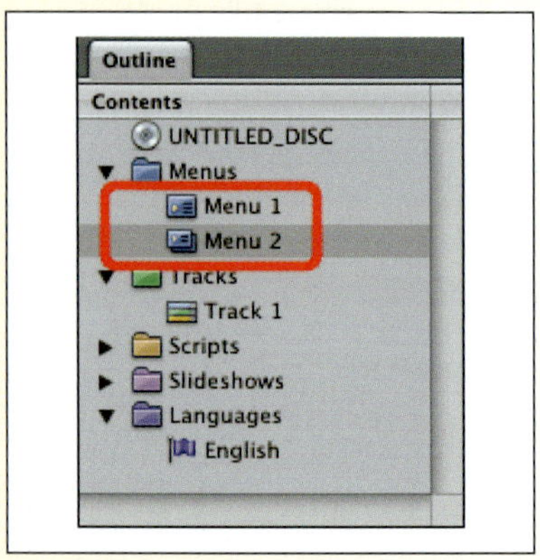

Figure 3-17. The new layered menu, which does not replace the default menu in a new project

1. **Create a Layered Menu.** Layered menus are quite distinct from standard menus. You'll need to add a new layered menu to the system. You can accomplish that easily by selecting Project → Add to Project → Layered Menu, or, if you prefer, by pressing Shift-⌘Y. Notice that now you have two menus under the Menu folder of your Outline view, as shown in Figure 3-17. Layered menus have a different icon, too.

After that, delete the first menu, as you don't need it. (You wouldn't be able to use it because it wasn't a layered menu.)

2. **Set the First Play and Remote Control Options.** Be sure to check the properties for the disc (using the Inspector) and set the First Play property to the newly added menu if it isn't set already. You also should verify the Menu and Title entries for the Remote Control property. Figure 3-18 provides a recap of these key disc properties.

3. **Add the Layered Photoshop File.** With the layered menu in place, just drag-and-drop the *quick_castle_tours_scaled.psd* file onto the menu and select the No Layers Visible option. (See Figure 3-19.)

   If you look at the Inspector for the menu, you'll see that now you can turn on the background layers for the menu. You need to turn on the background image and the title, as in Figure 3-20. To get started, you're also going to turn on the normal states of each button. You'll shut those off later, but you want to make sure you get the buttons' borders correct, and that's a heck of a lot easier when you can see them!

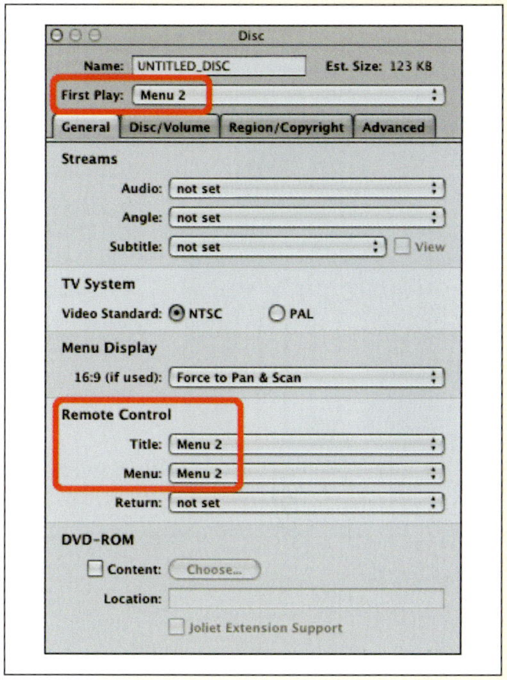

Figure 3-18. Updating the critical disc properties after deleting the default menu

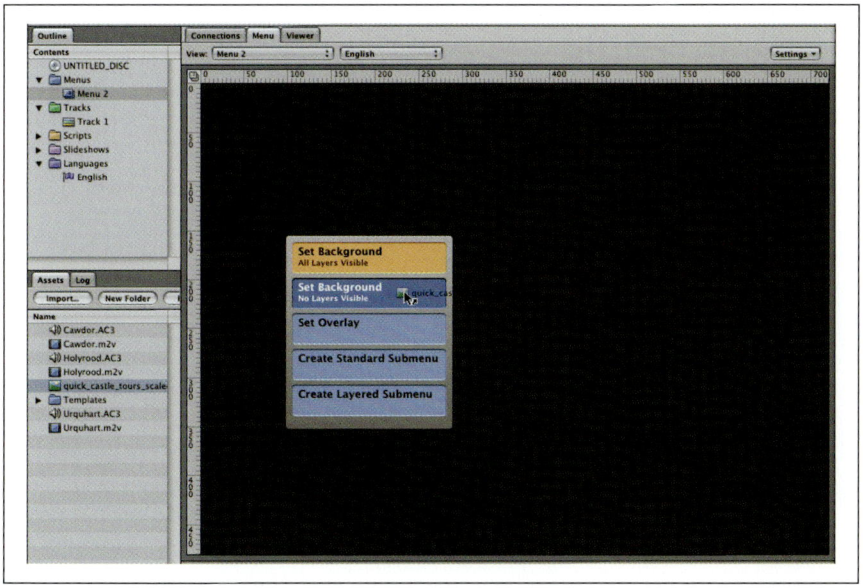

Figure 3-19. Setting the asset for a layered menu using drag-and-drop

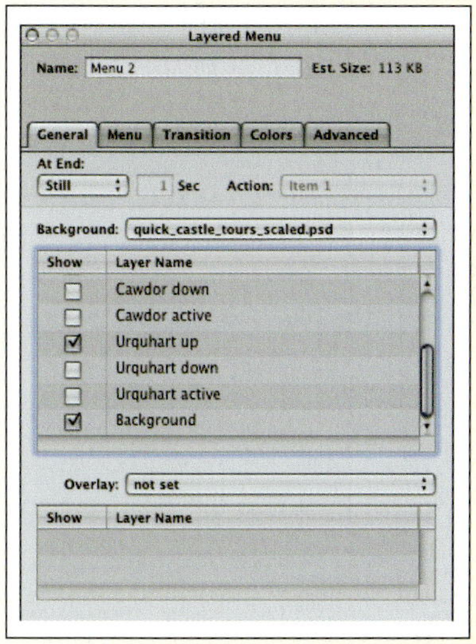

Figure 3-20. The Property Inspector for a layered menu

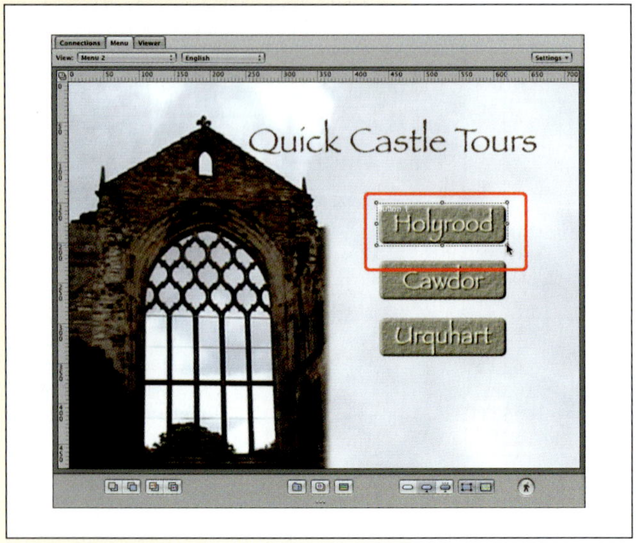

Figure 3-21. Dragging an outline to create the first layered button

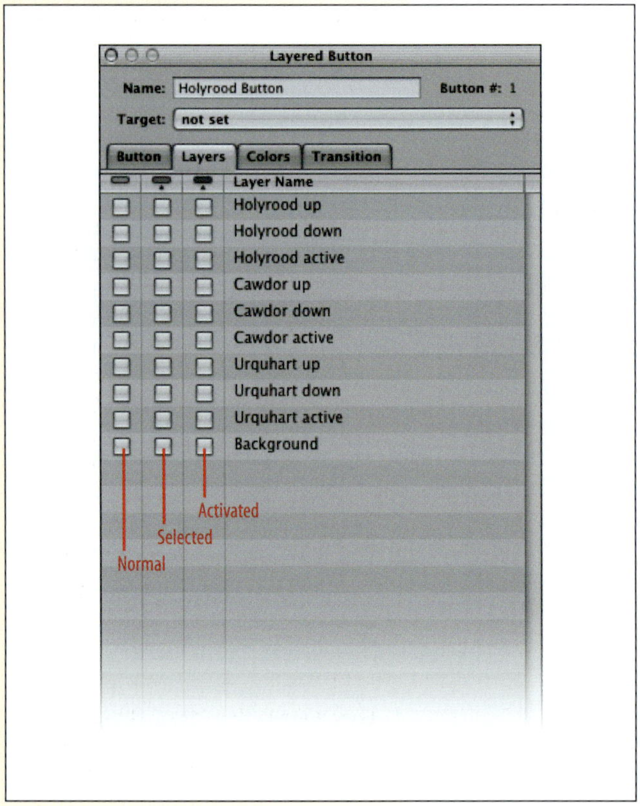

Figure 3-22. The columns for picking a button's normal, selected, and activated layers

4. **Create a Layered Button.** To create a button in a layered menu, drag an outline for the area the button will occupy, just as you do for standard buttons in standard menus. Figure 3-21 shows the first button outline.

5. **Match Each Button Layer to a Specified Button State.** The properties for layered buttons, however, are a bit more interesting. Figure 3-22 shows the Inspector view of our Holyrood button. Notice the matrix of layers. This is how you control which layer shows for each of the three button states. Just select the appropriate checkbox for each button state and make the next button.

   When you're done, you should have a menu similar to that shown in Figure 3-23. Now you can go back to the list of layers in the main menu (refer back to Figure 3-20) and shut off the Holyrood Up layer from the "background" of the menu.

6. **Add Video and Audio to Each Button.** You still need to hook up the video and audio for each button. First get the tracks ready. You can create three separate tracks simultaneously by dragging the video (*.m2v*) files to the Track folder in your

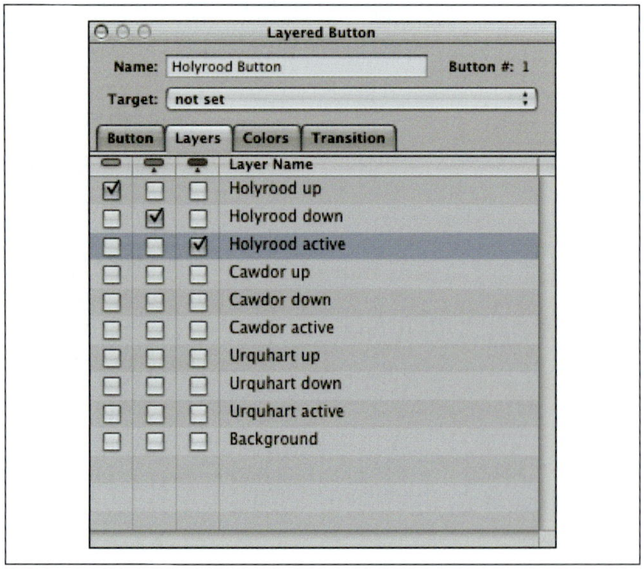

Figure 3-23. The specific layers you need to set for the Holyrood menu button

Outline view. Alternatively, you can create a new, empty track (Control-⌘T) for each clip and drag the video clips to the Track editor one at a time. Regardless of which method you use, you should check them briefly in the Track editor to make sure the audio (*.AC3*) files were properly associated with the video tracks. If you want, you also can use the Inspector to set the End Jump property for each clip to leave you back at the menu.

Now you can set the target for the buttons by simply right-clicking/Control-clicking each button in turn (see Figure 3-24) or by using the Inspector (see Figure 3-25).

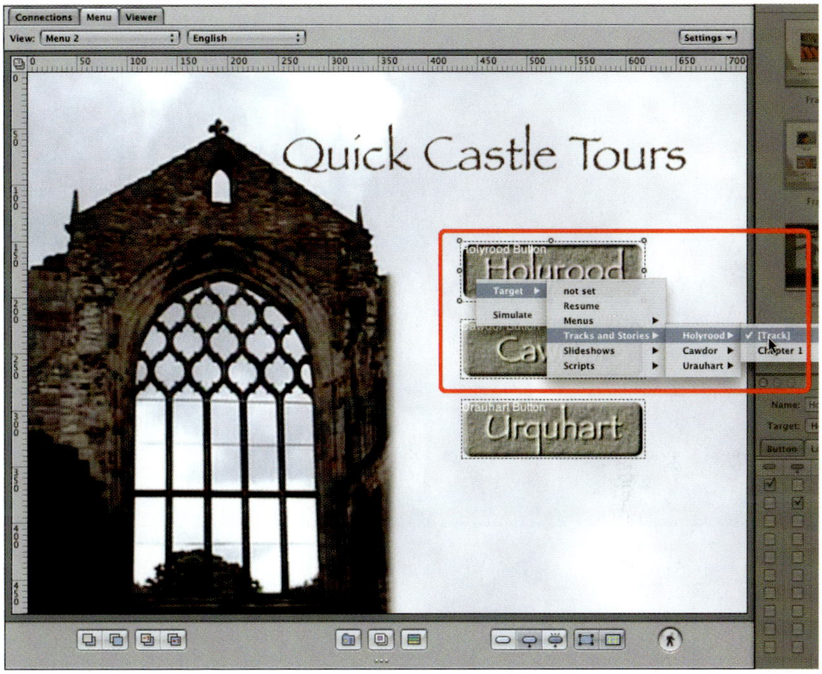

Figure 3-24. Setting the target on a layered button through context menus

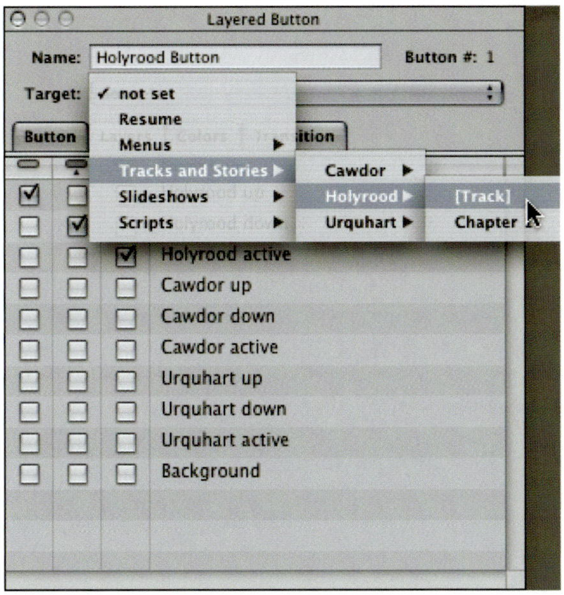

Figure 3-25. Setting the target on a layered button using the Inspector

Themeworks Music

Most of the music you hear in this companion DVD content was composed, orchestrated, and performed by Christian Hessler of Themeworks Music (*www.themeworksmusic.com*). The castle themes, Death and Transfiguration menus (in Chapters 4 and 8), and intro music are clips from his larger works for film, television, and multimedia. Although emerging sound-loop software and libraries fit the bill for most DVD projects, sometimes original composed music is what you need to take your project over the top. Find composers such as Christian in your local area who would be happy to be part of your next great opus. Not only will your DVD project stand apart from the rest, but also you'll be supporting local arts!

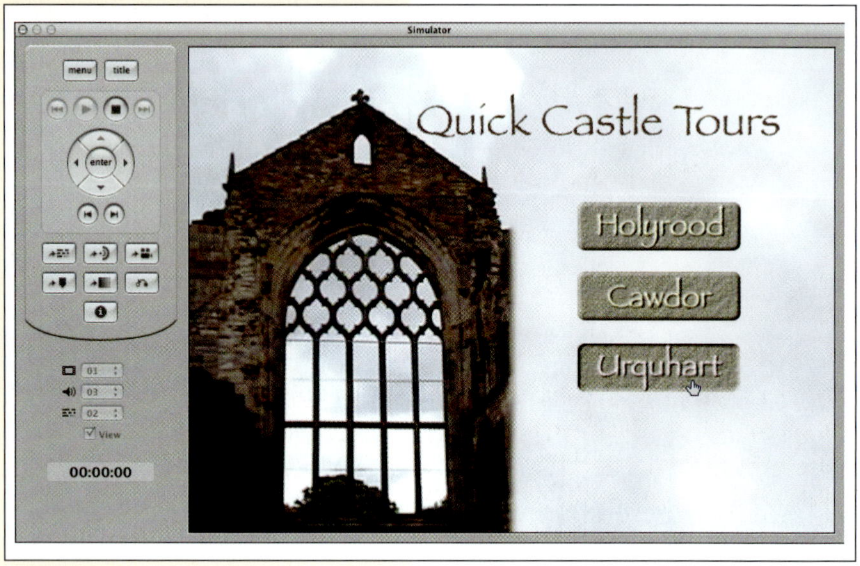

Figure 3-26. Our layered menu in the Simulator

That's it! Your project should be ready to simulate. With layered menus, it's much more important to simulate your project and test all the navigation possibilities. We had only three simple buttons with one layer per state, and we still had 10 layers to deal with. Larger, more intricate projects can have many cracks and crevices that you don't notice until the full menu is up and running in front of you.

Figure 3-26 shows our Castle Tour project in the Simulator. There's nothing special about starting the Simulator for layered menus; you just need to be more careful when checking to see if everything works.

STUDIO ADVICE

Knowing Your States

Be sure you shut off the "up" button states on your menu property sheet. If you don't, when you run this (and similar) menus in the Simulator, it will look like the buttons aren't being selected or activated. The "up" layers are currently part of the background and they stay on top of your buttons! They'll still work (you'll go to the right video clips), but you won't see what's happening in the menu. We have to admit to doing this about once per project!

A Hawaiian Vacation Menu

You've used layered Photoshop documents to create menus in DVDSP before, but how many layers have you used? Menu buttons can activate more than one layer at a time—without much more work than it takes to activate a single layer. This fact gives you more flexibility in designing and working with your menus. You can create fantastic menus, and you don't need to flatten everything in Photoshop before putting them to work in your DVD project.

Here, we'll build a menu for a Hawaiian Vacation DVD that creates a dynamic menu system. The menu consists of a picture of Hawaii and four buttons associated with four of the Hawaiian Islands. Navigating to a particular menu button will do three things:

- Highlight that button
- Highlight the associated island
- Display the island's name and a quick blurb about it

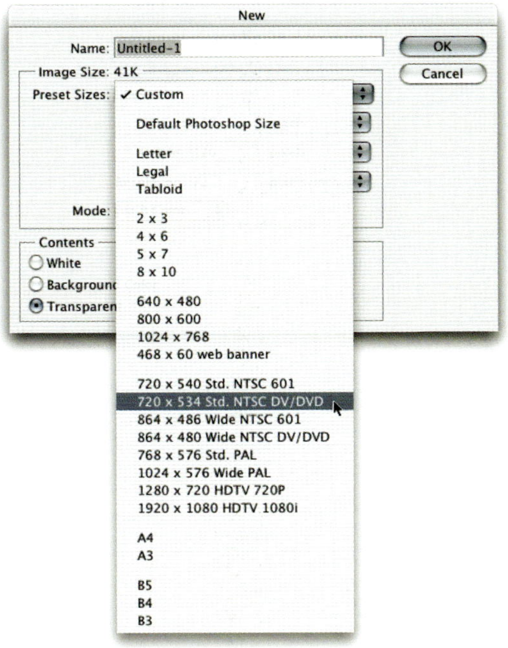

Figure 3-27. Photoshop 7.0 presets for use with DVD menus

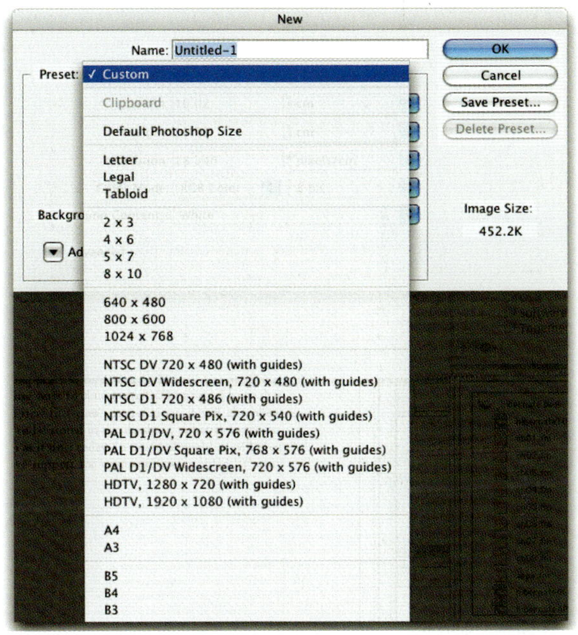

Figure 3-28. Photoshop CS template sizes including nonsquare DV templates

Start off with a regular NTSC/DV-size Photoshop document. Photoshop makes that easy with a preset you can use for new documents, as shown in Figure 3-27. Photoshop CS has an expanded section, as shown in Figure 3-28.

Build up your menu and don't be afraid to use layers! Remember that DVDSP loves layers. Figure 3-29 shows our main menu for the Hawaiian home-video DVD, and look at all the layers!

With DVDSP, it's easy to work with that many layers. Layers make tweaking and playing in Photoshop much more enjoyable and productive, so don't be stingy.

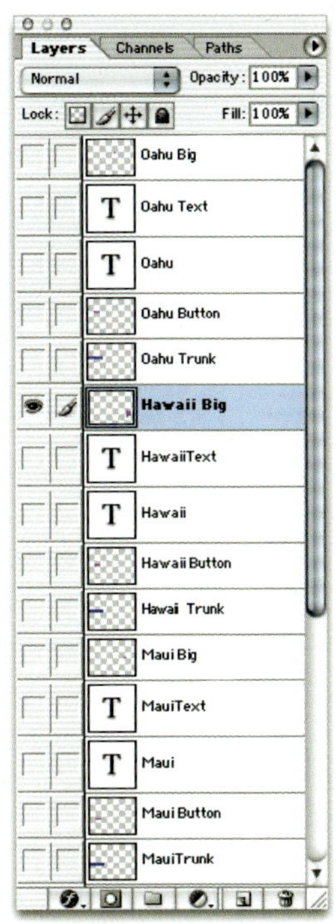

Figure 3-29. Some of the layers of our full menu

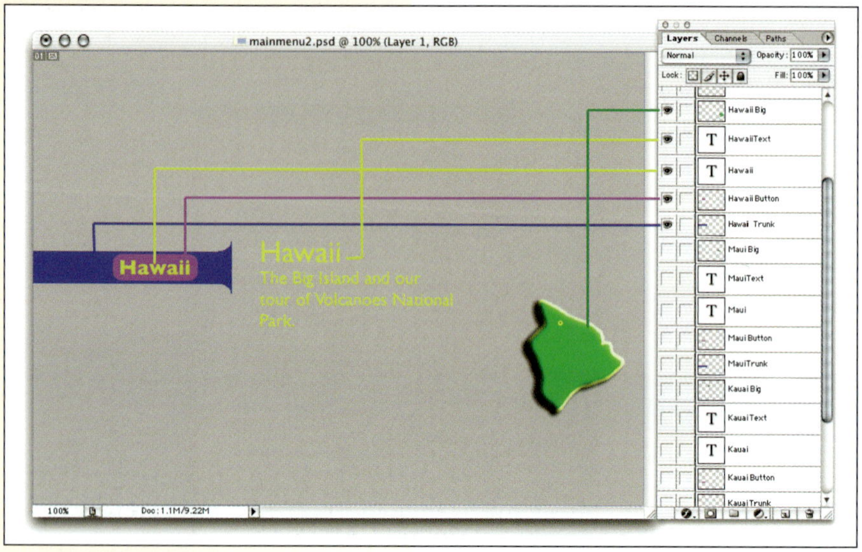

Figure 3-30. The specific layers for the Big Island button

For this menu, we want each island button to highlight the island itself and to bring up a quick description of the video segment. We simply make layers for each piece we're interested in. For example, our Big Island pieces break down as shown in Figure 3-30.

We repeat that for each button. Figure 3-31 shows what we expect to see when all the pieces of the Big Island content are put together with the background.

At this point, resize the image to 720×480 if you were working with square pixels. Be sure to shut off the Constrain Proportions option.

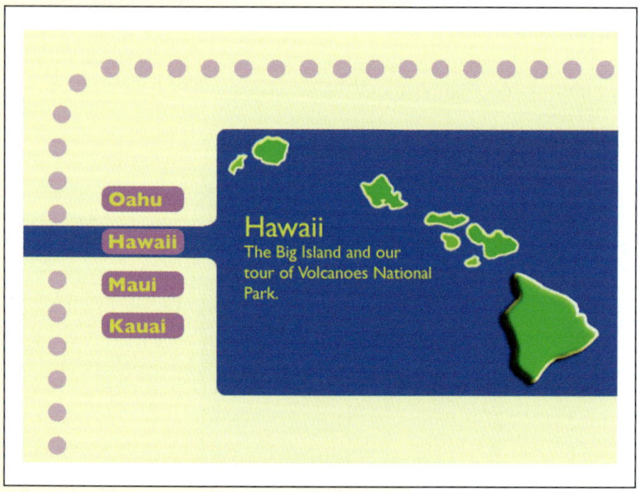

Figure 3-31. A completed menu with the Hawaii button active

Building the menu

Great, now that the menu graphic is ready, let's build the menu itself. Because we're concentrating on the menu for this project, you'll need to import only the following file:

- *tour_menu.psd*

1. **Create a New Layered Menu.** You can do this with the Shift-⌘Y keyboard shortcut. Go ahead and drag your file into the Menu editor with no layers visible. You'll need to turn on the following layers (you also can rename the menu if you want, but that's not critical):

   Ⓐ Background

   Ⓑ Islands

   Ⓒ Our Hawaiian Vacation

   Ⓓ Kauai

   Ⓔ Maui

   Ⓕ Hawaii

   Ⓖ Oahu

You should end up with a menu that looks similar to that shown in Figure 3-32.

The property sheet will show you all the possible layers, just as before. The fact that we have multiple layers selected for the background (see Figure 3-33) is no problem.

2. **Create the Buttons.** Go ahead and define the buttons for each island by dragging the mouse over each name on the left. You can hook up the jump target to the Under Construction video clip if you want. (See Chapter 1 for a refresher on hooking up buttons to jump targets.) Now that we've defined all our buttons, we'll go back to the Property Inspector to turn on all the right parts.

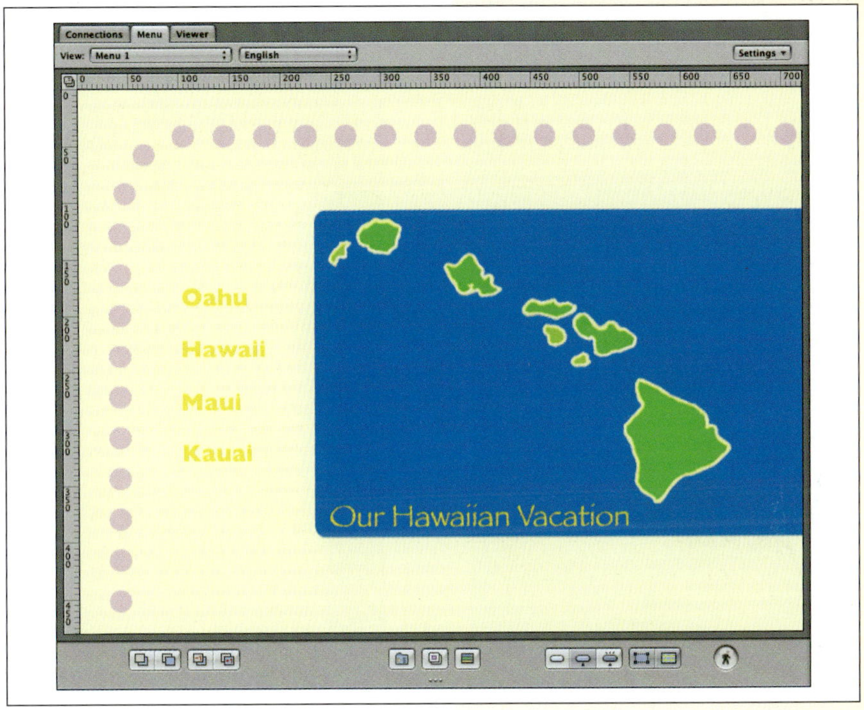

Figure 3-32. Our many-layered Hawaiian Vacation menu

Let's start with the Oahu button. As before, each button has three separate states: Normal, Selected, and Activated. For the Normal state, turn on the Oahu Button layer. That gives you the purple button behind the text. For Selected, turn on several layers: Oahu Trunk, Oahu Text, and Oahu Big. Now turn off the Oahu Button layer. You'll use the same layers for the Activated state, but you'll also use the Oahu Button layer to make it obvious when the user has activated a jump. And because you made everything a layer, it's easy! Figure 3-34 shows the matrix you should have now for the Oahu button.

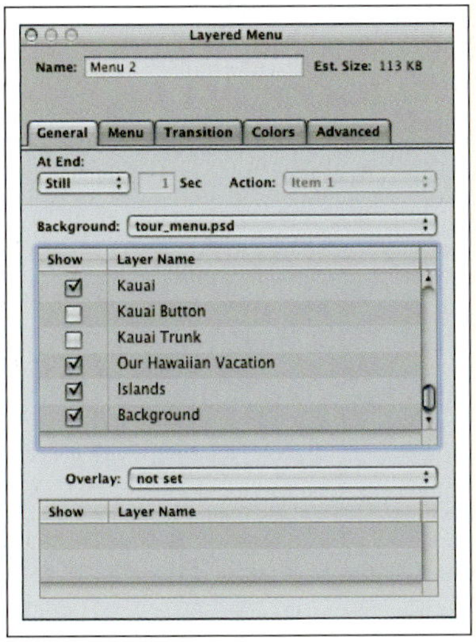

Figure 3-33. The layers list for our menu

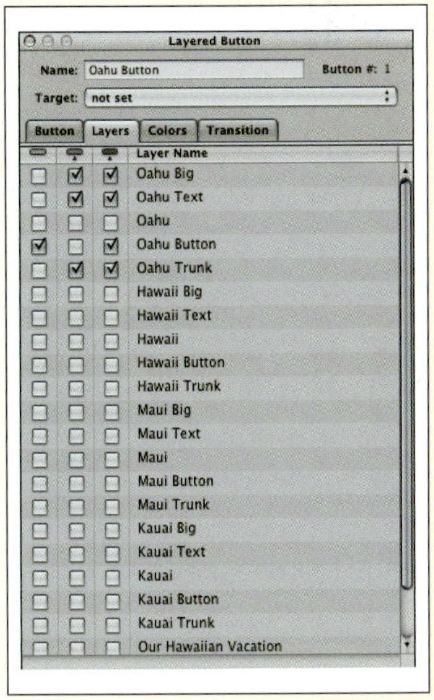

Figure 3-34. The layers list for the Oahu button

Do the same for each button. You'll end up with a fairly interesting matrix, as shown in Figures 3-35, 3-36, and 3-37.

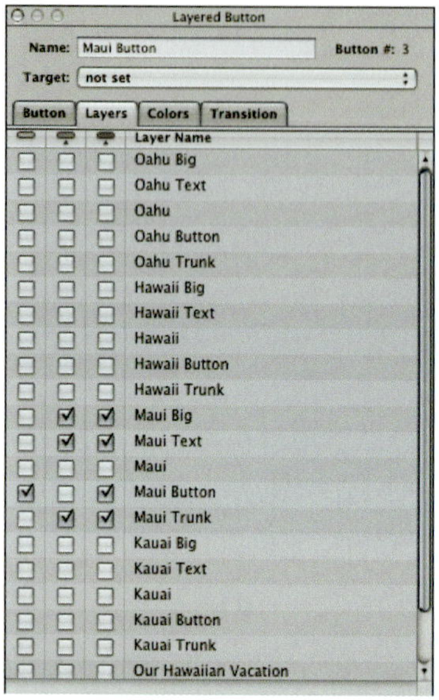

Figure 3-35. The layers list for the Maui button

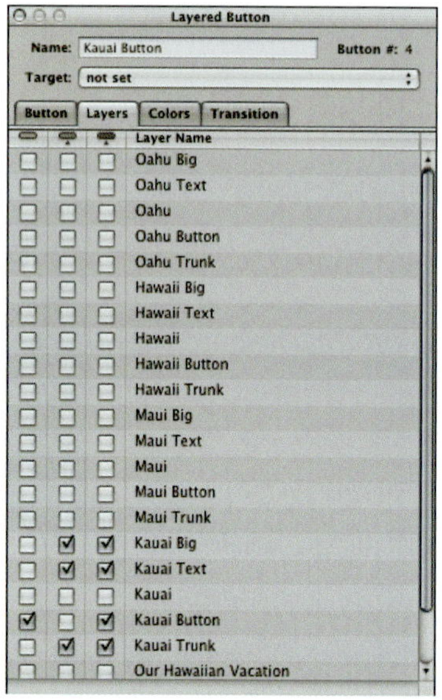

Figure 3-36. The layers list for the Kauai button

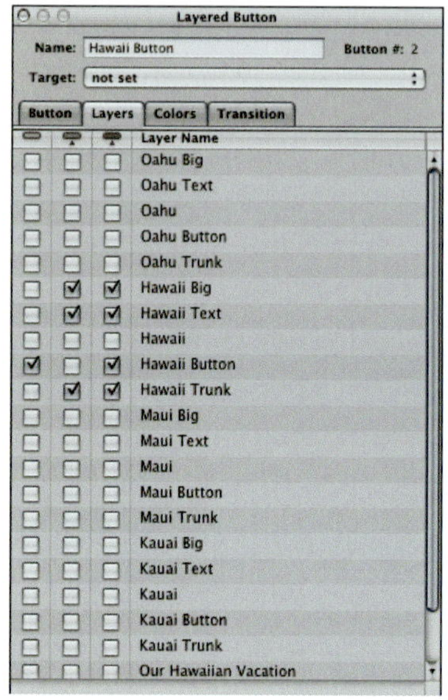

Figure 3-37. The layers list for the Hawaii button

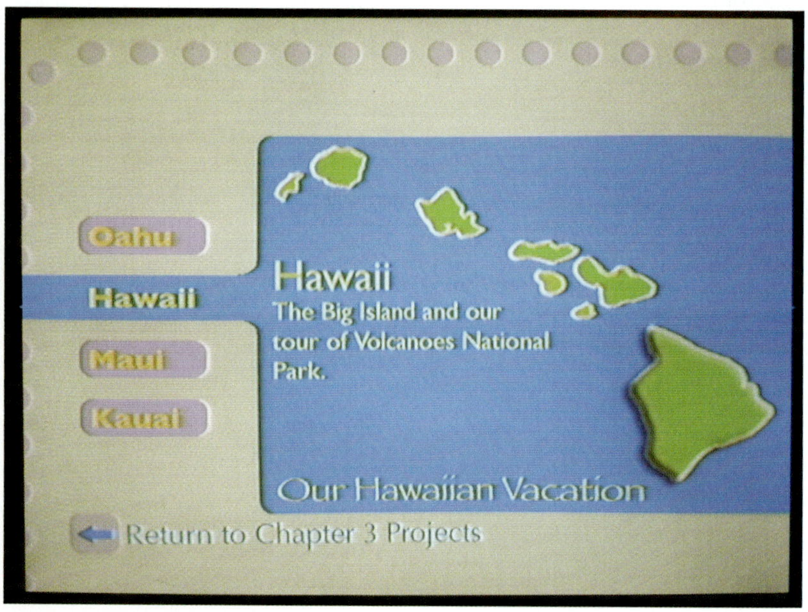

Figure 3-38. The completed menu in action on a TV set

3. **Save the Project.** You might have been saving your work all along, but make sure you do so now—just in case. You should be able to preview your menu and make sure all the layers are doing what they're supposed to do. Many times you'll catch stray layers during the preview. Don't worry—just stop the preview, fix the layer, and preview it again.

When you're happy with the menu, save everything and move on to the next part of your project. In our case, it's a no-frills family vacation slideshow, so we're ready to burn the disc! Figure 3-38 shows our menu in action.

An extra trick

DVDSP enables you to play some interesting games with the layered menus. If you can't decide between a few features (or you've got a client who wants to see several color schemes), you can substitute different Photoshop documents without losing your button and layer definitions. For that to work, though, every Photoshop document must have the same set of layers, in exactly the same order.

For example, our first attempt at the menu was to use a large, embossed version of each island as the appropriate button was selected. Let's say that several members of the review committee (which you often have, even at home...) have remarked that the 3D islands didn't stand out enough and broke with the clean design theme of the rest of the menu. It's hard to say who's right until you see it running, so we went back to Photoshop to create a copy of the existing menu. Then we tweaked the various Big layers (Oahu Big, Maui Big, etc.) to use purple islands instead. Figure 3-39 compares the results.

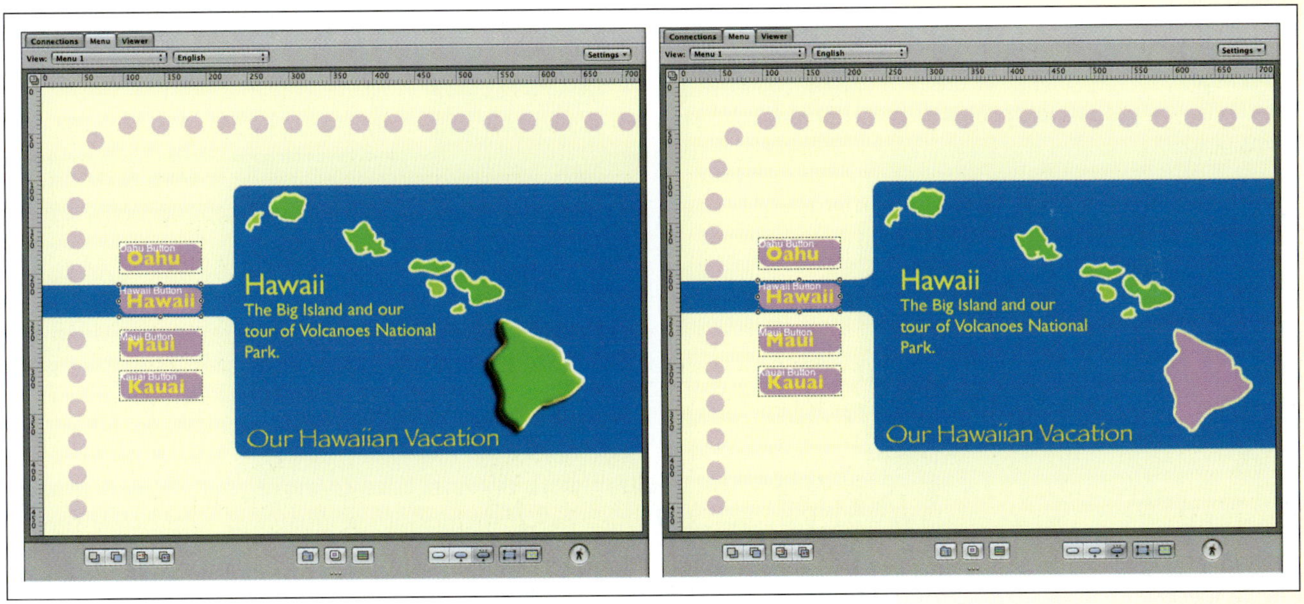

Figure 3-39. The original 3D highlights (left), and an alternative callout scheme (right) for the islands

Now we can load both Photoshop documents into DVDSP and flip back and forth between them with the property sheet for our menu. Simply change the image resource (see Figure 3-40) and reset the layers for the menu. All the button layers should remain intact.

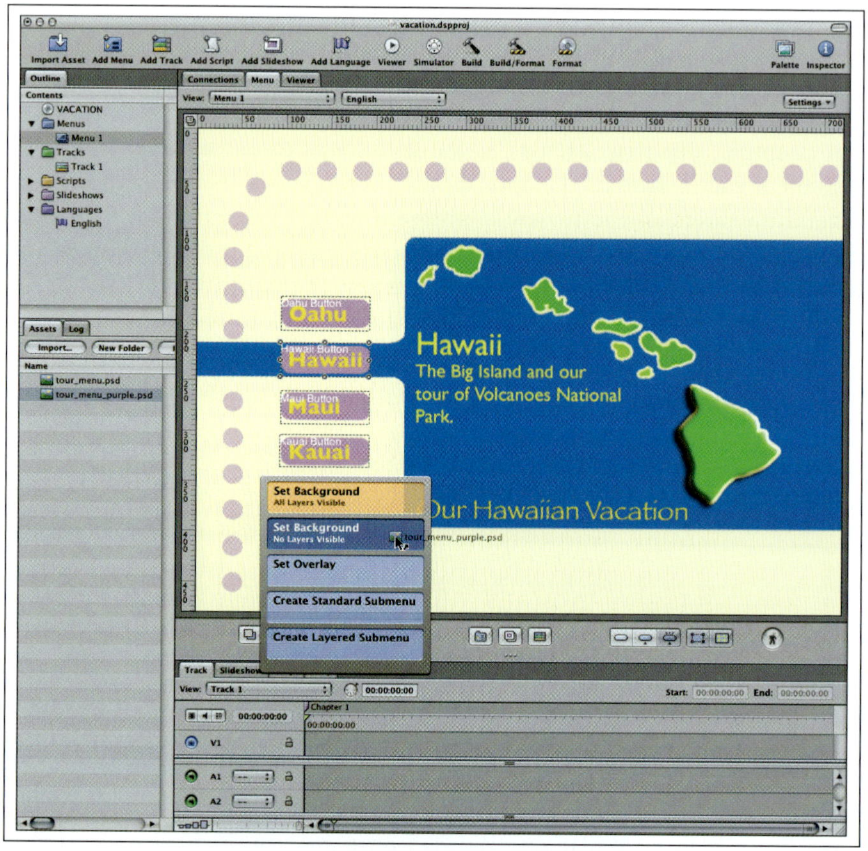

Figure 3-40. Changing the menu's Photoshop document

Note that this wouldn't work if you needed to move your buttons around, but for color- and effect-based themes, it's a great way to preview several variations without having to duplicate all the effort of creating menus. It also means you're free to edit or fix a given menu's document even after you've been using it inside DVDSP. Chapter 10 shows you how to use templates to manage bigger theme changes that can include relocating buttons.

Plan Ahead!

If you build many layered menu projects, you'll undoubtedly get into the habit of carefully planning the buttons. Building the layers required for a button is time-consuming and doesn't lend itself to on-the-fly design inside DVDSP.

For that reason, you might want to create a set of layers for any temporary or test buttons. One of our reviewers, Eric Bolden, found himself needing a button just to try out a quick feature. But he was using a layered menu and couldn't set up an "empty" button with no layers. DVDSP realized the button was empty and didn't make it part of the menu. Having a "spare" set of layers enables you to try quick things without having to go back to Photoshop to redesign your menu first. For the final DVD, you can just leave those layers unused.

Home Movies

Well, after all that work to set up the menus, where's the home movie? DVDSP 3 can make very quick, elegant work of the most dreaded of home-movie variations: the slideshow. Or rather, DVDSP 3 can make elegant slideshows. DVDSP always has supported slideshows, but until Version 3, it was just that: a set of individual images that advanced every few seconds or waited for the viewer to click the Forward button. DVDSP 3 added support for spiffy transitions, and that's what we're going to look at here.

If you're still using Version 2, fear not. The end of this section discusses using iMovie to accomplish the same effects. In fact, the iMovie discussion is relevant to everyone because it covers the way-cool Ken Burns Effect.* Even DVDSP 3 users should consider creating their slideshows in iMovie. But if "quick and easy" is what you're here for, check out just how seriously DVDSP 3 takes those terms.

Starting a slideshow

Starting a slideshow in DVDSP is very simple. You can click the Add a Slideshow button in the toolbar, right-click the outline, and add a new slideshow, or just press ⌘K. Figure 3-41 shows the resulting entries and the various tabs of interest to you when building your slideshow.

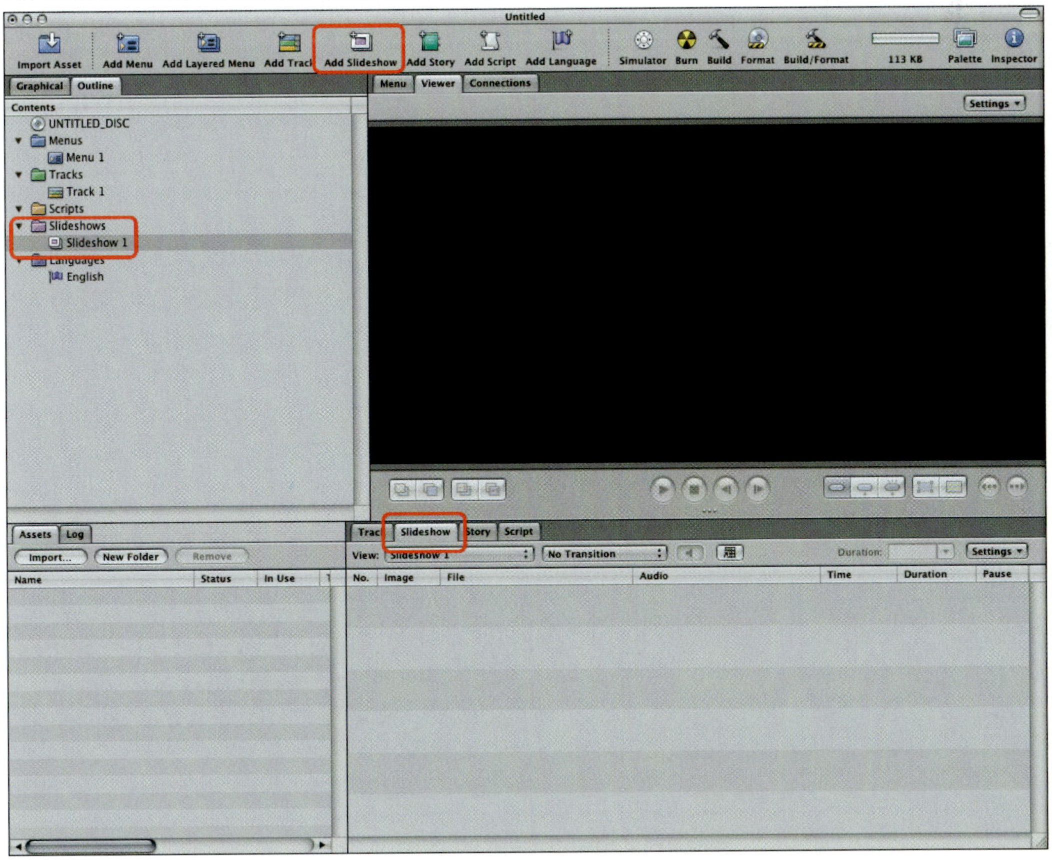

Figure 3-41. The relevant interface parts for slideshows

* Ken Burns is a documentary filmmaker renowned for such films as *JAZZ* and *Brooklyn Bridge* and for biographies of luminaries such as Thomas Jefferson and Frank Lloyd Wright. His eponymous effect in iMovie stems from his use of panning and zooming on still photos to add liveliness and action.

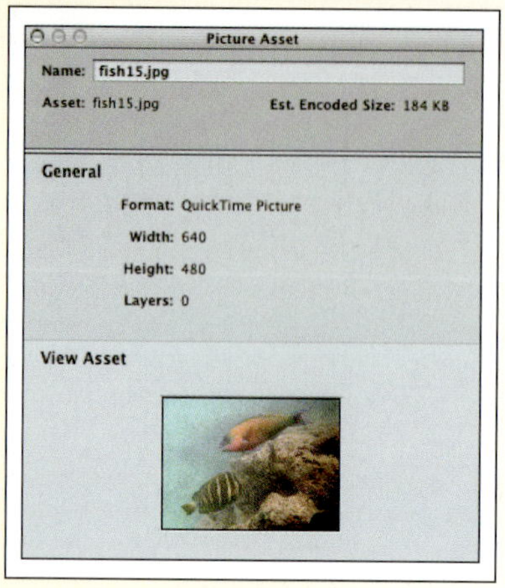

Figure 3-42. A slide asset in the Property Inspector

Notice that the Viewer tab is selected in the Main window. As you work with individual slides in the Slideshow tab at the bottom, they appear in this Viewer tab. Using the tab is not required, but sometimes it can make things easier when you can see the slide at a reasonable size.

You import slides as assets just as you would menu graphics or audio and video clips. And if you click one of the imported slides, it shows up in the Property Inspector just like other assets. Figure 3-42 shows an example. This is useful for reviewing your slides' dimensions. You don't have to make them all the same size—DVDSP will scale them—but it is still a good idea to make their aspect ratios all roughly 4:3.

Adding these assets to the slideshow is child's play. (No, really! We had a computer-savvy three-year-old assemble some pictures from her preschool. The dramatic arc of the slideshow might have been a bit amateurish, but we've certainly paid good money to see worse in a theater!) You just drag the slides from the Assets window to the Slideshow tab and drop them there. If you don't like the slide order, you can rearrange it, as shown in Figure 3-43.

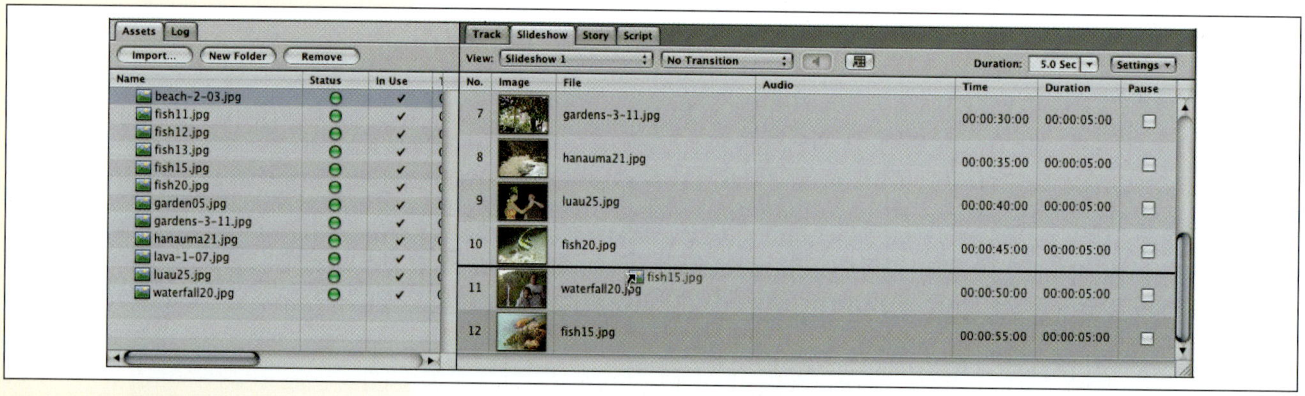

Figure 3-43. Arranging slides in the Slideshow tab

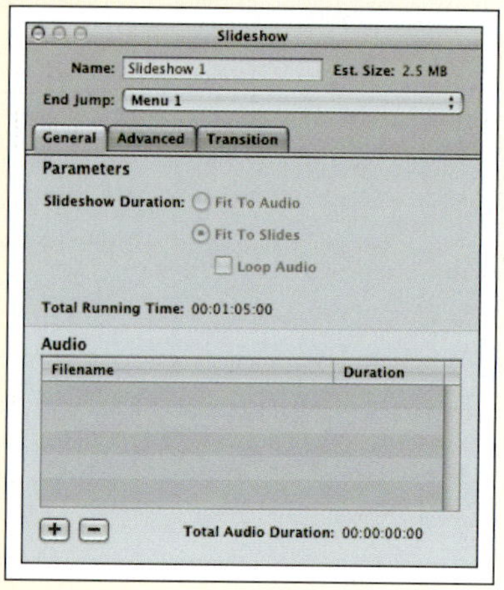

Figure 3-44. Our slideshow in the Property Inspector

Once you have all your slides in the slideshow, you can add audio if you want. The Property Inspector for the slideshow (you might need to select the slideshow in the Outline tab) is an easy place to add and arrange your soundtrack. Figure 3-44 shows the Inspector, where you also get a summary of the show's duration.

We didn't add any audio for this go-around, as we're concentrating on the visual aspects of transitions, but you can drag-and-drop audio assets into the Inspector or use the + button at the bottom to search for new audio files to use. (They will be imported automatically as assets for you.)

Adding transitions

The Inspector for slideshows also enables you to set the transition for the entire show. Plus, you can set transitions on individual slides if you want to use particular changes for particular images.

Slideshow transitions

If you want the same transition applied to the entire slideshow, you have two options. If you have the Inspector for the slideshow handy, just switch to the Transition tab, as shown in Figure 3-45. You can pick and preview transitions here.

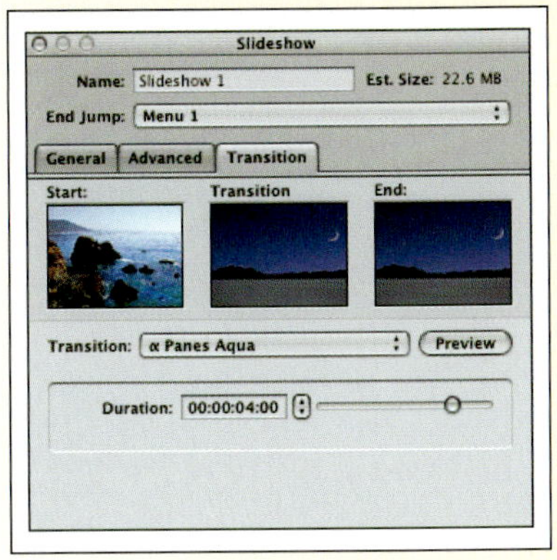

Figure 3-45. Setting and viewing transitions in the Inspector

You also have control over the duration of the transitions. Keep this time fairly short. Remember that the default length of a slide is only five seconds. A four-second transition will effectively cut down the slide's viewing time to three seconds. Put that same transition on both sides of the slide and you have only one second to view the slide!

In addition, you can set the transition for the slideshow using the pop-up list in the Slideshow tab, as shown in Figure 3-46. You can't preview the transitions

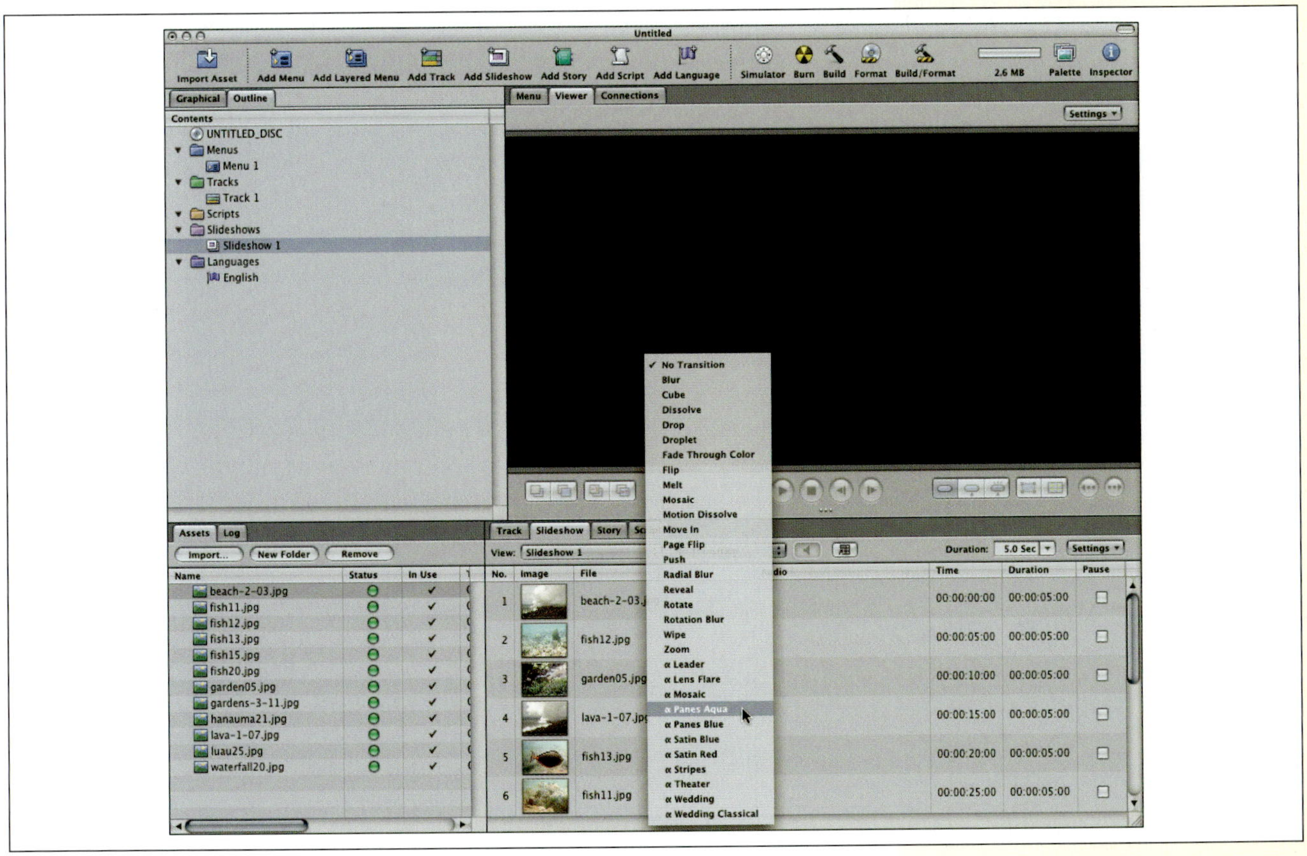

Figure 3-46. Selecting from the complete list of transitions in the Slideshow tab

from this list, though, so you might want to continue using the Inspector until you get familiar with the various transitions that are available.

You can't set the transition duration here either, but you can edit that easily in the Inspector (refer back to Figure 3-45) at any time.

Individual slide transitions

Another thing you can do is tailor the transitions to each slide. Obviously this takes more time, but it can make the slideshow more interesting with only moderate effort. Then again, going berserk and using too many different transitions can be distracting. As with all things in life, balance is key.

A quick way to set the transition for an individual slide is simply to Control-click the slide in the Slideshow tab and set the transition from the context menu you see in Figure 3-47.

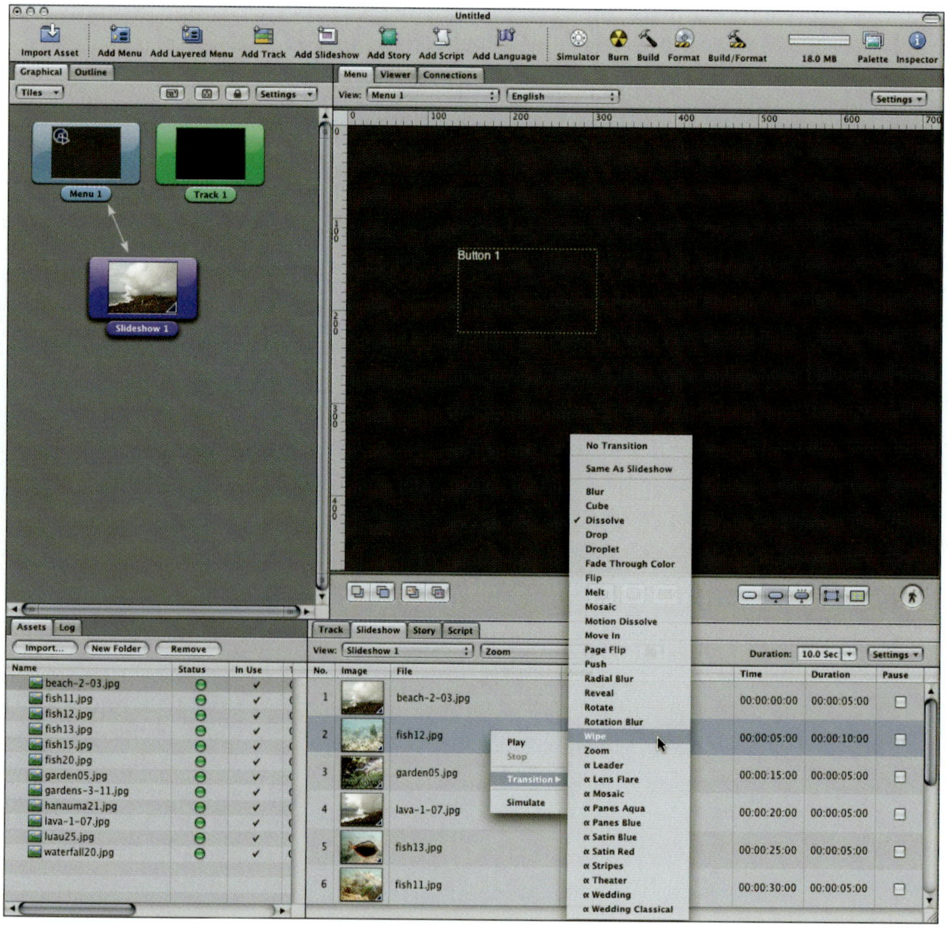

Figure 3-47. Setting the transition for one slide

If you don't set a transition for a slide, it will default to the transition set for the slideshow.

Of course, you can use a slide's Property Inspector to set its transition. Figure 3-48 shows the Inspector for one of our slides. As with the Slide Inspector, you can preview the transition and control its duration here.

Here, you also can configure any other transition features you have chosen. Some transitions enable you to specify things such as their starting point or the direction in which they move. Others don't. If you haven't played with transitions before, it's a good idea to just sit down and run through all of them to get a sense of what's available.

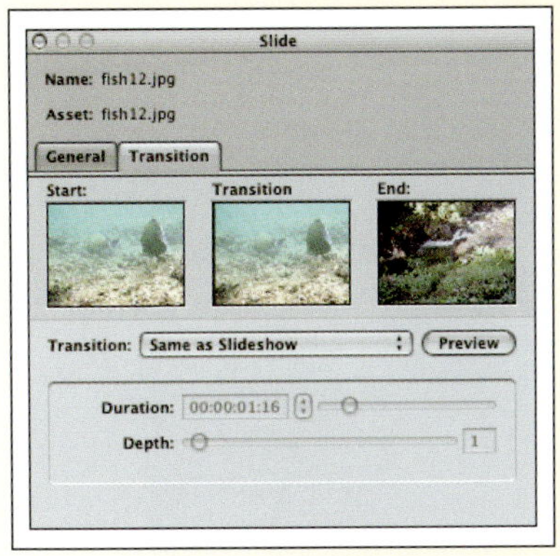

Figure 3-48. Fine-grained transition control for individual slides

Previewing transitions

When you're ready to preview your slideshow with all its glorious transitions, you can use the Simulator. Note that the transitions happen in real time. (When you go to burn the disc, the transitions will be converted into movies.) If your slideshow already is hooked into the larger project, you can just simulate the disc. You also can use the Graphical view to great advantage here.

Figure 3-49 shows the context menu you get when you Control-click a slideshow in the Graphical tab. You can select the Simulate option and the Simulator will start right on the slideshow. The show doesn't even have to be hooked up to anything. If you're using DVDSP 2, you can use the context menu from the Outline tab to accomplish the same thing.

All in all, not too bad for software meant to burn a DVD. But if this isn't enough, you should check out the next section on using iMovie to energize your slideshow like the pros do.

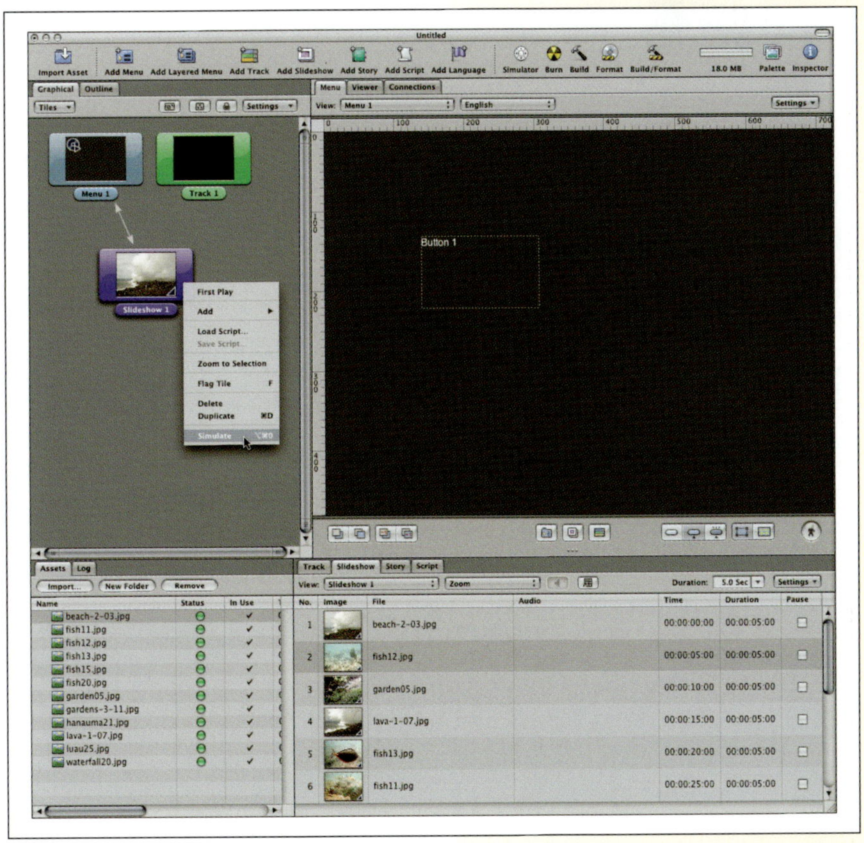

Figure 3-49. Simulating a slideshow from the Graphical tab

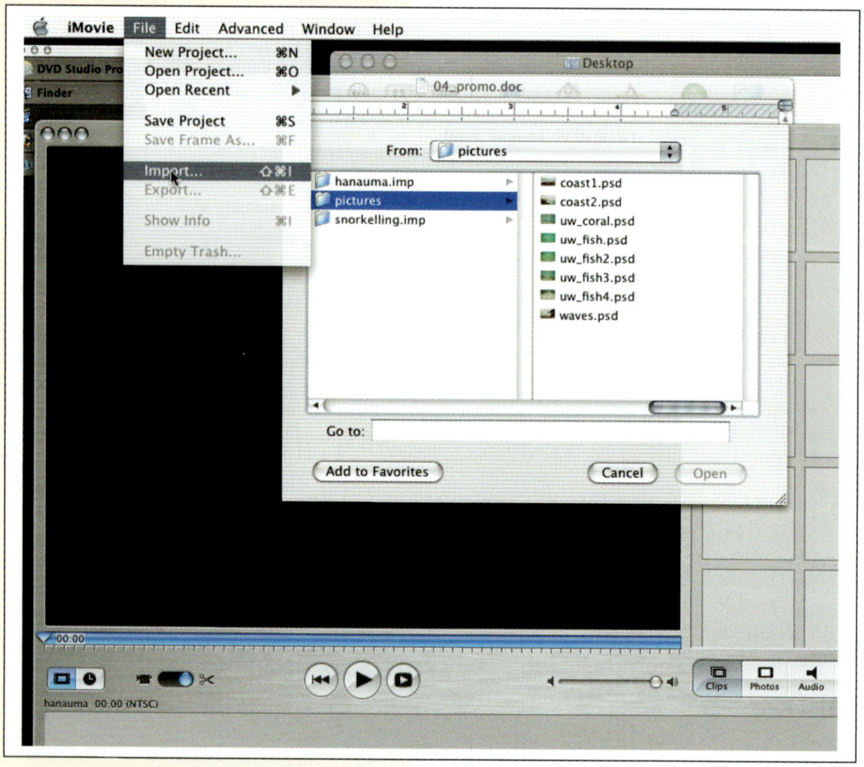

Figure 3-50. Importing slides into iMovie

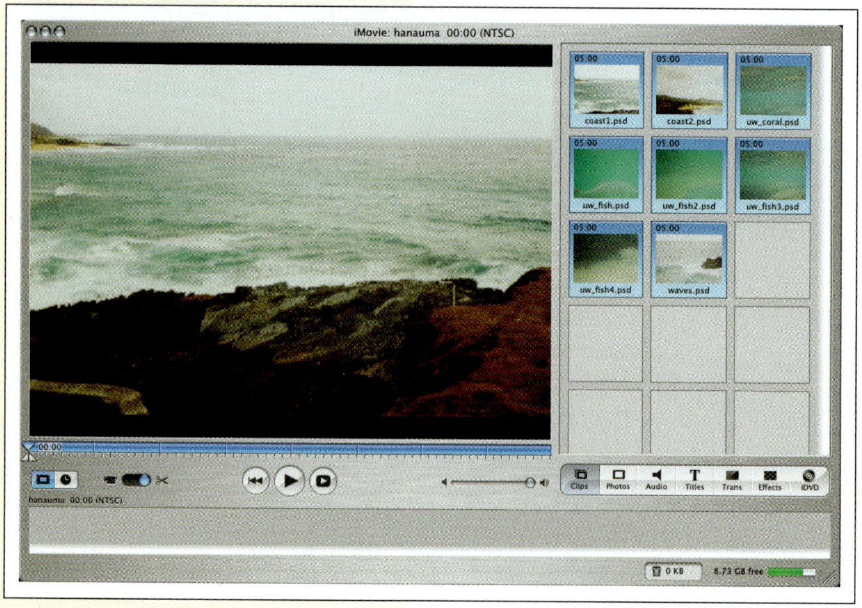

Figure 3-51. iMovie clips section after importing several slides

iMovie tangent

What about those of us still using DVDSP 2? Transitions are new to DVDSP 3, so slideshows in DVDSP 2 were limited to very basic slides. However, you still can put together a good-looking show with the help of other tools already on your Mac. (This is a good tangent for DVDSP 3, too!) iMovie is a free software package bundled with every Mac these days. It has many spiffy features for hacking together home movies. The feature we're most intrigued by here is the support for slides and the marvelous Ken Burns Effect.

1. **Import the Slides.** If you start up a new project in iMovie, you can import all your stills by selecting File → Import... (see Figure 3-50). Feel free to use some of your own images, or you can grab everything in the *projects/ch03/slides* folder from the accompanying DVD. The images should show up on the Clips tab (see Figure 3-51).

   Even though your still images end up in the Clips bin, the Ken Burns Effect is on the Photos page. The easiest way to work with this effect is to create the slideshow first, and then go through and apply the effect where appropriate.

2. **Place Your Clips on the Timeline.** From the Clips bin, drag your photos to the strip at the bottom of the page (see Figure 3-52). You can rearrange photos in that timeline, so don't worry too much about the ordering. Once the photos are in the timeline, switch to the Photos tab. (If you've ever imported photos with iPhoto, you'll see those images pop up on the Photos tab as well. You can ignore those for the purposes of this project.)

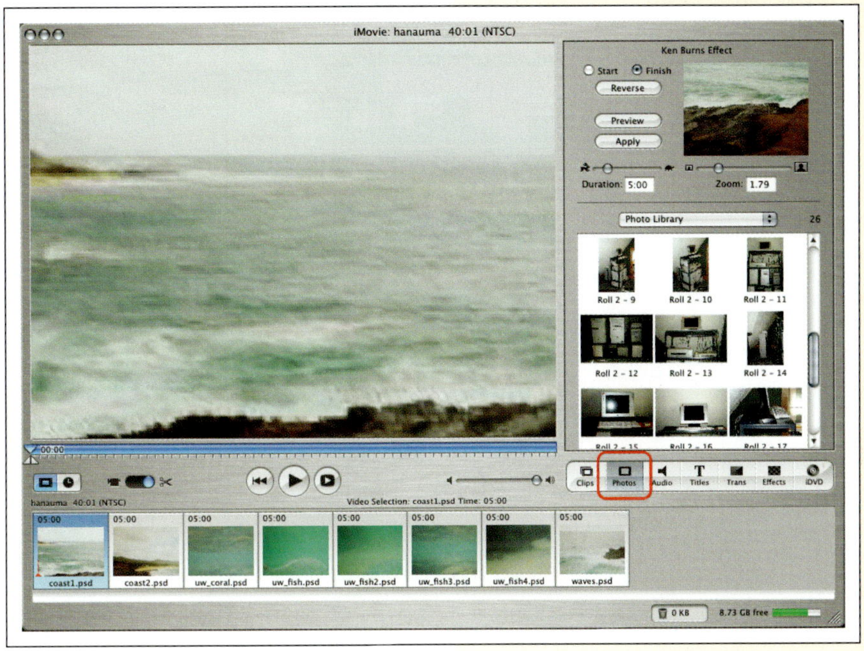

Figure 3-52. The Photos tab in iMovie

3. **Create the Effect.** Click any of the photos in the timeline and they should appear in the Ken Burns Effect portion of your screen.

   Creating the effect is very easy. Figure 3-53 calls out these steps:

   ① Click the Start button.

   ② Move the image of the photo around by dragging it. You also can play with the Zoom slider to focus on a particular spot.

   ③ Click the Finish button.

   ④ Zoom and move the image around to the place you want to end up.

   ⑤ Click the Preview button to make sure it animates the way you want it to.

   ⑥ Click the Apply button.

   Repeat for the next image!

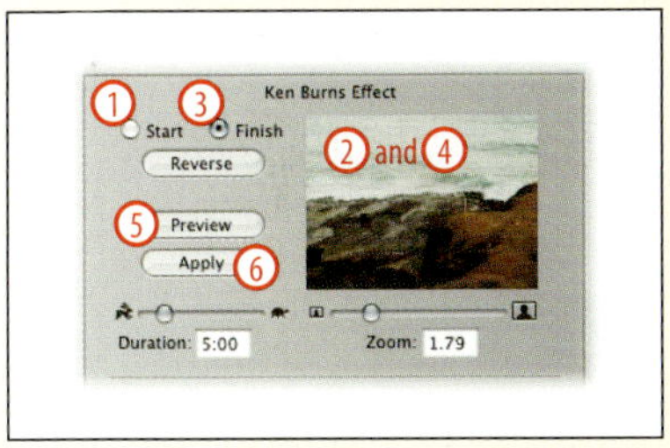

Figure 3-53. The Ken Burns Effect controls in iMovie

4. **Set the Duration.** You also might have noticed the Duration slider. By default, iMovie gives you a five-second clip for each imported still image. You easily can adjust that length here while setting up your Ken Burns Effect. When you click the Apply button, it will change the length of the real clip in the slideshow.

iMovie and Screen Size

After all that careful planning and worrying about nonsquare pixels in DVDSP and Photoshop, iMovie can throw you a curve. It uses the square-pixel dimensions for all its projects. So, if you want to use your images in iMovie, make sure they are 720×534, not 720×480.

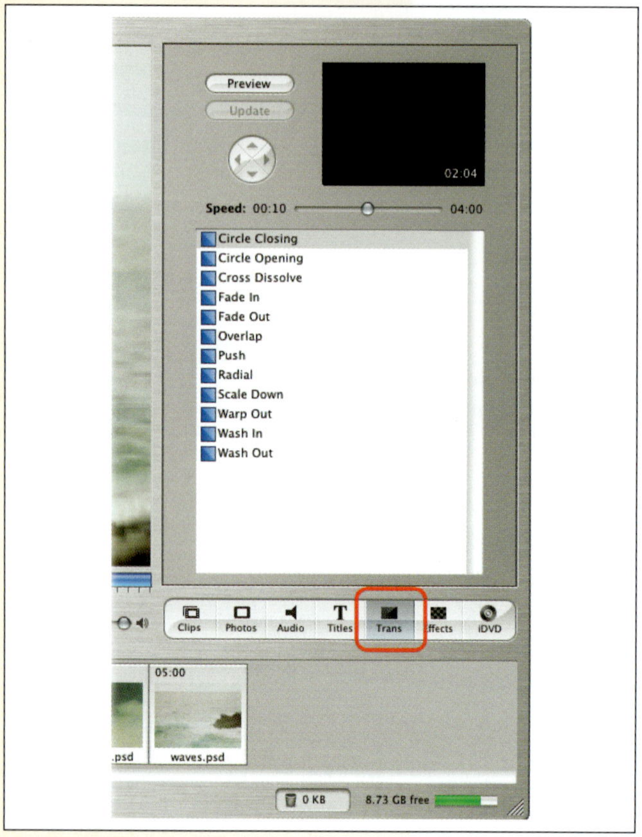

Figure 3-54. The Trans tab in iMovie

5. **Create the Transitions.** Now that each slide is a bit animated and more interesting, you can make switching from slide to slide a bit more enjoyable for the viewer. The Trans tab (see Figure 3-54) enables you to insert various transitions between slides.

   Select the first slide in your Timeline window by clicking it. Now you can preview the various transitions in the right pane by simply clicking each transition. Most transitions work best when applied between two slides. For the first and last slides, you can use the Fade In/Out and Wash In/Out effect. They switch from black (or white) to the selected clip—they aren't meant to switch between two clips.

   Once you have chosen a transition, you can set its duration. As you can see in Figure 3-54, transitions can last from 10 frames (about 1/3 of a second) up to 4 seconds. Set your duration, and then simply drag the transition down between the two clips and drop it there. (You pick up the transition from the list—grab its name or the little blue box next to it.) Figure 3-55 shows an example. That's all there is to it!

   Repeat that for each still. The Cross Dissolve transition is certainly the easiest and most common, but using a few variations can be fun. For example, a still using the Ken Burns Effect to pan left to right might lead nicely into a Push transition.

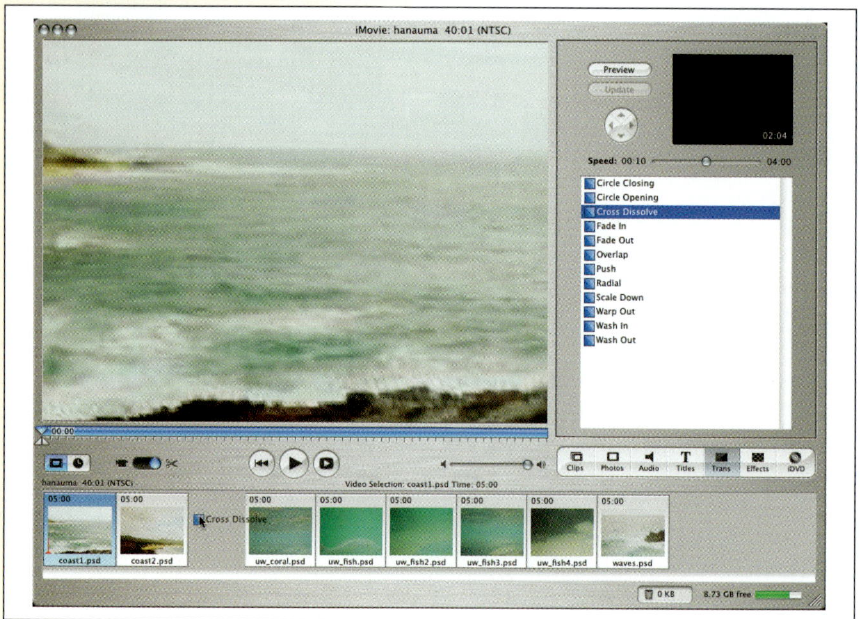

Figure 3-55. Inserting a transition in iMovie

6. **Add a Soundtrack.** The last feature we want to add while we're here in iMovie is the soundtrack. iMovie can import lots of different types of audio files. You can use the Import menu item or you can pull something from your

iTunes collection. The Audio tab also enables you to record audio on the spot and pull something from an audio CD.

If you have an audio file ready, drag it from the audio list on the right, down to your timeline. (If you were looking at the timeline in the "clip" mode, it will switch to the "time" mode automatically.) Now you have two audio tracks in case you want to narrate over your soundtrack or maybe add some sound effects. You can see the tracks in Figure 3-56.

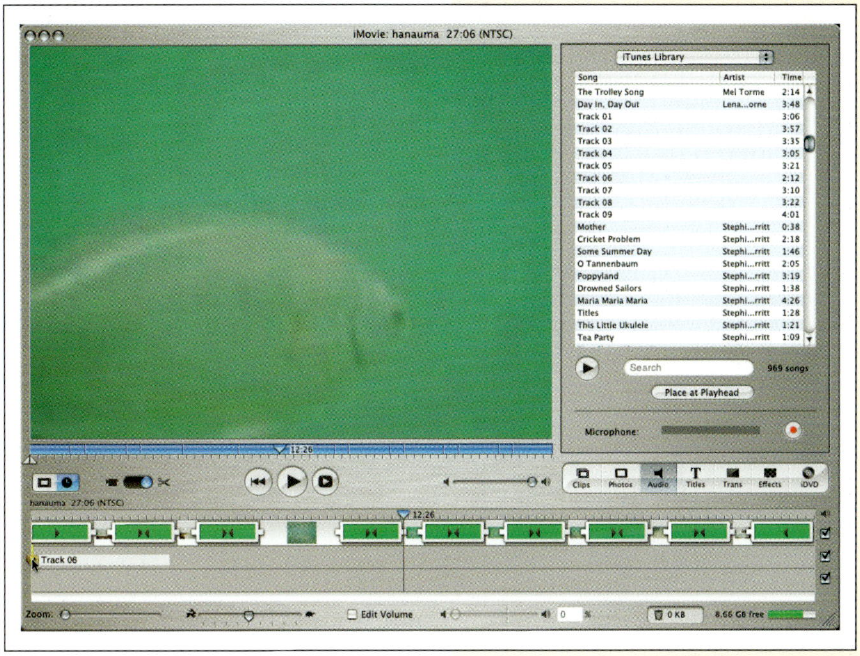

Figure 3-56. Inserting a soundtrack for our slideshow

You also can turn on the Edit Volume feature, which enables you to fade in and out at key moments. In our example in Figure 3-57, the song we chose is too long for the slideshow, so we trimmed it and faded it out.

7. **Export the Movie Back to DVDSP.** Exporting movies made in iMovie for DVDSP is not the easiest thing to do. iMovie is geared to play with iDVD. But it's not the end of the world. You have two relatively simple options, and each requires only a few steps.

   Figure 3-58 shows how to export the slideshow to a DV stream. Then you can convert that stream to MPEG-2 using DVDSP, Final Cut Pro, QuickTime Pro, or Compressor.

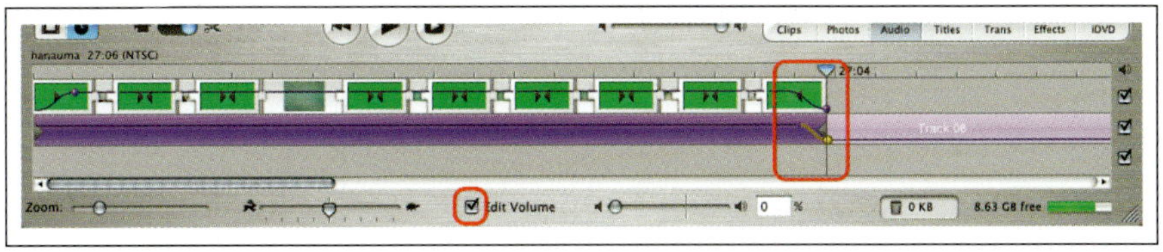

Figure 3-57. The audio editing tools in iMovie

> **STUDIO ADVICE**
>
> ## iMovie and Transition Rendering
>
> If you've had your Mac for a while, you might not have upgraded iMovie to the 4.x version available by purchasing the iLife suite. If you end up using iMovie for your slideshows on a regular basis, it might be worth the upgrade cost. iMovie 4 contains some great improvements in real-time rendering for transitions on Macs with enough power.
>
> But we want to stress that iMovie 3 still does a fine job. You'll just have to cultivate a bit of patience as you let the little red render lines march across your screen.

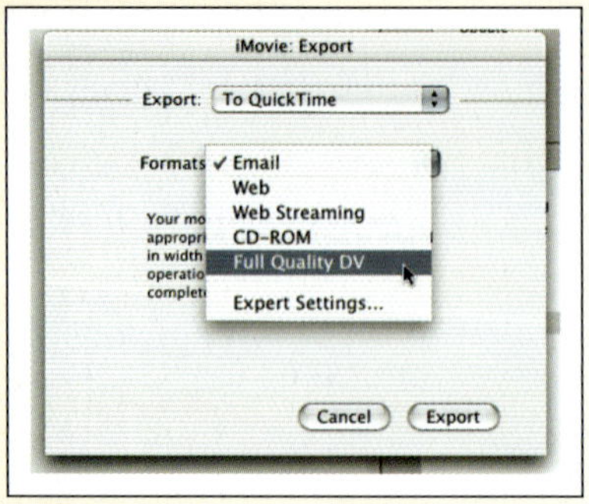

Figure 3-58. Exporting to DV from iMovie

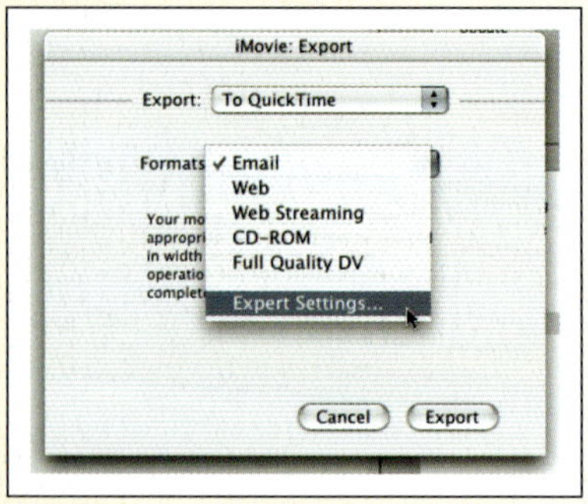

Figure 3-59. Exporting directly to MPEG-2, part 1

Your first option is to export the movie to DV and use Final Cut Pro. To do this, first set the Export option to QuickTime. Next, choose Full Quality DV for the Formats option (see Figure 3-58). Finally, import the new clip into Final Cut Pro and export to MPEG-2.

Your other option is to go directly to MPEG-2. *Direct* is a bit of misnomer, as the main export page doesn't contain an MPEG-2 option. But as shown in Figure 3-59, the page does include an Expert Settings option, and that option will do the trick.

Here, you must export the movie and audio separately. First, set the Export option to QuickTime. Next, choose Expert Settings for the Formats option (see Figure 3-59). Then, export the movie to MPEG-2, as shown in Figure 3-60. Export the movie through Expert Settings again, only this time, export with the Sound to AIFF option.

So, neither option is very difficult, but there's no one-stop shopping. Still, the overall quality of the slideshow definitely is worth this extra bit of effort.

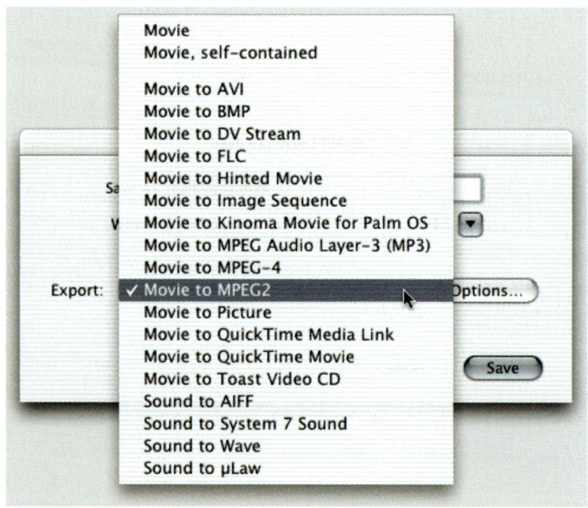

Figure 3-60. Exporting directly to MPEG-2, part 2

What? No Motion?

So, you say you've done all this work for a bunch of great-looking, *still* menus? Well, sometimes that's what the doctor ordered. The fancy buttons you can create with layered menus often can be more impressive than a video background. But DVDs can handle both types of menus, so Chapter 4 will tackle adding audio and video elements to your menus. That should complete your menu arsenal so that you can produce the best possible DVD for each project you undertake.

4 Great Menus: Motion

There's so much to menus in DVDSP that we had to include another chapter on them. This chapter tackles several zippy projects that include motion menus (with video and audio elements), using the built-in transitions from DVDSP 3. Chapter 11 covers custom transitions.

In this chapter

Video Transitions

On the way to making full-motion menus, DVDs allow for a nice intermediate step that doesn't require much effort to include in your DVDSP projects. When you switch between menus, you can add a transition much like we did between slides at the end of Chapter 3. In fact, you have access to exactly the same set of transitions. DVDSP treats the last frame of the current menu and the first frame of the destination as two slides and builds a transition between them.

However, as with slides, this feature is available only in DVDSP 3 and higher. But there's a fallback for these transitions, just as there was for the slides. You can create video sequences in your video editing software and squeeze them into menus. These menu transitions built outside of DVDSP are a bit more complex, so we put them in their own chapter—Chapter 11. You'll also see how to build your own custom transitions in that chapter if the existing set doesn't satisfy your creative needs.

In the meantime, those of you using DVDSP 3 are in for a nice treat. Transitions are simple to set up and they provide quick flash to your projects.

Creating menus using the graphical interface

Even though we're reusing content from a previous project, we're going to build this more or less from scratch again. This time, however, we'll rely on the Graphical view (also new to DVDSP 3) for our primary content canvas.

A graphical interface demonstration

Start by importing the same assets we used for the Castle Tours DVD in Chapter 3:

- *cawdor.m2v*
- *cawdor.AC3*
- *holyrood.m2v*
- *holyrood.AC3*
- *urquhart.m2v*
- *urquhart.AC3*

In addition, you can obtain the following new menu images from the *projects/ch04* directory:

- *cawdor_menu.pict*
- *holyrood_menu.pict*
- *urquhart_menu.pict*
- *quick_castle_tours_scaled.psd*

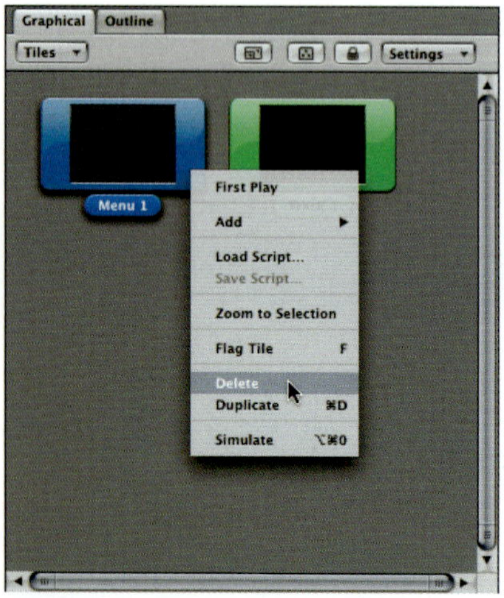

Figure 4-1. Deleting an item from the Graphical view

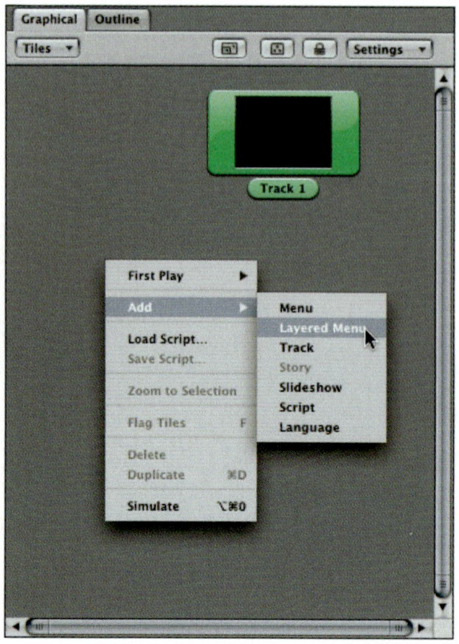

Figure 4-2. Creating a new item in the Graphical view

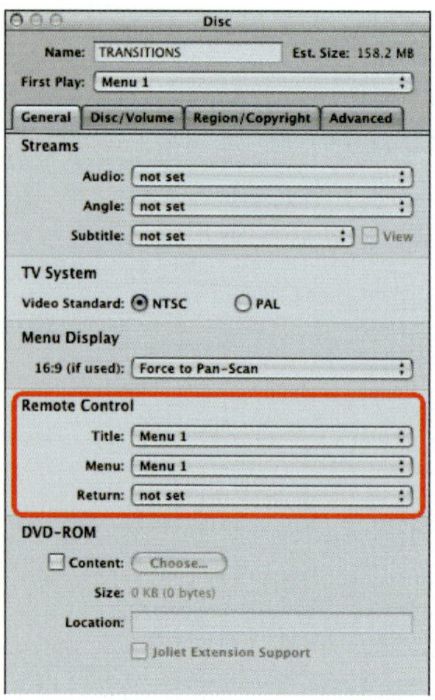

Figure 4-3. Resetting the Remote Control actions after creating the new main menu

Now perform the following steps:

1. **Open the DVDSP Graphical View.** You can open the Graphical view with the Control-⌘5 keyboard shortcut. Figure 4-1 shows that tab in its initial state. It also highlights the context menu (Control-click or right-click) for menus.
2. **Delete the Default Menu.** We want to delete the default menu. Go ahead and do that now. You also can delete the default track the same way.
3. **Add a Layered Menu.** With the default parts cleaned up, you can add your layered menu. You can do that using any of the techniques from Chapter 3, but Figure 4-2 shows the context menu in the Graphical view and how to use that for the new layered menu.
4. **Set the First Play and Remote Control Options.** As before, deleting the default menu will leave your First Play and Remote Control options unset. Figure 4-3 shows a recap of the disc properties you'll want to set in the Property Inspector for the disc. (Clicking anywhere in the background of the Graphical view will make the disc the focus of the Property Inspector.)

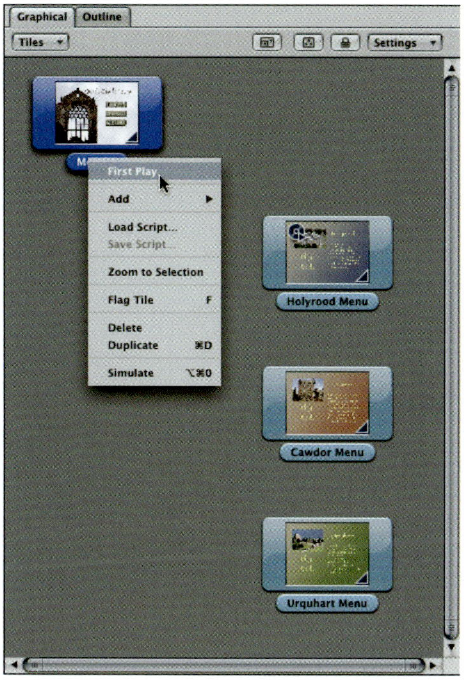

Figure 4-4. Using the Graphical interface to choose the First Play item

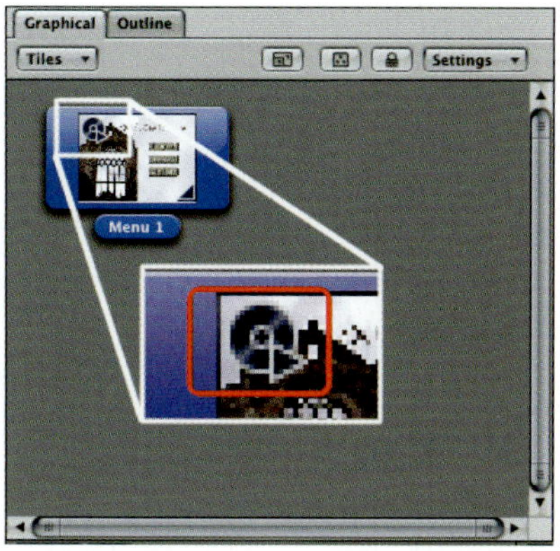

Figure 4-5. Identifying the First Play item in the Graphical view

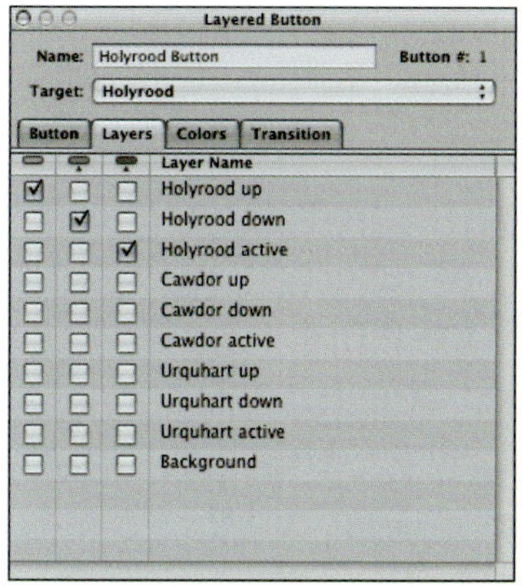

Figure 4-6. A reminder on setting up layers for buttons in the layered menu

You can set the First Play property in the Inspector or use the context menu in the Graphical view (see Figure 4-4) for the new layered menu. You can verify that your menu is the first play target by looking for the DVD-with-pointer icon in the upper-left corner. Figure 4-5 calls out that icon.

5. **Define the Three Buttons for the Castle Tours.** You do this in the Menu view just as you did in Chapter 3. You'll use the Inspector for each button to turn on the appropriate layers. Figure 4-6 shows a reminder for the layers you need.

6. **Add the Submenus.** To add a submenu to your project, grab its background asset from the Assets view and drop it in the Graphical view, as shown in Figure 4-7. We'll use the PICT assets from the *project/ch04* folder for our submenus. You'll have to do this one menu at a time. Dropping multiple image assets in the Graphical view creates a new slideshow!

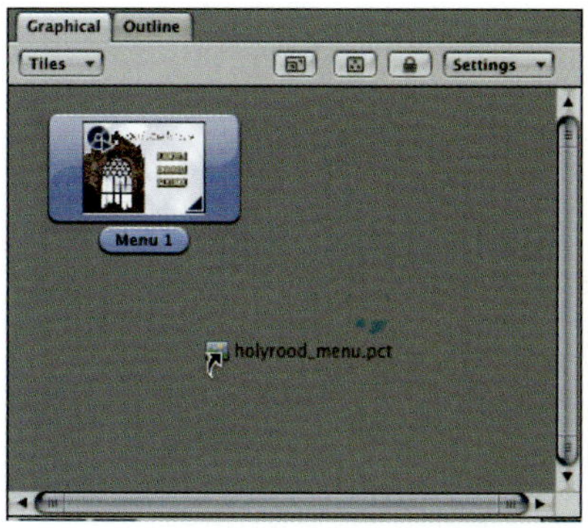

Figure 4-7. Drag-and-drop creation of a new menu in the Graphical view

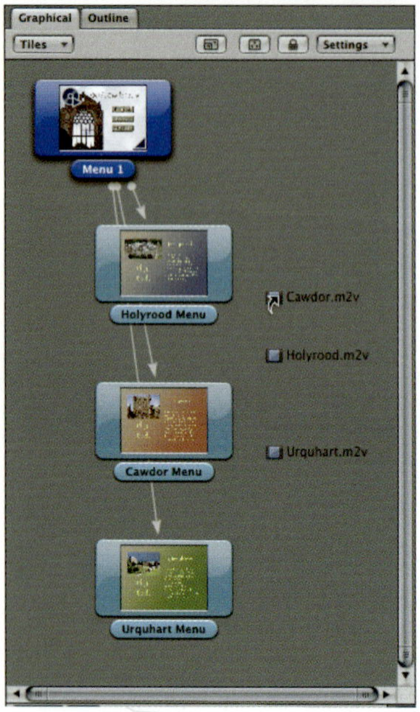

Figure 4-8. Adding multiple video tracks via the Graphical view

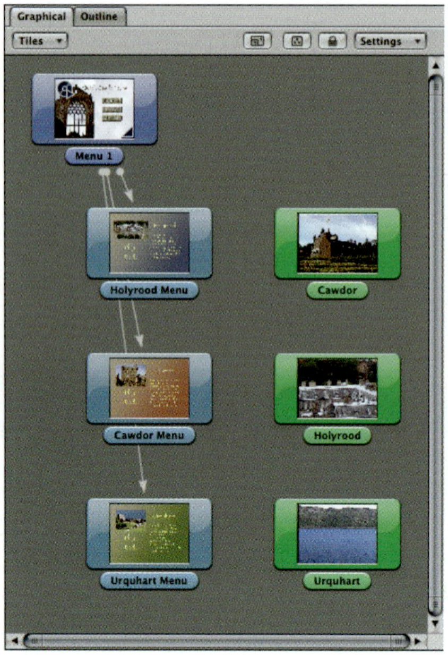

Figure 4-9. The visual result of adding our tracks

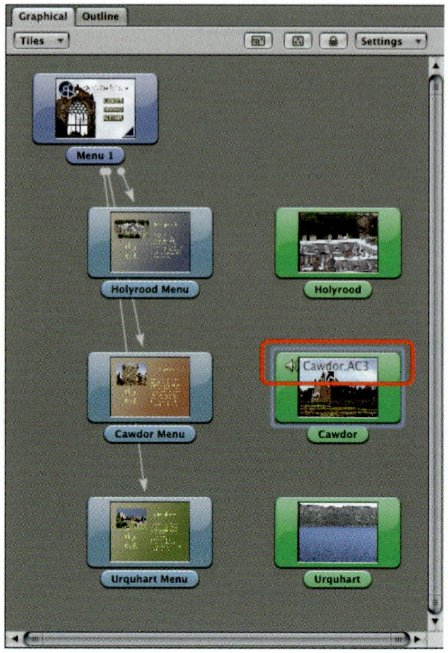

Figure 4-10. Drag-and-drop addition of an audio track in the Graphical view

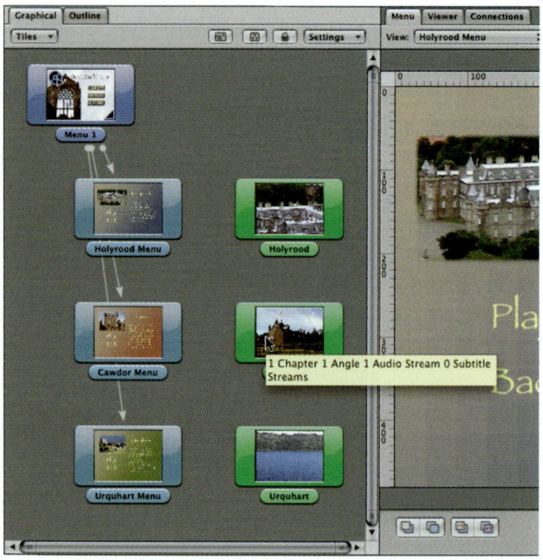

Figure 4-11. Some tool tip information provided in the Graphical view

7. **Add Tracks and Audio.** You can use the same drag-and-drop approach for the tracks. Although this time, you can grab all the tracks and drop them in at one time (see Figure 4-8). You'll probably get a stack of tracks that are too close together to be useful. Reorganize them however you want to make them easier to access. We ended up with the arrangement shown in Figure 4-9.

   Clicking any of the track items in the Graphical view should make that track the focus of the Property Inspector as well as make the track available in the Track view. If the audio was not attached automatically to the video track, you can do that quickly in either the Track or Graphical view. Figure 4-10 shows how to add an audio track to a video clip in the Graphical view.

8. **Check Out the Tool Tips.** Although this Graphical view is great for getting a good overview of your project, it doesn't show too many details on the items themselves. If you let your mouse hover over an item, though, you'll get a fairly verbose *tool tip* (small pop-up window) with information on that item. In particular, you can see how many video and audio tracks are attached. Figure 4-11 shows the tool tip for one of our video items.

9. **Link the Buttons to the Submenus.** Hook up the buttons from the main menu to point to their respective submenus. Each submenu has a Play and a Back button. Just define simple buttons for each. We chose to use the Arrow Point shapes in the Apple-supplied list. The Play buttons should point to the video tracks. Figure 4-12 shows the Inspector for the Play button on our Holyrood menu.

   When you're done, you should end up with the items and connections shown in Figure 4-13. You'll probably want to set the End Jump properties on each track to point somewhere useful. We send the user back to the submenu.

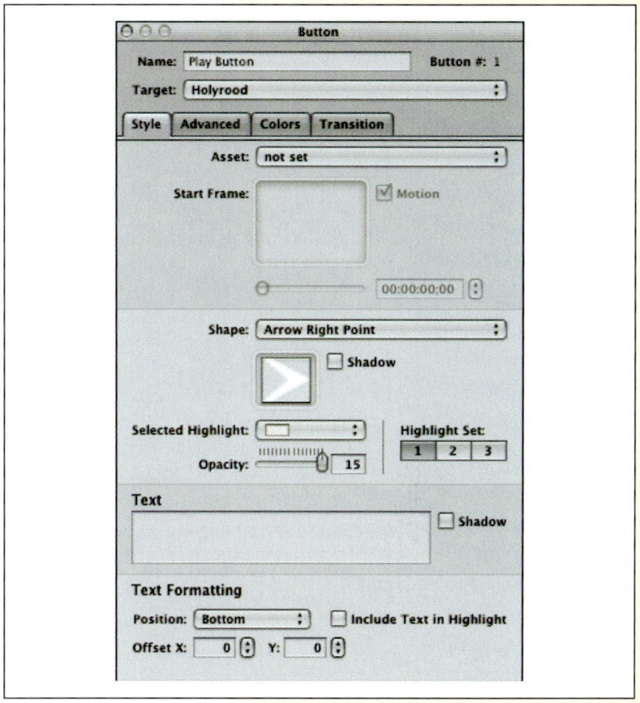

Figure 4-12. Property summary for our Play buttons

Specifying transitions

Now that you have the project flow completed, adding transitions happens quickly. Depending on how you want the transitions to behave, you can specify them in one of two places. A menu can have a default transition that applies to all its buttons. An individual button, though, can override that default with its own transition. We'll try it both ways for this project.

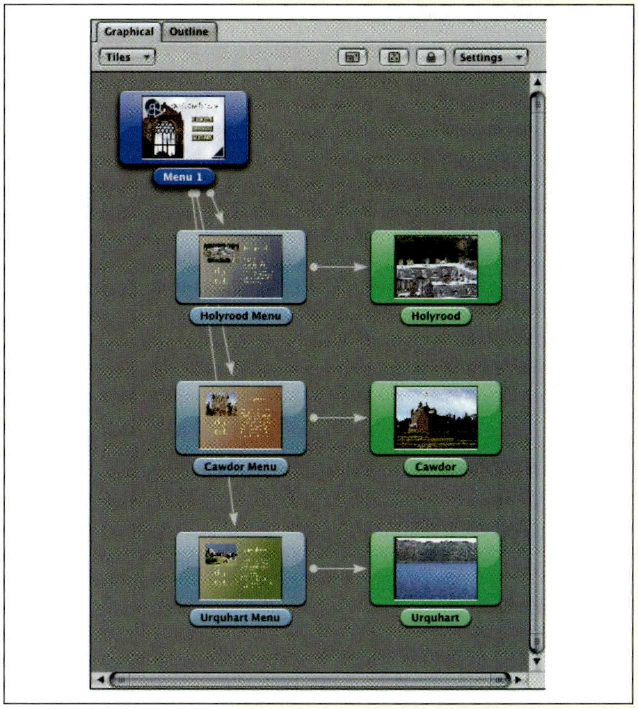

Figure 4-13. Our initial setup for the augmented Castle Tours DVD

Menu transitions

Figure 4-14 highlights the Transition tab of the Property Inspector for a menu. You can pick the transition from a pop-up list. Then you can tweak any parameters the

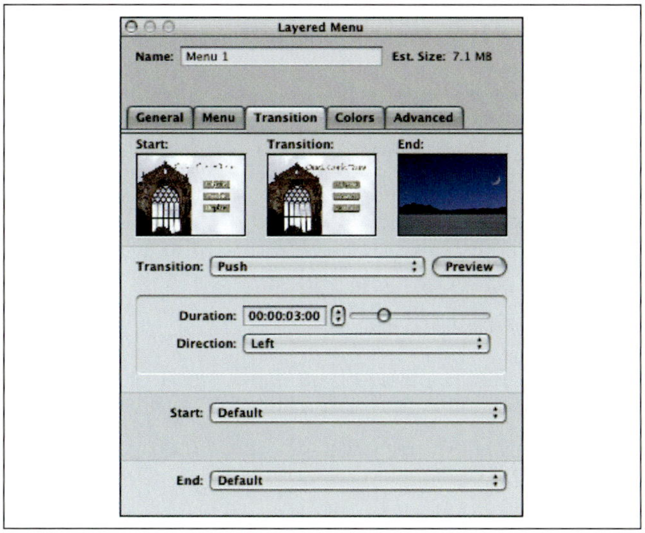

Figure 4-14. Setting the transition for the entire menu

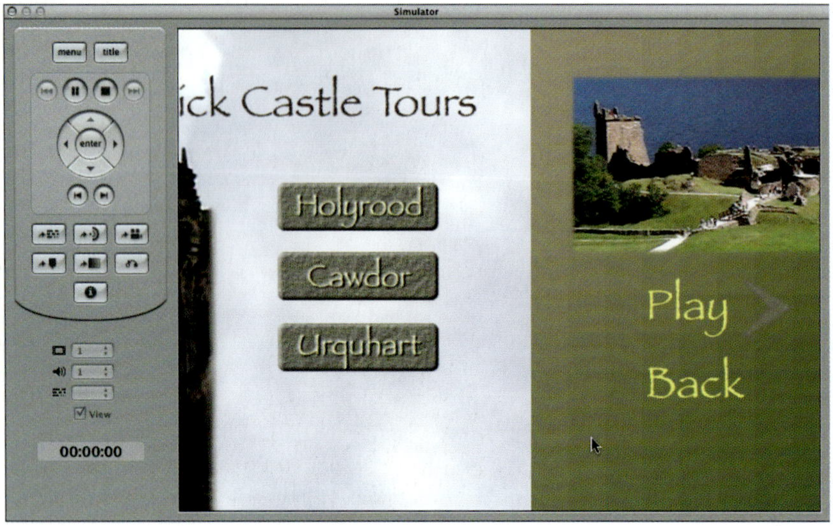

Figure 4-15. Our new menu in mid-transition

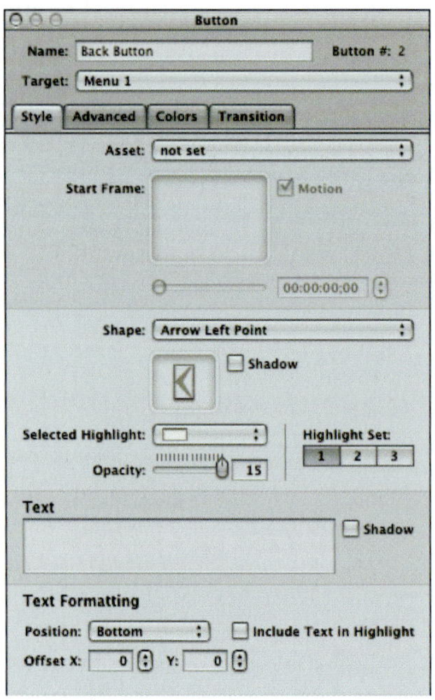

Figure 4-16. Property summary for the Back buttons

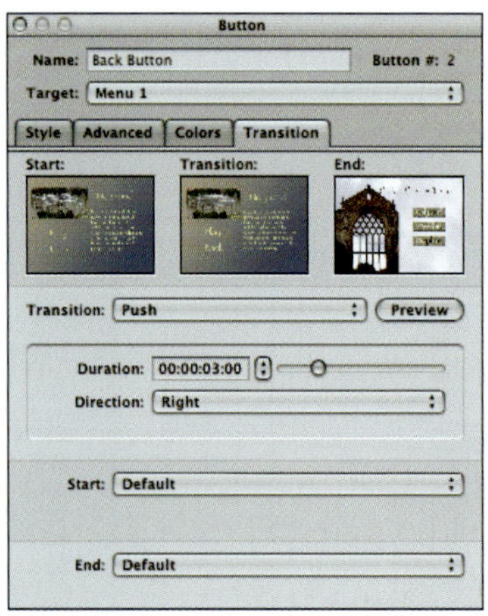

Figure 4-17. Setting the transition for one of the Back buttons

transition might have—just as you did for slides. We chose the Push transition and set it up to take three seconds.

Any button on this menu now should use the Push transition to get to its target. If you want to test things out to make sure this is working, just launch the Simulator. Figure 4-15 shows one of our transitions in action.

Button transitions

Button transitions are analogous to all the other transitions; they're simply specified in the Inspector for an individual button. Although it's often easier to slap a default transition on a menu, adding a transition to each button enables you to vary the attributes of your transitions or even to supply different transitions altogether. (But don't go overboard...too many transitions can be distracting from your overall project!)

Let's add per-button transitions to the Back buttons on our submenus. Figure 4-16 shows you the general properties for the shape and target of the Back buttons.

The transition we chose is shown in Figure 4-17. Notice that we use the Push transition again, but this time we push to the right instead of using the default of left.

When all is said and done, your Graphical view should provide a complete overview of the project, as shown in Figure 4-18. Try it out in the Simulator. Don't be too concerned if the transitions appear choppy or don't function in real time. They'll be rendered before the project is burned and should play quite smoothly in regular players.

One last note about the Graphical view: you can tell transitions are applied to an item by looking in the lower left-hand corner of the item's thumbnail. That little blue triangle means a transition exists. Again, you don't get details, but you do get a quick overview of everything that's happening.

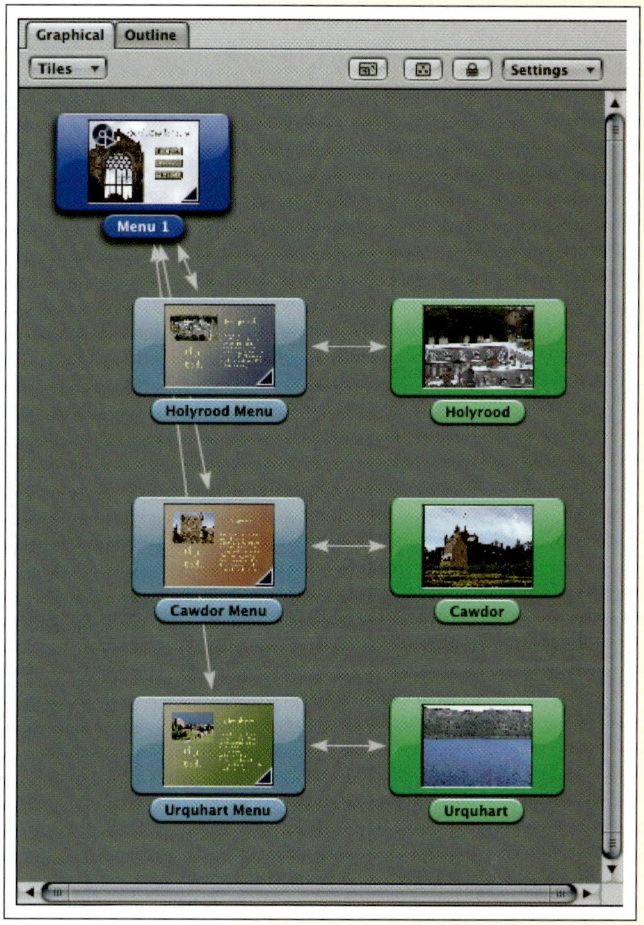

Figure 4-18. Our completed transition project

Motion Menus

The term *motion menus* is perhaps a misnomer. Motion certainly can be involved, but the menu doesn't have to be in motion to count as a motion menu. Any menu with video or audio content (as opposed to a static or layered image) is a motion menu. Background audio—even with a still graphic for the image—is considered a motion menu. Buttons that show a small video clip require a motion menu. Subtle, wavy backgrounds with still buttons also are motion menus.

Music, motion backgrounds, and movie buttons might seem like disparate things, but they are closely tied together in the world of DVD menus. They all involve a common element: a video for the primary menu asset.

We're drawing this distinction clearly because motion menus have some limitations. Although they can be arresting visually, things such as layered buttons do not work in motion menus. And remember, even if the only "motion" is an audio file playing in the background, you're in motion-menuland.

Menus and music

Of course, sometimes the audio and video assets are worth including in your DVD. As an example, we'll build a simple menu with an audio loop. Adding audio really can amplify the importance of an image. One interesting case of audio effects was studied with computer users looking at a still image of artwork. Half the users were shown the image on its own and half were shown the image with music playing in the background. The participants were asked about the image the next day. Many of the folks in the music group recalled watching an *animated* image.

Death and Transfiguration

For several of the remaining projects, we'll be "DVDifying" parts of the *Death and Transfiguration* anime series from Studio Anti-Thesis. It's an original series published online at *http://dnt.studioantithesis.com*. With AIs, huge corporate conglomerates, and peril-flaunting humans everywhere, it's a great read in its own right. But we're here to make some snazzy menus that reflect the quality of the series, so on with the task at hand....

If you have a simple still image that you use for a menu, you can add audio without much trouble—you can do it directly in DVDSP. In the first versions of DVDSP, you had to create a simple movie using your still image and the audio file in an editor such as iMovie or Final Cut Pro. That wasn't difficult, but it certainly was annoying.

This next project is really a detour; it's part of a larger project we'll be building in this chapter. But it's important to try it out and run it in the Simulator. You'll want to see (and hear!) the end results.

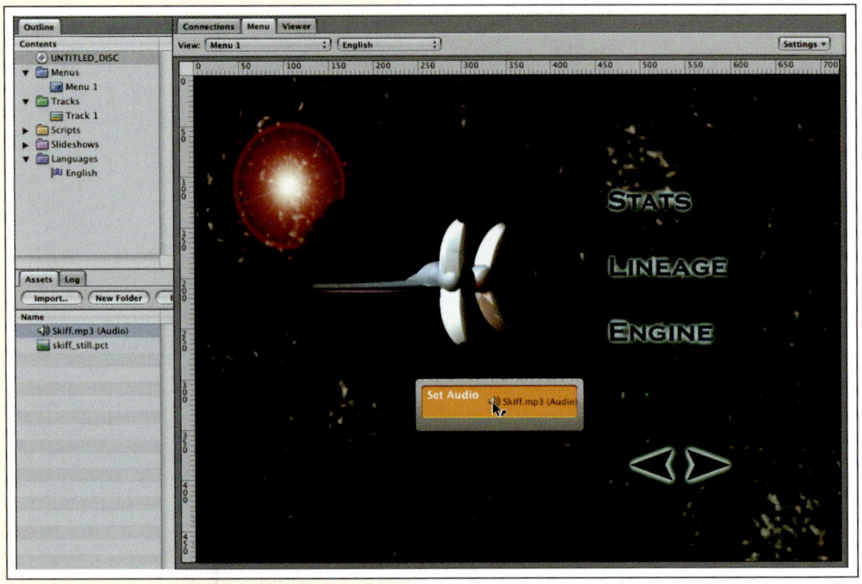

Figure 4-19. Specifying an audio file for the menu

An audio menu demonstration

First, grab the two assets you'll need for this test from the projects/ch04 folder:

- *skiff.mp3*
- *skiff_still.pct*

For this quick menu, you'll let DVDSP convert the audio file to the proper format. For larger projects, you need to make sure all your menu audio clips are encoded in the same way. (For example, on the DVD for this book, some of the other menus use AC3 audio, so we had to convert the MP3 for this project to AC3 as well.) You can set the image as the background for a new menu. Now we're going to add the audio. Ready? Drag the *skiff.mp3* file over to your menu and drop it there, as shown in Figure 4-19.

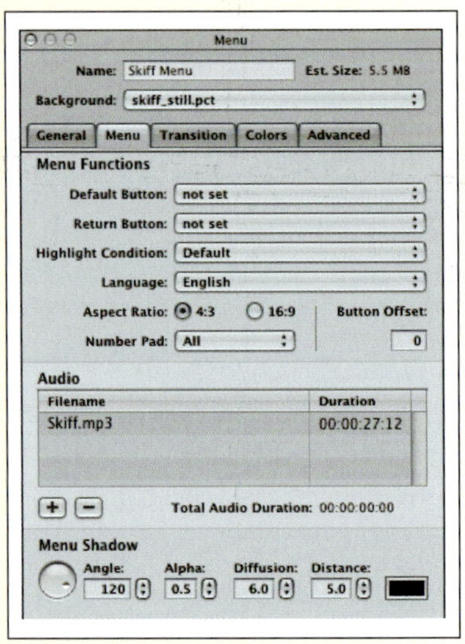

Figure 4-20. The properties of the new audio menu

You're done! At least, you've completed the business of adding audio. If you were setting up a real menu, you'd have to create and link the buttons, of course. But you were doing this to test the audio. Take a look at the Menu tab in the properties for the audio-enhanced menu in Figure 4-20.

There are a few interesting things to notice. First, you can see that an audio asset has been set. You can change the asset here later if you need to. Second, you can see that now the image has a start and end position. This menu uses a still image, so there's not much reason to play with these values. (As you'll see in the next project on quick video buttons, though, these values can be useful when playing with video assets.) DVDSP took the liberty of setting our values to match the duration of our audio clip—precisely what we want.

Go ahead and simulate your project and verify the audio is playing. You also should see another clue that you have a motion menu in the Simulator (see Figure 4-21). The counter is active—not stuck on 00:00:00, as it is for still menus.

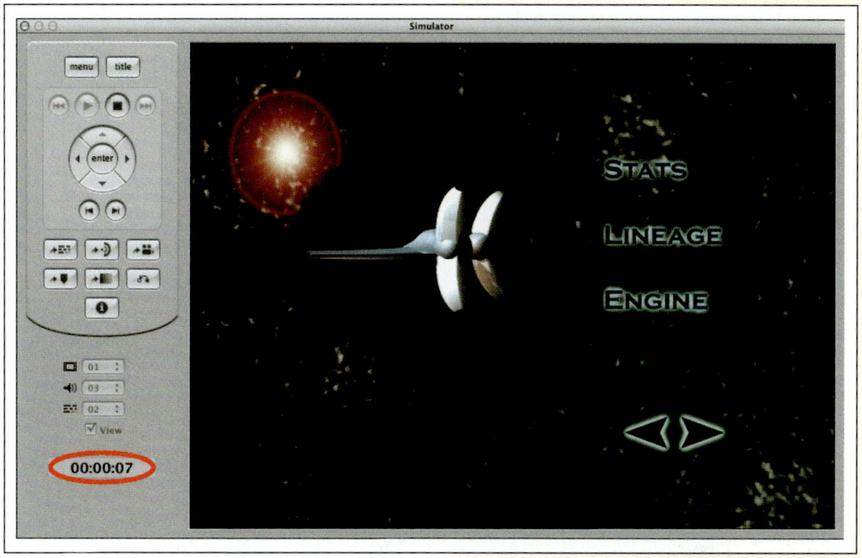

Figure 4-21. Our audio menu in action in the Simulator

End actions

Another default DVDSP chose for us that usually matches what we need is the At End property (highlighted in Figure 4-22).

Usually you just want to loop the audio when you're done. You have two other choices, though.

Still
: This stops the audio (or video) and waits for the user to make a selection. This can be useful if you have some intro text or images that aren't meant to loop.

Timeout
: This option stops the audio (or video) and jumps somewhere else. You can specify that action by picking a target similar to what you would do for a button. For example, Figure 4-23 shows us setting the end action for our Skiff menu to Timeout. One second after the audio stops, we default back to the main menu. This option can be a life-saver for kiosks in which you want to make sure you go back to your primary menu if the user walks away.

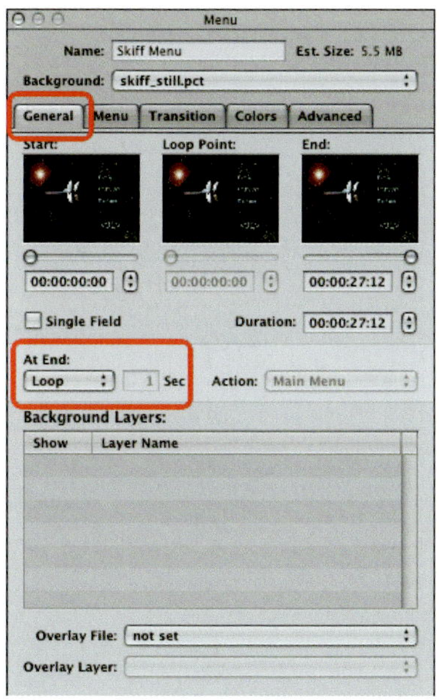

Figure 4-22. Making sure our menu loops properly

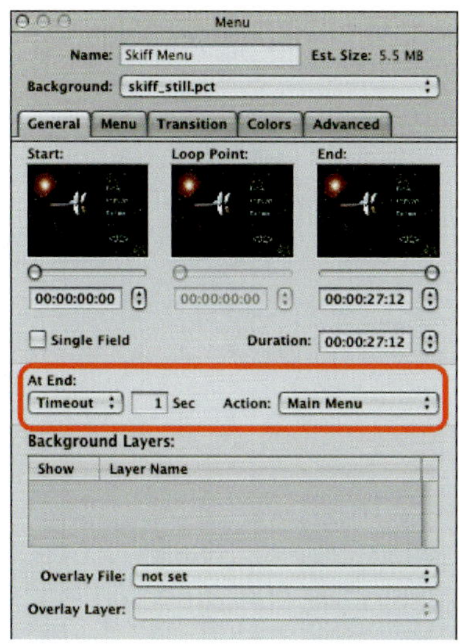

Figure 4-23. Using the Timeout feature for an alternate way to handle our menu

Quick Motion Buttons

Another quick way to get some dynamic content into your DVDSP projects is to use video assets directly for your buttons. You've actually done something quite similar when working with the templates

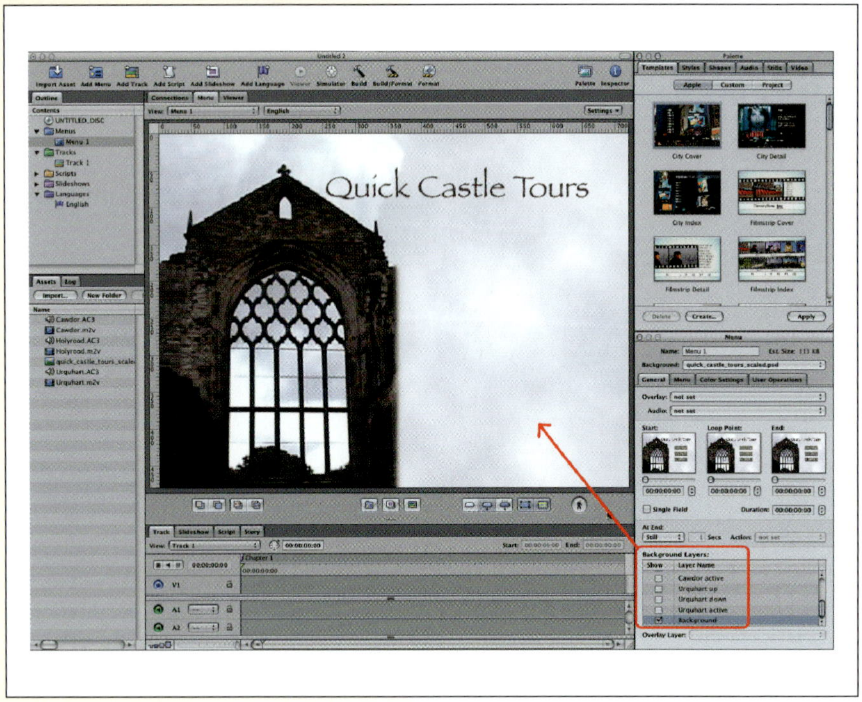

Figure 4-24. Setting up the background for the motion button menu

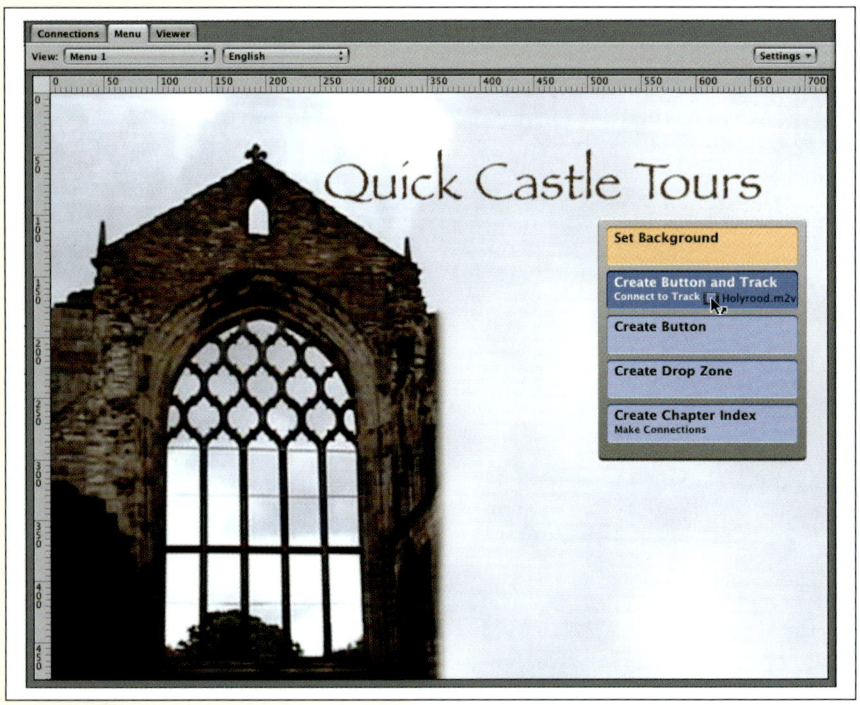

Figure 4-25. Creating a motion button (and video track)

in Chapter 1. You can do the same work in DVDSP with your own custom backgrounds.

Adding motion video buttons

Let's make another quick-and-dirty test menu to prove how easy it is to incorporate your own video buttons.

Motion video button demonstration

Start a new project and grab the following files from the *projects/ch04* folder:

- *castle_tours_scaled.psd*
- *cawdor.m2v*
- *cawdor.AC3*
- *holyrood.m2v*
- *holyrood.AC3*
- *urquhart.m2v*
- *urquhart.AC3*

Then, perform the following steps:

1. **Create a New Menu.** Use the *castle_tours_scaled.psd* file for the background of a *standard* menu (not a layered menu). Then, use the property sheet to turn on the background layer. See Figure 4-24 for an example. Notice that you don't have any buttons now. That's just fine.

2. **Create a Video Button.** Grab the *holyrood.m2v* asset and drag it over to the menu. Choose the Create Button and Track option, as shown in Figure 4-25.

   Voilà! You should have a video button. We moved ours into position along the right side and ended up with the menu you see in Figure 4-26. The default size is based on the video's aspect ratio. You can resize it as you want. DVDSP will scale and crop the video to fill the button while keeping the video's aspect ratio intact.

3. **Add Video Buttons for the Other Tracks.** As you add the other tracks you can reposition them in the menu. DVDSP will highlight alignment points with other objects in the menu. If you want, you also can steal some of the audio from one of the castles for the entire menu to utilize. The menu's property sheet will enable you to choose which segment of the music you want to loop. It also sets the duration shared by all the other buttons. These properties are highlighted in Figure 4-27.

   You can use the property sheets for the individual buttons to select which video segment is seen in the menu loop (see Figure 4-28). As noted before, the menu properties determine the button loop's duration (refer back to Figure 4-27).

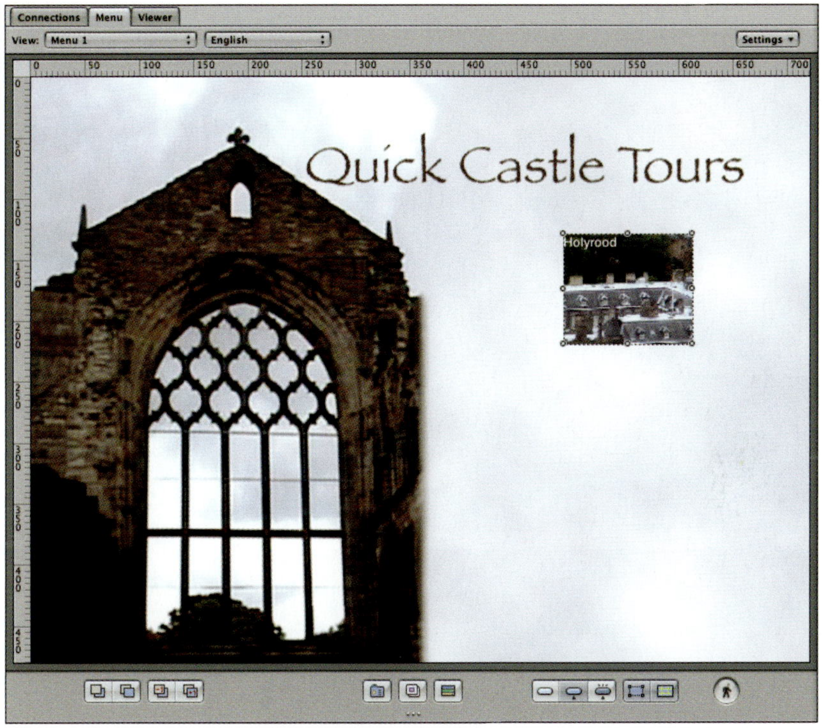

Figure 4-26. A video button

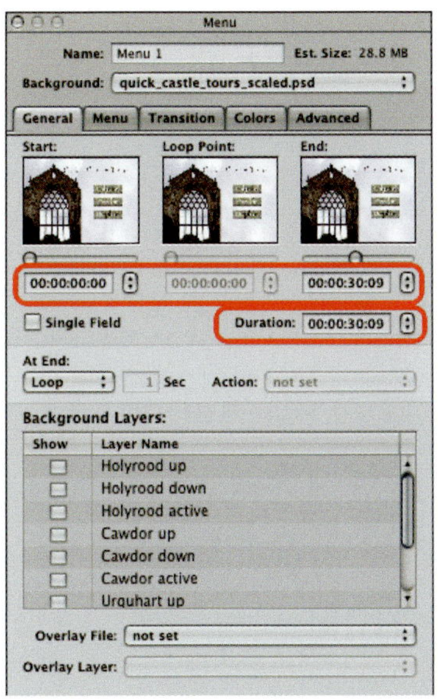

Figure 4-27. Setting an overall audio file on a menu with video buttons

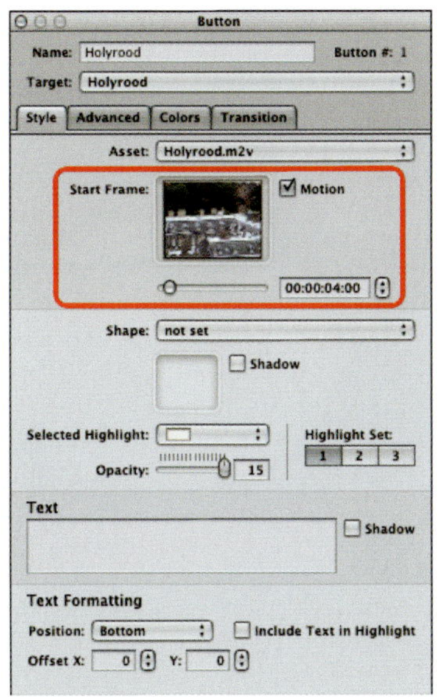

Figure 4-28. Picking the starting point for your button loop

Other Motion Menus

Background music is all well and good, but what about full-fledged video menus? The background is active, the buttons are active, there's good audio, and you still can create fancy, custom buttons to jump to some other part of the disc. It all can be yours; you just have to do a bit more work when creating the menu.

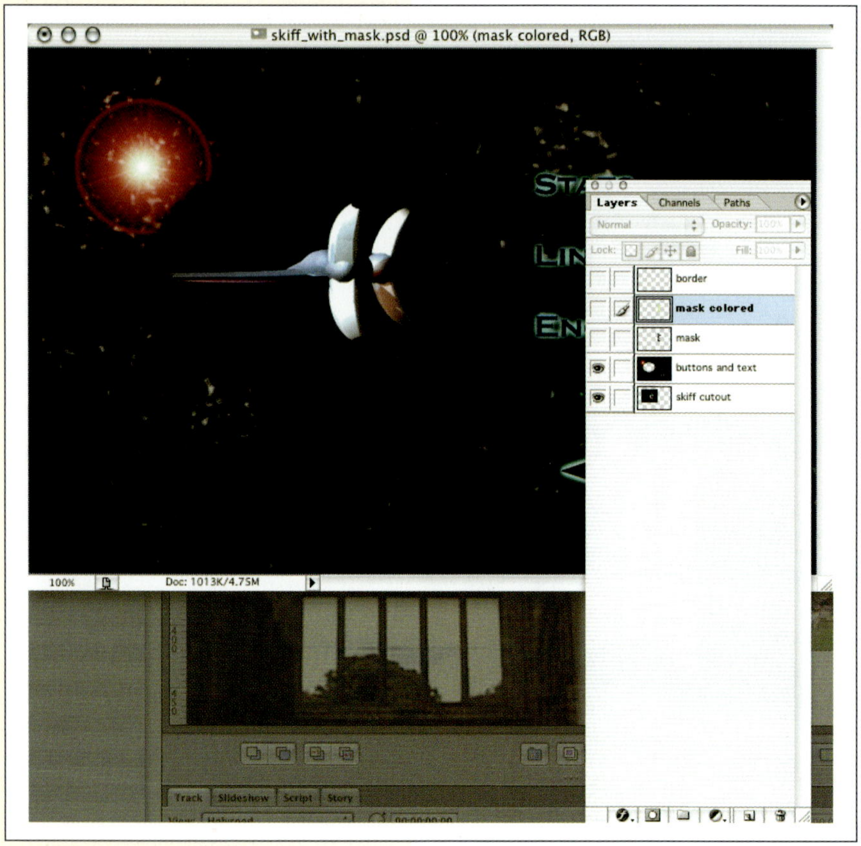

Figure 4-29. Photoshop menu construction

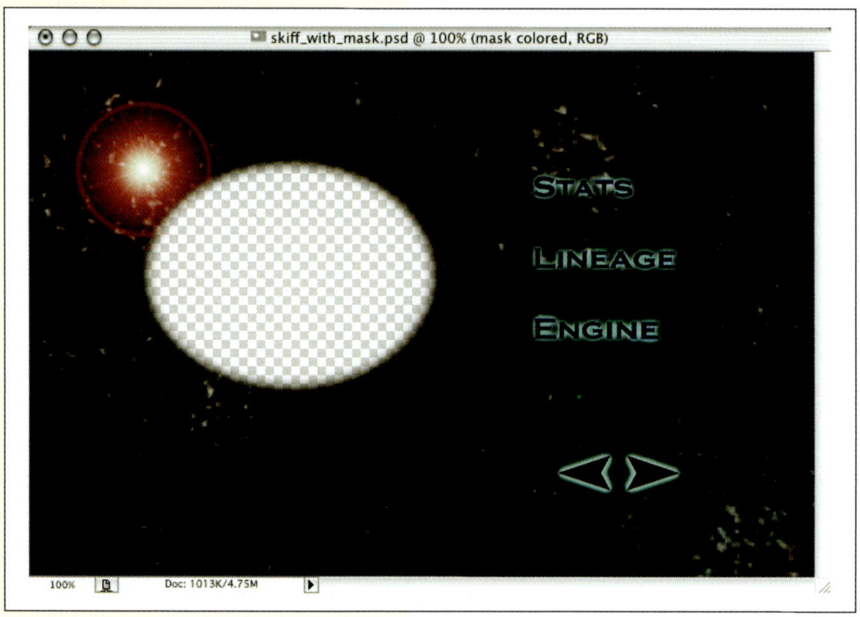

Figure 4-30. The skiff menu with a cutout for the video clip

It's definitely easier in DVDSP 2 and higher than in earlier versions. (Previously you had to do all the motion work in some other application, such as Final Cut Pro or After Effects.)

For motion menus with basic buttons (rectangular highlights) you can do your work in just about any application. The end result essentially will be the same as in the previous example. For this next example, though, let's move that skiff we worked on earlier.

Photoshop and Final Cut Pro

With Final Cut Pro, you can use some of your Photoshop skills to create motion menus that will be easy to deal with once you're back inside DVDSP. First off, you need some footage of one of the skiffs. You'll work with a rotating 3D clip of a skiff. On its own, this really isn't enough, so you'll call on Photoshop for some help.

Recall the still menu from the audio example. We used a flattened image. Let's go back to the original Photoshop document with all the layers available. Cut out the area where the skiff was and you've got a nice image for compositing with video. Figure 4-29 shows the candidate menu. Notice the separate layers.

To use this image in Final Cut Pro, you simply need to delete the still-frame layer. The hole left behind will be transparent, not just white or some other background color. You should be left with a single layer that looks like the one in Figure 4-30.

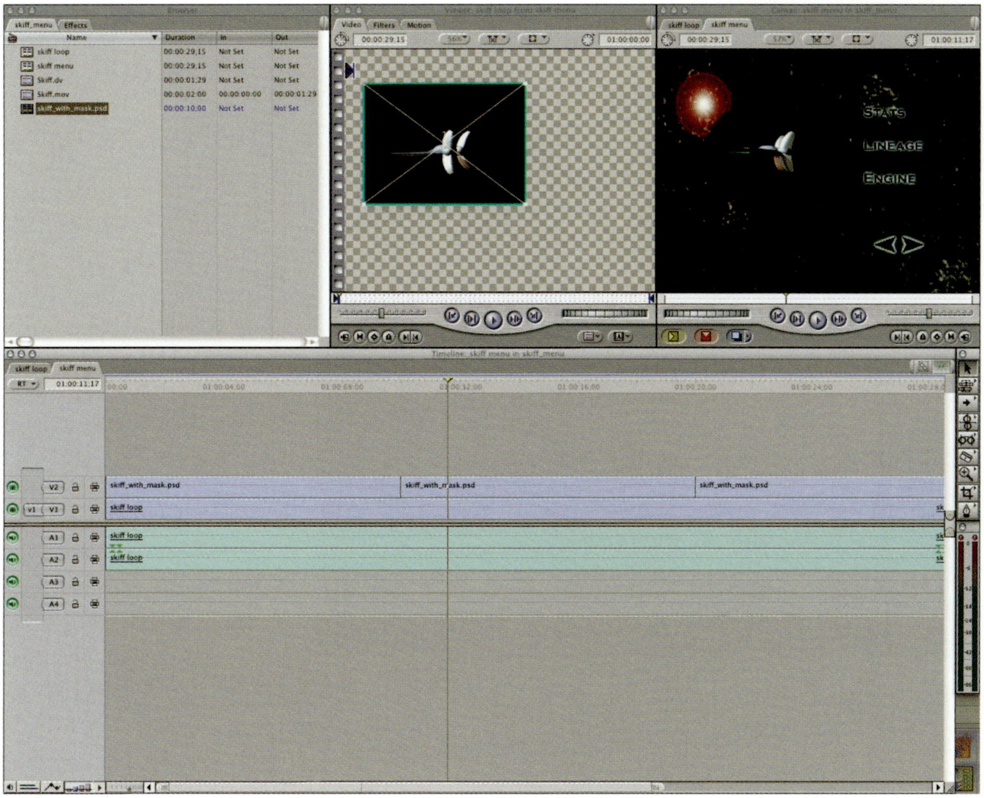

Figure 4-31. Compositing stills and video in Final Cut Pro

In Final Cut Pro, load the *skiff.mov* file and put in enough copies to make the loop last about 30 seconds (or as long as you want the loop to last). You should have something similar to the timeline at the bottom of Figure 4-31.

Now get that menu document into Final Cut Pro. You'll need to create a second video track and place the still image in that track. Extend the still to last as long as your video clip. Note that you also might need to manipulate the size of the video to get it just right. The transparent part of your menu should enable the real video clip to show through, as you can see in the upper-right corner of Figure 4-31.

STUDIO ADVICE

Photoshop Transparency and Final Cut Pro

Photoshop enables you to create some very subtle effects with transparency. You are not restricted to "on and off" transparency. Parts of your image can be translucent. You can create smooth antialiasing on a shape with semitransparent pixels around its edges.

Fortunately, Final Cut Pro is well aware of these shades of transparency (stored as alpha-channel information in Photoshop). Those translucent elements will appear just fine in Final Cut Pro. You can composite them over video in Final Cut Pro just as you would layer them over other still images in Photoshop. Play around with this feature. It's a relatively easy way to create some good-looking motion menus without constructing the entire menu in Final Cut Pro.

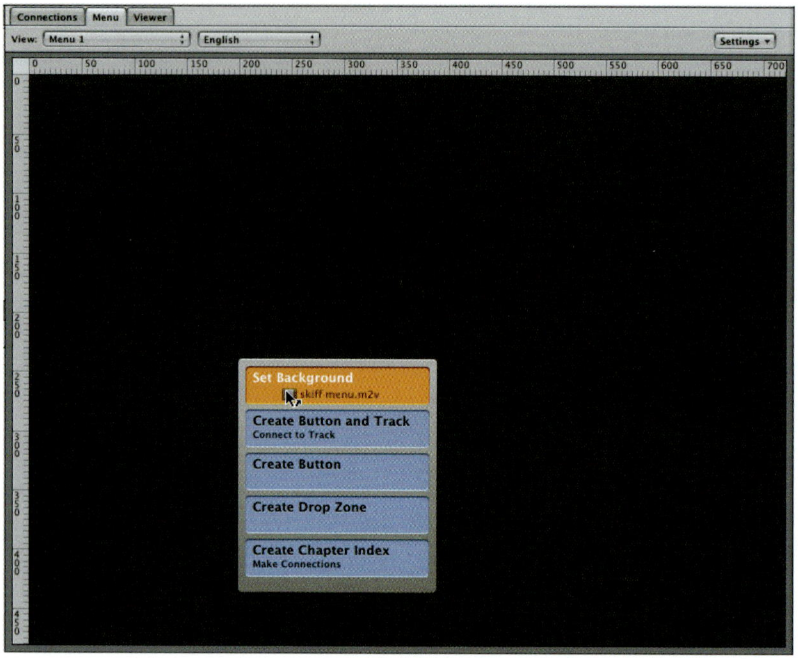

Figure 4-32. Adding a video background to our video menu

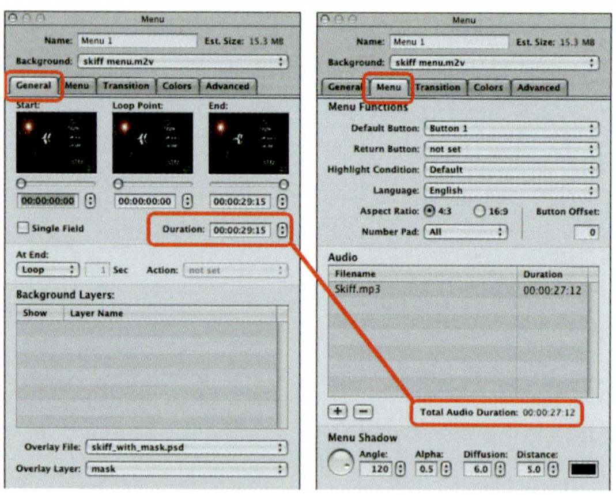

Figure 4-33. Matching up the video and audio loops

That's it for Final Cut Pro. After exporting the menu as an MPEG-2 movie, we'll jump back into DVDSP.

Adding buttons to video menus

As with the audio-enhanced menu from the previous project, you need to click the Add Menu button in DVDSP to get going. Now you can drag the *skiff_menu.m2v* file onto the untitled menu, as shown in Figure 4-32.

You also can set the audio at this point, just as you did for the still menu. You'll want to pay more attention to the loop points, though. Figure 4-33 shows the properties that make our video coincide with the duration of the audio. Notice that we're off by a few seconds. That's OK. As long as you aren't wildly off, DVDSP will fill in the gaps. It's easiest to have too much video—you just get a few seconds of silence in the audio track. If the video doesn't have an obvious loop point as ours does, you simply can line up the durations to match exactly.

We planned ahead in our Photoshop document, so we have some easy targets for creating buttons. Just drag the usual rectangle to the left of the text parts and choose a nice shape and a complementary color for the highlights. We picked the Simple Button shape and an off-white color at a transparency level of 5. Figure 4-34 shows the menu in Preview mode.

Woo-hoo! You've just created your first full-video menu. Well, OK, so it wasn't that different from the menu to which you added the audio, but to paraphrase the inimitable Homer Simpson, we stand by our enthusiasm. Woo. Hoo.

Overlays

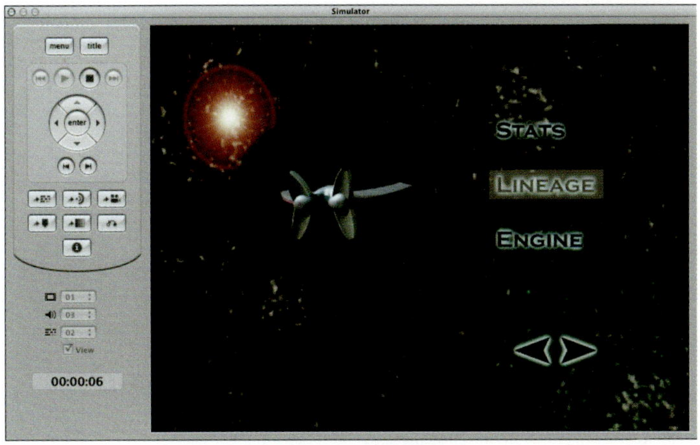

Figure 4-34. Our full-motion video in Preview mode

Motion menus can feel restrictive when you get to the point where you're creating the buttons that show up on the screen. Highlighting rectangles on the screen just isn't as fancy as you might want to get. The Apple-supplied shapes help, but even those can be a bit repetitive. Although you are restricted to using overlays when you have a motion menu, you can make some pretty creative custom overlays.

At its heart, an overlay is just a Photoshop layer that tells DVDSP what to draw when you select the rectangle that defines a button. The default overlay is a solid layer, so the highlighted rectangle shows up as just that: a rectangle. But you can make fancier overlays.

1-bit (black-and-white) overlays

If you want a shape other than a rectangle to show up, you can create a Photoshop document to aid in this endeavor. Start with the standard NTSC-DV size document. Make the background completely white (or transparent). Now draw shapes in black wherever you want a custom-shaped button to appear. We often start with a still image of the video we'll be overlaying as the background. Then we make the overlay layer partially transparent so that we can see where to draw our shapes. Check out Figure 4-35 for an example of creating skiff-shaped buttons.

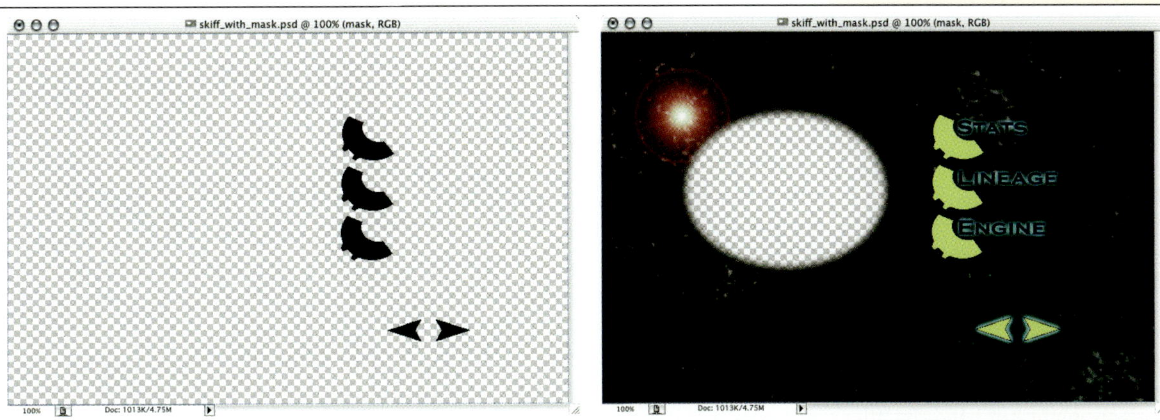

Figure 4-35. A 1-bit button mask (left), and the same mask laid over our menu image (right, with false coloring)

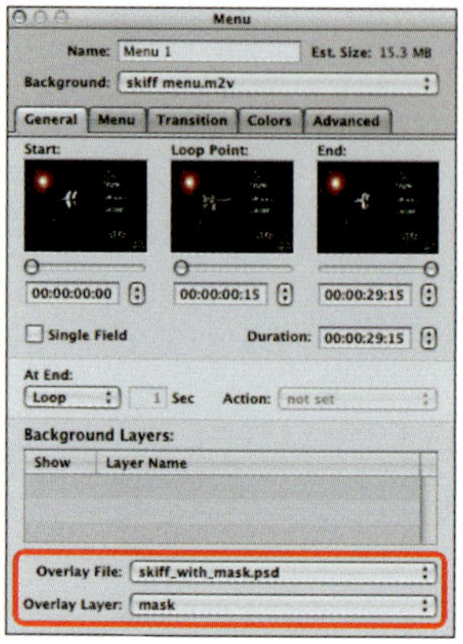

Figure 4-36. Using a mask with a video menu

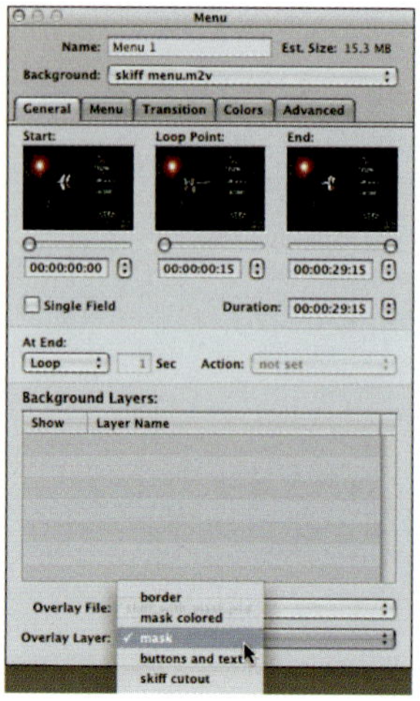

Figure 4-37. Setting the specific layer to use for our mask

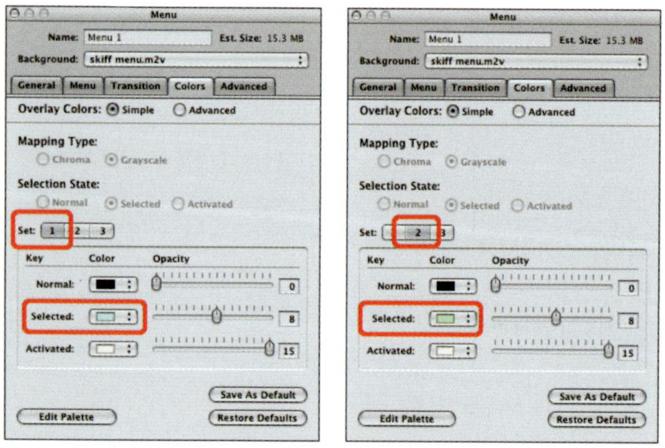

Figure 4-38. Highlight colors for the masked buttons

The black parts will show up as the buttons in DVDSP. But be careful: only the black parts will show up. Notice that we had to create shapes for the "next" and "previous" arrows in our menu, too!

1-bit overlay demonstration

Here are the steps to follow to create a 1-bit overlay:

1. **Create a Motion Menu.** Start by creating another motion menu using the same video clip you used for the previous project. You even will use the same number of buttons. This time, however, you're going to add the overlay layer for the skiff-shaped buttons.

2. **Import the Photoshop Document and Set the Overlay Options.** Import the *skiff_with_mask.psd* document into DVDSP. In the property sheet for the menu, notice the overlay options (see Figure 4-36).

   Set the overlay to use the *skiff_with_mask.psd* document. If you have more than one layer in the document, you'll also have to pick which layer to use. Figure 4-37 highlights this part of the Inspector.

3. **Pick the Highlight Colors.** You also should pick your highlight colors now. If you don't, you won't be able to see your buttons as you draw their bounding rectangles in the Menu editor. For our current menu, we'll use both sets of highlights. We'll use light blue for the skiff cutouts and light green for the next and previous buttons. Figure 4-38 shows the exact values we chose.

TIP

Looping

If you get to plan the audio or video from scratch (as we did for the previous two menu projects), you easily can create clips that "loop" well. Audio and video loops are meant to run continuously—and smoothly. You can create a clip that ends in a spot that coincides with the beginning of the clip.

If you use part of an existing audio or video clip for a menu, you might not have the ability to create a real loop. You can cheat a little, though, by fading in at the beginning of the clip and then fading out at the end. The loop won't be perfect, but it won't have any jarring jumps from the end of the clip to the beginning.

One other important thing to remember is that you should make your loops last at least 30 seconds. Most DVD players experience a small pause when they jump back to the beginning of the menu loop. During that pause, the viewer cannot interact with the menu. If your loop is only four seconds long and the jump takes up one second, the viewer can't use the menu a full 20% of the time.

4. **Define the Buttons.** If you haven't defined the buttons yet, go ahead and do that now. There's nothing fancy to do here; just set up the same old boring rectangular buttons. This time, though, make your rectangles cover the custom shapes you built in Photoshop. Figure 4-39 shows the results of outlining a masked area. Table 4-1 lists the exact coordinates for the buttons in this menu.

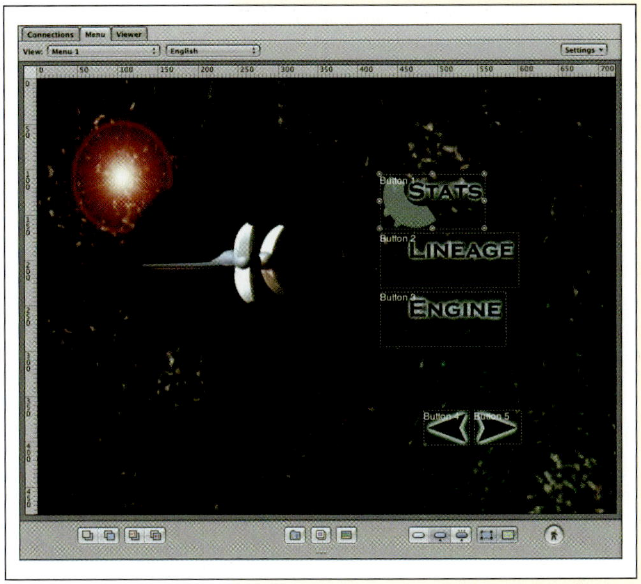

Figure 4-39. Creating buttons with an overlay mask

Table 4-1. Button coordinates for the skiff menu

Button	Left	Top	Right	Bottom
Stats button	427	106	558	166
Lineage button	427	171	600	231
Engine button	427	235	584	295
Prev button	481	366	538	405
Next button	544	366	603	405

All is good so far. When you preview the menu, you should see shiny new skiff buttons, as shown in Figure 4-40.

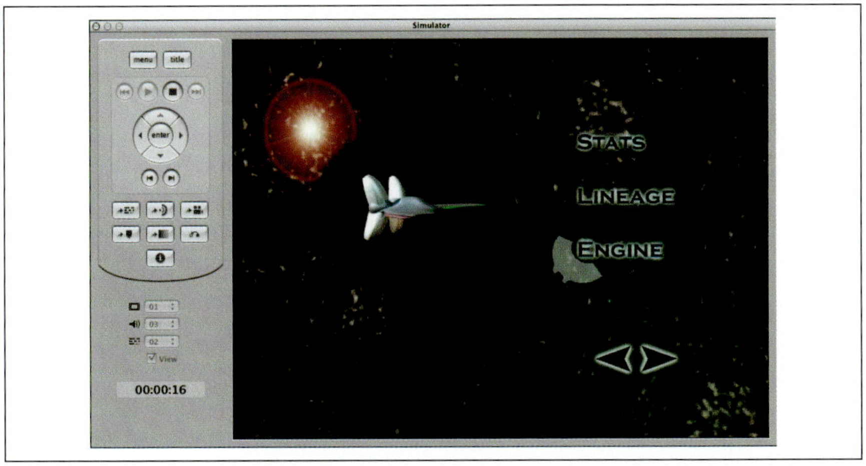

Figure 4-40. Our overlay mask buttons in Preview mode

You can use this same trick to create underlines, text, picture frames, or any other shape to use as a highlight. It obviously takes more planning than the simple highlights, but the results usually are worth the effort.

2-bit (grayscale) overlays

Although 1-bit overlays are easier to deal with, you can, in fact, use a 2-bit overlay with black, white, and two shades of gray (66% and 33%, respectively). Figure 4-41 shows these grays against a blue background for proper contrast.

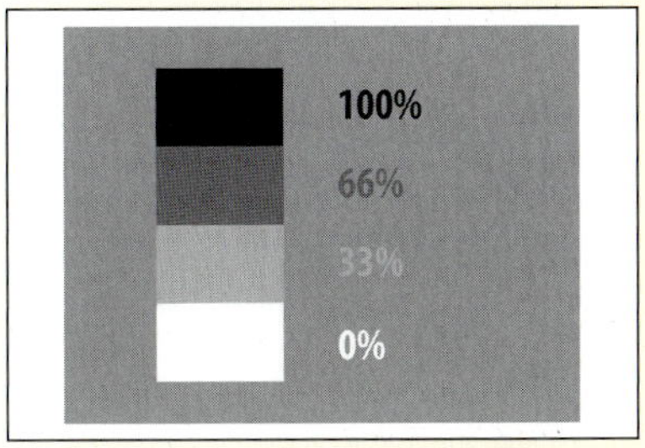

Figure 4-41. The four shades of "gray" you can use in a 2-bit mask

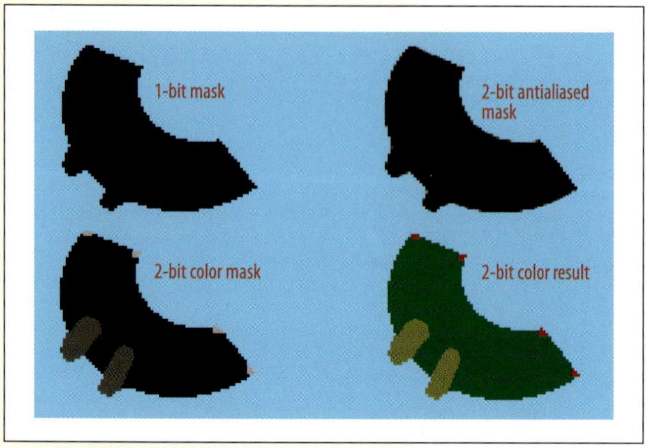

Figure 4-42. Variations on 1-bit and 2-bit masks

You can use the shades of gray in one of two ways: as translucent variations of the highlight color, or as four separate colors. See Figure 4-42 for an illustration.

The translucent variations on the highlight color enable you to do things such as antialias your highlight shape. That can provide some smoother, more professional highlights.

The multicolor trick works even better, but requires a great deal of attention to details. You have to draw with the shades of gray, but you have to think in terms of the highlight colors. (Usually you'll leave white as the completely transparent background.)

Let's try this for ourselves. For the 2-bit overlays, we'll work on the main skiff menu for *Death and Transfiguration* (which we covered earlier in this chapter). It has four screens that lead to the detail pages for each of the four skiffs. We want to highlight each screen in a separate color as the user navigates around the menu.

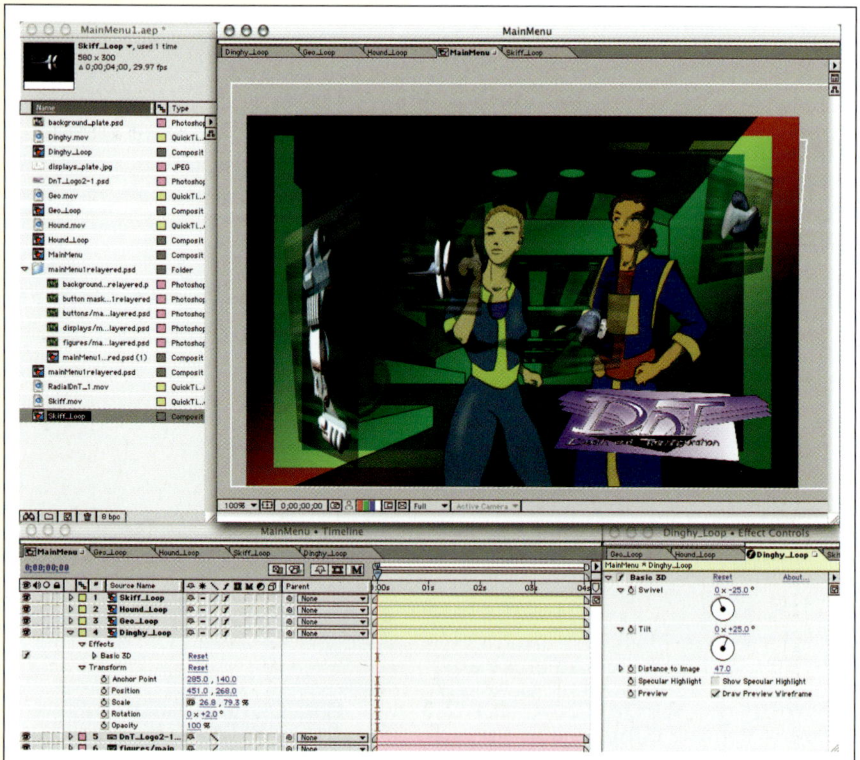

Figure 4-43. An After Effects project to create a fancy motion menu

Another After Effects detour

First we start off by making our video clip for the menu. The creators of *Death and Transfiguration* built a control room with several video screens, one for each of the four skiffs. We'll put a little spinning version of each skiff in its respective screen. Because they wanted to play with perspective and blinking lights and whatnot, the creators used After Effects to create this motion menu. Figure 4-43 shows the basic menu with and without the video elements.

We don't have time to go through all the details of making the video in After Effects, but you can look at the complete project on your own. The *dnt_main.aep* file contains this project as well as the complete animation. The upshot is that we included four video loops that were imported as QuickTime movies. We ran the loops continuously for 30 seconds and exported the final product to DV, then used Compressor to get our MPEG-2 resource.

We could have done similar work with video drop zones in DVDSP, but the odd perspectives of the video clips and their short durations made it easier to build the loops in an external program.

Creating the overlay

Figure 4-44 outlines the steps we'll have to go through to get our video menu and overlay ready for primetime.

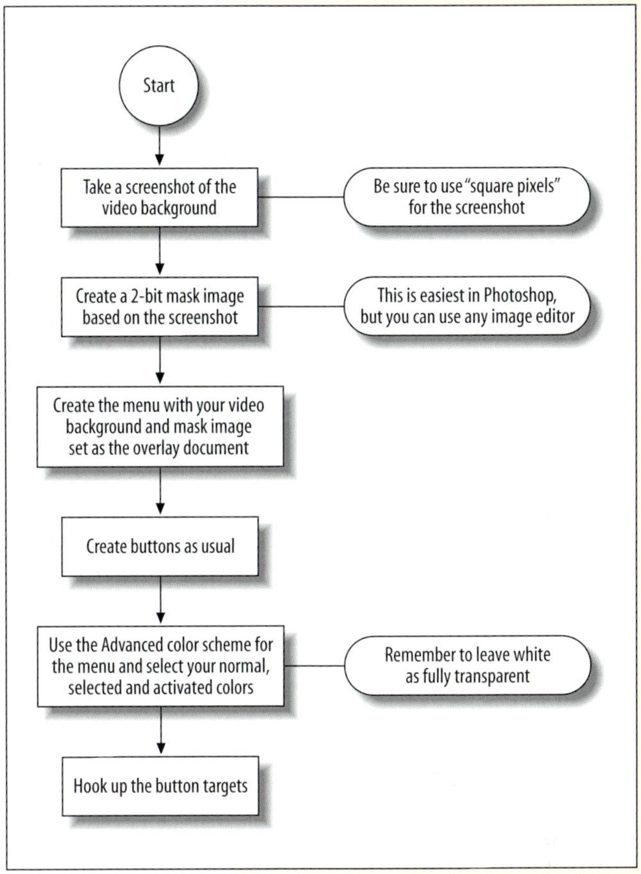

Figure 4-44. Flow for creating a button overlay for a video menu

Now that we have the video, we can create our overlay. Using the same basic menu background from the After Effects project (leave it at 720x480), start a new Photoshop document. Alternatively, you can make a screenshot of your menu at 100% in After Effects. That's cheap, but at least you get a picture of the whole menu to play with. Create a new layer with a transparent background. (We'll change it to a white background as our last step.) We'll create borders around the various floating screens containing the skiffs, and we'll use the different shades of gray to achieve different border colors for each skiff.

For each skiff screen, use the Polygon Select tool to create an outline similar to the screen, but a bit bigger. Figure 4-45 shows an example for the first skiff.

Fill that selection with one of your three grays (black, dark, or light).

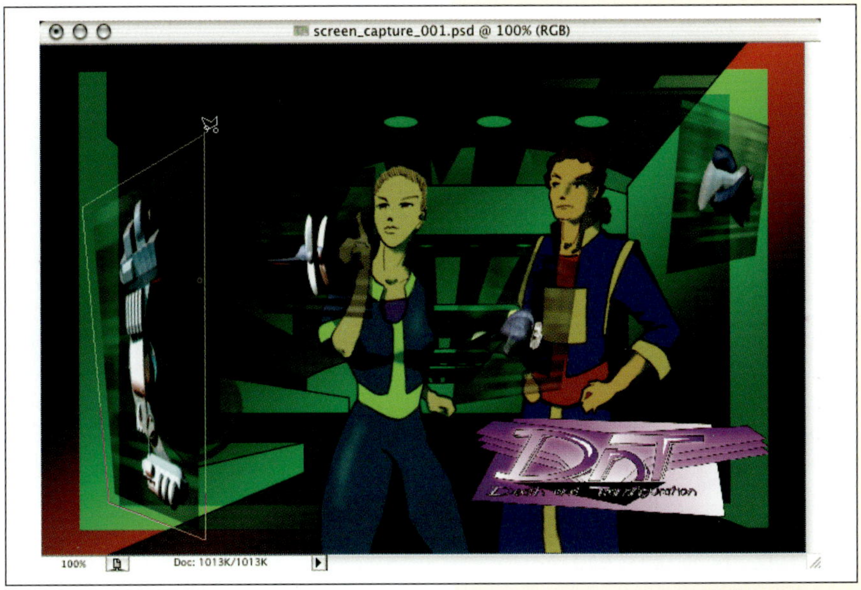

Figure 4-45. A polygon selection around our proposed button (far left)

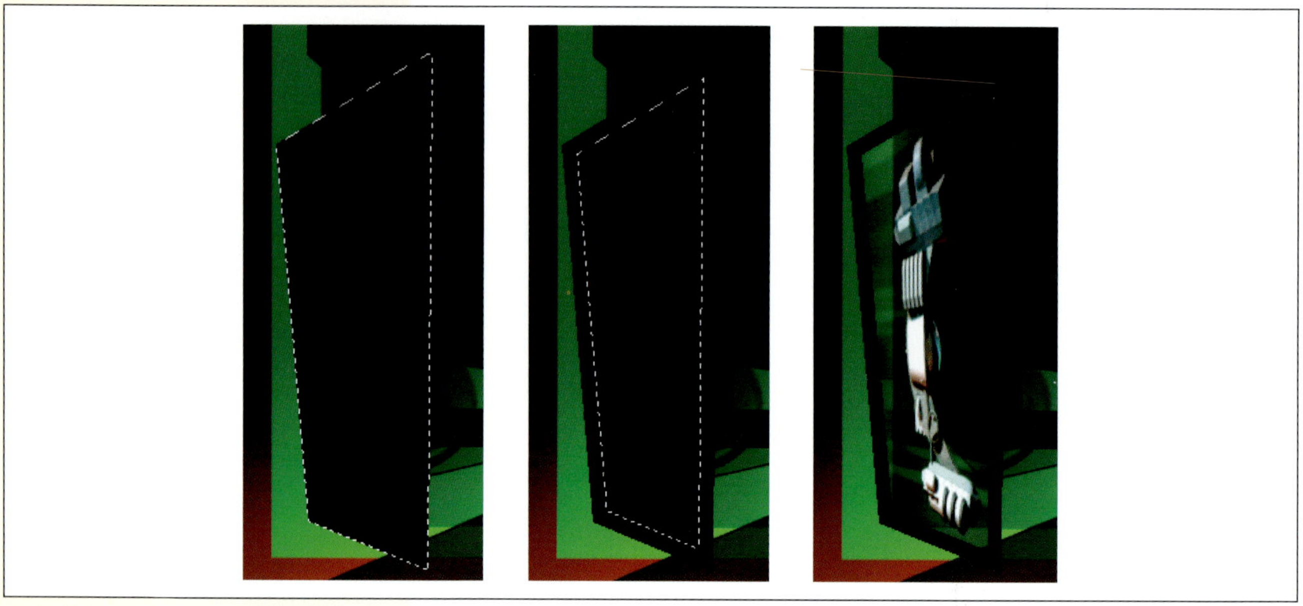

Figure 4-46. Creating a button frame: the first fill (left), the contrasted selection (center), and the final, cleared result (right)

Then choose Select → Modify → Contract to make the selection shrink by about 10 pixels. Now select Edit → Clear to remove that selected area. Now your screen should pop through a nice gray border. Figure 4-46 shows the results for the first skiff. Figure 4-47 shows the borders for all four skiffs and the *Death and Transfiguration* logo.

Before you save the file, be sure to fill in the background of your mask layer to make it white. Save that file and import it into DVDSP. You'll need to specify the overlay document and layer, just as you did before. Now edit the menu to create buttons over each of the four video screens in the

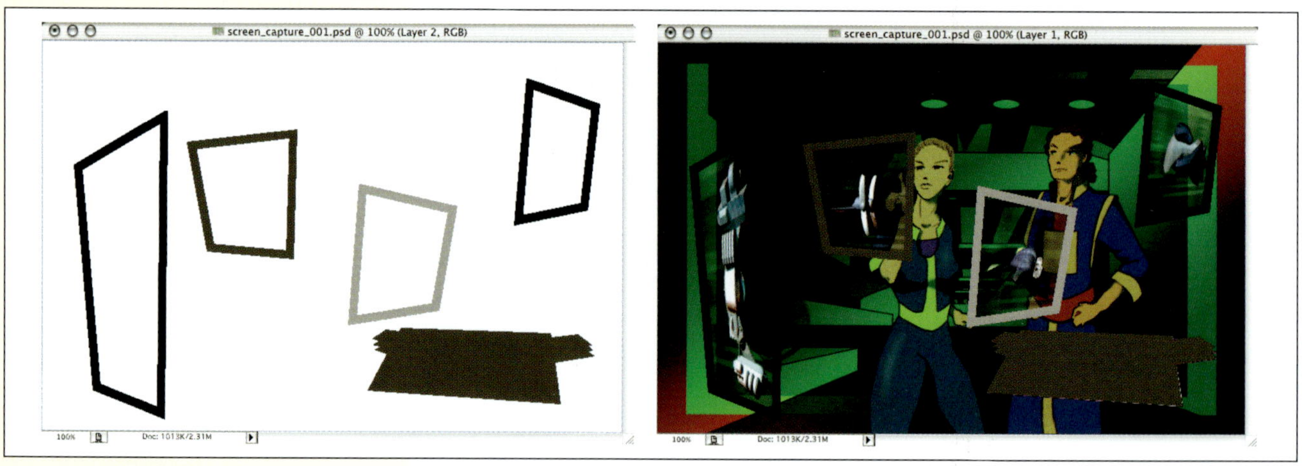

Figure 4-47. The completed masks (left) and a preview of where they'll land on the video (right)

menu. Figure 4-48 shows a completed menu. You can see that our outline shows up nicely for the currently selected button.

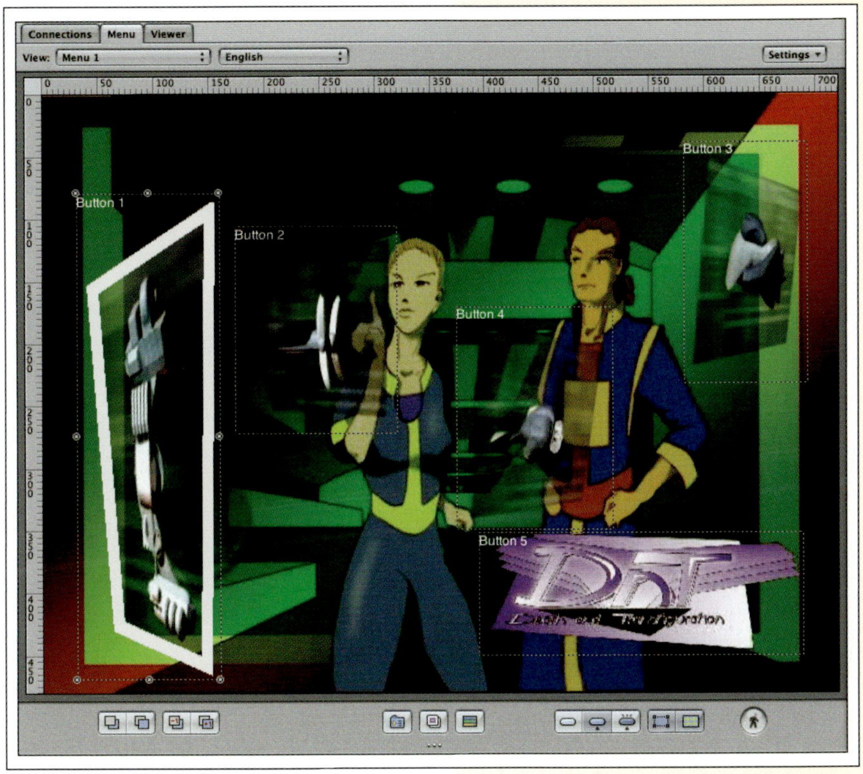

Figure 4-48. The outlines as buttons in DVDSP

Fancy overlay schemes

Back in the property sheet for our menu, you'll need to modify one more thing. For the overlay, you need to use a custom color scheme, so you can't use the "simple colors." Set the Overlay property to Advanced. You should see several new properties. We'll pick some nice, garish colors and set their opacity to 10 (see Figure 4-49). For the Activated state, bump up that opacity to 15.

Set the 100%, 66%, and 33% options for Set 1 to yellow, magenta, and light green, respectively. Each color you use for the outlines should be 60% opaque for the Selected state and 100% opaque for the Activated state. You certainly can play around with those values to suit your tastes. (For example, you could set the Selected states to 100% opaque and then have all the Activated states set to 100% white.)

Notice that there are two sets of colors. You have three color schemes available for your buttons. Each scheme can have its own three colors plus a transparent option. We used the second set to give us five distinct colors for our outlines, plus the logo. It's all a matter of planning out what you need to

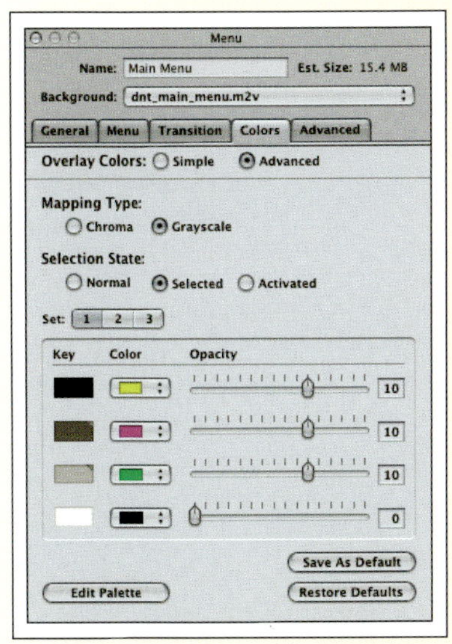

Figure 4-49. Nonsimple overlay color scheme for our menu

> **TIP**
>
> **Overlays and Still Images**
>
> You can use a custom overlay with still menus just as you do with video menus. This can be useful if your project is pushing the limit of available space. With a layered Photoshop document for the menus, each combination of layers creates its own full image. That can start to get out of hand if you support multiple languages for your menus.
>
> Using an overlay with a still image menu is no different from using overlays with video. You just specify the overlay to use in the menu's property sheet and place your buttons accordingly.
>
> One other benefit of overlays: they're fast. With layered buttons, the DVD player actually has to switch to a new screen every time you navigate to a new button. No such shift is required with overlays.

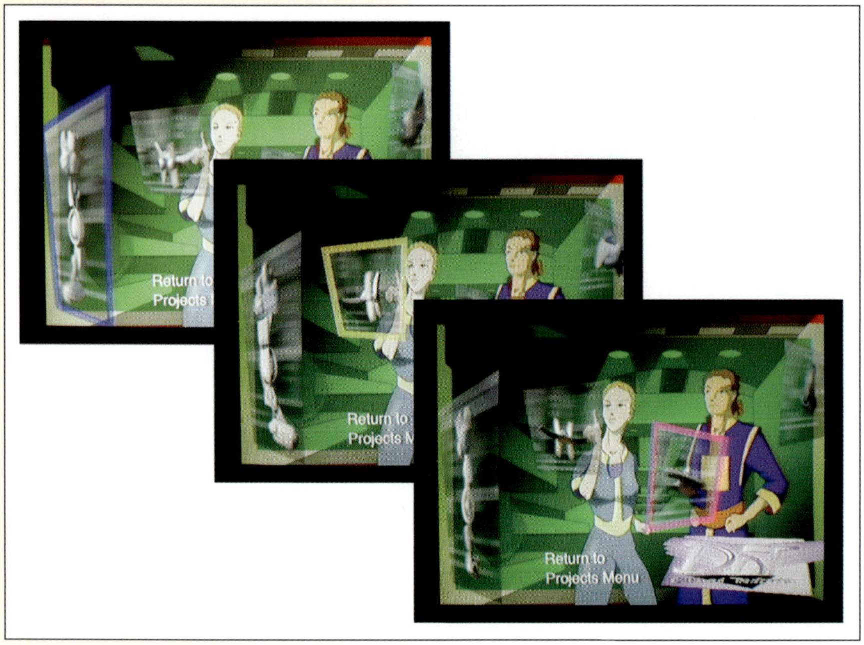

Figure 4-50. Mask buttons with fancy overlay in action

accomplish. If you've watched many DVDs lately, you can see that many designers aren't that ambitious. But don't let that stop you! Well-defined color overlays can create great-looking menus that exhibit great performance, too. Figure 4-50 shows some shots of our fancy overlay menu as seen from a television set.

Multicolor Buttons

Let's take a look at one more quick project with 2-bit overlays. Instead of gaining access to six effective colors for buttons, let's make one interesting, multicolor overlay button. We'll head back to our Hawaii tour over the volcano and turn out a quick menu with fancy buttons.

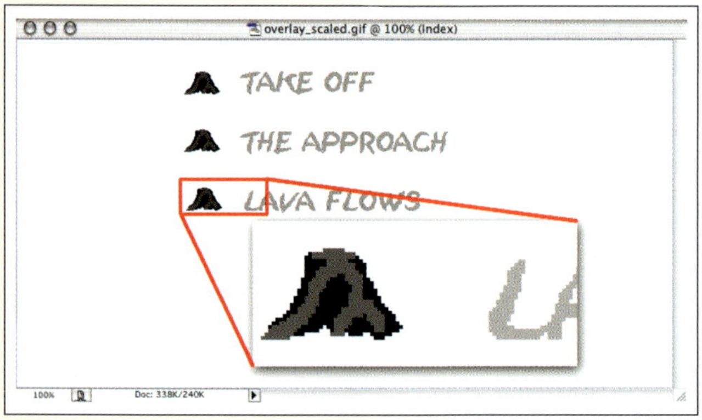

Figure 4-51. A 2-bit mask for a multicolored button and some text

The mask of the red death

Well, this isn't as horrific as Poe's mask, but we're going to use our mask to produce some nice red lava on a volcano. And for a change, we're going to use the mask for the button text as well. Figure 4-51 shows the grayscale version of the mask.

We intend to use the mask for all the menu options here. This isn't common, but it enables us to be remarkably flexible in our background choices. We can use a layered menu, a standard menu, a standard menu with music, or even a full-video background. In fact, let's go the full-video route.

> **NOTE**
>
> *One other thing to remember when building your mask is that fonts should not be antialiased if you're using your shades of gray for distinct colors. Having the occasional red pixel show up around the edges of your light-blue text is not often a desirable thing!*

Multicolor button demonstration

Start a new project in DVDSP and import the following files from the *projects/ch04* folder:

- *overlay_scaled.gif*
- *volcano.m2v*
- *volcano.AC3*

Notice that we didn't include any menu files except for the mask document. That's on purpose!

Now follow these steps:

1. **Create a Menu with the Video Clip as the Background.** Create a new menu as you've done before. This time, however, drag the *volcano.m2v* file over and set it as the background, as shown in Figure 4-52. If you want, go ahead and add in the audio clip as well.

2. **Set the Loop Settings.** We set the loop for about 30 seconds over the fairly constant, dried lava flows. The exact times are shown in the property sheet in Figure 4-53. Be sure to set the loop point at the starting point. The buttons don't show up until the loop kicks in. That's great for skipping over intro video that isn't part of the loop, but until the loop point, the user won't be able to see what he's selecting.

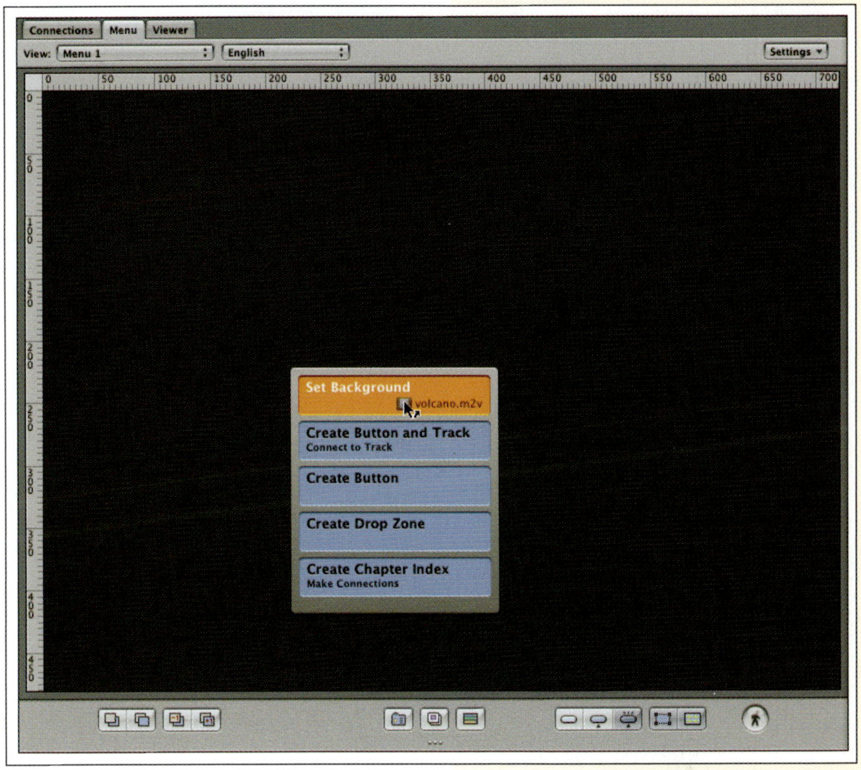

Figure 4-52. Setting up the background for our multicolored buttons menu

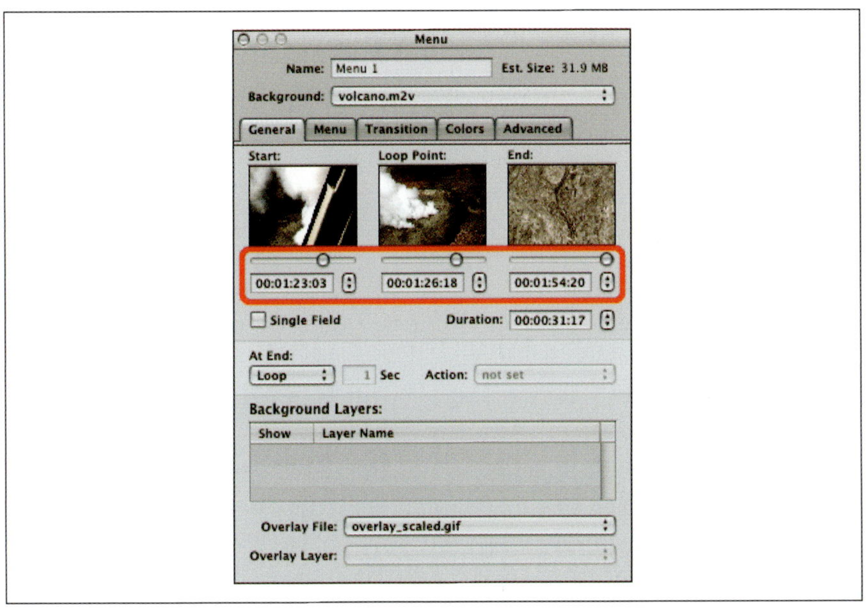

Figure 4-53. The loop settings for our video background (with a 3.5-second video lead-in)

Saving and Scaling Masks

You have to be careful about scaling your masks in Photoshop. Photoshop tries to "do the right thing" when resizing images, and it does some interpolation to keep the image looking good. That interpolation messes up your carefully crafted 2-bit image.

The easiest way to get around that is to save your scaled image for the Web and pick the GIF option. Then you can set it to a four-color palette and be guaranteed to get the right colors in your mask.

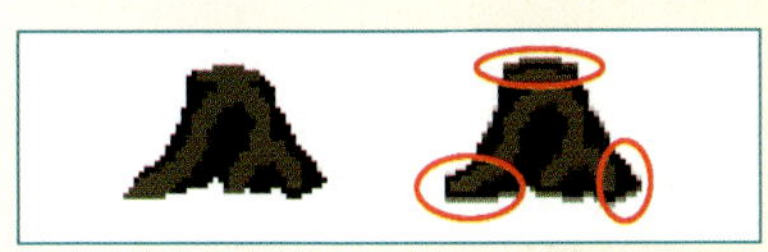

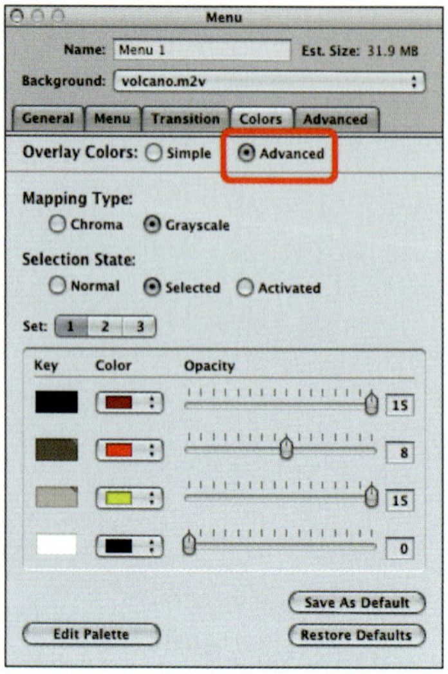

Figure 4-54. Using the Advanced color options for our multicolored buttons

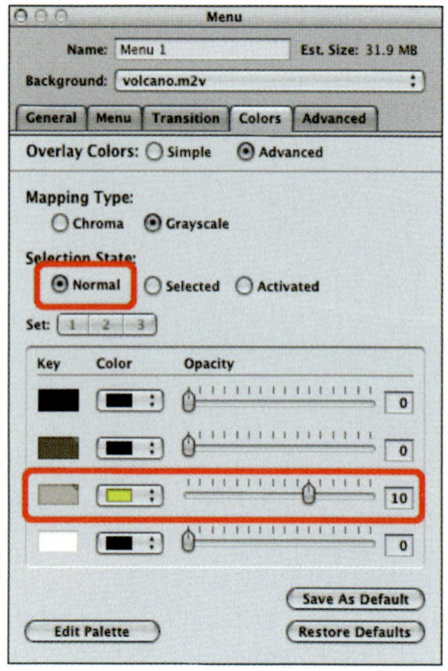

Figure 4-55. The normal settings for our multicolored buttons

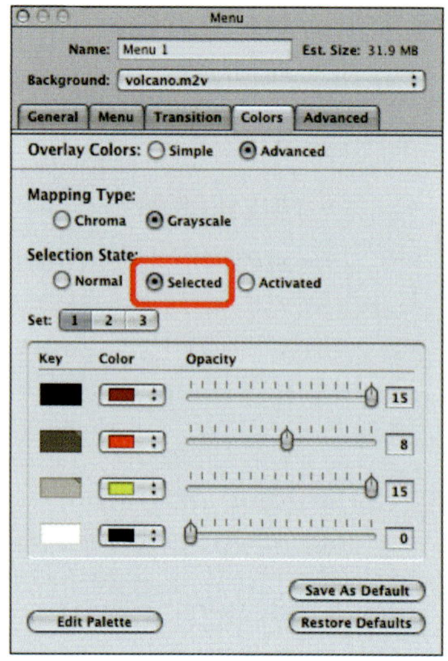

Figure 4-56. The selected settings for our multicolored buttons

3. **Set the Overlay Document to the Photoshop Mask File.** Only one layer should exist, so you don't have to worry about which one to pick.

4. **Set the Mask Colors.** Before you create the buttons, get the colors set up. In the menu's Property Inspector, take a look at the Colors tab. You want to use an "advanced" scheme that enables you to choose everything about the mappings. Figure 4-54 shows the options available for the Advanced color scheme.

   Notice that you can set different colors for each of the three states. You also can build three separate "sets" of buttons so that different categories or groups of buttons can have their own color mappings. We need to work with only the first set, but we want to make different colors for the three button states, as follows.

   The normal state

   Our normal state is, well, not quite normal. The text for our buttons appeared in the mask, so we need to make sure the text shows up even when the button is not selected or activated. The normal state is usually all transparent. We're going to leave the two colors for the volcano transparent, but we'll alter the color for the text (see Figure 4-55).

   The selected state

   For the selected state, we want the text to stand out and we definitely want the volcano icon to show up. For the text, we set the transparency on the text color to completely opaque. For the volcano, we pick a brown and a red and make them at least partially opaque. Figure 4-56 shows these settings.

The activated state

For the final state, we want to give the user confirmation that the player understood his selection. We'll make the lava in the icon completely opaque and change the color of the text just for good measure. Those settings are shown in Figure 4-57.

5. **Create the Buttons.** Nothing new here; we just drag an outline over the button area, as we did before. This time, however, the buttons should appear as we're dragging! Figure 4-58 shows this spiffy feature in action.

   Notice that the colors (and opacity levels) we picked for the normal state of the buttons made the text show up even before we selected it as part of a button.

Hook up the buttons to the chapter markers, as you normally would, and take your new multicolor buttons for a spin. Figure 4-59 shows our menu running in the Simulator.

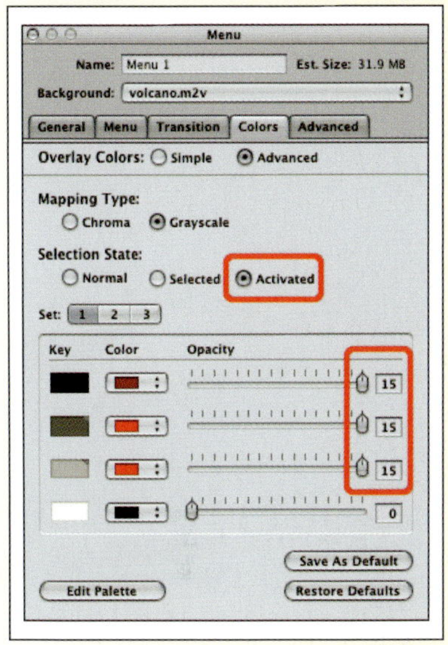

Figure 4-57. The activated settings for our multicolored buttons

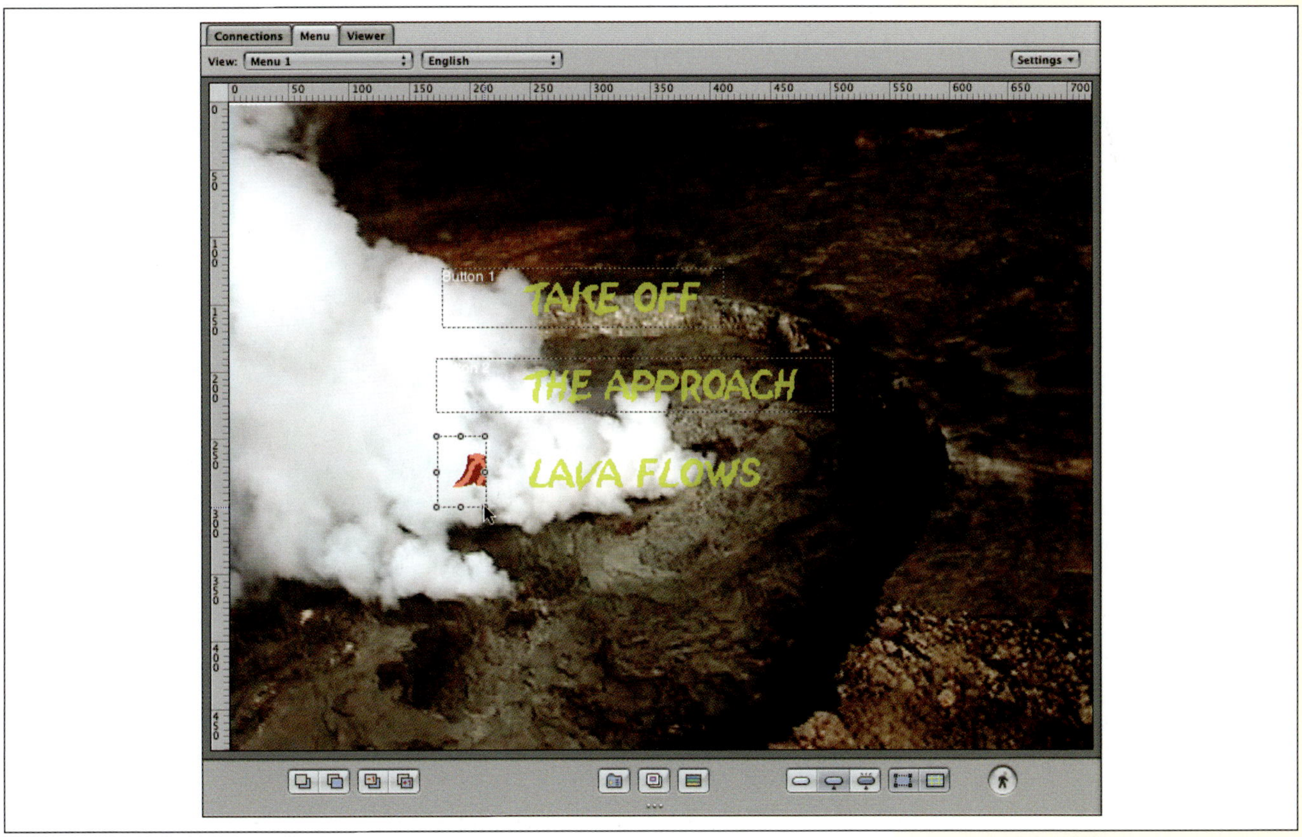

Figure 4-58. The button (including its colors) appearing as we drag the new outline

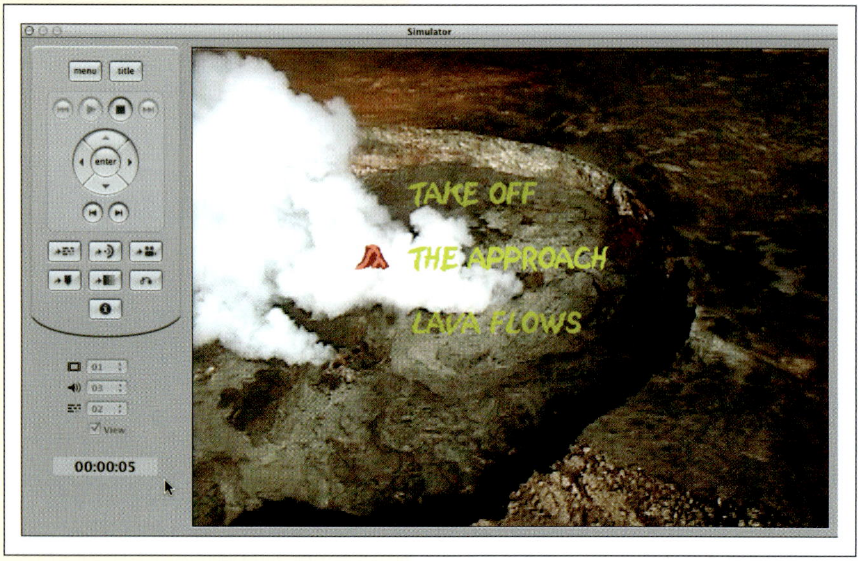

Figure 4-59. Our multicolored volcano buttons in action

Clever use of advanced mappings

Just a quick parting note: you can do some clever things with these 2-bit masks and the advanced color schemes. One very clever trick is to use the three shades of gray to build three unique icons. They have to occupy the same space, so you can't go completely crazy, but you can do additive things to the icons.

We won't bother to build an entire menu project for this snippet, but Figure 4-60 shows an example of such a dynamic icon. The only trick to using this in DVDSP is to make sure you keep the unwanted colors fully transparent in the normal and selected states.

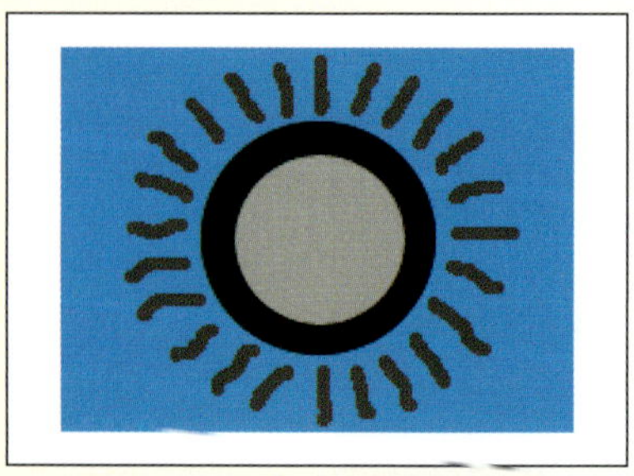

Figure 4-60. A complex multicolored button

In this example you could set the following colors and transparency levels for the different shades of gray, as shown in Table 4-2. Remember that a transparency of 0 is fully transparent, and a transparency of 15 is fully opaque.

Table 4-2. Highlight colors for a complex overlay (as shown in Figure 4-57)

	Normal	Selected	Activated
Black	Blue, 15	Blue, 15	Blue, 15
Dark gray	Black, 0	Black, 0	Yellow, 15
Light gray	Black, 0	Yellow, 15	Yellow, 15
White	Black, 0	Black, 0	Black, 0

Appendix B has some more notes on color combinations with respect to menu buttons. The DVD included with the book also has some onscreen tests with various colors so that you can judge for yourself what combinations are appealing.

Layered Versus Standard Menus

Now that you've seen all the possibilities for both layered and standard menus, we thought we'd recap their features. Table 4-3 contains a few pros and cons to keep in mind when you're planning your project.

Table 4-3. Pros and cons of the menu types in DVDSP

Layered Menus		Standard Menus	
Pros	**Cons**	**Pros**	**Cons**
Fine-grained control over background and buttons.	Requires extra planning; "new" buttons mean editing original graphic elements.	Buttons and menu structure are very dynamic; "new" buttons can be created on-the-fly, often without penalty.	Static background (even if it's video). All information required for the menu must be present in that one background element.
Completely custom buttons with control over normal, selected, and activated states.	Slower response to user input.	Buttons are overlays which are managed by the player hardware—very good button response times.	Limited to (effectively) three colors per button.
Great integration of dynamic graphic elements.	No video or audio elements.	Allows video and audio elements.	Takes up more space on the disc.
Works great with Photoshop.	Requires Photoshop.	Works with Photoshop as well as other graphics editing programs.	Did you read that three-color limit on buttons?

Beyond Menus

Although menus certainly constitute a vast majority of the "work" that goes into a DVDSP project, the physical menus are only part of the possibilities. Now that you have the basics for all the various types of menus under your belt, we'll look at some of the other features you can pack into a DVD. We'll start by adding support for multiple languages and move on to other fun topics such as scripting and games. But don't forget the lessons from these menu chapters—every project you build will have at least one menu!

Le DVD Entwurf Internacionale (An International DVD Project)

5

One of the most exciting and truly revolutionary aspects of DVDs is their ability to carry multiple languages (both spoken and subtitled) for a given video segment. This is probably the most publicized feature of DVDs. Commentary tracks and international releases would not exist without it.

In this chapter we'll tackle an international project. When handling multiple languages in a project, you have four primary areas of concern:

- Will the menus have multiple versions?
- Do you need to support multiple spoken languages?
- Do you need to provide subtitles?
- How do you enable the viewer to pick the menu language?

With the exception of that last area, we'll look at each aspect for the project and build it up step by step. The last bullet point requires a bit of scripting, but don't worry, we'll discuss that in Chapter 7. For now we'll just make sure you can choose through the menus whatever combination of audio and subtitle languages you want. So, let's get going!

In this chapter

Multilingual Menus

If you expect an international audience will be using the DVD you're creating, you might want to support menus with the same languages as those spoken (or subtitled) on the disc. Fortunately a number of mechanisms in DVDSP are available to help you out.

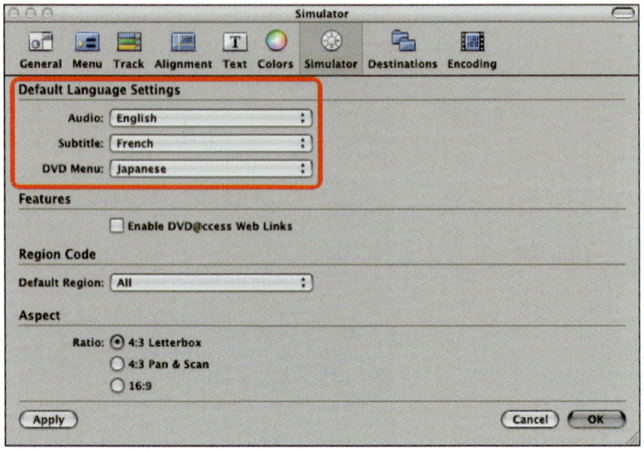

Figure 5-1. Simulator preferences that you can set to trigger other default languages for things such as menus and subtitles

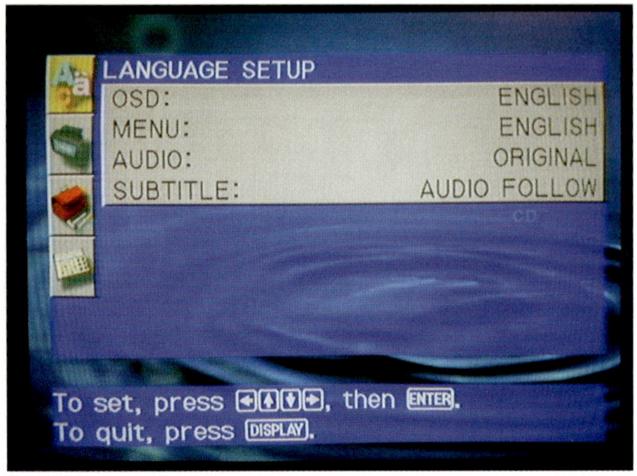

Figure 5-2. Setup screen for a Sony consumer player

Getting started

The first thing you have to do is figure out in which language the user wants to see the menu. We're going to start with the simplest approach: use the player's default. Figure 5-1 shows the Simulator preference in DVDSP for the default language. Most home players have a similar setup option (see Figure 5-2 for the setup of a Sony player), though they usually are set at the factory.

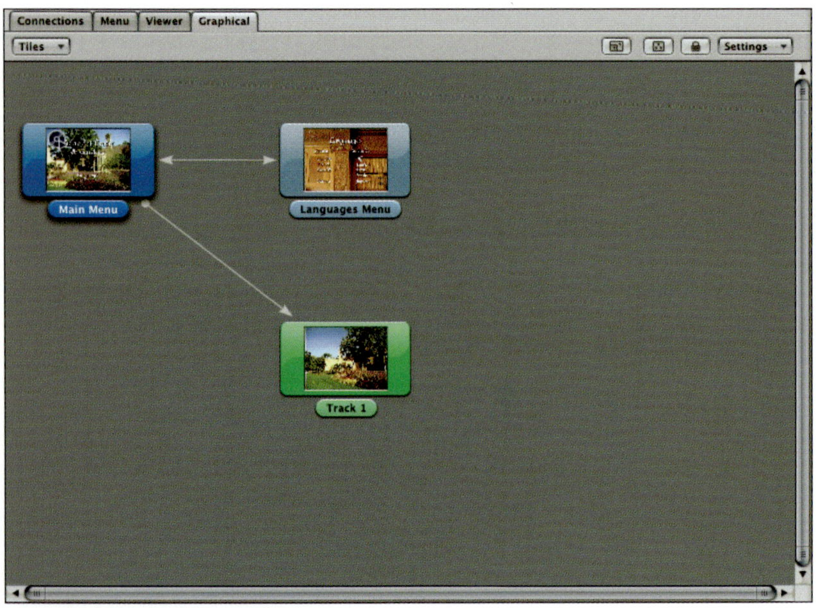

Figure 5-3. The only connections made, even with multiple languages supported by the menu

It's a bit tedious to switch the preference on the Simulator, but doing so enables you to verify that your menus will find the player's language correctly.

From each menu variation, you still can go to a common video segment. Figure 5-3 shows the connections for our first project menu in the Graphical view. Here the video clip has only a classical music soundtrack, so no special subtitling or audio track considerations exist. Even though we have several versions of the menu for our various languages, one Play button launches the clip.

Menu restrictions

If you plan to support multiple languages in your menus using the player's language, you have to live with a few restrictions:

- You must use the same background layers across separate files. For layered menus this means separate files for each language. Each file must have exactly the same set of layers (number and order).
- The buttons apply to every language, so be careful how you lay them out. That means the boundaries have to encompass any text or characters that make up the button in any language variation (i.e., make them big enough for the "biggest" language).

Because of these restrictions, it's easiest to begin with single-layer menus and then pick which file to use for which language. You can, indeed, use layered documents, but the background layers have to be the same for each language. (This feels like an oversight, but layer selections apply to the entire menu, not just to the current language.)

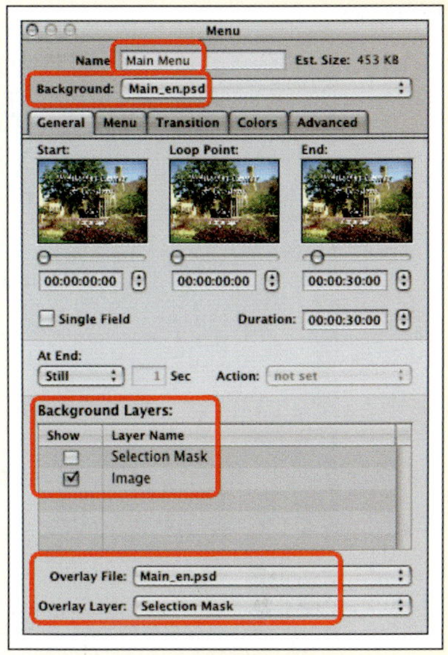

Figure 5-4. The Background and Overlay properties for our menu

Our first multilingual menu

Let's start with a simple main menu. Grab the following image resources from the *projects/ch05* folder and import them into a new project:

- *Main_en.psd*
- *Main_fr.psd*
- *Main_ja.psd*
- *Main_sv.psd*

Then do the following:

1. **Create the Main Menu in English.** We'll start the project by setting up the main menu in English. When that's working, we'll add the other languages and tailor the menu to work with all of them. Go ahead and drop the *Main_en.psd* file on the menu screen. Make sure that only the Image layer is turned on in the Menu Property Inspector. Use the Selection Mask layer for the Overlay layer. (You'll have to specify *Main_en.psd* again as the overlay file.) See Figure 5-4 for a look at our initial properties.

   Notice that the Outline view (Figure 5-5) has one language defined. Our default is set to English. You can, of course, alter that in the Preferences for DVDSP.

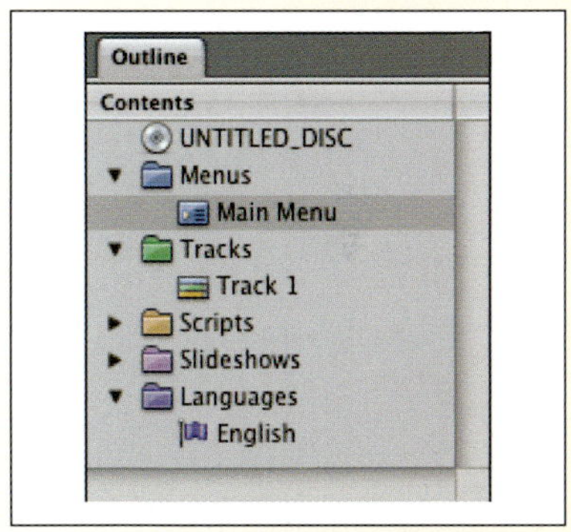

Figure 5-5. The first language added to our project

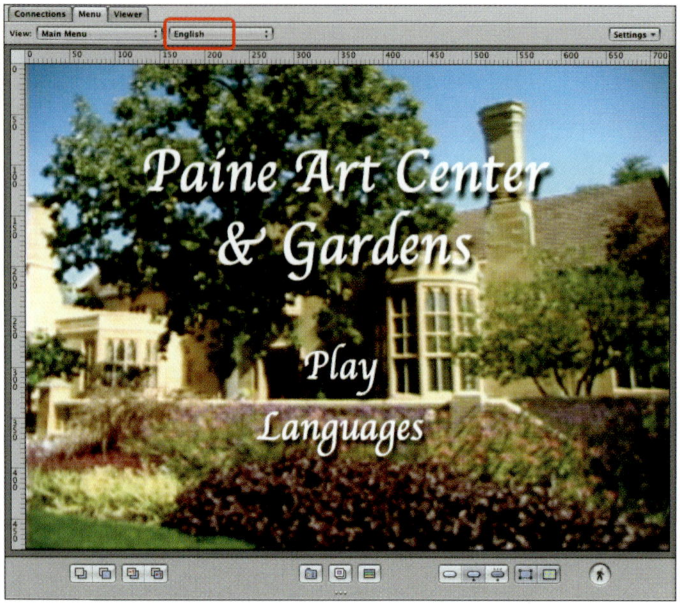

Figure 5-6. The language indicator in the menu screen

So far we haven't had multiple languages in our projects, so it has been safe to ignore the Languages folder altogether. That's no longer true in this chapter. The menu screen also carries language information, as highlighted in Figure 5-6.

Again, we're starting with one language, but that Choice button will become vital in a few moments. The Property Inspector for your menu should reflect the choice you make in the menu screen. See Figure 5-7 for an example.

Now that we know we're working on the English menu, we'll go ahead and set up the buttons and masks. The Selection Mask layer is a simple black-and-white mask, so you don't need to configure any fancy color schemes. The mask should have styled icons book-ending the Play and Languages labels. Drag the buttons over the text to create a menu similar to that shown in Figure 5-8.

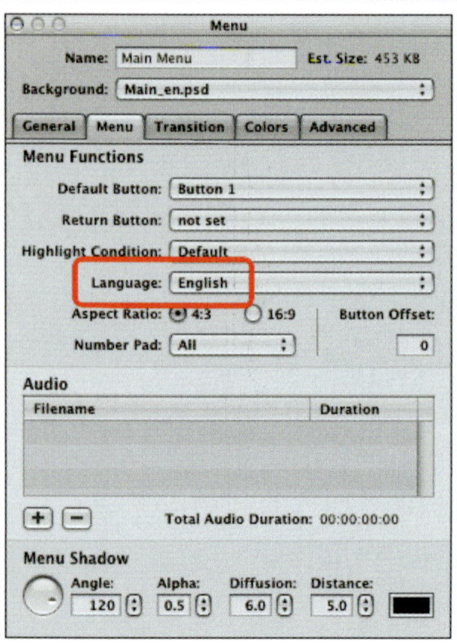

Figure 5-7. The language property for the Menu Inspector

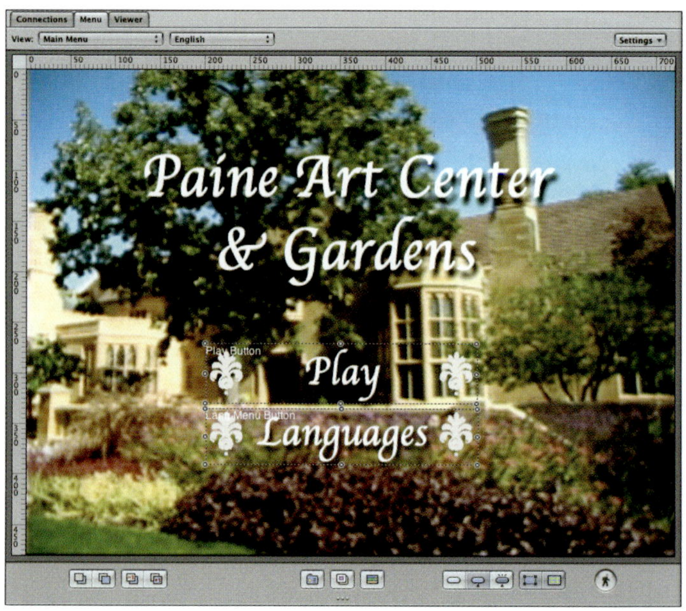

Figure 5-8. Our first buttons defined for the multilingual menu

2. **Add Audio and Video.** Add the following audio and video tracks to the project:

    - *paine.m2v*
    - music.AC3

    The Play button should do just what it says: play the track. Create the track with the video (and audio) files. Go ahead and set the track as the target for the Play button.

3. **Create a Languages Menu.** For the Languages button, you need to set up another menu. Import the remaining image resources for this project:

    - *lang_en.psd*
    - *lang_fr.psd*
    - *lang_ja.psd*
    - *lang_sv.psd*

    Create a new menu by pressing ⌘Y or using your favorite shortcut. Use the *lang_en.psd* file for the English version background. You'll need both the Image and Languages layers turned on for the background. You should use the Selection Mask layer for the overlay, just as you did for the main menu.

    The buttons are pretty much where you would expect them to be. For now, we want to hook up only the Back button, as in Figure 5-9. The target should take us back to the main menu. (We'll hook up the other buttons in the remaining projects in this chapter.) We just want to make sure the basics are in place.

    Go ahead and simulate the project to test it out. You should be able to go back and forth between the two menus and play the video clip. Make sure you work out any problems with the navigation now so that you have fewer things to worry about as you add the other languages.

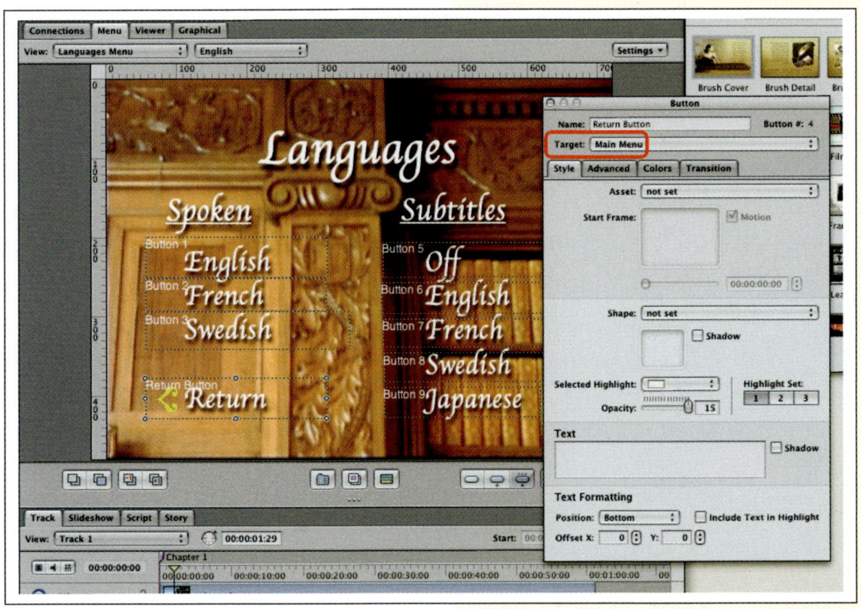

Figure 5-9. Quick setup for our Languages menu

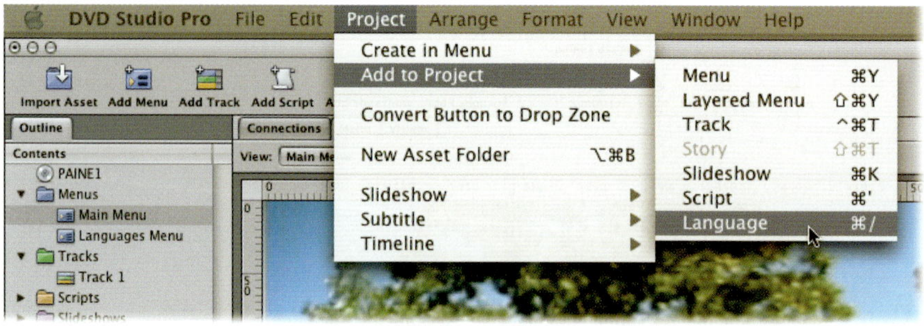

Figure 5-10. The menu and shortcut option for adding a language

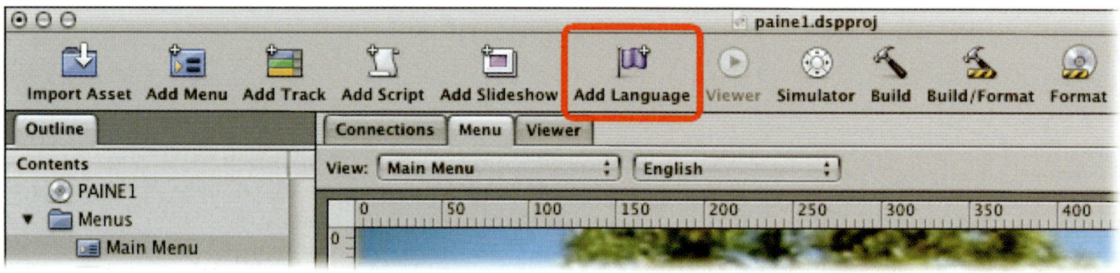

Figure 5-11. The Add Language toolbar button for adding a language

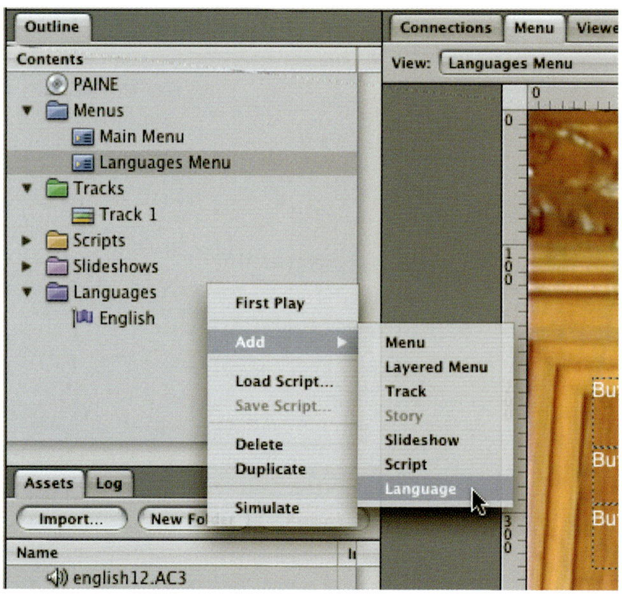

Figure 5-12. Adding a language through the Outline tab

4. **Add a Second Language.** Now comes the fun part. Let's add another language to the project. You can do that in one of several ways:

   - By selecting Project → Add to Project → Language or by pressing ⌘/ (see Figure 5-10)
   - By clicking the Add Language button (see Figure 5-11)
   - By Control-clicking the Add → Language option in the Outline tab (see Figure 5-12)

   You should end up with a new language in the Outline tab, similar to Figure 5-13.

5. **Rename the Language in the Outline Tab.** Select the new language in the Outline view and take a look at the Property Inspector, as shown in Figure 5-14.

You can pick any name you want. Typically, you'll use the name of the language, but you can be more creative, too. You might specify both the language and the audio encoding (e.g., English 5.1 or English Stereo). In any case, the important part is the language pop-up. You can define up to 16 languages for your project. We used French for our second language, as you can see in Figure 5-15.

Now the menu screen should be a bit more interesting. Figure 5-16 highlights the language choices for our main menu.

6. **Set the Language and Background Image for the New Menu.** Go ahead and select French as the language and then bring up the property sheet for the menu. The property sheet contains several items that apply to only the current language. Other items apply to all languages. The different categories are shown in Figure 5-17.

   You need to change only the background image. Go ahead and select the *Main_fr.psd* entry from the property sheet pop-up, as in Figure 5-17. Figure 5-18 shows the result—a menu in French!

   Now you should be able to use the Menu editor's Languages button to bounce back and forth between the English and the French menus.

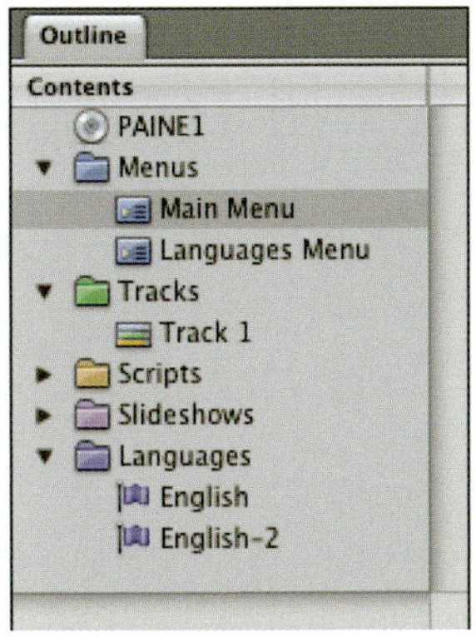

Figure 5-13. A newly added language in the Outline tab

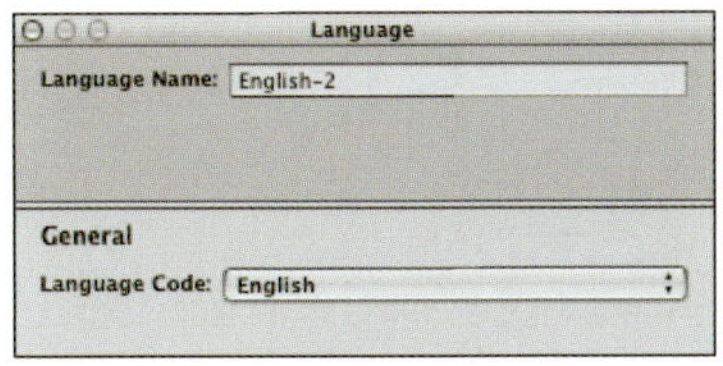

Figure 5-14. The Language property sheet in the Inspector

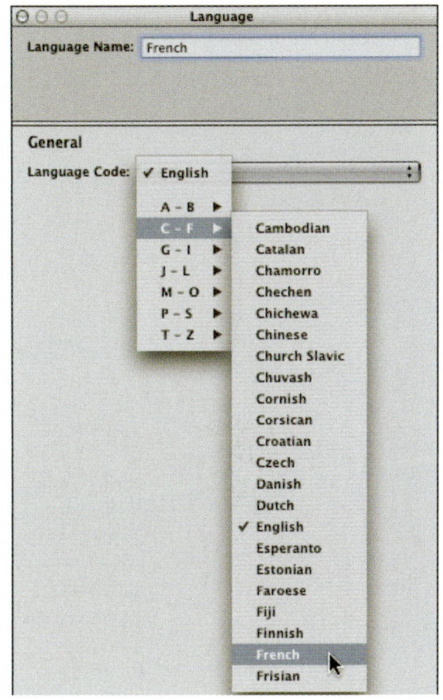

Figure 5-15. Setting the name and language code for our second language

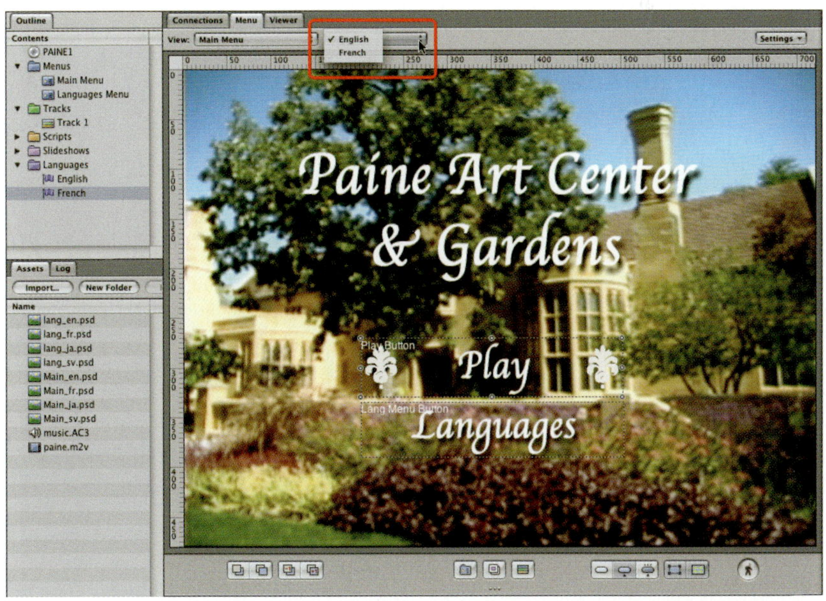

Figure 5-16. The language choice button for menus

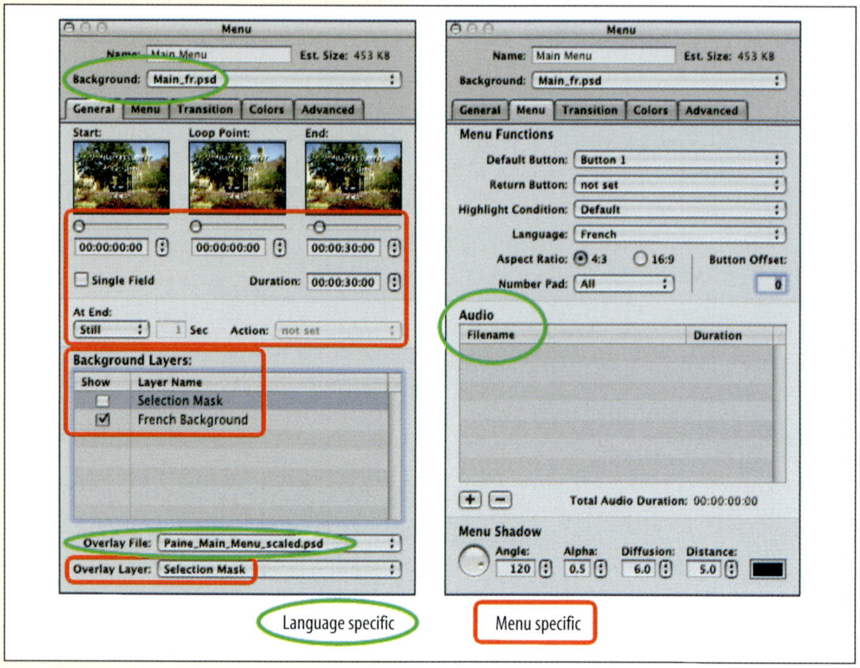

Figure 5-17. Language- and menu-specific properties

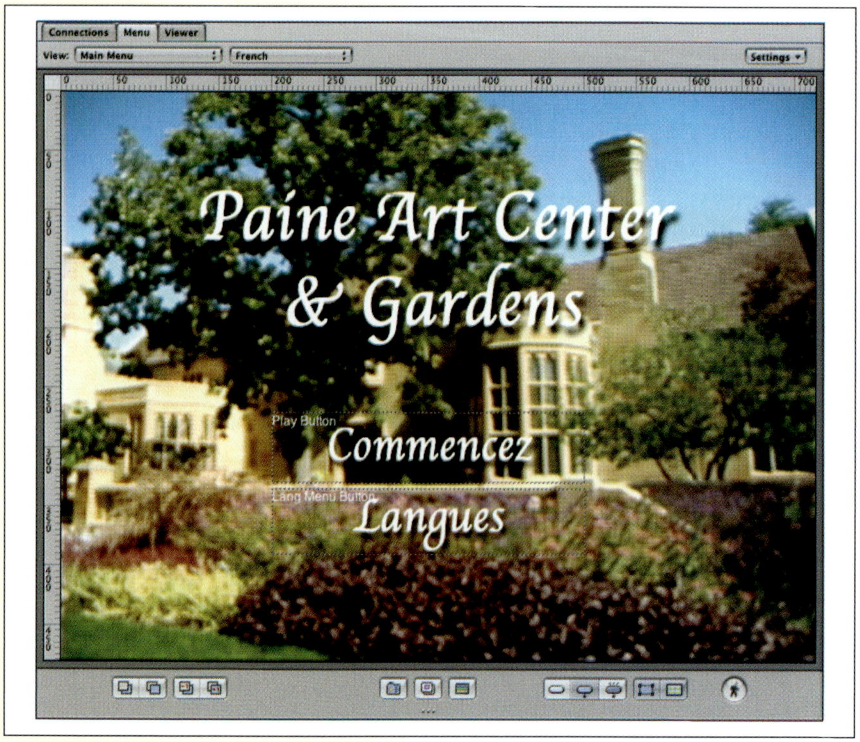

Figure 5-18. Our main menu with buttons in French

7. **Reset the Button Targets.** Note that while you were viewing the French menu, the buttons' targets didn't change. That's OK. Activating the Play button on the French main menu still should start the video track.

   Go ahead and select the Languages menu in the menu screen and set up the French version of this menu as well. This should involve changing only the background asset to *lang_fr.psd* in the Menu Property Inspector. Figure 5-19 shows the result.

8. **Add Other Languages.** You should be able to repeat the same steps for the Swedish (sv) and Japanese (ja) languages: add a new language, then update the property sheet for each menu. You should end up with all four languages in the Outline view, as shown in Figure 5-20.

   And all four languages should be available in your menu screen. You can test to make sure your menus come up correctly by altering the Simulator's default menu language. Open the DVDSP Preferences and look at the Simulator tab (also shown in Figure 5-21).

   You'll have to pick each language in turn to test out the menus. For example, pick French, close the Preferences window, and then simulate the project. Do the same for Swedish and Japanese. Just for fun, try selecting a language you did not add to the project. What happens?

You should find that the Simulator will select the first language if the disc does not have a language matching the player's setup. For this project, that means defaulting to the English menus.

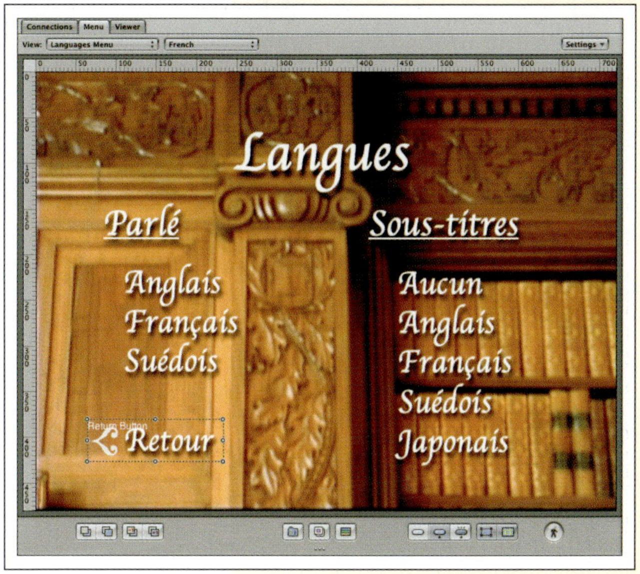

Figure 5-19. The Languages menu now in French

Plan ahead!

The big moral of this story is that you need to plan ahead for any international project. The usual workflow does enable you to get things working in one language and add more languages at a later time, but knowing what's coming around the corner can make things a lot easier.

If you're completely organized, you can define all your languages before beginning the menus. But it's best to be flexible. Oftentimes you'll end up adding or removing a language after the project has started.

Adding Multiple Audio Tracks

We've already looked at the basics for getting audio attached to your video clips in DVDSP. Feel free to refer back to Chapters 1 and 2 if you need a refresher. Including additional audio tracks to support multiple spoken languages is not much harder on the execution side—you just drag-and-drop them onto your video clip.

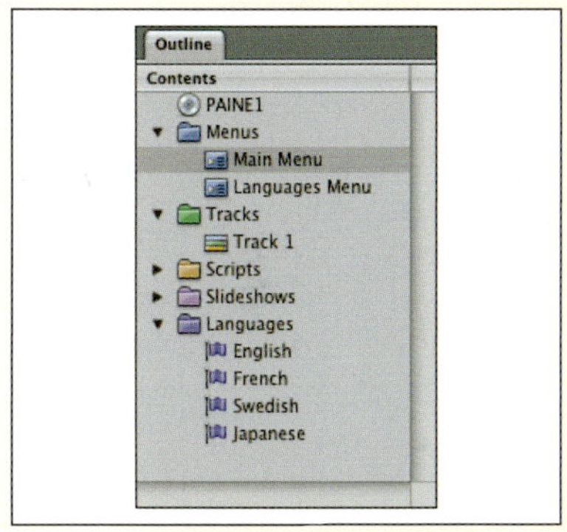

Figure 5-20. All four languages in the Outline tab

When adding multiple audio tracks in DVDSP, you mostly just have to make sure you obey the normal audio encoding rules for each individual track. You can have up to eight audio tracks. Of course, each new track does take up space on the disc, so you need to be more careful when creating long projects than you do when creating short projects.

Oh, and it probably goes without saying, but each audio clip should be the same length, too.

Soundtrack detour

Our previous project had a simple orchestral soundtrack. That soundtrack was built with, well, Soundtrack. Soundtrack is a fantastic application from Apple that enables you to put together your own music. It comes stocked with tons of audio loops that you can use to build an incredible array of music. You get a copy of Soundtrack with Final Cut Pro 4, but you also can purchase Soundtrack directly from Apple.

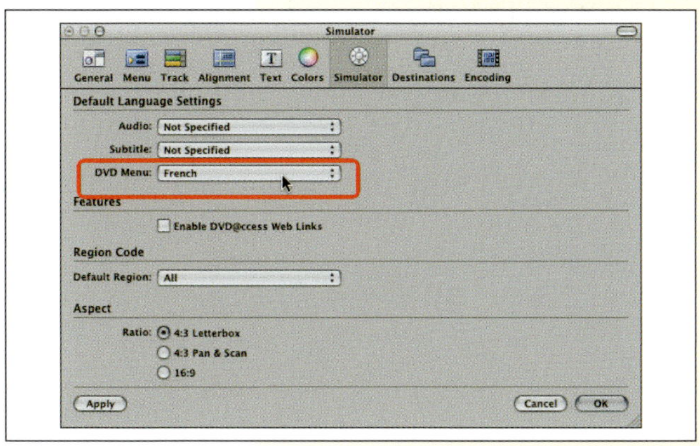

Figure 5-21. Setting the menu language for the Simulator

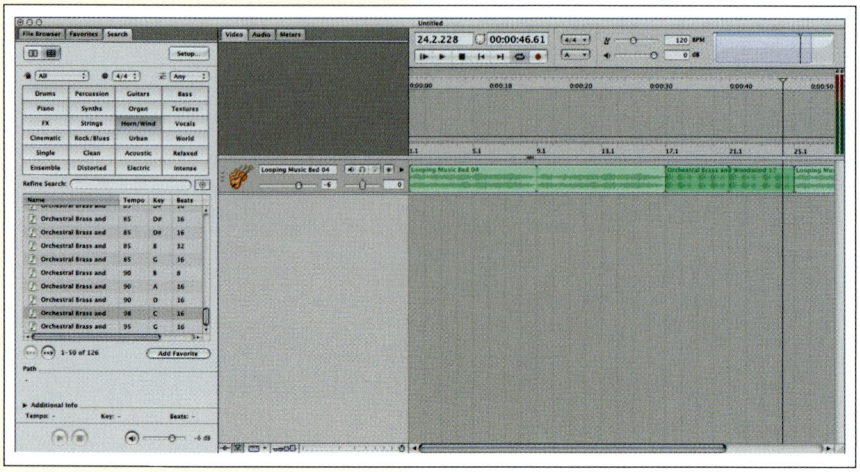

Figure 5-22. The Soundtrack interface

Figure 5-22 shows Soundtrack's main interface and the two loops we used to build that first soundtrack.

It's that easy. You just drag-and-drop loops into the soundtrack area. You can layer the loops to create more complex tracks. We'll be doing a bit of layering ourselves for the voiceovers in this project.

It's all in the timing

The audio loops have a sense of timing in and of themselves. You can search for different loops in the panel on the left of the Soundtrack interface by switching to the Search tab, as you see in Figure 5-23.

You'll notice different beats and different keys attached to each loop. Play around with these and listen to several clips to get a sense of the variations. One of the really useful things in Soundtrack is the ability to have a video loop as well. You can use the File Browser tab in the left panel to navigate to

TIP

Layered Documents

You can use layered documents in multilingual menus. You just have to ensure that all your menu documents have the same number of layers. They have to be kept in the same order, but they don't have to be named the same.

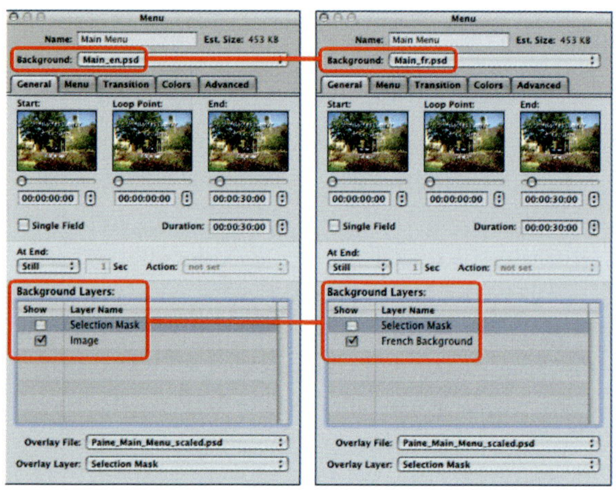

It's tempting to use the same document but different layers for each language. Regrettably that behavior is not supported in DVDSP. Changing the visibility of a layer applies to the entire menu, not just to the current language.

STUDIO ADVICE

Parting Photoshop Tip

Because keeping all the layers in the same order across multiple files can be difficult, at best, we usually make just one file with different layers for each language. Rather than use copies of that file in DVDSP, though, we just export PNG (or TIFF) files of each language, and one for the overlay mask. Photoshop makes that easy by saving only the visible layers to any new file you create.

For example, here are the layers we used for the Swedish menu, along with the resulting PNG file:

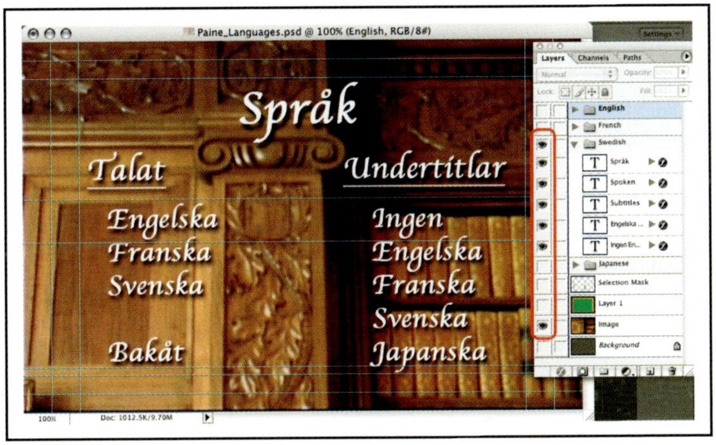

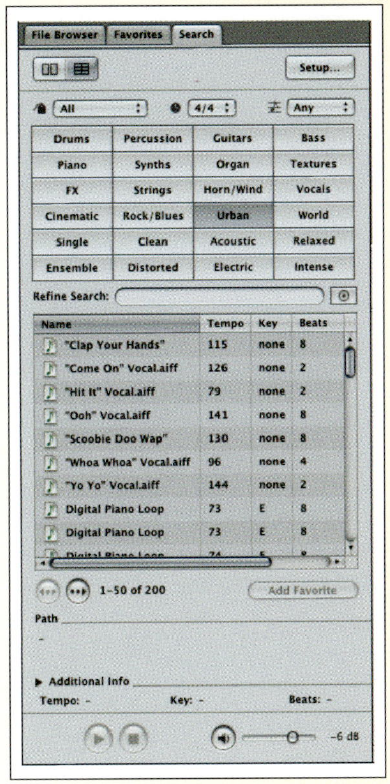

Figure 5-23. The Search interface for audio loops in Soundtrack

your clip and then just drag it over to the Video window in the main interface. Figure 5-24 shows the process with our Paine Art Center clip.

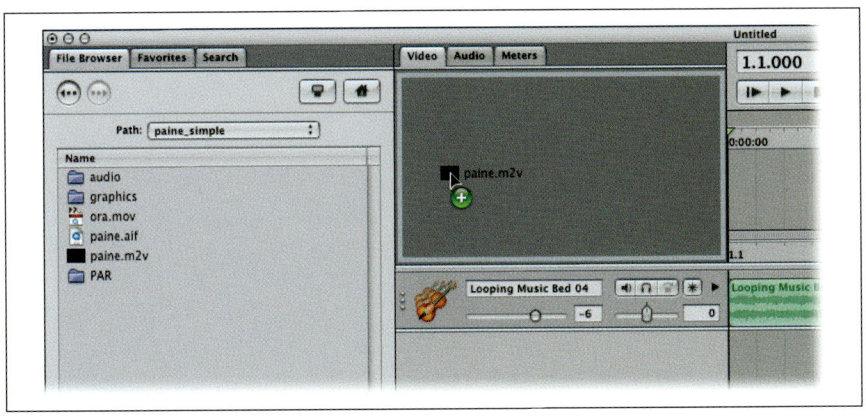

Figure 5-24. Adding a video clip to a Soundtrack project

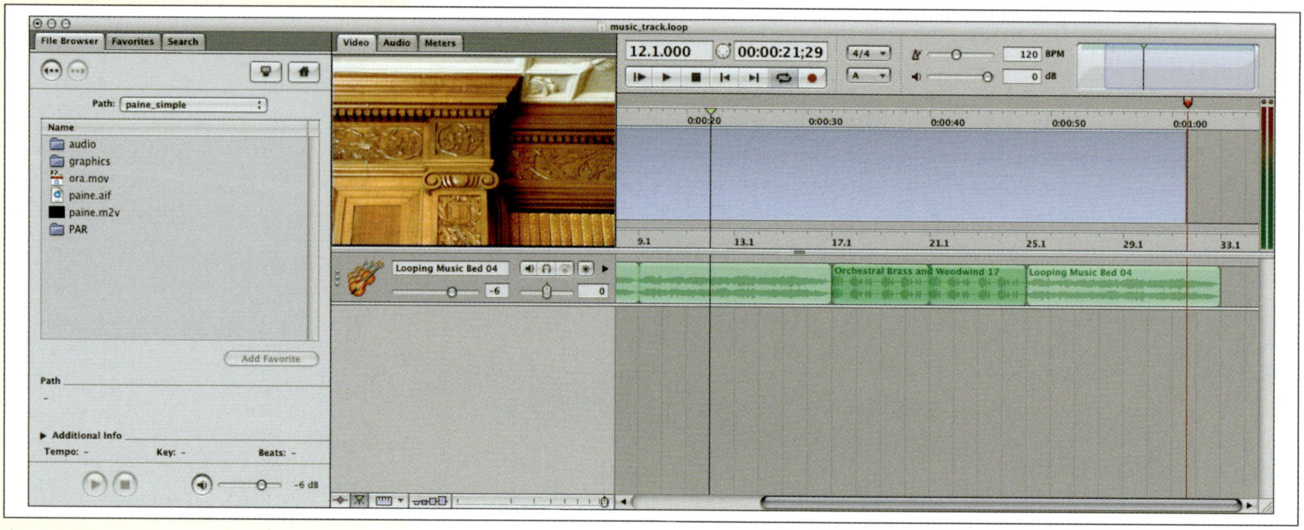

Figure 5-25. The video and audio tracks might not line up initially

Here's the fun part. Scroll or resize the window so that you can see the end of the video as well as the end of the audio track. Without any adjustments, our tracks look like Figure 5-25.

You can adjust the beats-per-minute property of your Soundtrack project to make the clip match your video more closely (see Figure 5-26).

You probably won't get a perfect, frame-accurate match on the soundtrack, but that's OK. If you notice in Figure 5-26, we left the video just a smidge longer. When we actually go to export the audio mix, we can set the length to match the video clip. A tiny bit of silence will be added to the audio track and everything will line up nicely once we bring it into DVDSP.

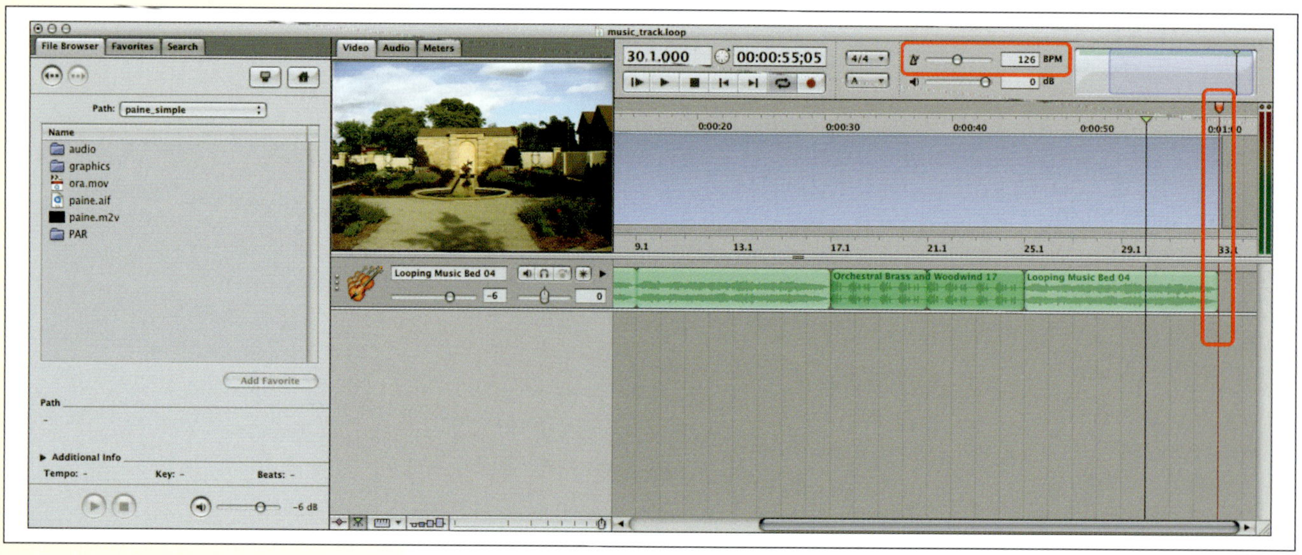

Figure 5-26. Changing the beats per minute to make the soundtrack match the duration of the video clip

Exporting and sample rate warnings

One thing to remember about the audio track is that we eventually want this on our DVD. We need to make sure the sampling rate is 48kHz. You can do that easily in the Project → Sample Rate menu, as shown in Figure 5-27.

When you open this project in the future, you'll probably get a warning about sampling rates that says "Playback will be sampled at 44.1kHz." You safely can ignore that warning. When the project is exported to disk, the sample rate you chose will be saved. Figure 5-28 shows this error, just for reference.

In fact, let's go ahead and export our current music track. From the File menu, select the Export Mix... option, as in Figure 5-29.

From here you should be able to choose a filename and folder for the AIFF export. Notice in Figure 5-30 that we checked the "Mute audio track from video" checkbox. Although our video didn't have any audio this time, checking this checkbox is a good habit to get into. The duration should match the longest clip (our video clip) but will not contain any audio from the video.

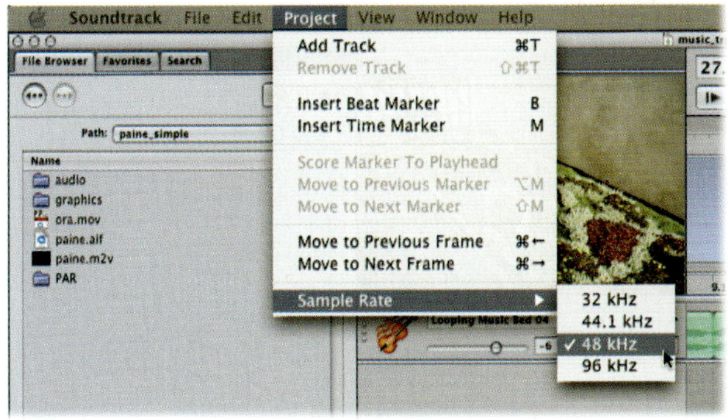

Figure 5-27. Setting the sample rate for a DV project

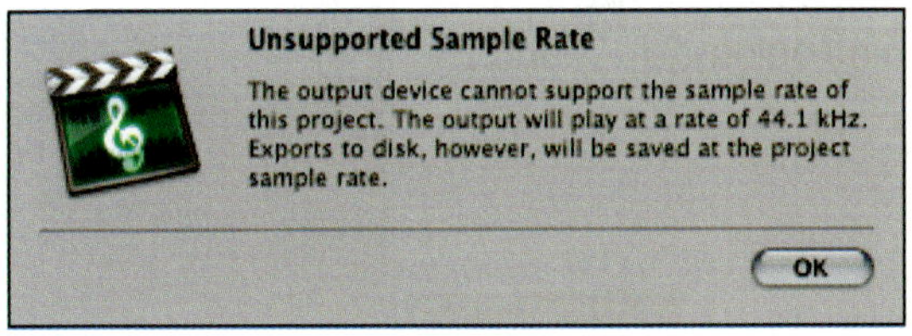

Figure 5-28. The preview-versus-actual sample rate warning in Soundtrack

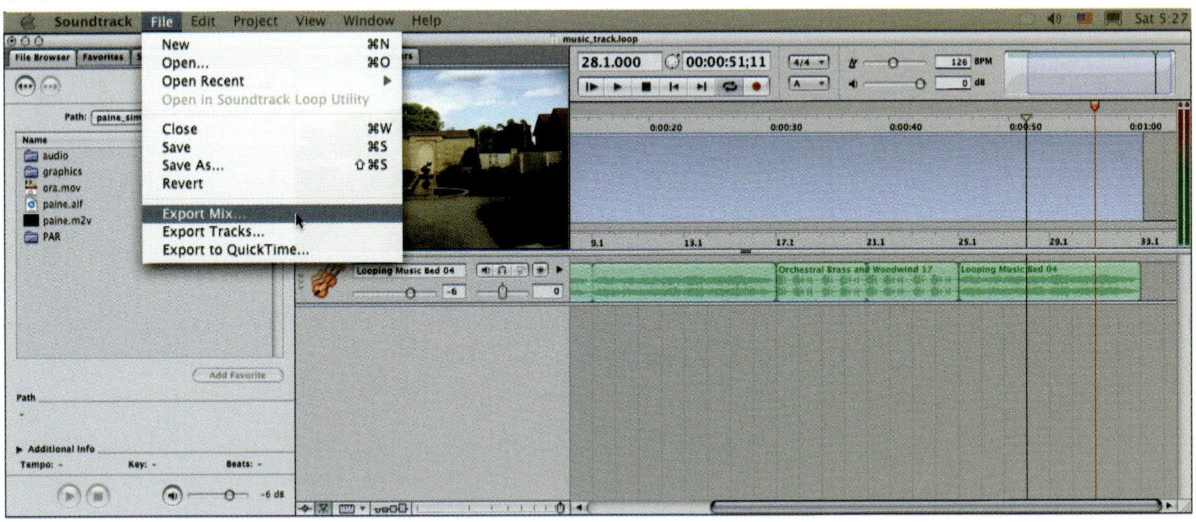

Figure 5-29. Exporting our first soundtrack

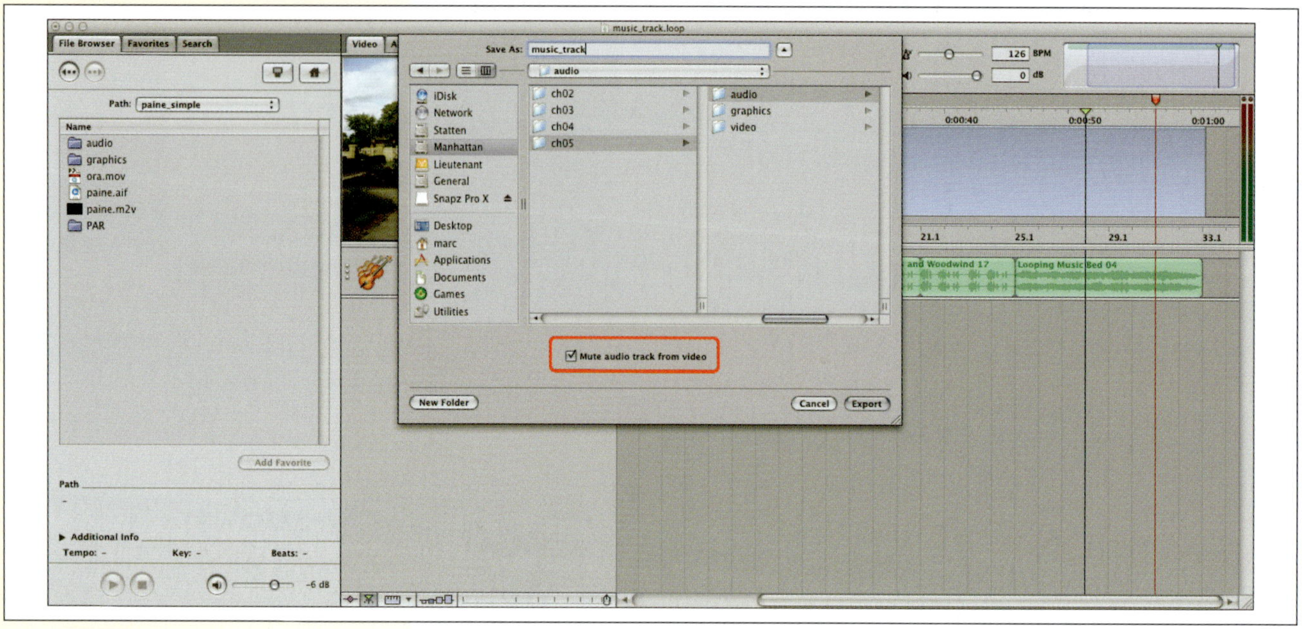

Figure 5-30. Ensuring we export only our custom mix (we don't want the video's audio track mixed in)

One-take recording

Hooray! You just re-created the audio track that you used for the first project. But we're really here to look at recording the voiceover tracks for the different spoken languages we plan to support (English, French, and Swedish). Soundtrack can make this fairly easy, if you keep it informal—and if you can do it in one take! Well, you always can throw things out, but Soundtrack really is a mixing tool, not an editing tool, so single takes are best.

To get a voice recording going, you need only one thing: some audio input hardware (i.e., a microphone attached to your computer). When you make a recording, Soundtrack automatically will create a new track for you and place the recording there.

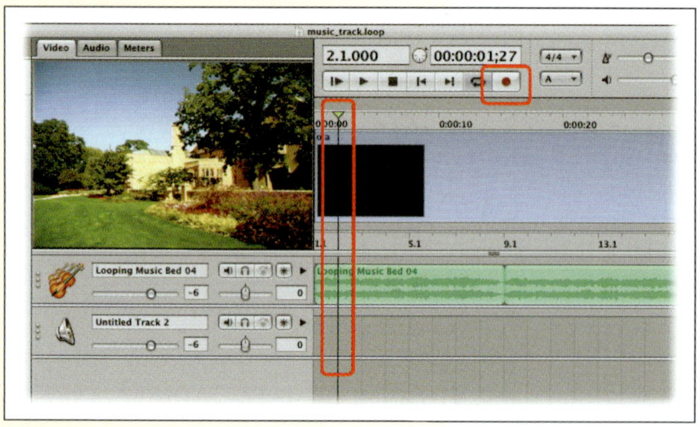

Figure 5-31. Starting a single-take recording

For the hardware, we used the iMic from Griffin Technology. This isn't an endorsement or anything, but boy does the iMic make sound-in easy. It's a USB interface with separate 1/8-inch jacks for a microphone and headphones. Those "hands-free" headsets work great for quick jobs such as this. You can tell the computer to play the current mix in the headphones while you're recording via the microphone.

Now you can start recording. Just move the green playhead to the beginning of the track (or wherever you want to start, really) and click the record button. Figure 5-31 highlights both components.

You should end up with a recording dialog similar to the one you see in Figure 5-32. Notice that you can pick the name of the clip as well as the input and monitor sources. You get a stereo recording by default, but you can make it mono later if needed. If you don't want to listen to the other tracks while you're recording, you also can mute the project before clicking the record button.

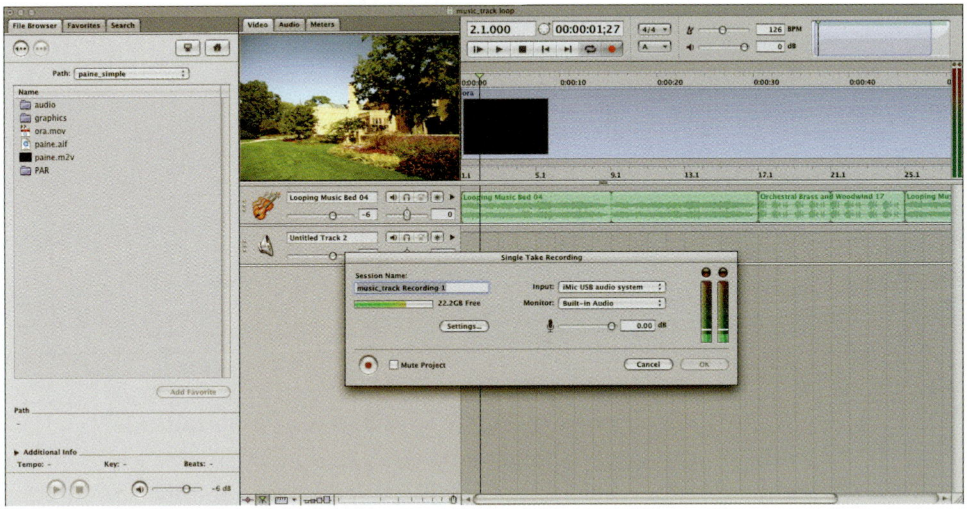

Figure 5-32. The Single Take Recording interface

The video and audio (unless muted) should run in the background while you record your voiceover. If you let the audio play, you should play it back through headphones so that it doesn't get "re-recorded" as part of the voiceover track. That can cause lots of problems later on—one tweak to the soundtrack part and you have to redo all the voiceovers!

When you're done, click the stop button (actually, the record button changes to become the stop button) and your new track should appear similar to Figure 5-33.

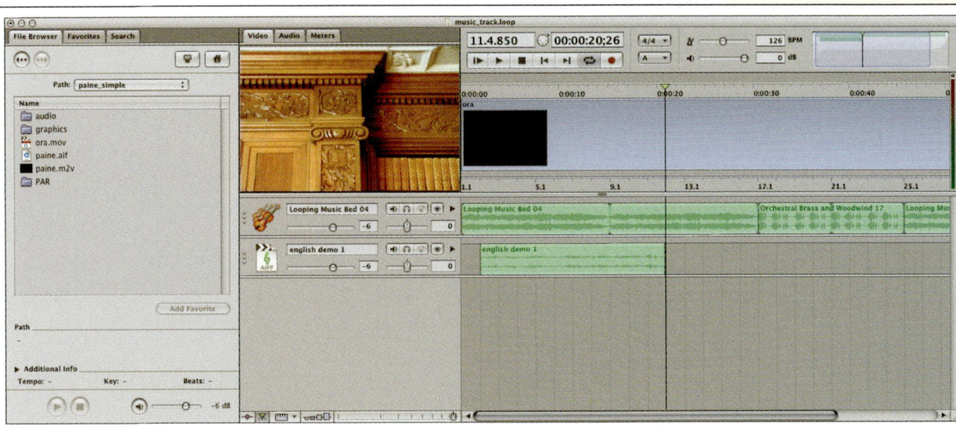

Figure 5-33. Our first voiceover in Soundtrack

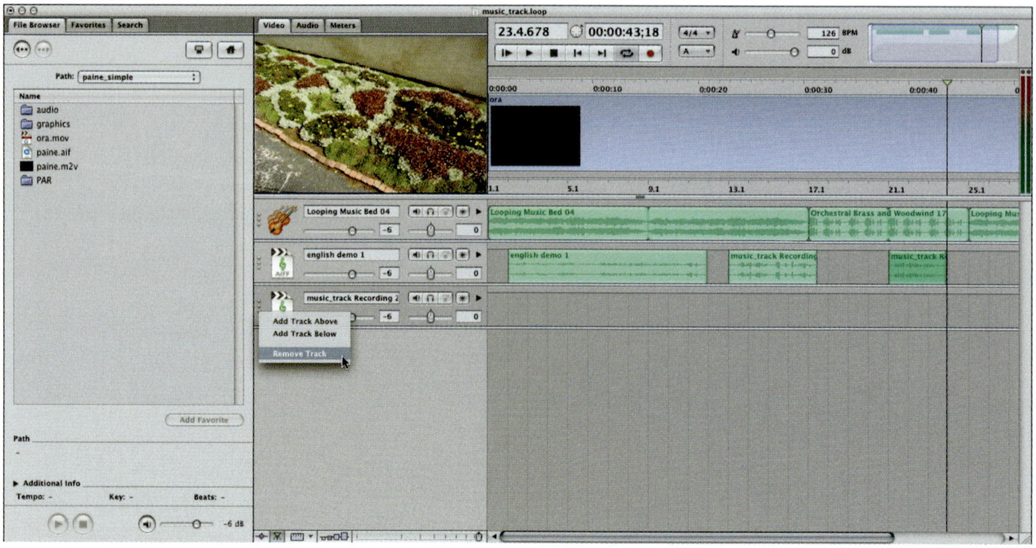

Figure 5-34. Combining recordings and discarding empty tracks

Just repeat that process for each sentence or paragraph of your voiceover. If you're experienced in such things or if it's a candid environment such as a director's commentary, you can do the entire track in one take. Otherwise, break it up into small pieces and then combine the clips later. Figure 5-34 shows our English track with multiple recordings. You also can see the Control-click menu we used to remove the empty tracks after combining our voice clips.

Multiple languages

For the other languages in this project, just repeat the preceding process. One of the nice things about Soundtrack is that you can mute particular tracks and still listen to others. We used this muting feature to keep all our language recordings in the same project, as you can see in Figure 5-35.

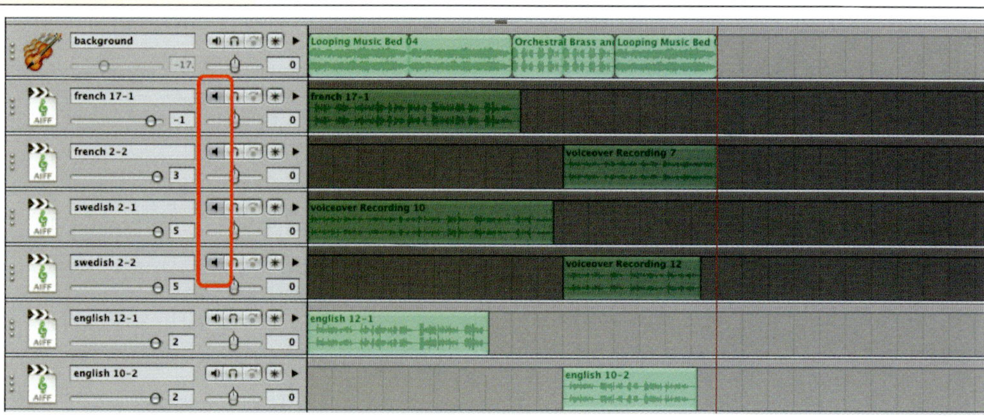

Figure 5-35. Muting unwanted tracks temporarily or for an export

Before you export the mix, just make sure you have the right set of tracks enabled. Muted tracks will not be exported.

Exporting and encoding

You should format the files you export from Soundtrack as 48kHz stereo AIFF. You can use that format directly, but as we discussed in Chapter 2, you should encode the files with A.Pack. The bit rates you pick don't have to match from language to language; you just need to balance your audio throughput with your video. Audio tracks are not interlaced and, therefore, not additive. You just need to look at the maximum video bit rate, add that to the maximum audio bit rate for any one audio track, and make sure that sum* stays smaller than 10Mbps.

The layouts you pick for each audio stream also can be independent. You've probably seen discs with audio tracks labeled "English 5.1," "English Stereo," "French Stereo," etc. Again, just make sure you've got room on the disc for the added tracks.

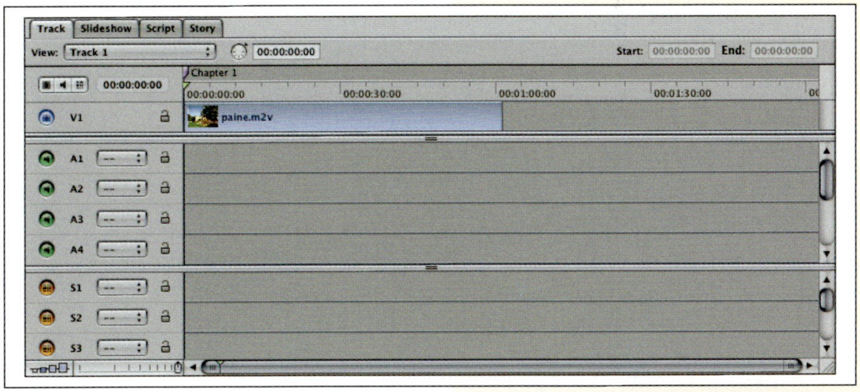

Figure 5-36. The first step in adding multiple audio tracks: adding the video track

Adding audio tracks in DVDSP

Working with multiple audio tracks is remarkably easy. You just drag-and-drop each language on its own line and you're done! Well, you should specify what language each track uses, but that's also fairly easy.

Audio track demonstration

To add an audio track in DVDSP, do the following:

1. **Place a Video Track in the Track Editor.** Start off with a video track and drop it in the Track editor, as shown in Figure 5-36.

2. **Add Several Audio Tracks.** Grab your first audio track and drag it over. You should see a result such as that shown in Figure 5-37.

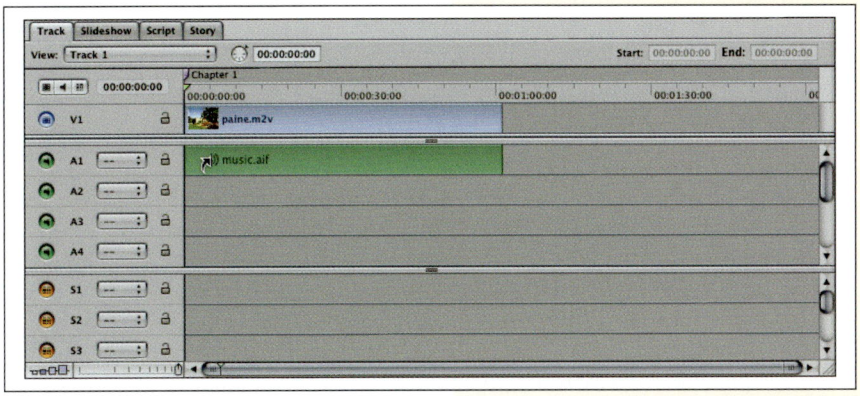

Figure 5-37. Adding the first (of many) audio tracks

* Technically, you need to leave room for any subtitle tracks, but those tracks are really tiny—about 0.01 Mbps each.

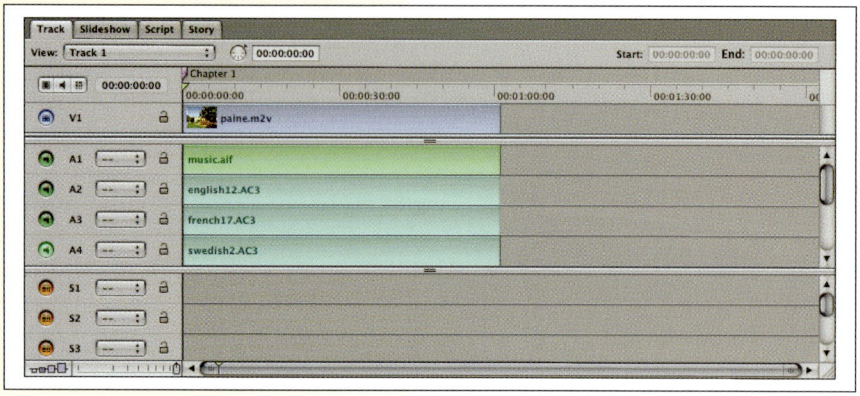

Figure 5-38. A video track with four associated audio tracks

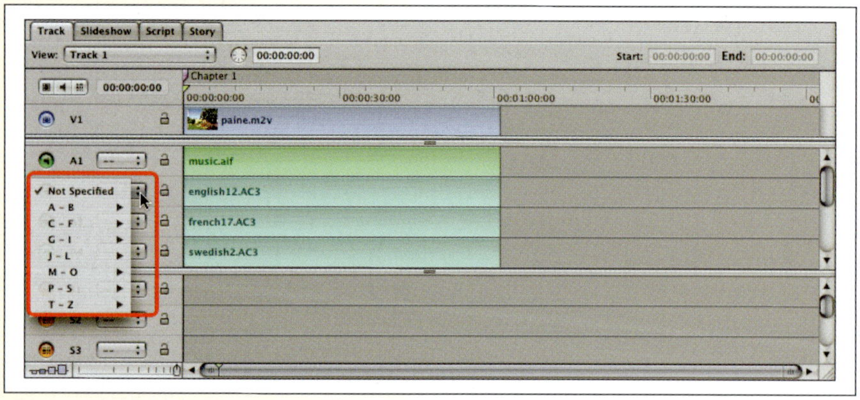

Figure 5-39. Setting the language code for an audio track

Go ahead and do it again for any other audio tracks. You might end up with a track list similar to that shown in Figure 5-38.

3. **Assign Languages to the Audio Tracks.** The big step you have to take now is to assign a language where appropriate. Each audio track has a language prefix pop-up menu to the left of the track. For instrumental tracks, you can use the default of Not Specified (see Figure 5-39).

   For most tracks, however, you'll want to follow the alphabetized, cascading menus and pick a language. Figure 5-40 shows the end result of our quick multilingual demo.

   Note that you also can adjust this value in the Property Inspector for the audio track. Just click the track you want to adjust and open the Inspector (Shift-⌘I). Figure 5-41 highlights the language selection pop-up.

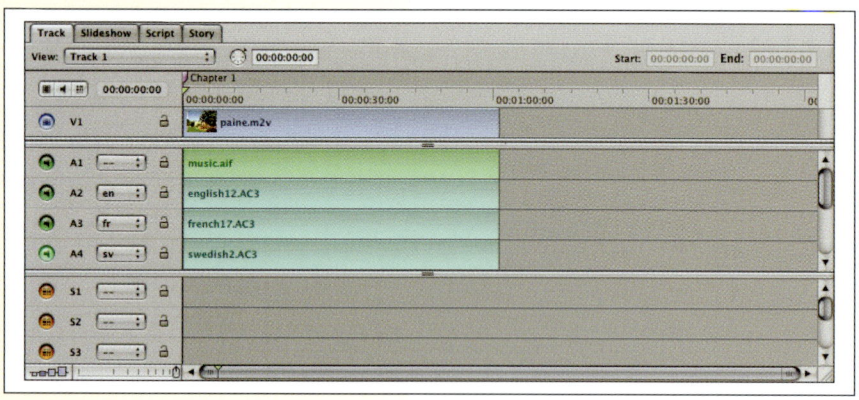

Figure 5-40. The language codes for all three speaking tracks

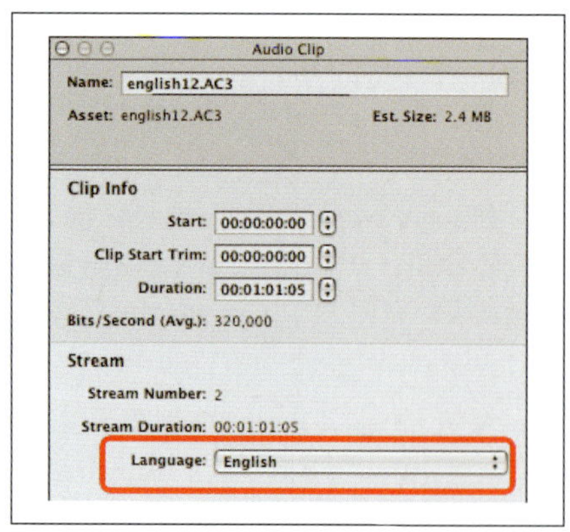

Figure 5-41. The Language property in the Audio Clip Inspector

A Multilingual Audio Project

So, let's set up the Paine Art Center project with real, live audio clips for English, French, and Swedish. We'll start with the previous project and build on the work we did there to keep the multilingual menus in place.

A multilingual demonstration

The first thing to do is throw out the music-only track. (You could keep it around, but that just adds more mess to the menus. You want less mess for right now!) Go ahead and import the following language files from the *projects/ch05* folder:

- *english12.AC3*
- *french17.AC3*
- *swedish2.AC3*

Then perform the following steps:

1. **Prepare the Audio Clips on the Track.** Drag those clips to the audio track section of the Track editor. Assign the appropriate languages to the tracks as well. You should end up with a screen similar to the one shown in Figure 5-42.

   You can simulate the project right now, even without hooking up the menu options. Figure 5-43 shows the Simulator running with the French language selected. You can "manually" switch the audio language using this pop-up. This has the same effect as using the audio button on consumer remote controls.

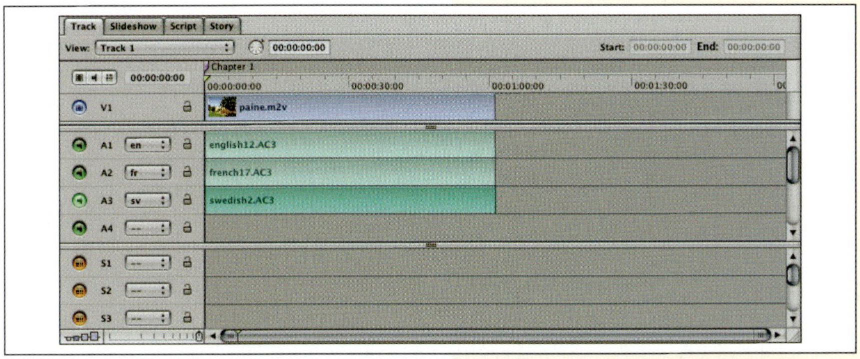

Figure 5-42. The Paine clip with English, French, and Swedish audio tracks

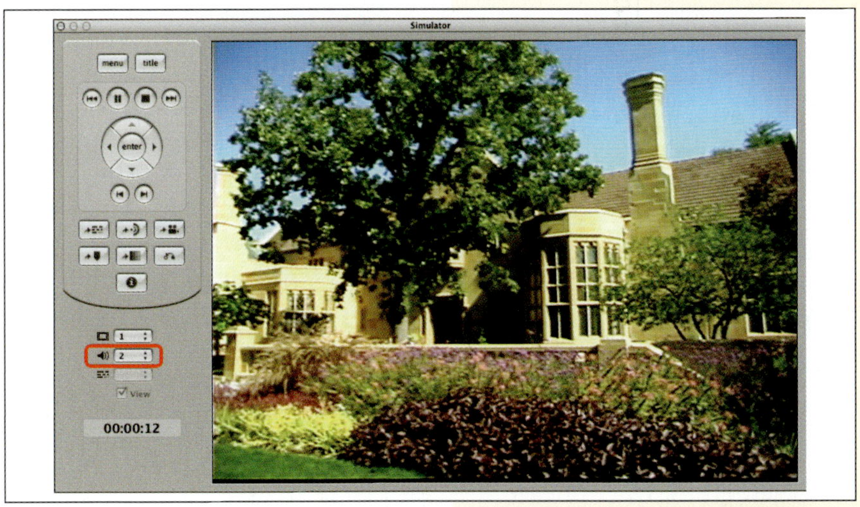

Figure 5-43. Choosing the audio track in the Simulator

> **NOTE**
>
> *If your player has a preferred audio language and a matching language is available on the DVD, that language will be chosen and used unless the user manually changes the audio track.*
>
> *If no matching language is found or the player doesn't have a preferred language setup, the first audio track (A1) will be used. Keep that in mind when dropping your audio clips into the editor. After the first language, no special properties exist for subsequent tracks.*

2. **Build Menus to Support the Tracks.** Getting the audio tracks into the project is the easy part. Getting your menus built to support them is more interesting. Look at Figure 5-44 and recall that our Languages menu screen had a column for spoken languages.

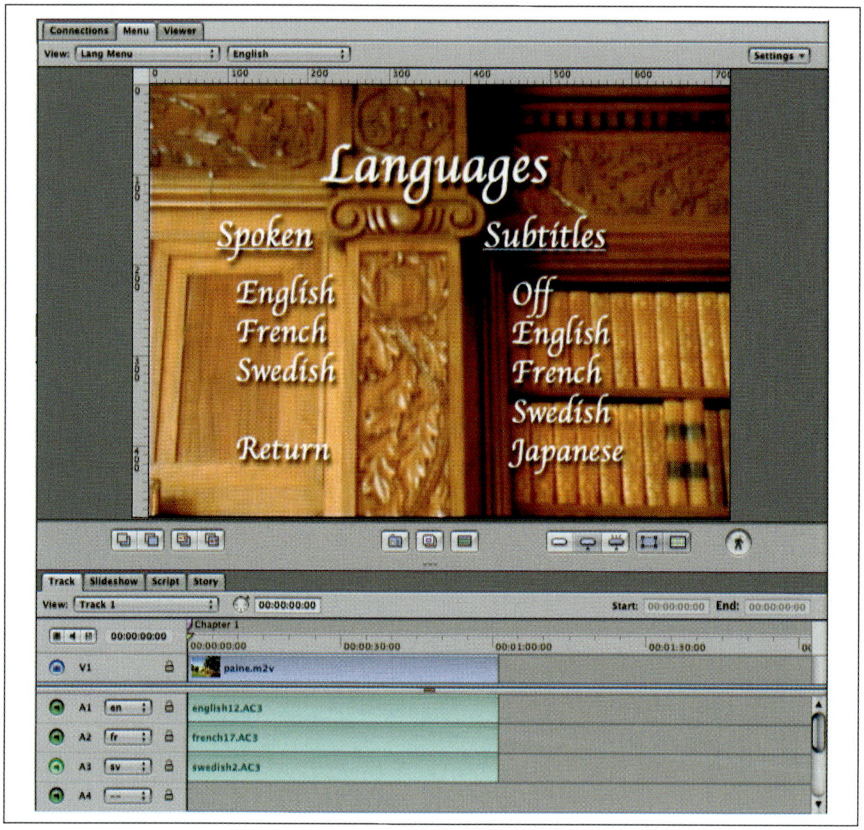

Figure 5-44. Back to our Languages menu to set up the audio buttons

You want to set the language buttons to switch to the selected audio track. Depending on your specific project, you can have the buttons go ahead and jump to the subtitle buttons, go back to the main menu, or just leave the selected language alone. Without setting up a more complex menu system, there's no easy way to indicate that you selected a particular language. The typical trick for "confirming" the fact the user pressed something is to switch to another button on the same page.

For this project, let's go ahead and define all the audio buttons. We'll have the audio buttons jump to the Return button. (We'll worry about the subtitle buttons in the next section.) Your menu should end up looking something like that shown in Figure 5-45.

For the button targets, then, you need to do two things. First, set the target on the button, the same way you did previously. The actual target

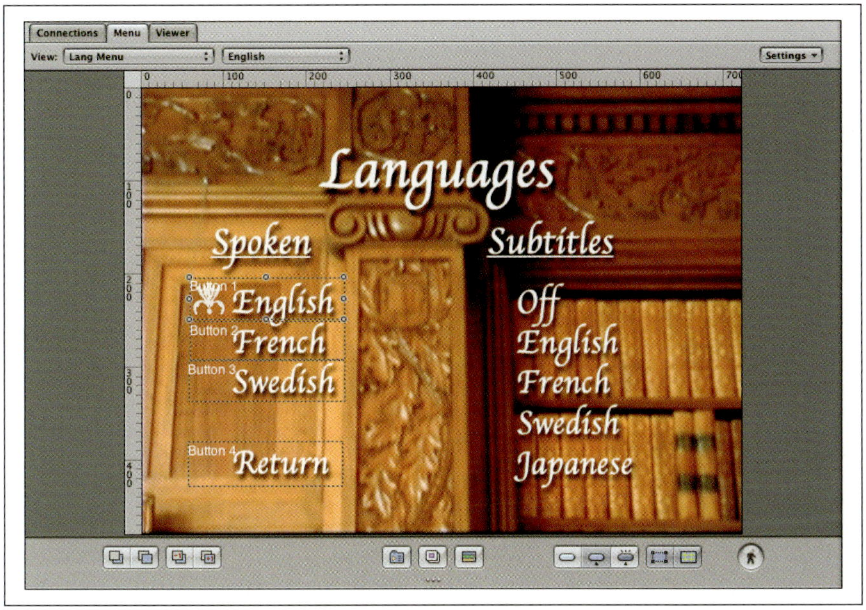

Figure 5-45. Our Languages menu with buttons for the spoken language options

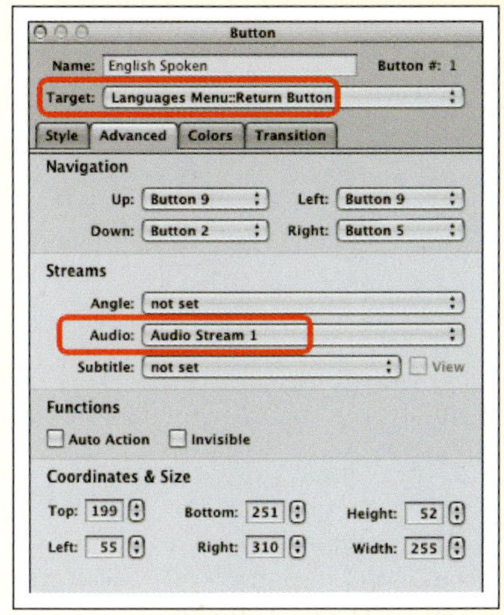

Figure 5-46. Setting a button to alter the audio track

will be the Language menu's Return button. (You might have labeled your menu and buttons differently.) The second step is to set the Audio property on the Advanced tab so that it forcibly uses Audio Stream 1 (our English audio track). Figure 5-46 highlights those properties for the English Spoken button.

You should be able to simulate your project now and use the Languages menu to select the spoken language before you begin playing the video clip. Once the clip starts playing, you'll still be able to flip through the different audio tracks with the remote control buttons.

Remote control support

We've mentioned the remote control a few times. For consumer players, supporting the remote control is essential. Fortunately, the default behavior in DVDSP for remote control operations is pretty smart. Or rather, DVDSP stays out of your way and most consumer remote controls have good default behaviors.

You can alter the behavior, though, if that's appropriate for your project. For example, you might want the Audio button on the remote control to take you to the fancy menu for selecting languages that you spent all this time configuring. Just open the Inspector for the Disc (you might need to select the Disc item in the Outline view first) and check out Figure 5-47.

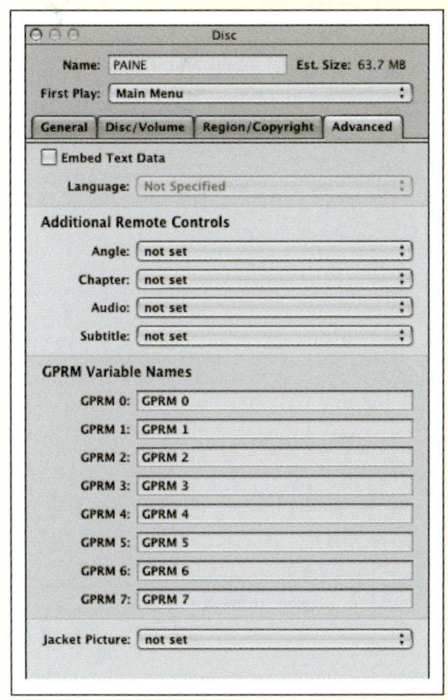

Figure 5-47. The Advanced properties on the Disc Inspector

You can change the Audio button's target here if needed. Most of the time you'll leave these properties in their default "not set" states.

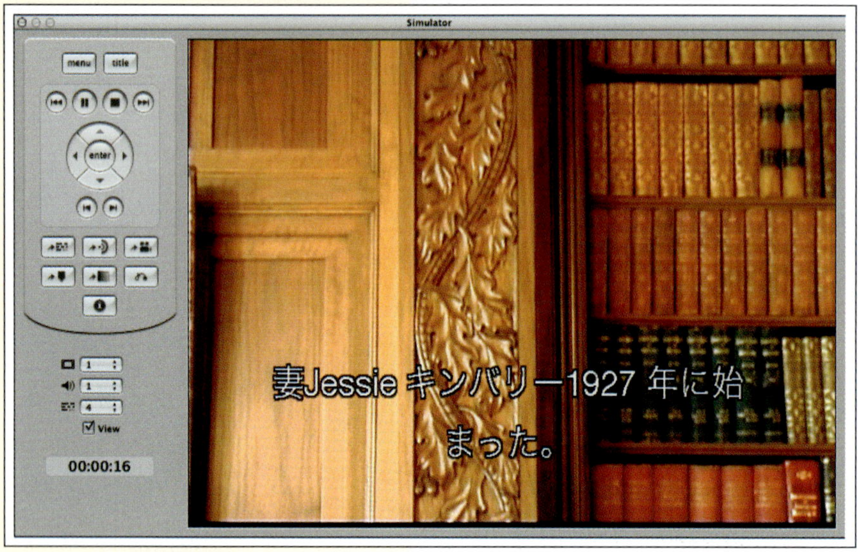

Figure 5-48. Japanese subtitles in action

Adding Subtitles to Video Tracks

Even without the intention of supporting multiple languages, you might end up subtitling some projects for use in bars, airports, or other loud, public spaces, or for assisting the hearing-impaired. We'll tackle getting several subtitle tracks onto our newly narrated video clip. Just to prove it works, Figure 5-48 shows our clips with Japanese subtitles active.

Subtitles can go a long way toward internationalizing your video. Even without alternate soundtracks, you can have multiple subtitle tracks and expose your work to a wider audience. Plus, DVDSP 2 made some great leaps forward in helping you create good-looking subtitles quickly. All the same great features are in DVDSP 3, too.

> **NOTE**
>
> *We need to start the subtitles section by asking forgiveness from anyone who actually speaks French, Japanese, or Swedish. Our translation team is, well, a bunch of computers on the Internet. We tried cleaning things up a bit, but really, we are under no illusions that these are real translations. They illustrate the point at hand, so at least we have that.*
>
> *If you're really doing multilingual projects, you no doubt have real translators. DVDSP supports a variety of subtitle document formats that make it easy to get help from outside sources. We'll go through an example of such a format in this chapter.*

To add subtitles to your DVD, you have two options:

- Do it all in DVDSP.
- Create an importable file in some other application.

We'll look at both approaches. After all, we're going to write subtitles for four languages!

First, let's tackle English using DVDSP directly. This is amazingly simple and only slightly tedious. (Well, slightly tedious is like saying Mount Everest is only slightly tall, but hey, we're trying to put a positive spin on this!) The good news about doing this in DVDSP is that you can get away with little to no planning.* You just jump in and do it.

For our other languages (French, Swedish, and Japanese), we'll go the importable file route. DVDSP supports four different subtitle document formats:

SCR
: Daiken-Comtec Laboratories Scenarist format

SON
: Sonic Solutions format (bitmap-based)

* OK, OK. You never can get away with absolutely no planning. If you're doing a serious subtitling job, you definitely have to plan. What we mean to say is that you can get the actual subtitles into your project right away. The content and timing of those subtiles certainly should be planned—extensively, if possible.

STL
: Spruce Technologies format

TXT
: Plain text files (that still must be formatted as STL)

We'll use the simple but flexible STL format and prepare our subtitles in the TextEdit editor.

Adding subtitle tracks with DVDSP

You've probably noticed the subtitle tracks in DVDSP sitting below the audio tracks (see Figure 5-49). They've been there all along; we just haven't used them—yet. Let's do that now.

Adding subtitle tracks

Here are the steps for adding subtitle tracks to your project:

1. **Add a Generic Subtitle to a Subtitle Track.** If you're going to use DVDSP directly for creating your subtitles, all you have to do is double-click or Control-click in a subtitle track and insert a subtitle. Figure 5-50 shows an example of the menu you get from the Control-click method.

   Notice that you have a few options. You can insert a subtitle at the playhead or wherever it fits next. You also have an import option, but that's for later. Go ahead and choose the Add Subtitle option. You should see a nice yellow box similar to that shown in Figure 5-51.

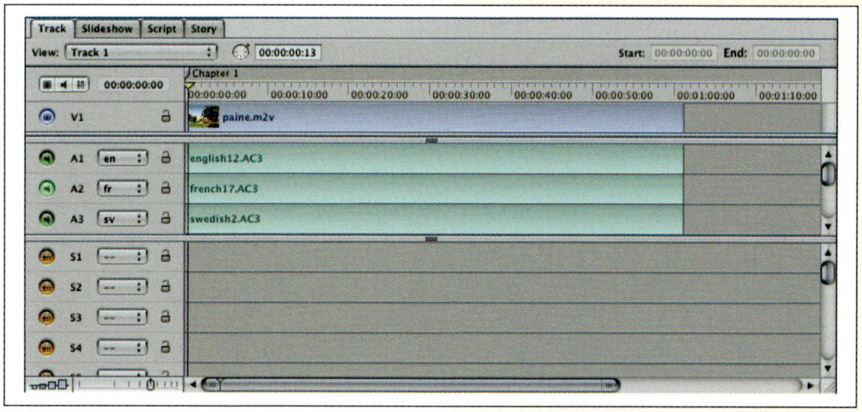

Figure 5-49. The until-recently-ignored Subtitle tracks in the Track tab

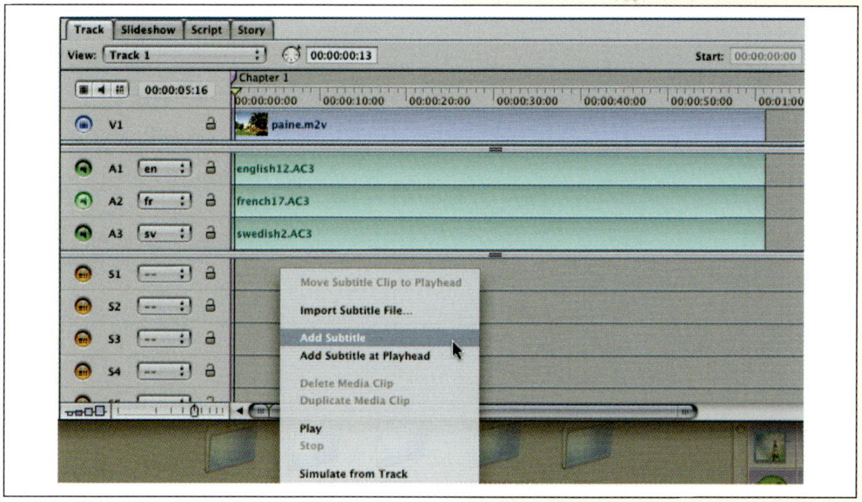

Figure 5-50. The Control-click menu for adding subtitles

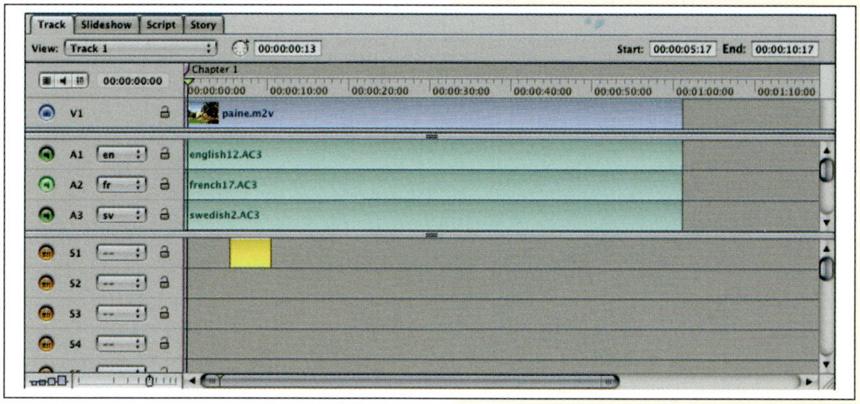

Figure 5-51. Our first (empty) subtitle

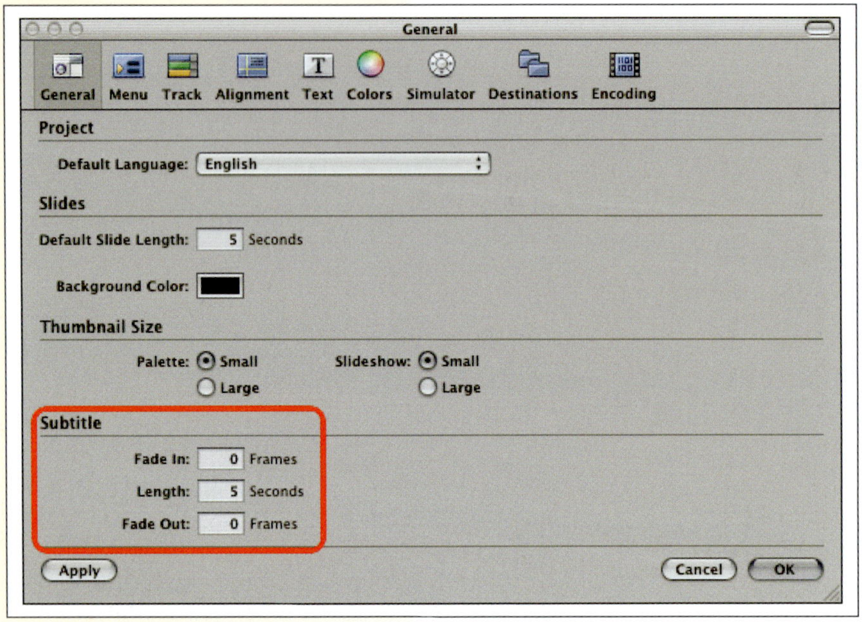

Figure 5-52. Subtitle preferences in DVDSP

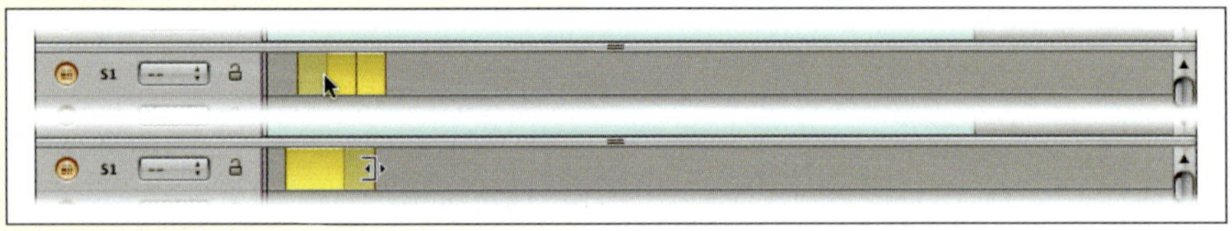

Figure 5-53. Relocating a title (top) and adjusting the duration (bottom)

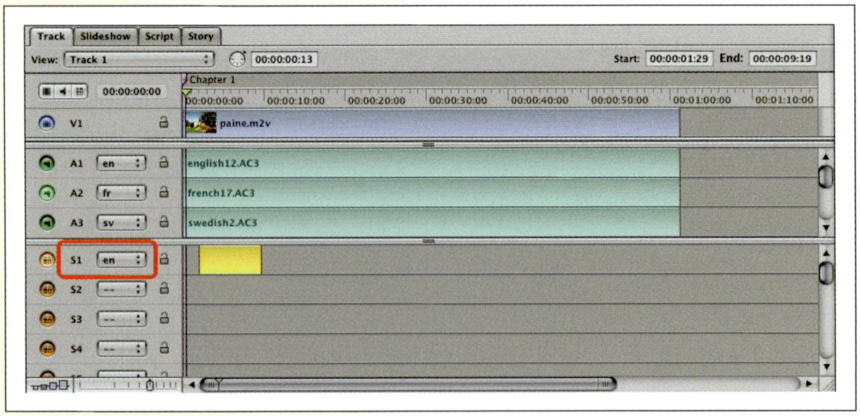

Figure 5-54. Setting the language code for subtitles on the track

2. **Set the Subtitle Position and Duration.** The default duration is five seconds, but that's something you can control in the DVDSP preferences, as shown in Figure 5-52. The Text and Color portions of the preferences also enable you to specify the default font and color for subtitles. But don't worry too much about the duration of the subtitle; you'll probably be adjusting that to fit the actual audio.

   Fortunately, DVDSP makes moving and adjusting subtitles as easy as drag-and-drop. Just drag any subtitle to its new location by clicking the middle of it. Clicking near the edges enables you to "resize" the subtitle, affecting its duration. Figure 5-53 shows examples of such adjustments.

3. **Set the Subtitle Properties.** Beyond the location and duration of subtitles, you'll want to be aware of several other properties. First and foremost is the language setting. You can choose the language this subtitle track represents in two places. Figure 5-54 shows the easy one, to the left of the track itself. Figure 5-55 shows the other place you'd expect to access such a capability: the Property Inspector.

   The Inspector shown in Figure 5-55 also has several other properties worth looking at. You can control the start and stop times, the duration, the fade in/out effects, the horizontal and vertical position and orientation, and the colors.

4. **Enter the Subtitle Text.** Once you have the little yellow box, how do you enter text? Easy: double-click the yellow box in question, and start typing. (You can do the same via the Inspector.) You should see your subtitle text appear as you type. Figure 5-56 shows an example.

   Once you have typed in the text, you'll probably want to verify the placement and duration against the video. Just click the timeline a few seconds prior to the start of your subtitle clip and press the spacebar. That should play the video and show the newly added subtitle clip. Adjust the placement and duration to fit the video.

When the first clip is done, just repeat the same process for the rest of the clips. Again, this is not the speediest way to go, but it is easy and you get to see immediate results. Be sure to save your work often, just to make sure you don't lose any subtitles in case something goes wrong. You might be

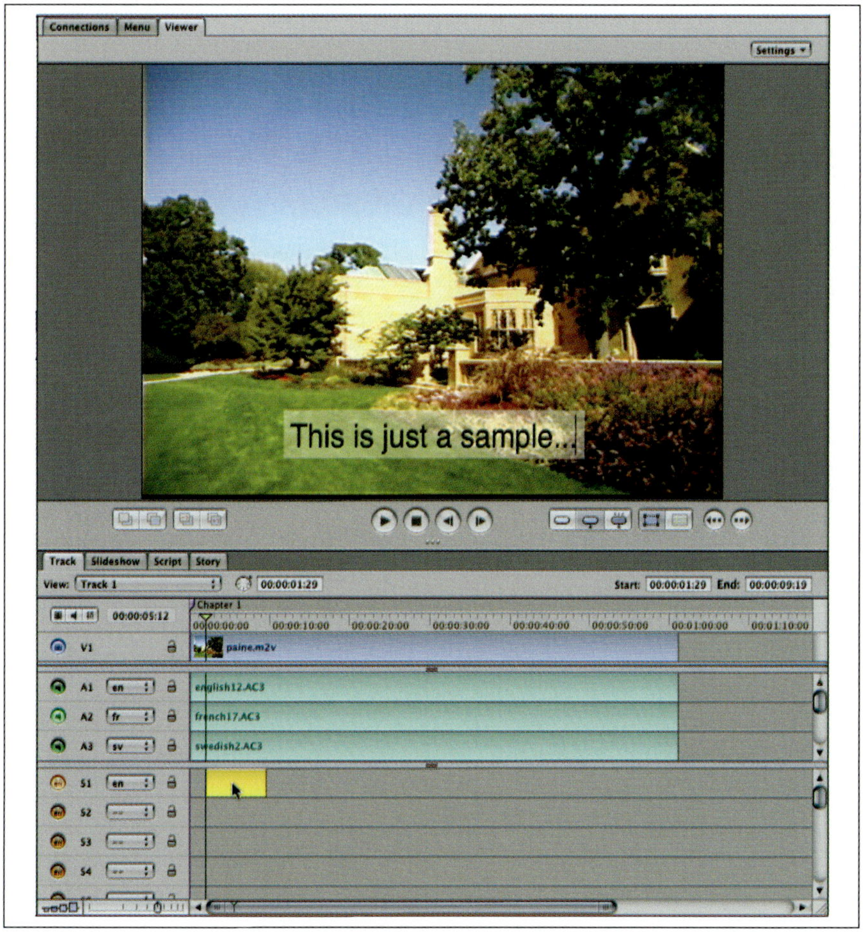

Figure 5-56. Typing the contents of our first subtitle

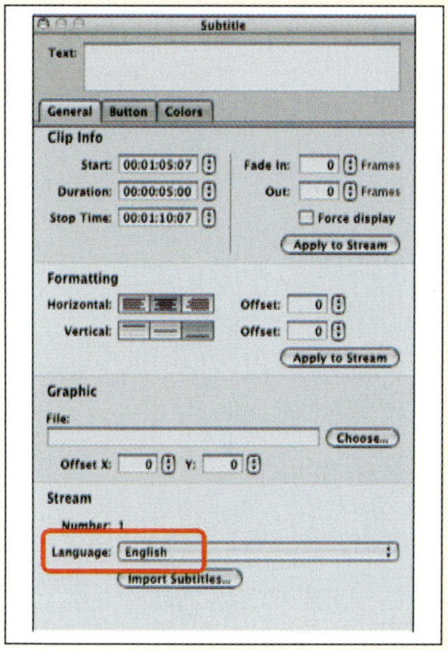

Figure 5-55. Using the Inspector to change the language code for a particular subtitle track

Arbitrary Subtitle Placement

Timecode, for those of you who have never seen it, is an interesting experiment in filming whereby four single-take shots are shown simultaneously. Each shot takes up its own quadrant of the screen. The director brings certain conversations to the foreground with volume control, but you are free to pay attention to any one shot if you so desire.

Although it requires some manual intervention on your part, you still can subtitle such a film. You set the vertical alignment to the top of the frame and set the offset value so that the first line of subtitles falls in the middle of the screen. You place the subtitles for the first two shots by left-justifying the text and manually inserting spaces between the two subtitles. Then, just add a bunch of blank lines (by pressing the Return key) to get you to the bottom of the screen. There, you can add text to the third and forth shot, just as you did for the first two.

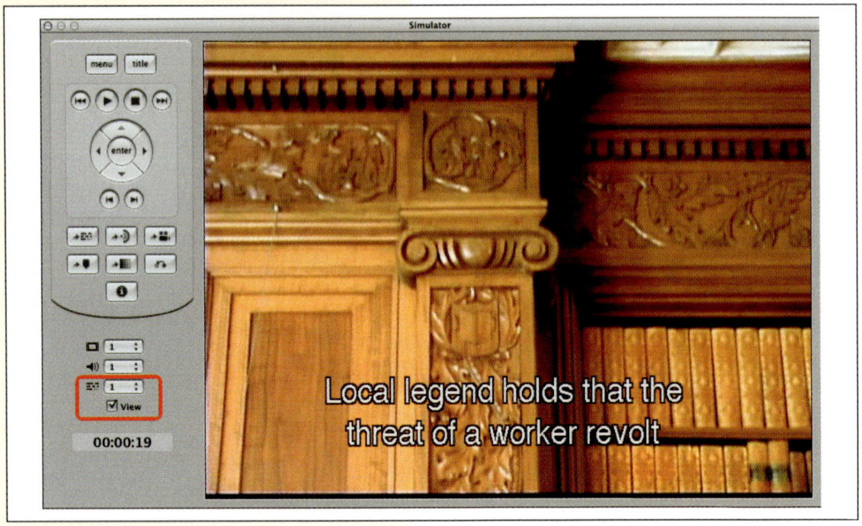

Figure 5-57. Using the Simulator to view our subtitles

surprised at how frustrating it can be to re-create a batch of titles you just finished. (Or maybe you wouldn't be surprised, in which case we empathize greatly.)

When you're done, save your work and simulate it. You should be able to turn the subtitles on and off in the Simulator. When they're on, you should see them come and go just as you placed them. Figure 5-57 shows our first project running in the Simulator.

Multiple subtitle tracks

Not much magic is involved in adding multiple tracks. Just pick the next track (up to 32 tracks), set its language, and go. Although this is an easy process, performing it manually for a feature-length movie would get pretty boring. Plus, you'd have to learn how

STUDIO ADVICE

Formatting Subtitles

Here is the dialog that DVD Studio Pro 3 uses to format subtitles. You've got the standard text editing options available in the Subtitle window. You can choose your font face, size, and color. You can set the text alignment (centered is most common, and also the default). You also can set a few visual attributes on the font:

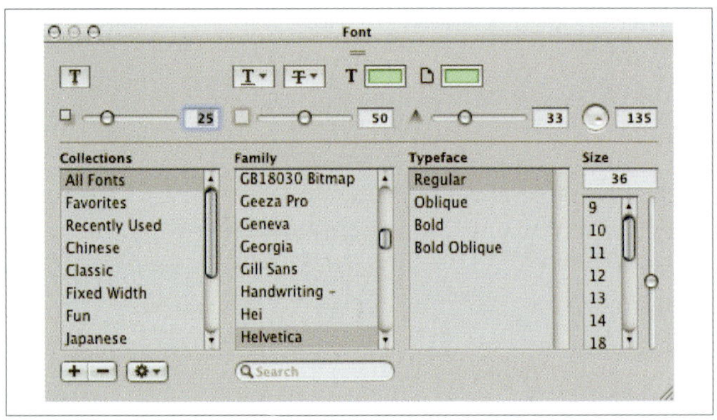

Be careful about using colors. Although you can use multiple colors (for different subtitles or even within the same subtitle), keep your variations to a minimum. Most consumer DVD players have a limited capacity for handling subtitle colors. Exceeding that capacity probably will mess up your subtitles in unpredictable ways.

to type all the characters that show up in French, Spanish, Swedish, etc., and those are just the Western languages. Typing CJK (Chinese, Japanese, Korean) characters on a standard Western keyboard would prove trying to the most patient of souls. That's where subtitle imports come in handy.

The Spruce Technologies format

This format is straightforward and enables you to control most of the interesting bits of formatting for your subtitles. You can set things such as the font family and color as well as provide the actual content of the subtitles (including the location and duration of each subtitle clip).

The basic idea is to provide a few format commands and then provide the subtitle clips. If we were to write as an STL file the English subtitles we built in the previous project, the file might look something like this:

```
// We could set font stuff here, but we're just going to rely
// on the defaults
// First paragraph
00:00:00:13,00:00:04:24,The Paine Art Center and Gardens is located in Oshkosh
00:00:05:00,00:00:08:01,and is considered one of America's castles.
00:00:08:07,00:00:12:18,Construction was started in 1927 for Nathan Paine
00:00:12:19,00:00:17:07,and his wife Jessie Kimberly (of Kimberly-Clark fame).
00:00:17:12,00:00:21:11,Local legend holds that the threat of a worker revolt
00:00:21:17,00:00:26:02,kept the Paine family from taking up residence.
00:00:26:18,00:00:30:10,It was eventually opened to the public in 1948.
// Second paragraph
00:00:38:11,00:00:43:11,The Paine offers tours of the interior and gardens.
00:00:43:17,00:00:46:17,The well maintained gardens include a Reception Garden,
00:00:46:18,00:00:50:07,a Contemporary Garden, a Shade Garden and a Rose Garden.
00:00:50;14,00:00:55:14,The Rose Garden rounds out the Tudor construction
00:00:55:23,00:01:00:23,and invites guests to spend time outside as well.
```

Timing in import files

If you're going to have multiple languages translated and bundled into an STL file, you should figure out the timing on the English subtitles and use that as a basis for the translated tracks.

If you're starting from scratch, just pick an interval and common duration for each clip. Use the normal import facility and then just move and adjust each clip to fit the video. Quite frankly, if we had the first subtitle project to do over again, that's how we'd do it. You can type a lot faster in a typical word processor. Plus, you can use the spellchecker to avoid any embarrassing typos that clients inevitably notice.

You can include an empty subtitle track, but most players these days have an "off" option for all subtitles, thus rendering the empty track redundant.

SMPTE Timecode

The time markers you see in the STL format show up all over when dealing with audio and video. They follow the Society of Motion Picture and Television Engineers (SMPTE) format of hh:mm:ss:ff. The hours (hh), minutes (mm), and seconds (ss) parts are straightforward. The frame (ff) part is more interesting.

First off, you have to know what medium you're dealing with. NTSC uses 29.97 frames per second. PAL uses 25fps, and film is 24fps. Although you won't be dealing with the film rates for output to DVD, you might need to convert incoming timecode entries if they originally were taken from film.

The other thing to note is that unlike markers, you are not restricted to particular frames for subtitles. The upshot is that you can place subtitles wherever you want, and their duration can be frame-accurate as well.

TIP

Accented Characters in Mac OS X

If you're responsible for typing your own translations (lucky you!), you can produce most of the accented characters that show up in Western (European) languages directly in OS X. Most times you can accomplish this with a key sequence. For example, you can produce the é character by typing Option-e and then e. For á, you'd type Option-e followed by a. Here's a table of other common entries:

Character	Key Sequence	Character	Key Sequence
ü	Option-u u	ñ	Option-n n
à	Option-` a	ê	Option-i e
å	Option-a	ç	Option-c

Of course, on a Mac, you can also pop open the Character Palette and select from, well, everything your fonts have to offer:

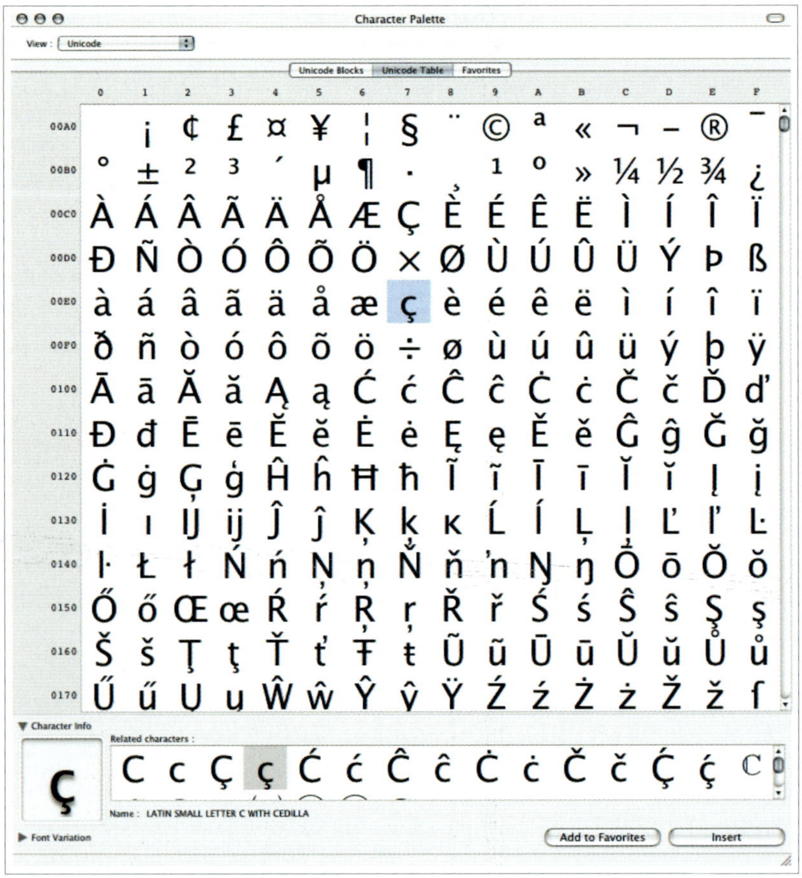

Importing subtitles

Our timing in this project isn't as strict as it would be for actual dialog, so we just used the import-with-interval approach. Go ahead and Control-click subtitle track two and import the *commentary_fr.txt* file. You should see results similar to those shown in Figure 5-58.

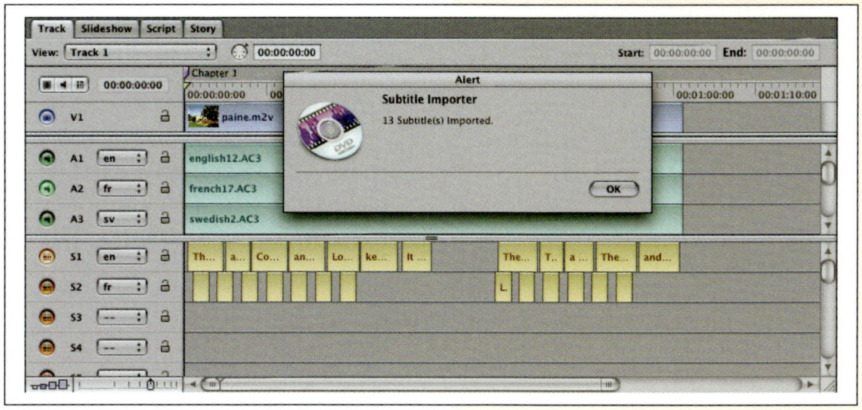

Figure 5-58. Importing an STL file for subtitles

Set the language for the track to French (fr). Now repeat for Swedish (*commentary_sv.txt*) and Japanese (*commentary_ja.txt*). We adjusted the clips based more or less on the spoken languages and how much text there was to read. You should end up with a set of tracks that look something like that shown in Figure 5-59.

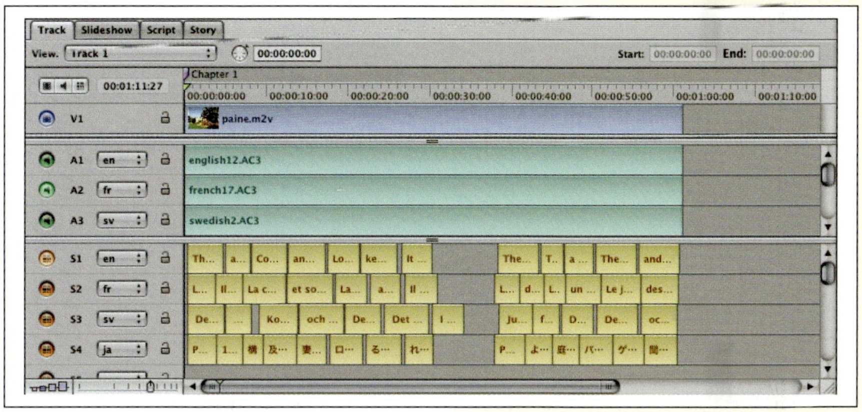

Figure 5-59. All four subtitle languages imported and sized

When you're done importing the tracks, run your project through the Simulator. You should be able to bounce around between the various subtitle clips. Figure 5-60 shows the Swedish and Japanese subtitles.

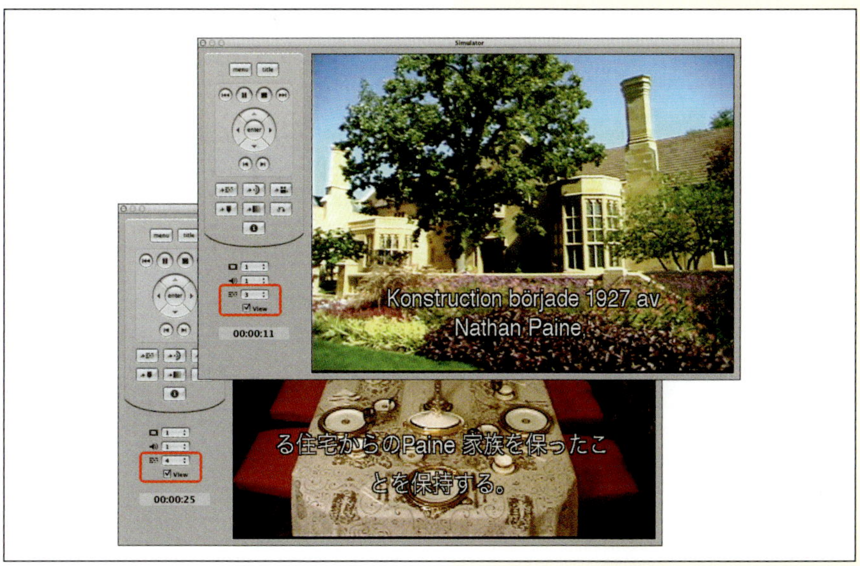

Figure 5-60. Swedish and Japanese subtitles in action

Selecting alternate subtitle tracks from menus

Our last task for the moment is to make the various subtitle and audio tracks available to users for configuration. Users might want to hear the French audio track but see English subtitles. You can accommodate that easily enough in DVDSP. You just have to make a menu with the right buttons and tell them what to change.

Our previous project went through the steps for attaching buttons to particular audio tracks. Subtitle tracks work in much the same way. One of the other track options in a

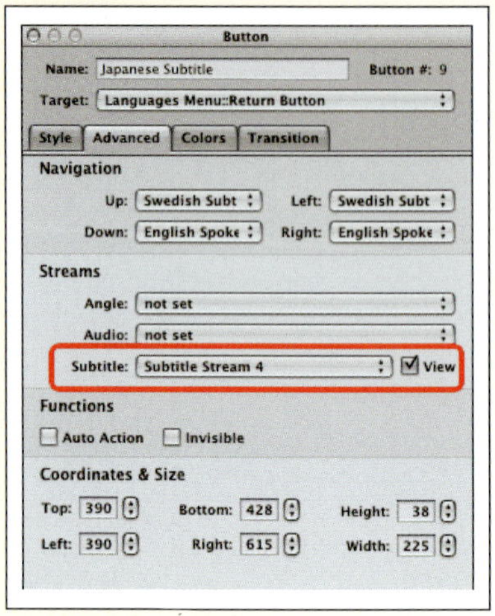

Figure 5-61. Configuring a button to change the subtitle stream

button's Property Inspector is the Subtitle track. Just set that to show the subtitle you want, as we do in Figure 5-61.

Notice the View checkbox on the right. To make the subtitle visible, you have to check that checkbox. Unchecking it enables you to shut off the subtitles. Our Off button selects Subtitle Stream 1 but unchecks the View checkbox. You could switch to an empty subtitle track or just rely on the player's "no subtitles" option, but unchecking the View checkbox is a somewhat cleaner approach.

We'll also handle the destination for the subtitle buttons in the same way we handled the audio buttons. We'll alter the audio buttons so that after selecting an audio track, the cursor jumps to the Off subtitle option. After the user selects a subtitle track, the cursor will jump to the Return button. Figure 5-62 traces the flow for our audio and subtitle buttons.

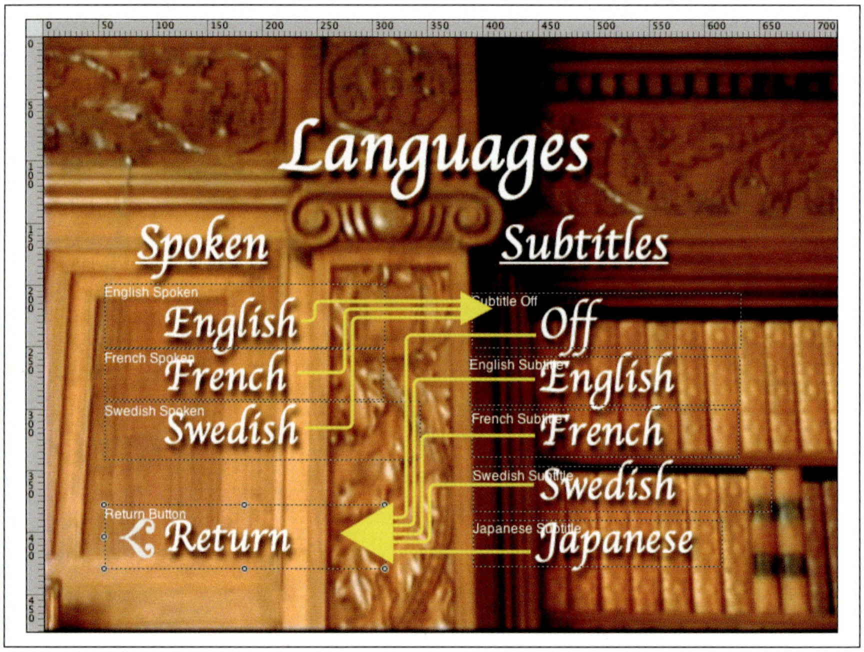

Figure 5-62. The flow of our audio and subtitle buttons

Launch your project in the Simulator. You should be able to set the language and subtitle tracks in the Languages menu. Then when you play the clip, you should see (and hear!) your choices.

Remember that you can use the audio and subtitle buttons on your remote control to switch tracks dynamically. That can be a quick way to check that all the options got placed on the disc properly. You still need to verify that all your configuration buttons work, but you don't have to watch the entire video in every language. That's what your Quality Assurance department is for, right?

Unicode Characters

If your subtitles include CJK or other-bit characters, you need to make sure you save your text file with the proper encoding. In TextEdit, you make that decision in the Save Dialog, as shown here. The easiest choice is UTF-16. This should keep everything intact as you import the subtitles into DVDSP.

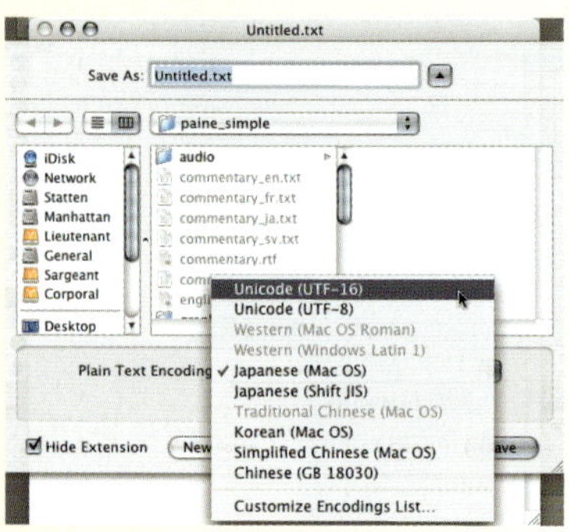

The Director's Cut

6

So now that you have your home movies on a DVD, what do you do? Well, you start adding bonus features, of course! This chapter recaps some of the common bonus features, such as multiple audio tracks and deleted scenes. It also goes into depth on how to create *stories*, which allow you to program alternate endings or PG-13 edited content right into your main video clip.

In this chapter

DVD Video Features

Multiangle Video

Multistory Movies

DVD Video Features

You've seen the powerful tools available for presenting audio and text in DVD projects. Let's turn our attention to the video side of the house. You have even more options in this area.

We'll be looking at two primary video features: multiangle video and multistory video. Multiangle video is commonly touted as "bonus" video on directors' cut discs. This feature enables you to provide multiple views of a scene. You can have rough cuts, greenscreen backgrounds, animatronics, or just a different angle stitched up together to enable the viewer to pick a version as the disc plays.

Multistory video is subtler, but arguably more powerful. This feature enables you to build different versions of a film and arrange the various parts in such a way that each version plays seamlessly from beginning to end. A simple example of this involves presenting an alternate ending to a movie. But you don't have to stop there! You can weave together an entirely separate director's cut and offer it on the same disc as the theatrical release. Plus, you can add a deleted scene and make it look as though it was planned all along.

Notes on what you already know

In earlier chapters you saw several features you often find on special-edition discs. For instance, you can present screenplays and production photographs as a slideshow (as we discussed in Chapter 2). An audio commentary by the director or other participants is simply the same thing as an alternate language (as you learned in Chapter 5). You can present deleted scenes and bloopers as regular clips from a menu (and you learned how to present regular clips in Chapter 3).

Before we get into new video features in this chapter, we want to go over a few things you should consider concerning the features we've already discussed. None of these is a requirement, mind you; all we're providing are recommendations and considerations.

Audio commentary tracks

One of the nifty side effects of supporting multiple languages is that you can do anything you want with the audio tracks. You can provide several tracks of a particular language, or even a track without any language at all. For example, the special-edition release of *Legend* comes with an isolated score by Tangerine Dream. This is the same technique you'll use to add an audio commentary track.

If you're planning to add such a track, you need to keep a few things in mind. First, unless you have the entire commentary track scripted, you should consider using the main audio track as a lower-volume background. You can bring up the volume of the main track during pauses in the commentary (this is known as *ducking*).

Another thing to remember is that you don't have to use the same quality for all your audio tracks. If you have several languages plus an audio commentary (or several audio commentaries), you can encode the commentary tracks at much lower bit rates. Speech requires much less bandwidth than a full score with dialog.

Scripts as slideshows

Several discs include early copies of a movie's script for your reading pleasure. We don't have anything too exciting to point out here; you just present each page as a slide. You should enable the user to choose when to go to the next page, as reading speeds vary widely. You can pause individual slides using the Pause column (see Figure 6-1). Or you can select a group of slides and click the Manual Advance option in the lower-right corner (also shown in Figure 6-1; we shuffled the column a bit to the left so that you can see both the column and the Settings drop-down menu).

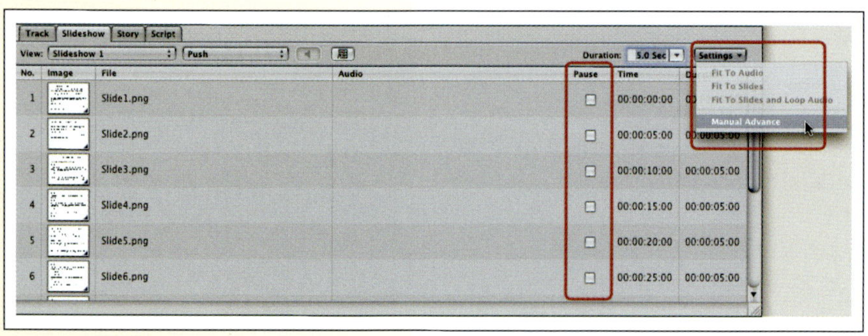

Figure 6-1. Two ways of setting the manual option on your slideshow

The important thing to consider for these types of pages is the font (typeface and size) that you use for presenting the pages. TV screens have a much lower resolution than DVDs can store. You should test various fonts to see which ones work best.

To that end, a sample project with script pages in various fonts and sizes is in the *projects/ch06* folder. Notice what happens with the fancy serif fonts. At small sizes the font's decorative details don't show up well on the screen. You'll often see a shimmer around thin lines, which can make the text very hard to read (see Figure 6-2).

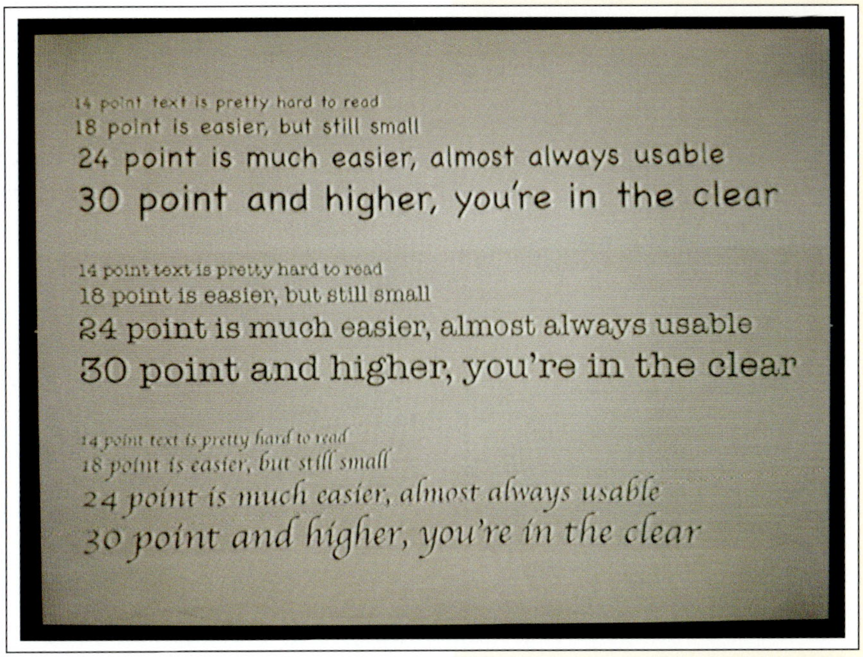

Figure 6-2. A test of readable fonts

Exporting slides

Many applications have a nifty "export for web" option that enables you to request a series of TIFF, JPEG, or GIF files. DVDSP understands all those formats. It even understands Adobe PDF files, but it will show you only the first page of a multipage document.

If the application you are using for the script or other text you want to present doesn't support a graphic output option, fret not. You're in for a bit of tedium, but you still can export. In Mac OS X, you can save everything as a PDF. Just pretend you're going to print your document. You should see a Save As PDF option at the bottom similar to that shown in Figure 6-3.

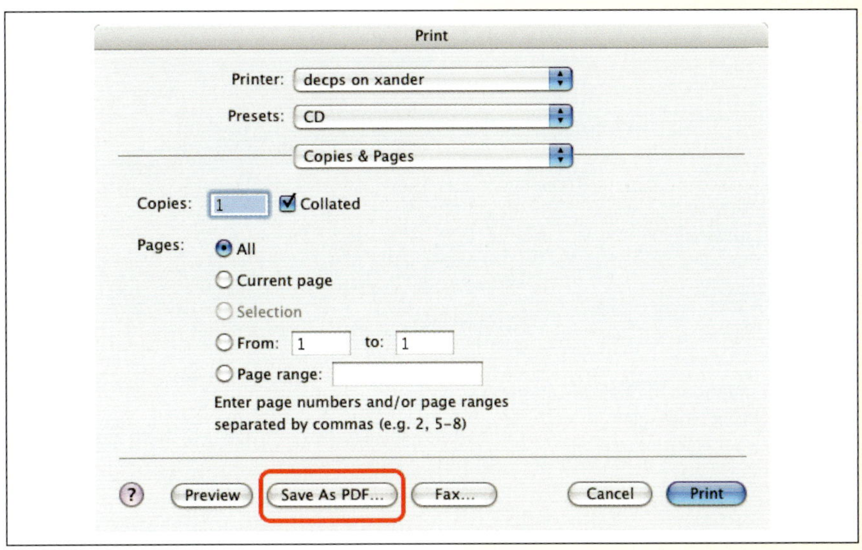

Figure 6-3. Saving an arbitrary document to the PDF format

Now comes the tedious part. The Mac Preview application can save each page as a JPEG or PNG, or whatever you prefer. But it will save only one page at a time. So, you have to save each page as you go. Save the pages into a common folder and use sequential names such as *page001.jpg*, *page002.jpg*, *page003.jpg*, or something similar. Of course, if you do this regularly, you should check out some of the commercial

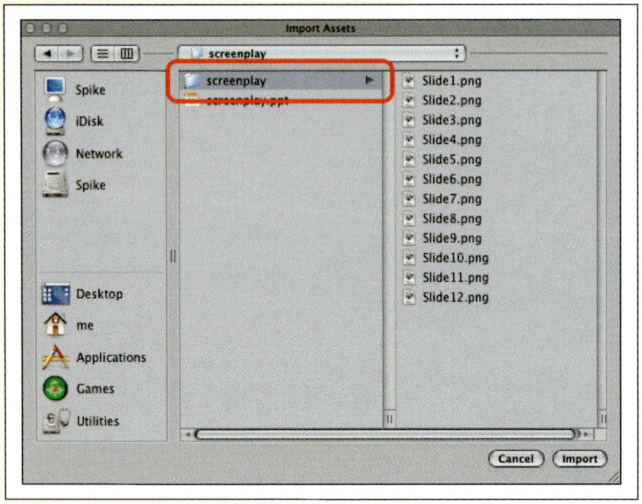

Figure 6-4. Importing an entire folder into your Assets tab

PDF tools such as Acrobat or PDFPen to automate this for you! Visit *www.planetpdf.com* for a big list of PDF tools.

One nice thing for such additional material is that DVDSP knows what folders are! You can import an entire folder of pages in one shot. Figure 6-4 shows how to select an entire folder when importing assets. You select the folder, not one of its contained files. The folder and it constituents will appear in your Assets tab.

You even can drag the entire folder from your Assets tab to the Slideshow window (see Figure 6-5). It keeps the pages in alphabetical order and applies the default duration to each slide.

You should be left with a complete slideshow. At this point you can reorder things as you need (for example, our slides needed a bit of adjustment to account for slides 10–12 appearing before slide 2). Then you can select them all (as shown in Figure 6-6) and turn on the Manual Advance option.

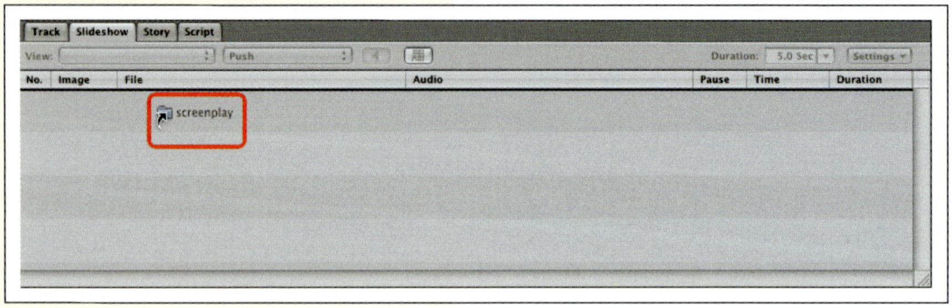

Figure 6-5. Adding an entire folder to a slideshow

Note also in Figure 6-6 that the duration of each slide still applies, even if you select the Manual Advance option. The pause simply will kick in after the slide has played. You can, of course, reduce the duration of the slides if you want to give the user more control.

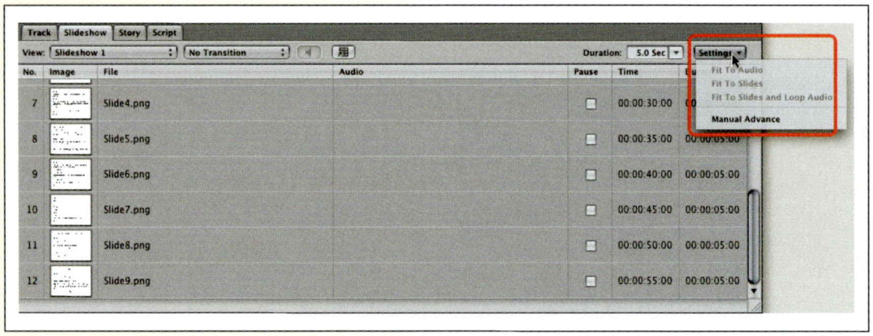

Figure 6-6. Adjusting the Manual Advance on all slides in our script show

Deleted or extended scenes

Adding deleted scenes to a movie disc is straightforward. You just add them to a menu as you would normally. You'll probably want to set their End Jump properties to lead back to the deleted scenes menu and select the next scene in the list.

One thing you see on many discs is an option to "play all" the deleted scenes. You can accomplish this behavior with scripts—which we'll cover in the next chapter.

Another fun thing that many discs are including these days is commentary on the deleted scenes. Again, this is no different from adding multiple languages to a main movie. You just need to make sure you provide buttons to activate the commentary version of the audio. The decision to listen to commentary on deleted scenes is usually spontaneous, so you don't want to make the user go to a separate menu to activate the commentary option and then come back to the deleted scenes menu.

The menus for such deleted scenes often look like the menu shown in Figure 6-7, which is from the independent film *Remaining Class*.

Notice the two "play" buttons for each clip. Pressing either button will play the deleted scene, but each button picks a specific audio track before starting the clip. Figure 6-8 shows an example of the properties for the two play buttons for the first clip.

Be sure you specify the audio track on every button. Otherwise you run the risk of the user bouncing around and getting the wrong audio track.

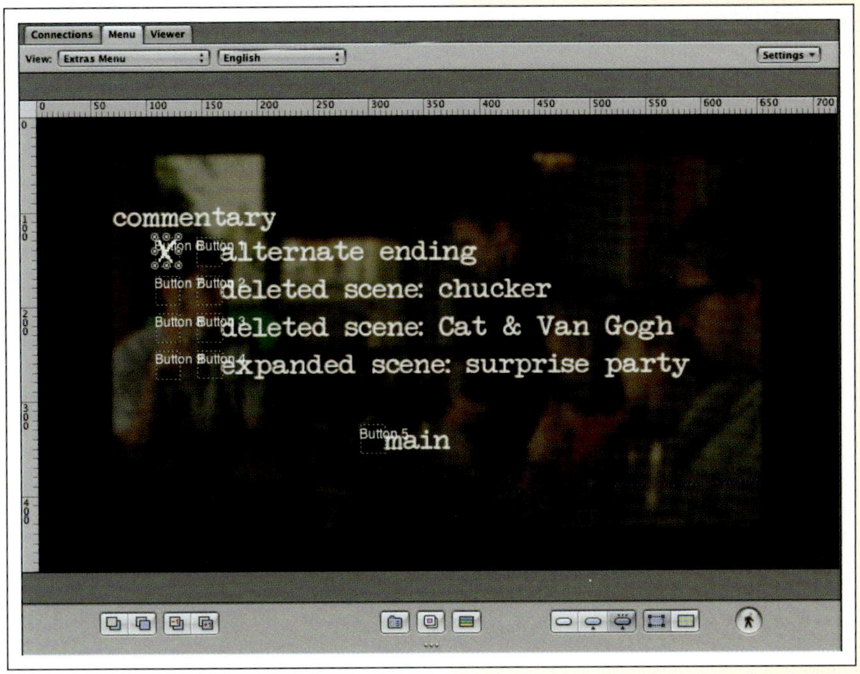

Figure 6-7. An example of a deleted scenes menu

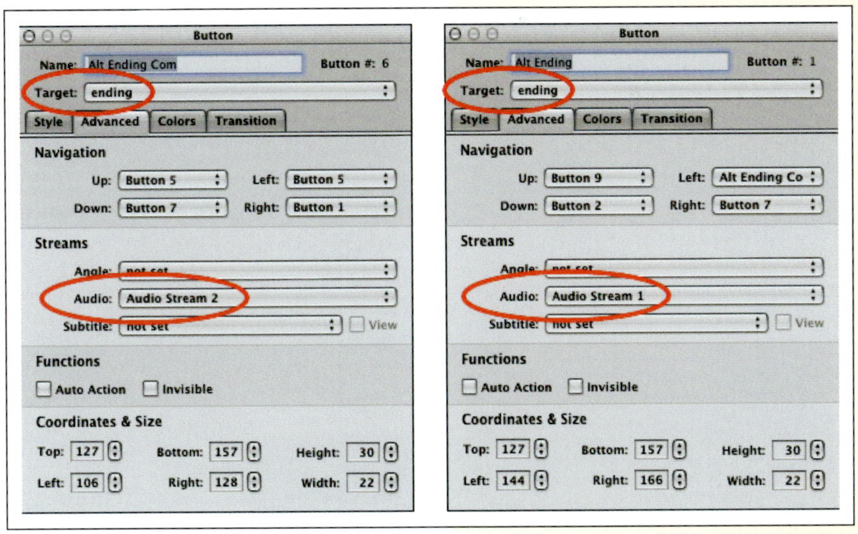

Figure 6-8. Configuring a button to start a clip and select the initial audio track at the same time

Multiangle Video

We've seen several examples now of multiple audio tracks in use. You can do the same with video tracks. However, this requires a lot more effort—both on the production side (creating two hours of video is much more work than creating one hour) and on the DVD playback side. Because of that, you'll have to keep some restrictions in mind.

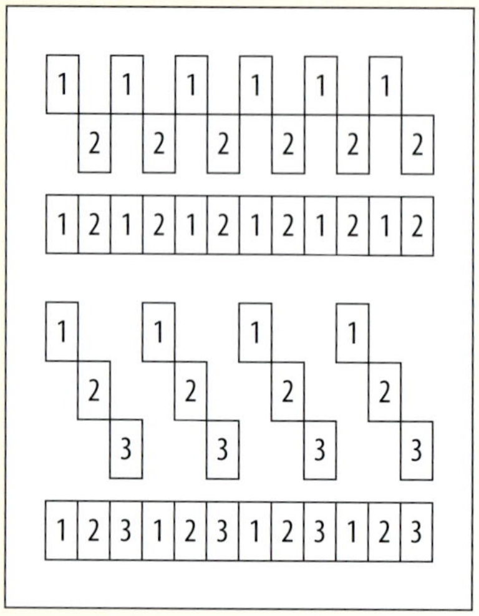

Figure 6-9. Two- and three-angle streams shown individually and interlaced

Multiangle video restrictions

To understand why you have restrictions on the content used for multiangle video, it helps to understand how the video is laid down on the disc. Frames from each angle of a multiangle clip are interleaved on the disc. Figure 6-9 shows an example of three clips as they logically would appear on the disc.

Because of this interleaving, you have two big restrictions on your video clips:

- They have to use the same bit rate—and this means VBR encoding is out.
- They have to obey throughput limits:
  - 2–5 angles: ~8Mbps max per clip
  - 6–8 angles: ~7.5Mbps max per clip
  - 9 angles: ~7Mbps max per clip

And, of course, these clips take up space, so unless you use a really low bit rate, you won't be able to have 5 angles in your 100-minute feature film.

Multiangle uses

Some of the most effective uses of multiangle video are for instructional purposes. One cool use in the world of special-edition DVDs comes from showing the construction of complex, composited shots. You can show the actors against a greenscreen in one angle, the greenscreen content in another angle, and the final composite in a third angle.

Computer animation shots can benefit from the same approach. You can show the timing stills, the wireframes, the basic shaders, and the final shot in their own angles to illustrate how the animation clip was put together.

The other instructional arena in which multiangle tracks work well is corporate training. Video of the presentation can be interleaved with video of the presenter to enable the viewer to switch back and forth at will as he would in the actual presentation.

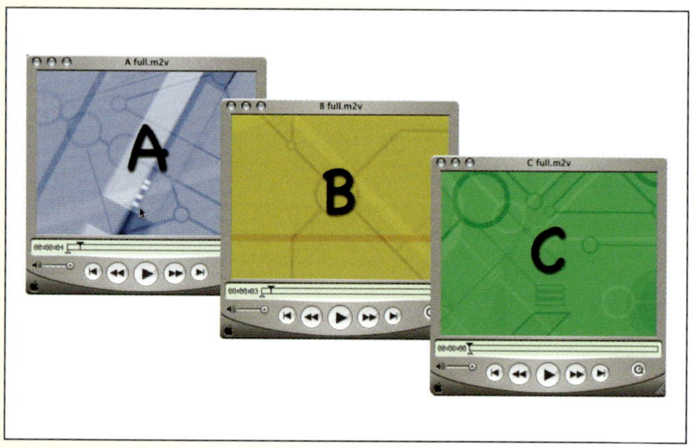

Figure 6-10. The three separate tracks we'll be using for our multiangle project

A multiangle project

For our multiangle project, we're going to start with some very simple images to make sure we can tell which track we're on. We built the images in Photoshop and composited them in Final Cut Pro over a motion background loop from Digital Juice. Figure 6-10 shows our three tracks: A, B, and C.

To start the project, import the following tracks, along with a simple Soundtrack audio track and the menu from the *projects/ch06* folder:

- *A full.m2v*
- *B full.m2v*
- *C full.m2v*
- *abc_audio.AC3*
- *abc_menu_scaled.psd*

Then perform the following steps:

1. **Add the Video Tracks.** Add the first track (*A full.m2v*) and the audio file (*abc_audio.AC3*) just as you normally would for any video track. The only interesting thing about this clip is that you can see we used basic VHS quality (~4 Mbps) for encoding the video to make sure we have room for the other angles.

   Adding a second angle isn't much more difficult. You might need to adjust your window to show the other video tracks. Once you can see as many tracks as you need, you just drag the next video track to your Track tab (see Figure 6-11).

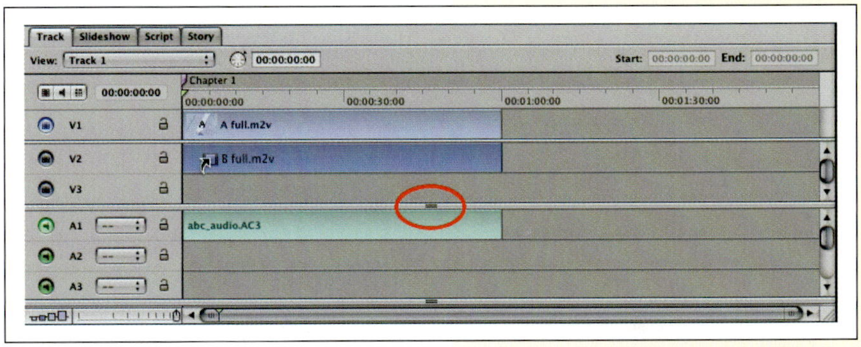

Figure 6-11. Adjusting the Track tab to show more video tracks (gray border highlighted in the center), and adding a second track

   You can continue adding tracks up to your limit (either the maximum of nine, or slightly lower, depending on your encoding bit rate). Our project should end up with three video tracks, as shown in Figure 6-12.

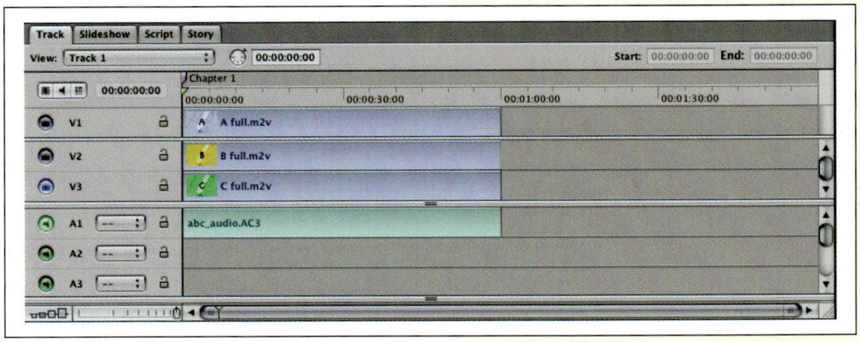

Figure 6-12. Our finished Track tab with three video tracks

2. **Create a Video Angle Menu Selection.** Selecting video angles from the menu works much like choosing audio streams. You can pick the video angle to show, as well as a target. Similar to our multilingual project in the previous chapter, you can stay on the same menu after selecting an angle, or you can jump right into the video.

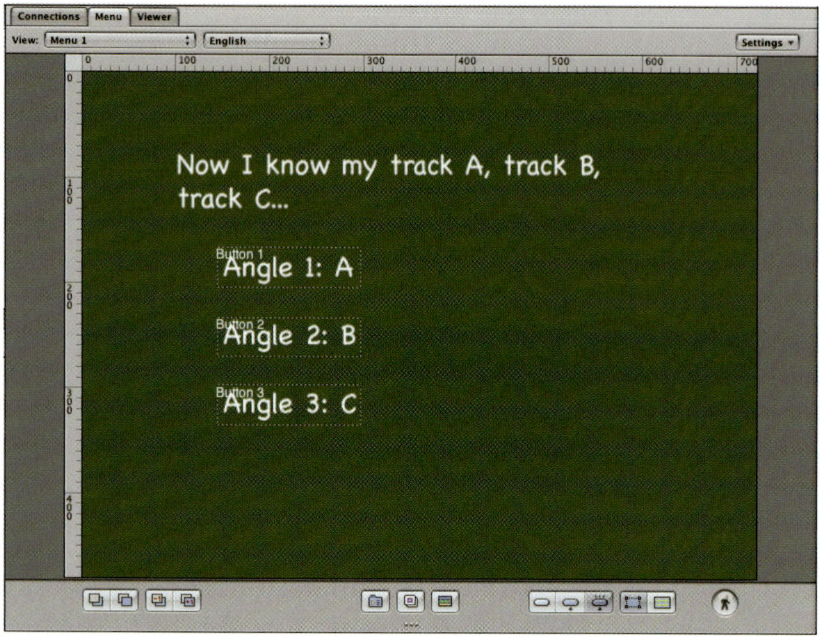

Figure 6-13. The first stage of our multiangle menu

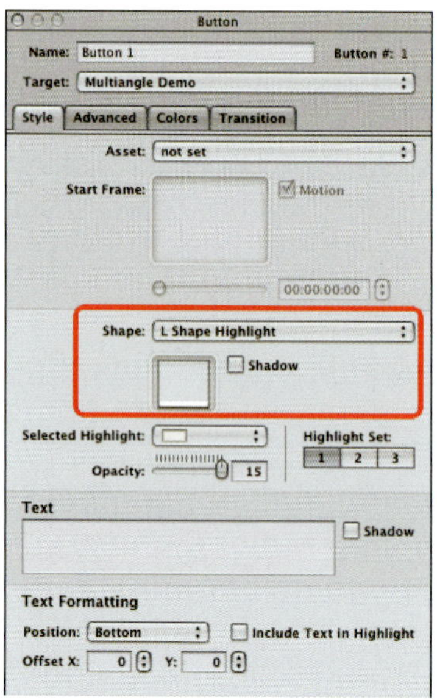

Figure 6-14. Setting the shape for a menu button

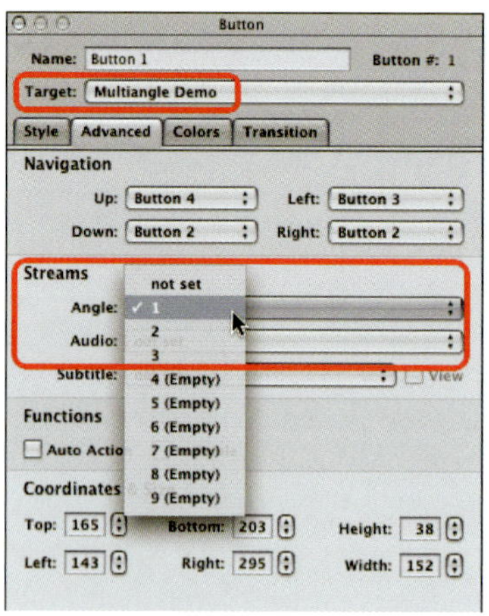

Figure 6-15. Setting a button to jump to a video track and picking the initial angle

Because our multilingual project goes the menu route, let's try the jump-right-in approach. Go ahead and create a new menu with the *abc_menu_scaled.psd* image for the background. Create three buttons around the Angle text entries. You should end up with a menu similar to that shown in Figure 6-13.

To make the buttons stand out when selected, we'll set their Shape attribute to L Shape Highlight, as shown in Figure 6-14.

3. **Set the Button Targets for the Appropriate Video Angles.** To hook up the video angles and targets, select a button and then open its Property Inspector. Figure 6-15 highlights the options we'll be configuring.

   Do the same for the other buttons. Each will select a different angle, but will use the same track for their target. Selecting a button during simulation should take you directly to the angle you picked.

Figure 6-16 shows our Simulator running the B angle.

Notice the highlighted camera button. This button enables you to switch streams dynamically, just as the other buttons did for audio tracks and subtitles. Most consumer players have similar buttons on their remotes as well.

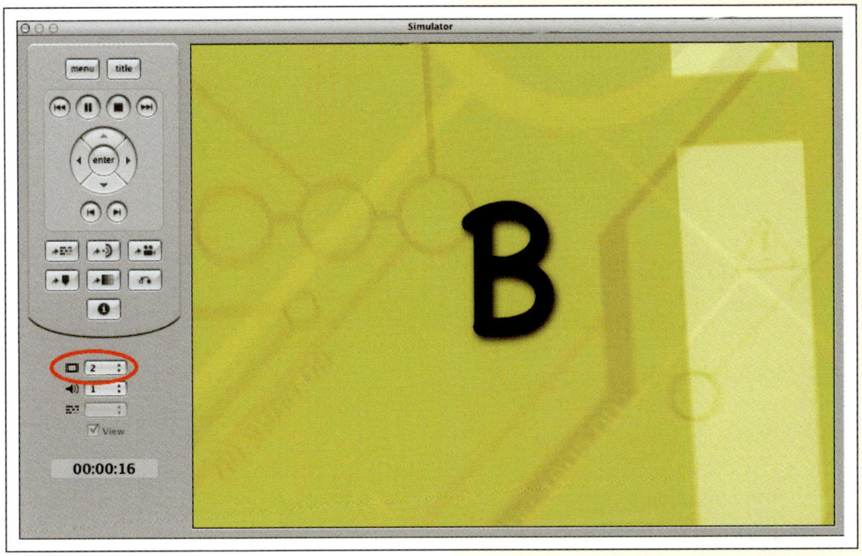

Figure 6-16. A multiangle project running in the Simulator; you can switch angles using the controls highlighted at left

Mixed versus multiple angles

You can use multiple angles to create a particular type of multistory film. You can mix single- and multiple-angle segments to create a full-length video that looks like multiple story tracks. As a sneak preview, Figure 6-17 shows the final setup.

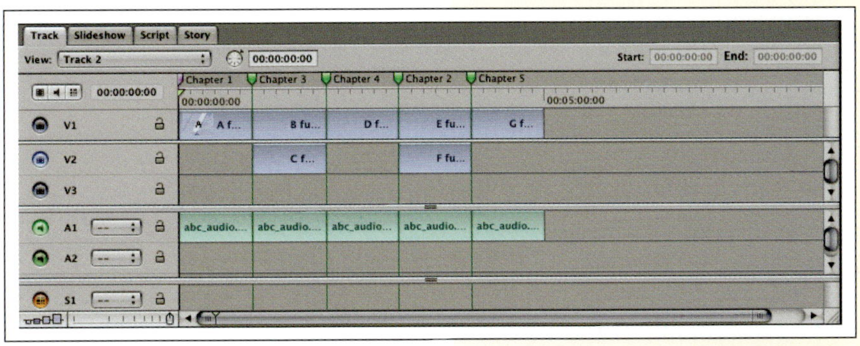

Figure 6-17. A mixed-angle project with two video tracks

This type of mixing enables you to cut down on the space required for the disc. You include only the multiple-angle shots for the segments in which the second and third angles actually exist. You still can use the menu to select video track 2. When no track 2 is available, track 1 will be shown.

If you have three or more tracks, however, each track must have its own clip wherever you place a mixed angle. For example, you can't have a clip in tracks 1 and 3, but not in 2. Because it takes so much work to plan mixed-angle presentations, though, this is rarely a problem!

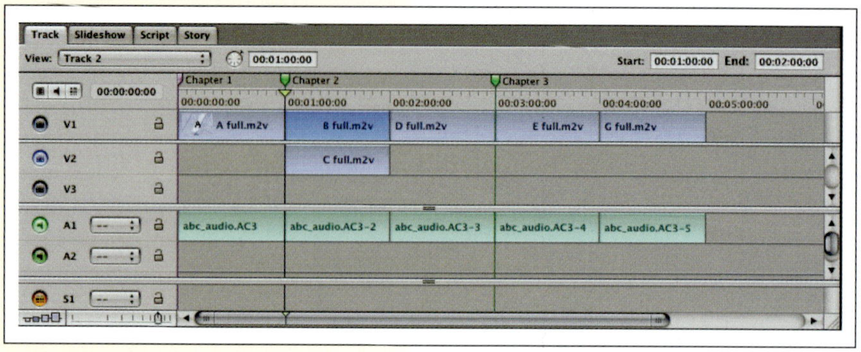

Figure 6-18. Starting a mixed-angle project with two new markers (labeled Chapter 2 and Chapter 3)

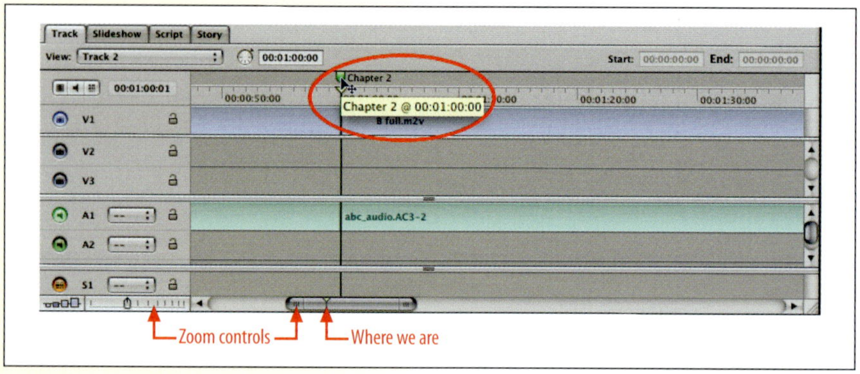

Figure 6-19. Zooming in to place a new marker in the timeline

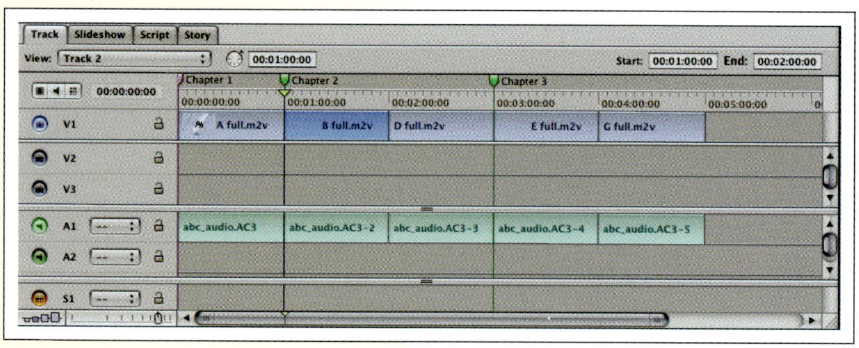

Figure 6-20. Our first mixed-angle clip

To mix single- and multiangle shots in the same project, follow these steps:

1. **Set Up Markers.** Choosing where to include the mixed-angle clips depends entirely on your project. The important thing to remember when working with those clips in DVDSP is to set up the markers. Figure 6-18 shows our main track and highlights the initial markers of interest to us. We put together a quick track consisting of the A, B, D, E, and G clips. We just repeated the audio clip five times—hooray for Soundtrack loops!

   We use the markers for alignment. Adding the marker in DVDSP has another salutary effect: it places markers only on good boundaries. Because of the way MPEG-2 video is encoded, not all frames are suitable starting points for a second (or third, etc.) video track. You'll need a marker at the end of the mixed-angle clip as well, but you can add that after the clip is in your timeline.

   If you need it, you can zoom in on the timeline to make sure you are where you want to be. Figure 6-19 shows the zoom controls and the pop-up information you get if you hover over a marker. That pop-up can be useful for verifying the placement of things.

2. **Drag a Clip to a Marker in Another Video Track.** Now drag down the C clip and drop it in the V2 track. If you drop it around the Chapter 2 marker, it should snap into place. It's best to make sure the beginning of the clip lines up with the marker. Now you should have a track window similar to that shown in Figure 6-20.

3. **Set the End Marker.** This one is easy. Control-click the C clip and select the Add Marker to Clip End option, as shown in Figure 6-21.

   Now both markers should be set. Figure 6-22 shows the result.

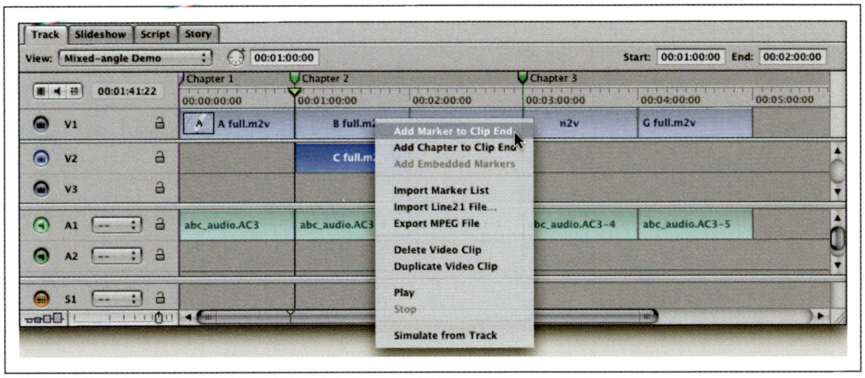

Figure 6-21. The right-mouse button menu for adding markers

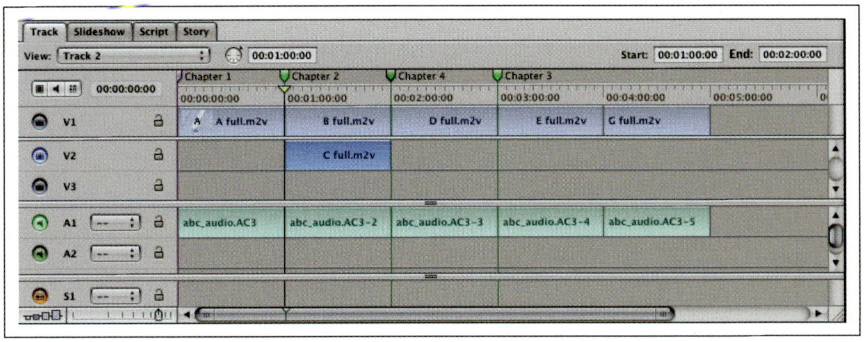

Figure 6-22. Both markers in place for our first mixed-angle clip

4. **Repeat for the Remaining Clips.** They all need their own markers at the beginning and end. This is easier in our case because we are creating a track from multiple segments. You can do the same work with regular tracks; you just need to remember to add the extra angles on marker boundaries.

Multistory Movies

A similar but quite separate notion in movie presentation is that of multiple stories. Where you can use multiple angles to show different versions of a particular scene, you can use multiple stories to rearrange the order—or even the existence—of the scenes. Want your disc to include the theatrical release alongside the director's cut version? A multistory disc is in your future.

Multistory discs might or might not have multiple angles. Consider the idea of a single disc with both the theatrical cut and the director's cut. Unless

Markers Versus Chapters

Many times your mixed-angle segments will represent complete chapters in your story. If that's the case, you certainly can use the chapter markers you inserted in your editing software. In fact, if you look at the Property Inspector for a marker (pop up the Inspector window [⌘I] and click a marker), you'll notice you can specify whether a marker represents a chapter.

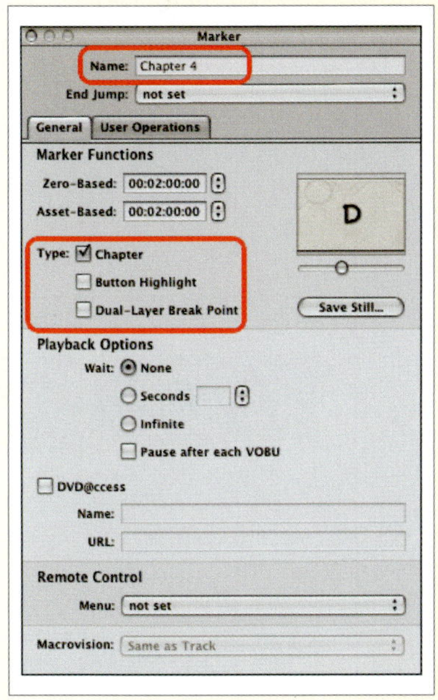

You also can rename the marker here, in case you find Chapter 2 (or whatever the next chapter number it used by default) to be a bit misleading!

we're talking about a short film, you couldn't fit both versions as separate tracks on the same disc. Instead, you can use stories to create two separate arrangements of the various chapters in your movie. Figure 6-23 illustrates the concept of multiple stories.

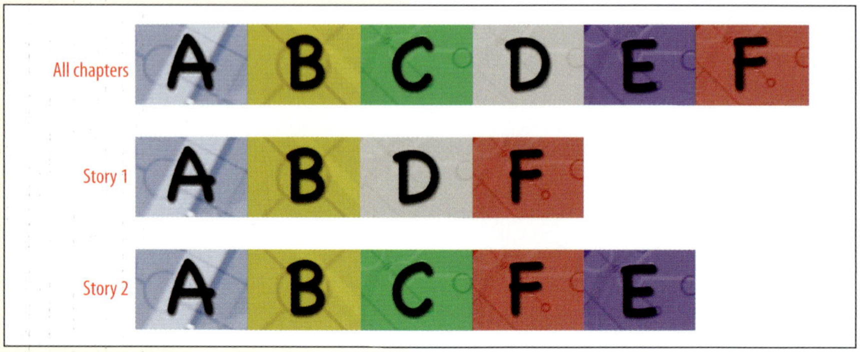

Figure 6-23. The basic concept of a multistory project

Indeed, you can take this to the clever extreme used in the movie *Memento*. The original release of the film was presented as a series of scenes shown in reverse chronological order. On the DVD, you can watch the film as it was released originally, or you can watch the scenes in chronological order, all without duplicating any of the video content.

Multistory restrictions

Now, multistory discs do have some limitations. The biggest limitation is the "distance" between clips. You want each story to run as a smooth, continuous movie. To do this, the next chapter in a sequence needs to be "close" to the current clip. How close depends very much on your bit rate. High bit rates can jump only a few seconds' worth of data. For standard NTSC-quality video, you can jump a few minutes—more if you use even lower encoding rates. Of course, all of this still depends a bit on your player.

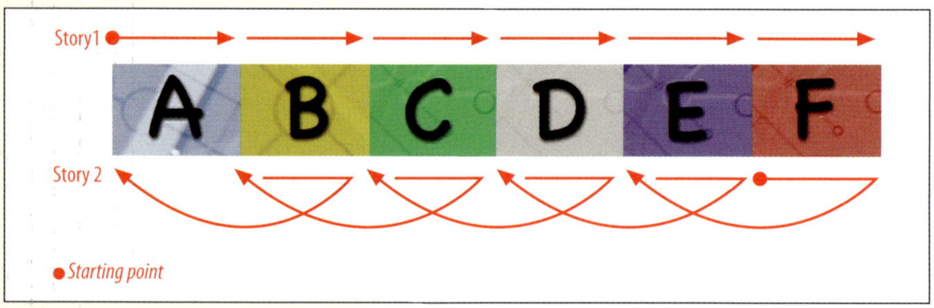

Figure 6-24. A possible story path to reverse the order of scenes

So, how would you pull off a *Memento*-style complete reversal? Figure 6-24 shows one possibility.

The Story tab

The Story tab consists of two parts, shown in Figure 6-25. The right side lists all your "options." It should be a familiar list of chapter markers for your video clip. You can drag, drop, and arrange those chapter markers on the right side to create a story.

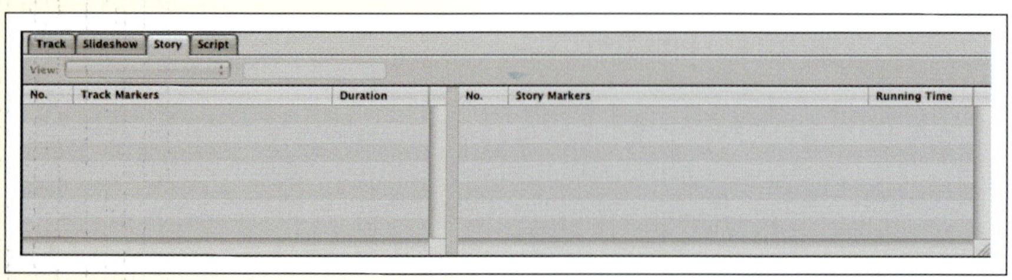

Figure 6-25. The Story tab in DVDSP

You can create multiple stories for any clip. In fact, you can have 98 stories if you want! They don't even take up noticeable room—each story is just a list of chapters in a particular order. No new video content is required to create a new story.

Another trick for multiple stories that fits nicely, even with high bit-rate projects, is the notion of the "airplane edit." That's where you take an R-rated film and make it presentable on airplanes. (Some films could get away with the occasional three-second edit. Others might need more... .)

A multistory project

For our multistory project, we'll head back to our alphabet clips. Grab the following files from the *projects/ch06* directory and import them into a new project:

- *ABCXYZ.m2v*
- *abcxyz_audio.AC3*
- *story_menu_scaled.psd*

Then follow these steps to create a multistory project:

1. **Add the Video Track.** We have only one video track, so go ahead and set that up. The video track contains chapter markers that we set in Final Cut Pro. Those should show up similar to the screenshot in Figure 6-26. Go ahead and add the *abcxyz_audio.AC3* audio track as well.
2. **Create the Menu.** Create the menu using the *story_menu_scaled.psd* file. Don't worry about hooking anything up just yet, though. Figure 6-27 shows the menu. We used the same L Shape from the previous project for our buttons.
3. **Add a Story.** You get one story—the video, as is, with the chapters in order—for

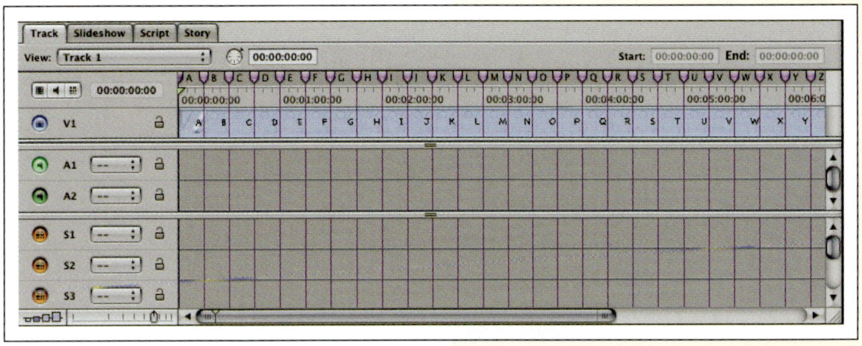

Figure 6-26. Our alphabet track and its chapters

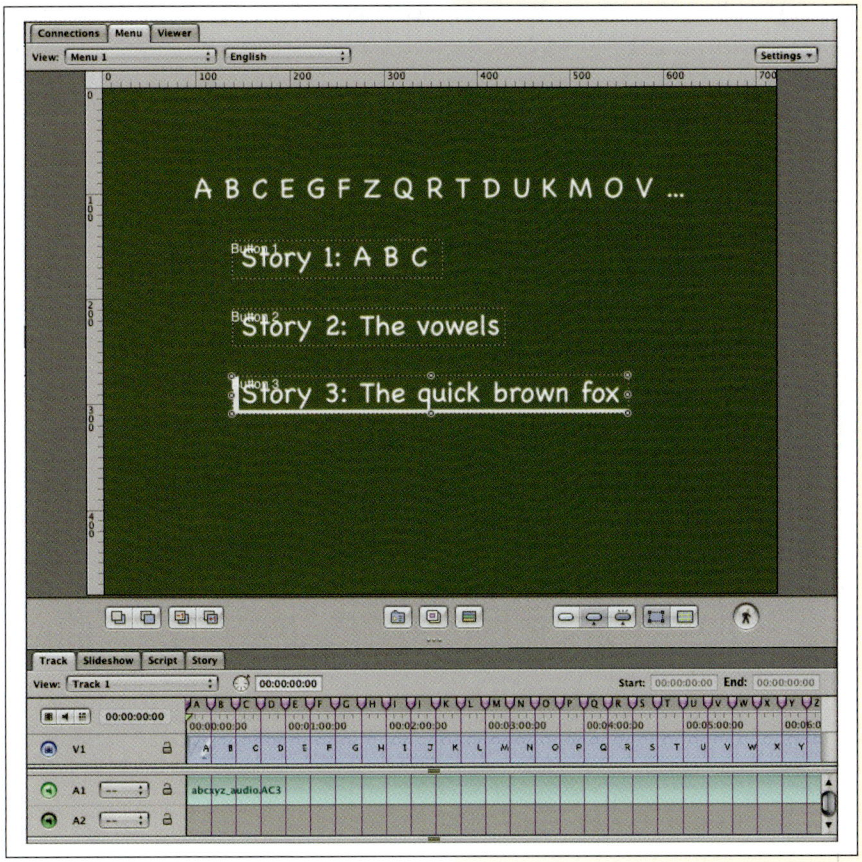

Figure 6-27. Our multistory project menu

free. We've been relying on that fact for every project up to this point in the book. But now we want to add a deliberate story. To do that, start by selecting the video track in the Track tab or in the Outline tab, as shown in Figure 6-28.

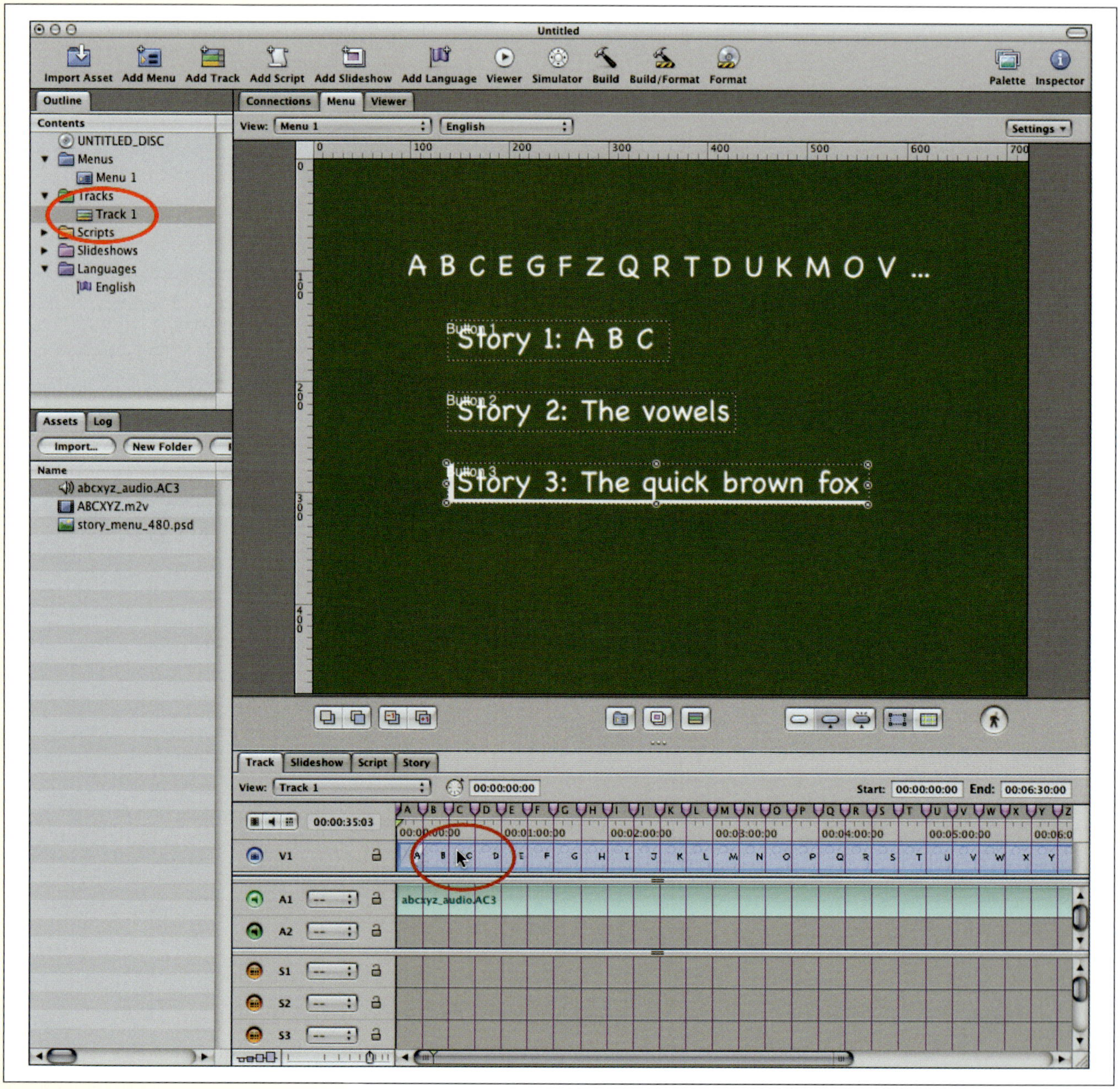

Figure 6-28. Two options for selecting the track to which we want to add a story

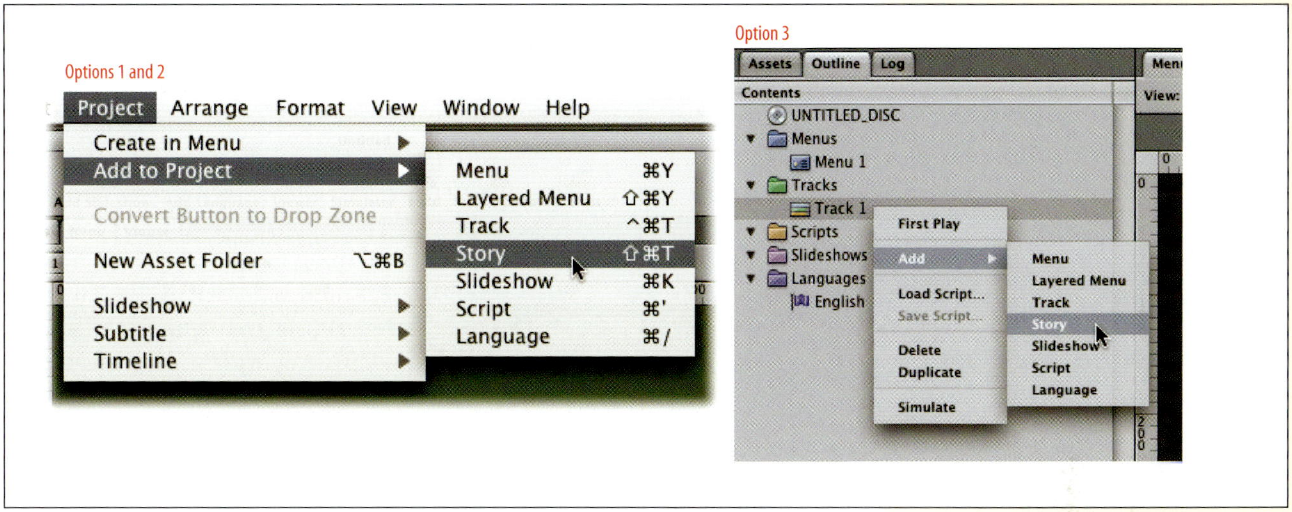

Figure 6-29. Story creation options

Now you can add a story using any of the following techniques (highlighted in Figure 6-29):

- By selecting Project → Add to Project → Story
- By pressing Shift-⌘T
- By Control-clicking the Outline tab and selecting Add → Story

Take a look at the Story tab now. You should see something similar to Figure 6-30.

We'll start with a "vowels only" story. Drag the A marker to the right side of the screen, as shown in Figure 6-31.

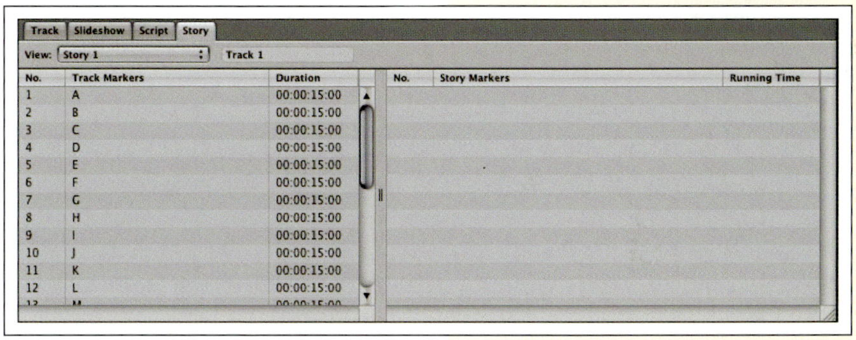

Figure 6-30. A new story, ready for populating

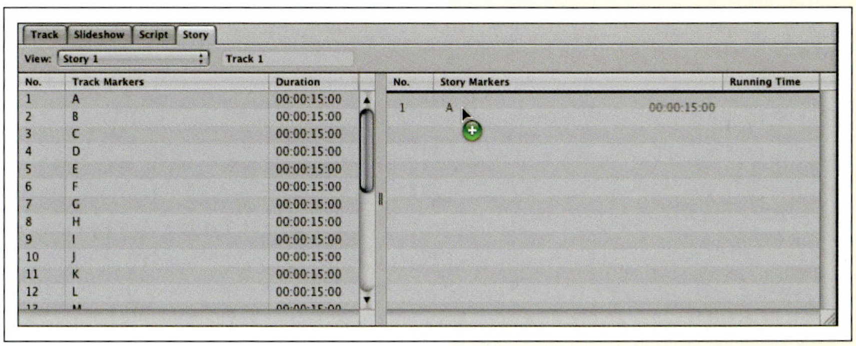

Figure 6-31. Dragging a marker to build a story

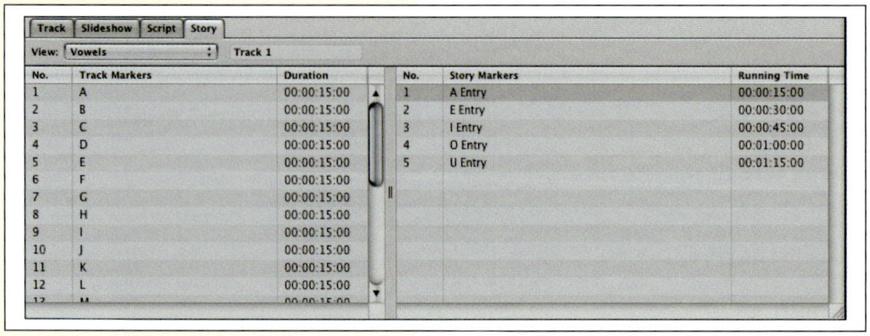

Figure 6-32. Our completed Vowels story

Then drag the remaining E, I, O, and U markers from the left side and drop them on the right side. (Of course, you're free to add Y if you want.) Now Story 1 should look like Figure 6-32.

If you want to rename your story to something more descriptive, now's a great time to do that. Make sure your story is selected in the Outline tab and bring up the Property Inspector. Figure 6-33 highlights the properties you most often edit for stories.

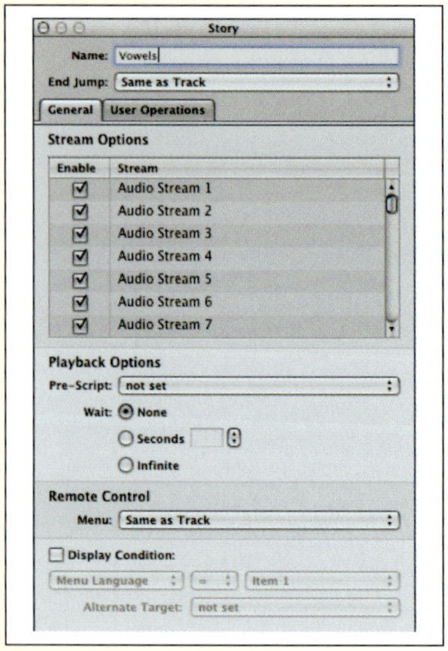

Figure 6-33. The Property Inspector for a story

Where's My Scrollbar?!?

A small bug that exists in early versions (through 2.0.4, as far as we can tell) of DVDSP 2 left the marker list in the Story tab without a scrollbar—even if you had many more markers than space to display them. You can get your scrollbar back by quickly resizing the window containing the Story tab.

The last version of DVDSP 2, 2.0.5, seemed to fix the problem. DVDSP 3 is happily free of this tiny bug as well. So, if you're still on an old version, you might want to consider upgrading!

4. **Hook Up the Stories in the Menus.** Let's go ahead and hook up the first two buttons on our menu, now that we have our track and the first story. Story 1 is just our track, so we set it up as usual. Figure 6-34 shows the target menu. For our Vowels story, we use the same basic menu, but we should have our Vowels story entry available, as shown in Figure 6-35.

   You can test these targets in the Simulator if you want.

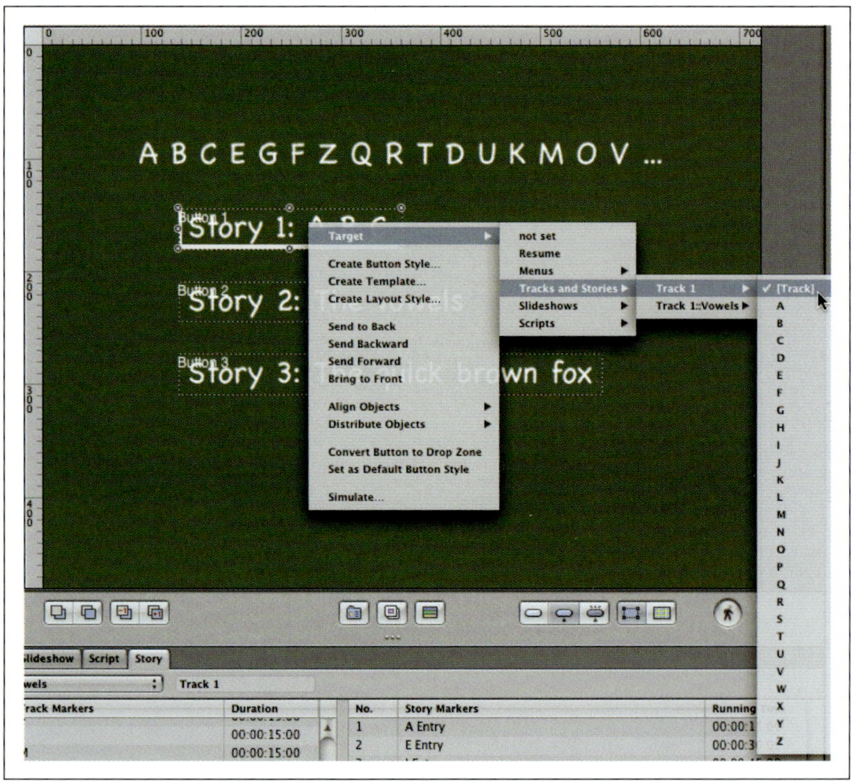

Figure 6-34. Selecting our normal track target

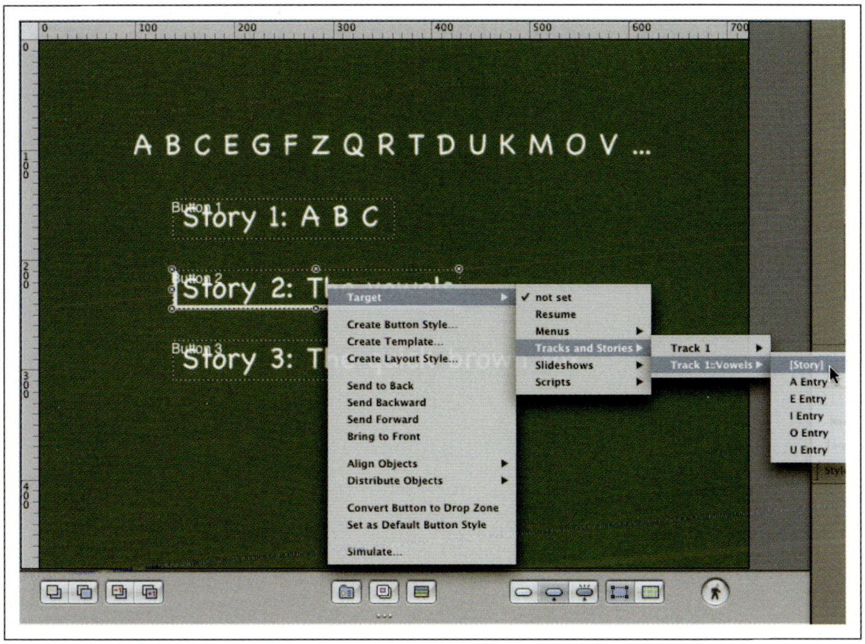

Figure 6-35. Selecting our first story target

5. **Add Another Story.** To add another story, just follow the same steps as before. Give it a descriptive name and drag your chapters over to the list on the right. Our second story spells out the phrase "The quick brown fox." Notice in Figure 6-36 that we hit the O chapter twice. That's perfectly legal. Whether that makes sense in your projects depends entirely on the projects. For some training presentations, repeating an overview or key concept might be quite reasonable. For movies, this is obviously less likely.

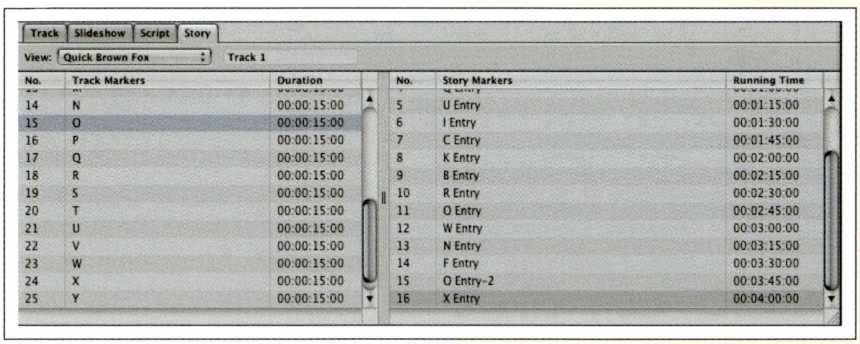

Figure 6-36. Our Quick Brown Fox story

6. **Add Alternate Audio Tracks.** Notice that we had to cheat on the audio track for this project. The audio loops along with the background so that you don't hear any interruptions when jumping from chapter to chapter in the story. That won't always be the case.

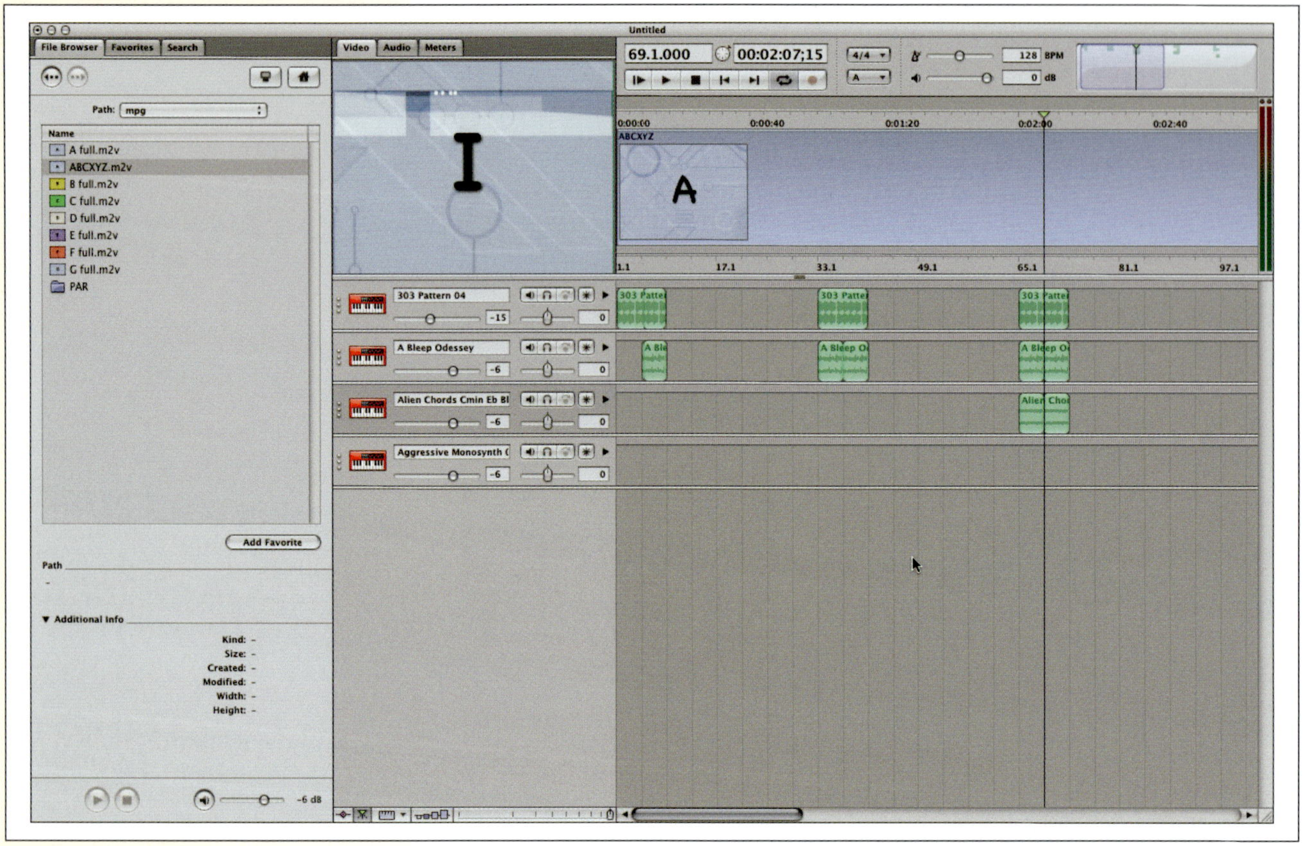

Figure 6-37. A tailored soundtrack for our Vowels story

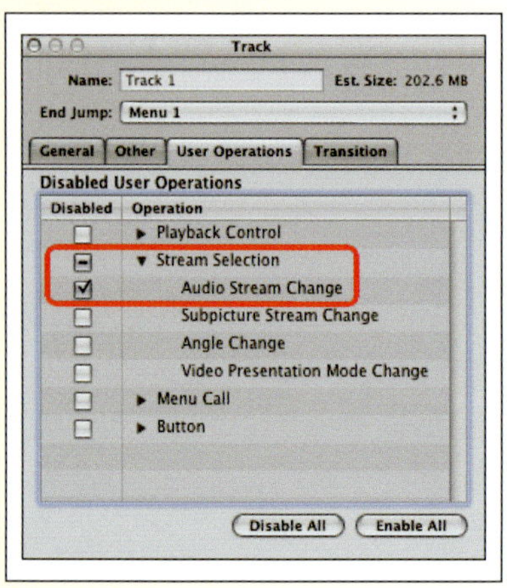

Figure 6-38. Blocking the viewer's ability to switch audio tracks

Consider a seamless but nonrepetitive audio track for the full alphabet. Jumping around to show the vowels would be quite jarring. You can create a second audio track for use with the Vowels story. The only trick is that the second audio track has to be the same length as the full-chapter video track. That means really disparate stories (such as our Vowels story) could end up with tailored audio tracks that look like Figure 6-37.

If you build these alternate audio tracks, you might want to make sure the viewer gets the right track. DVDSP provides a few mechanisms for locking out other audio tracks if they aren't appropriate for a given story.

One simple mechanism is to disallow the use of the Audio button on the remote control. That effectively stops the viewer from switching away from the current audio stream. Figure 6-38 highlights this approach in the Track Property Inspector.

Another, more flexible approach is to enable particular tracks for a given story. This can be handy when you have multiple audio tracks that are appropriate. Figure 6-39 shows the Property Inspector for a story and highlights the audio track options.

If you disallow stream switching (or if only one stream is available), the Simulator notices. Figure 6-40 highlights the audio button and shows it disabled for our story with only one valid audio track.

Although a director's cut certainly qualifies as a special edition for most films (and home movies!), you can do more things on a DVD. Scripting plays a large role in these added features, so grab a fresh source of caffeine and dive into the next chapter!

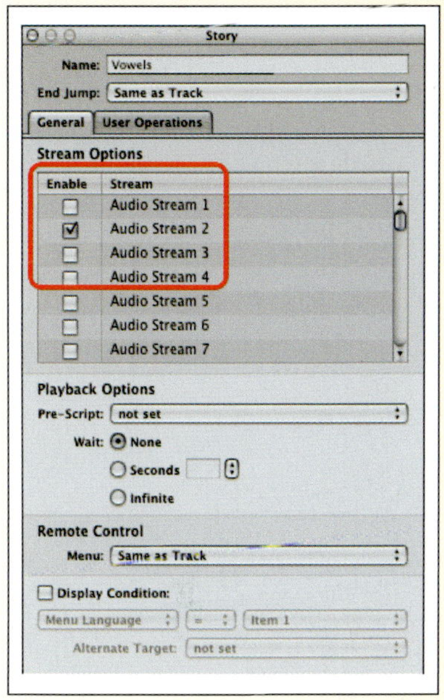

Figure 6-39. Enabling particular audio tracks for a story

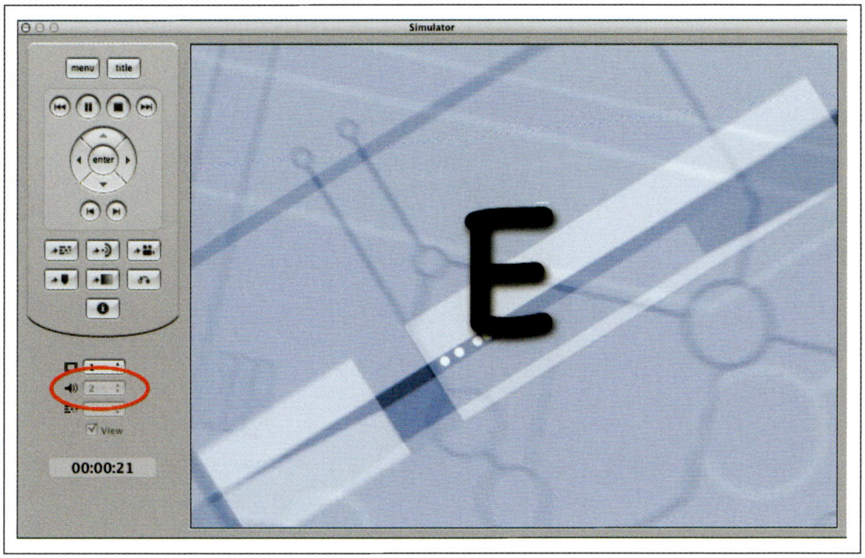

Figure 6-40. Our Vowels story playing with its own (and only) soundtrack

7

DVD Scripting

One of the biggest elements that sets DVDs apart from other video entertainment media is its interactivity. A DVD can appear to be "smart," remembering where you left off, or asking you questions and then tailoring its presentation to your answers. Scripting opens the door to these fancy behaviors. In this chapter, we take a look at using DVD scripts to make your DVDs a bit more mature and to increase their usability. (In the next chapter, we'll use scripting for a few tasks that are more lighthearted.)

In this chapter

DVD Scripting Features

DVDs support features beyond the great audio and video tricks we've looked at so far. DVD *scripting* allows you to expand the user's experience with your disc in very clever ways.

What are scripts?

The DVD specification enables you to control some basic player features via a simple scripting language. Only a few low-level operations are available for now, but you still can accomplish some surprising interactivity.

In DVDSP 1.5, you actually typed in a script that the player translated to the low-level operations for you. You could use a generic text editor and import the script for use in your project, or you could use the rudimentary built-in text editor.

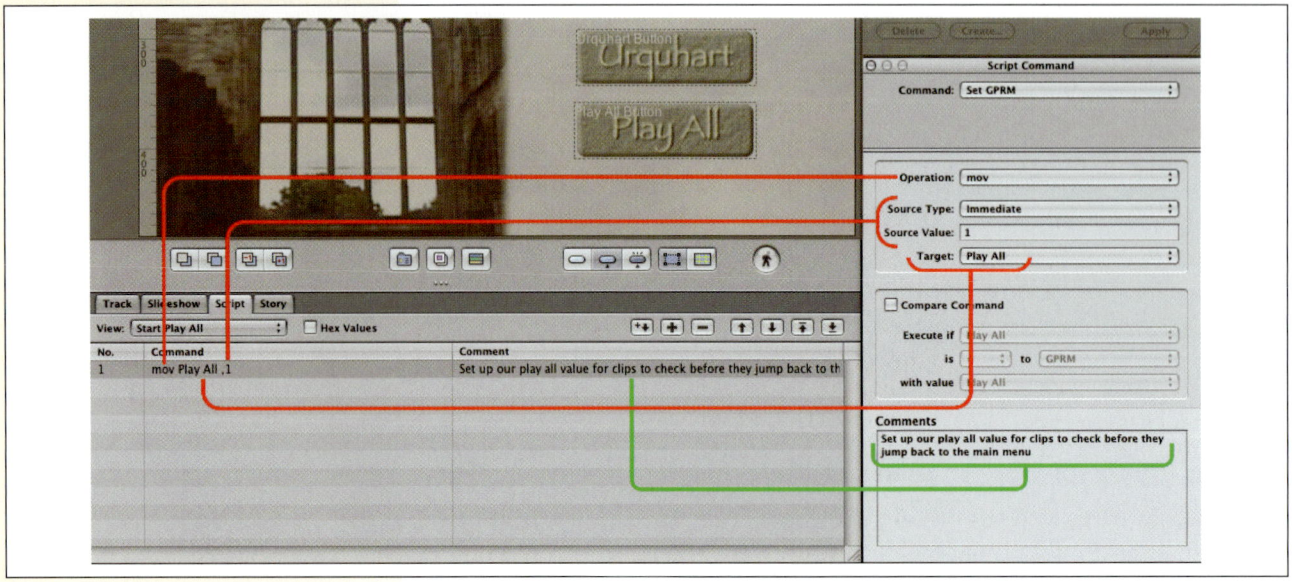

Figure 7-1. The script viewing and editing facilities in DVDSP

In DVDSP 2 and 3, however, you use a much more sophisticated interface. You still produce a script, but the process is more guided and takes advantage of tools you're already familiar with. Consider Figure 7-1, which shows the interface for editing a script command.

The script appears on the left, with line numbers and comments. You use the Inspector to edit one line of the script at a time. You can use the script toolbar to add, remove, and rearrange the lines, as shown in Figure 7-2. You also certainly can import and export scripts, but those files are large XML files meant for machines—not humans. Still, importing and exporting are valid options for "template" scripts or for copying a script to another project.

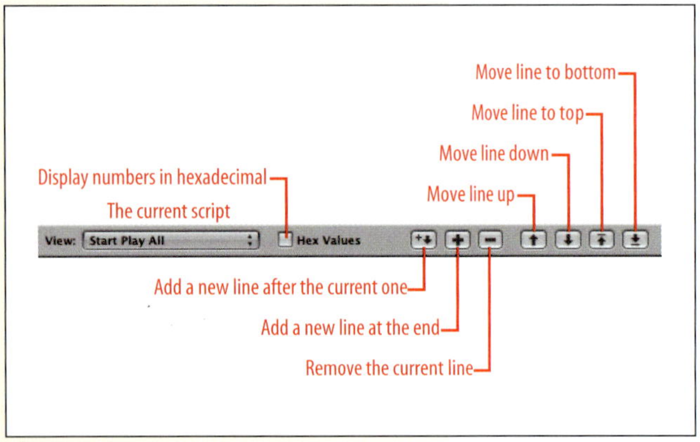

Figure 7-2. The script toolbar for organizing your commands

You can switch scripts with the View selector at the left, just as you switch tracks and menus. The Hex Values checkbox can be useful to programmers familiar with bitmask programming, but it's not vital to script operation.

Although seasoned programmers might find this interface somewhat cumbersome, it's a good compromise. They can accomplish everything they need to, while designers new to programming can get around and edit existing scripts without much trouble. Another thing to remember about this interface: scripts can be only about 120 lines long. So, this interface stops you from making syntax or length-limit errors. You're still free to make logic errors, but more on that later.

A simple play-once-on-insert project

Have you ever watched a DVD with a fancy, 20-second intro video that plays before every menu appears? It looks great the first time you see it, and you're glad to be alive in a world where such wonders cost only $18.99. But the second time it's not as fun. By the eighth time you're definitely interacting with the menus, but for the record, no consumer players (as of yet) respond to angry voice commands.

You can use scripts to alleviate this problem. Indeed, a fairly simple script can handle it and give you the exact behavior you want to see: play the intro the first time, but after that, just enable the user to move on to the menu.

The play-once demonstration

Let's go back to the Paine Art Center for this project. We'll simplify the menus to highlight the new scripting work. Import the following files from the *projects/ch07* folder for this project:

- *intro.m2v*
- *intro.AC3*
- *Languages_scaled.psd*
- *Main_menu_scaled.psd*
- *paine.m2v*
- *paine.AC3*

Now follow these steps:

1. **Set Up the Main Menu, Video, and Audio Track.** Do this in the same way you did in Chapter 5. This time, though, you need English only, so turn on the following background layers:

    - Title
    - English
    - Image

    You can go ahead and use the Selection Mask layer for the button overlay. Next, set up the *paine.m2v* video, along with its audio track. You should see a screen similar to Figure 7-3.

120?

Why limit script lengths? Well, scripts occupy some very particular areas of a DVD. You can set a pre-script that runs just before a video asset, for example. Remember that DVD players are not meant to be computers. Lots of compromises have been made to make sure players can be manufactured cheaply. (And how! You can get players for only $30 these days.) Limiting the complexity of scripting and navigation controls is one of those compromises.

Plus, you get stability. You know a script will fit within a certain size (1K, in fact: 128 commands x 8 bytes max/command), and that makes them easier to manage. The actual number you see in DVDSP reflects some behind-the-scenes work by Apple. You might notice that adding one command to your script reduces the number of available commands by two or three. That's normal—and you really don't have to worry until you're at the very edge of that limit.

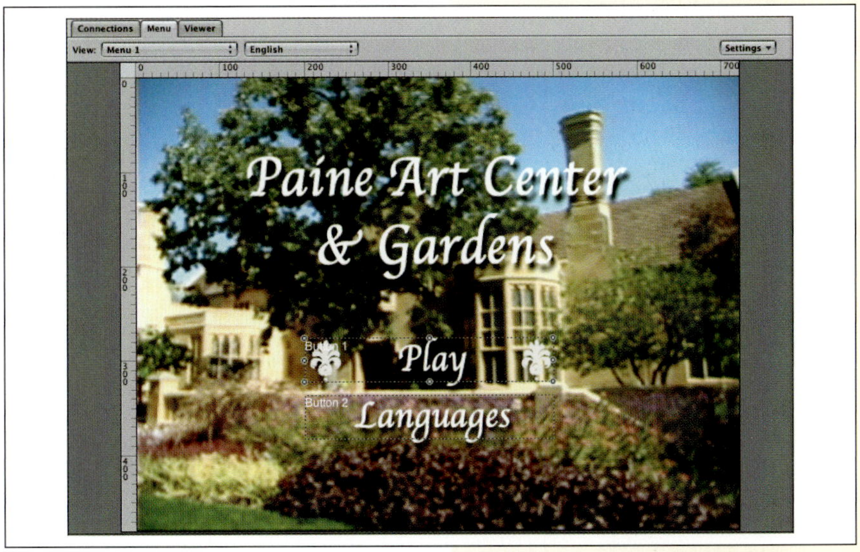

Figure 7-3. The main menu for the play-once project

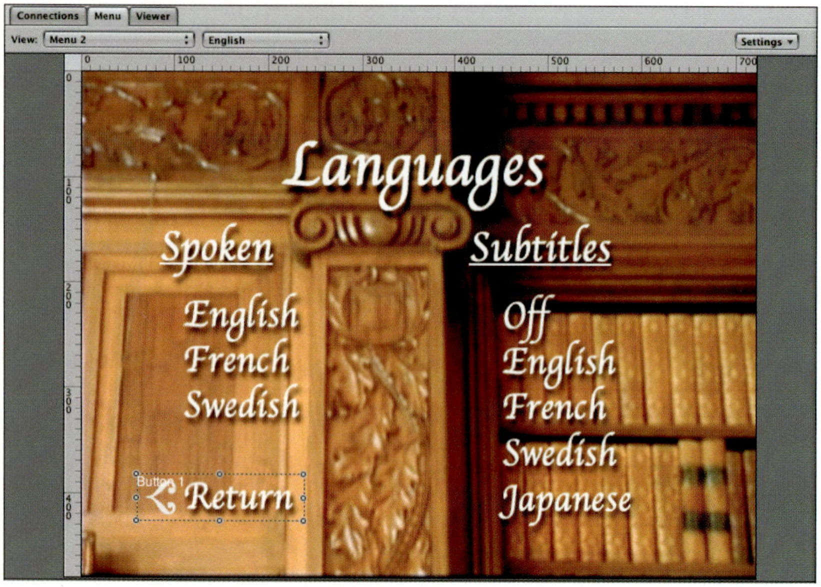

Figure 7-4. A simplified Languages menu for the play-once project

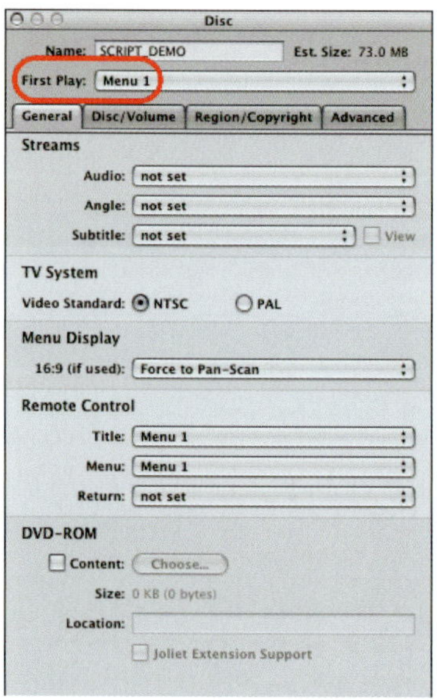

Figure 7-5. Making sure the menu is the default action for the disc

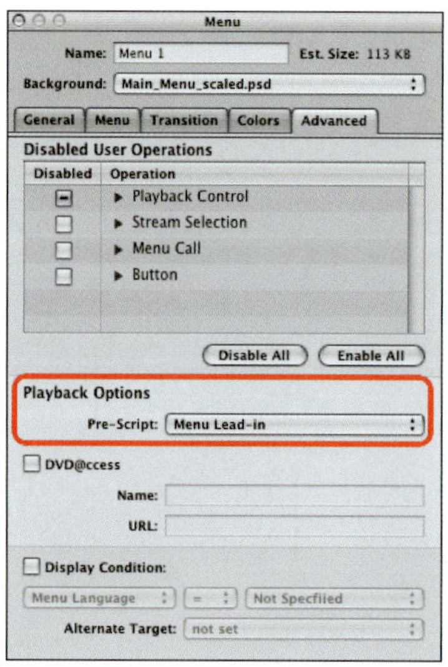

Figure 7-6. The pre-script area for menus

2. **Set Up the Languages Menu.** For the Languages menu, all you want to do is hook up the Return button to test the script. (If the script works, you should go right back to the main menu. If it doesn't, you'll see the intro clip again.) You'll need the following layers turned on:

   - Languages
   - Image

   Use the Selection Mask layer from *Languages_scaled.psd* for the button overlay. Your second menu should end up looking like Figure 7-4.

3. **Set the First Play.** Check the Disc Inspector sheet and make sure the menu is set up as the default action. Figure 7-5 shows a reminder of where to set that property.

4. **Create a Track for the Intro.** Now the interesting work starts. Add the *intro.m2v* video and its accompanying audio to a track called Intro. Don't worry that they aren't quite the same length. You'll use this clip as the pre-script entry for the menu. Next, bring up the Inspector for the main menu (see Figure 7-6) and take a look at the Playback Options section on the Advanced tab.

   We have ignored the pre-script pop-up before, but we're going to put it to good use now. Let's get started with one of the simplest scripts possible.

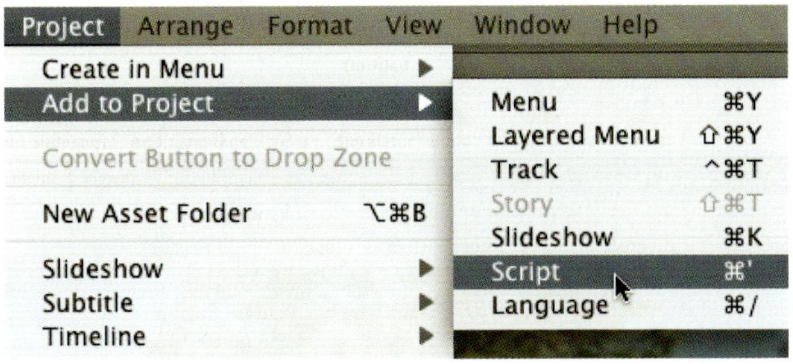

Figure 7-7. Two options for adding a script to a project

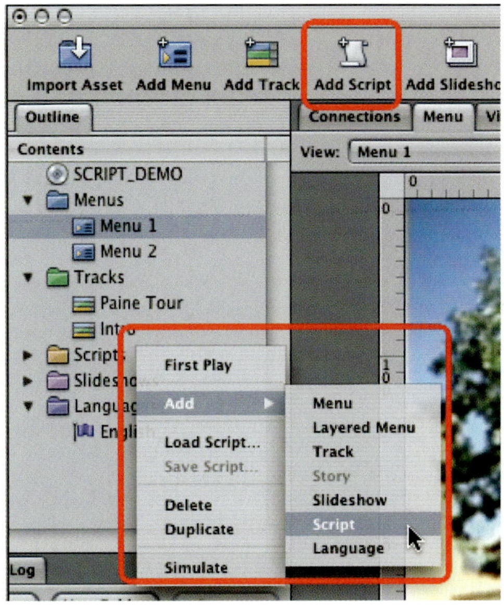

Figure 7-8. More options for adding a script

5. **Create a Simple Script.** You can add a script to your project in the following ways:

   - Select Project → Add to Project → Script (see Figure 7-7)
   - Press ⌘' (Command-apostrophe; see Figure 7-7)
   - Right-click or Control-click the Outline tab and select Add → Script (see Figure 7-8)
   - Click the Add Script button if you have it on your toolbar (see Figure 7-8)

   The initial script is empty—and quite boring. You should see something similar to Figure 7-9.

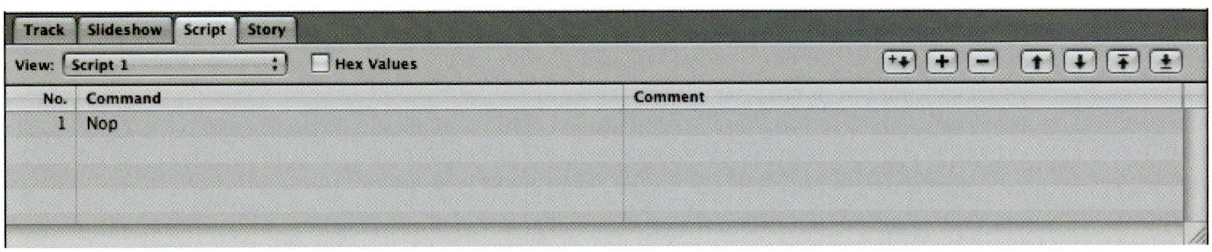

Figure 7-9. The Nop ("no operation") command, which all scripts start with

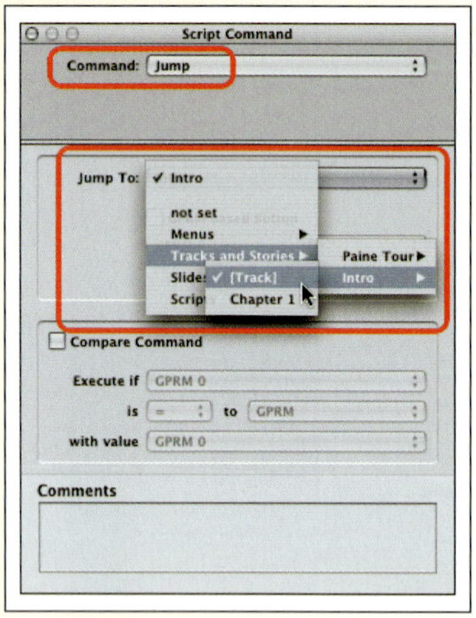

Figure 7-10. Configuring the first command for a jump

General Purpose Register Memory

You no doubt noticed those eight GPRM entries on the script property sheet in Figure 7-12. You can use a General Purpose Register Memory (or general parameter) as you would a programmer's variable. Indeed, the Inspector enables you to rename the GPRM 0-style entries so that they even look more like traditional variables.

Programmers (and now DVD designers!) use variables to store information for later use. For example, you can store the user's chosen language or the number of times he has accessed a particular clip. What you do with variables depends entirely on your project.

The DVD specification allows for 16 GPRMs, but only 8 are available through the DVDSP interface. Some DVD players might have other, fancier features available, but you might not be able to access those features through DVDSP.

Now you can work the magic that makes the best DVDs stand out from the pack. Your first script simply will start the intro clip. It uses the Jump command. To change the Nop command to the Jump command, click the Nop line and bring up the Inspector. You should see a property sheet similar to that shown in Figure 7-10.

When you're done with your changes, your Script tab should look like Figure 7-11.

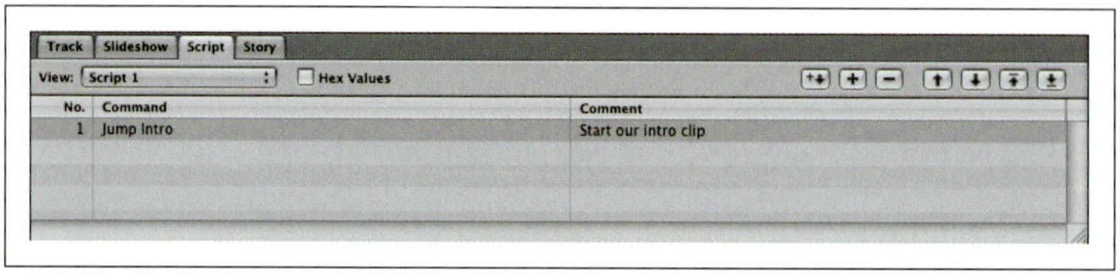

Figure 7-11. A very simple script with one Jump command

To make things easier on you later, you'll want to name the scripts something useful. We chose Menu Lead-in and set that in the Property Inspector for the script, as you can see in Figure 7-12. (You might need to select your script from the Outline tab to get it to appear in the Inspector.)

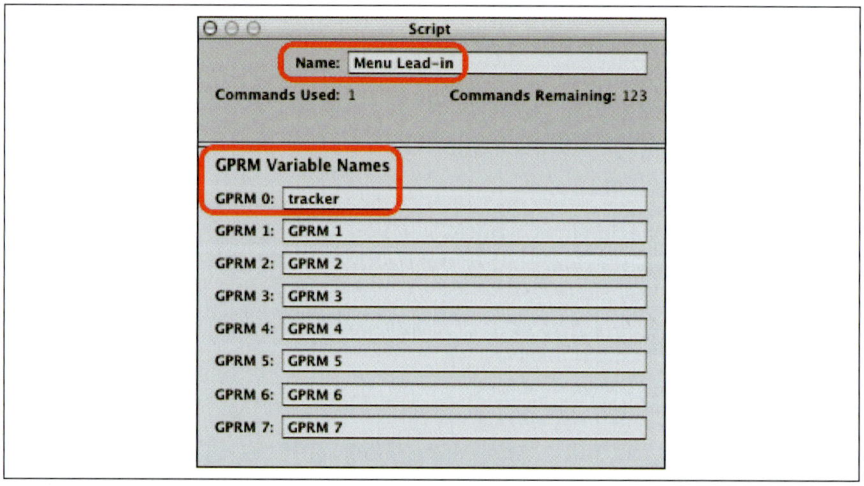

Figure 7-12. Naming the script (and the global variables, if needed)

Save your project and run the Simulator. Figure 7-13 shows the first thing you see when you start the Simulator.

That's not the menu! Which is exactly what you wanted. You chose the video introduction as the pre-script for the menu. Every time you go to this menu from here on out, that 20-second clip will play. Try it! Go to the Languages menu and then click the Return button. Or start the video clip and then press the Menu button on the remote control. It won't take long for that to become annoying, will it? Let's change that so it plays only once.

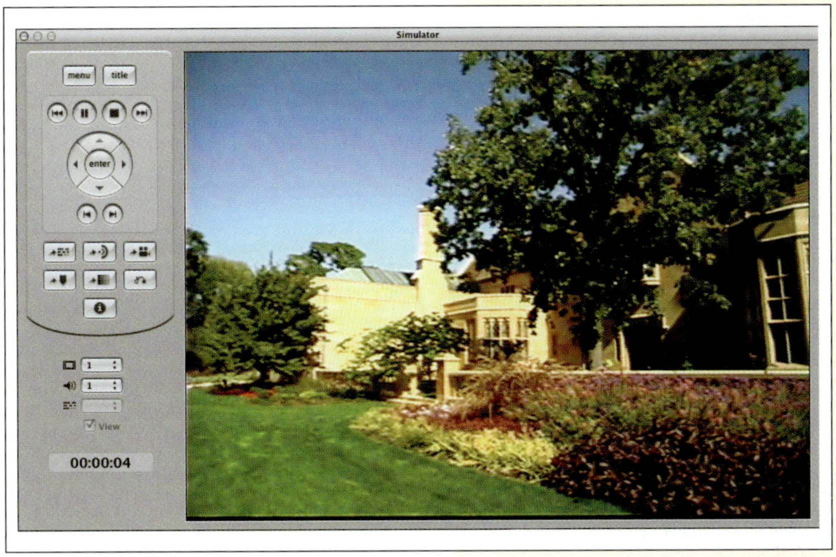

Figure 7-13. Our menu's first pre-script in action

A smarter script

Instead of blindly playing that video sequence every time you show the menu, you can modify your script so that it remembers whether the introduction has played already. If it has, it can direct the player to jump immediately to the menu.

Look back at those GPRM entries in the Script Property Inspector (see Figure 7-12). Those are the keys to remembering the state of the intro clip. You'll store a value in one of those entries and check it before playing your intro. The new script will look like Figure 7-14 when you're done. If you're just passing through this chapter, peruse this figure and skip to the next section. If you want the details, continue reading.

Go ahead and add two new lines to the script. Move the existing Jump line to the bottom. Your script should look like Figure 7-15.

You also can rename GPRM 0 to "tracker"—if you want to, that is. (Recall that you can do this in the Inspector for your script.) Now let's edit the first line. You want this to exit the pre-script if you've already played the intro. If the tracker variable is 1, that means you've seen the intro and should go on to the menu. So, start out as though you want to go on to the menu. Your command will be the Exit Pre-Script command, as shown in Figure 7-16.

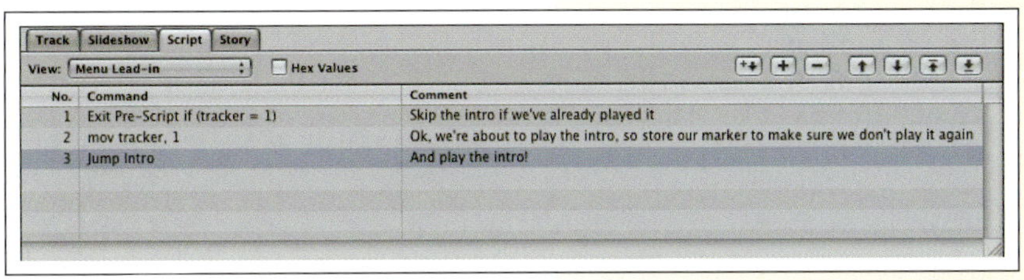

Figure 7-14. Our new, smarter play-once script in its final form

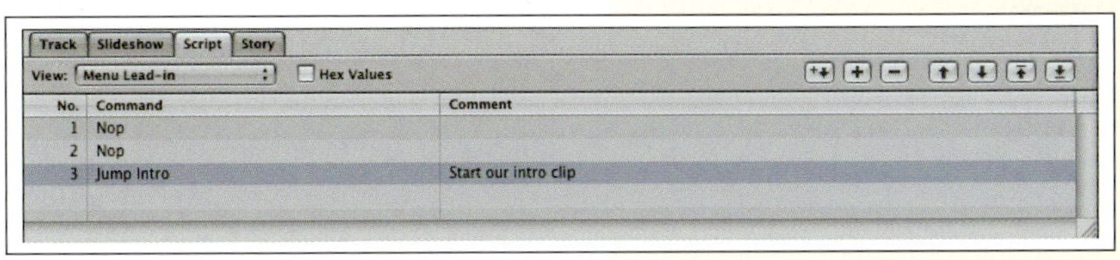

Figure 7-15. The first step to a smarter script: adding lines

But you want to exit only if you've seen the clip, so use the conditional portion of the Inspector. You'll jump only if tracker is equal to 1 (see Figure 7-17).

Notice the Immediate type in Figure 7-17. That enables you to use "immediate" values (e.g., numbers) in the command. These can be useful for counting, for picking stream numbers, or for "marking" things, as you're doing here. You also can compare against another GPRM or one of the System Parameters (SPRMs). But more on those later.

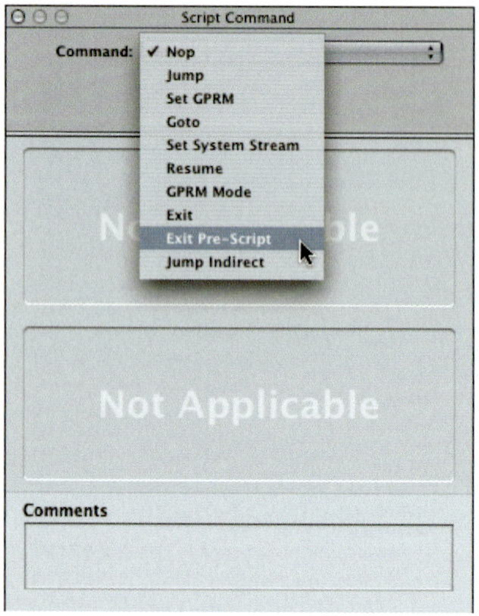

Figure 7-16. Selecting the Exit Pre-Script command in the Inspector

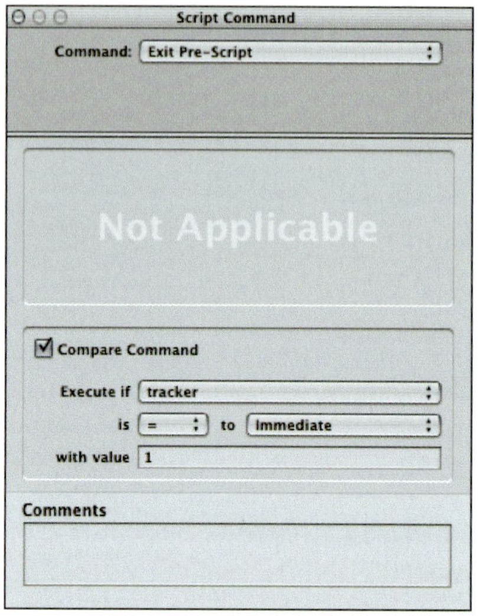

Figure 7-17. A conditional script command

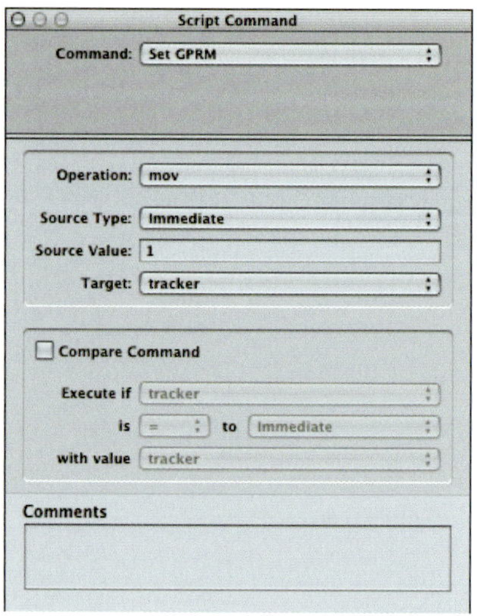

Figure 7-18. Setting a value for a variable

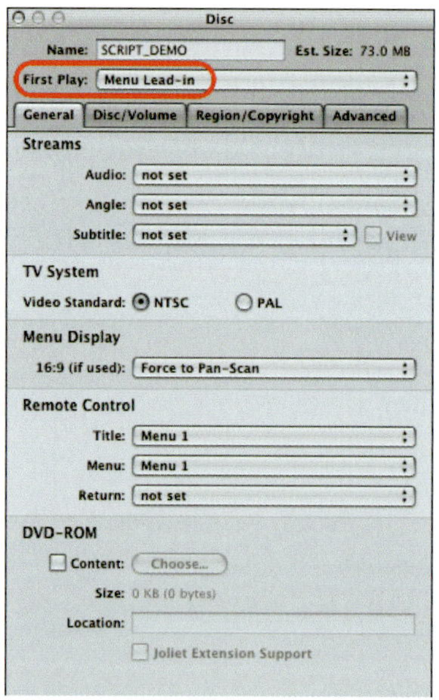

Figure 7-19. Choosing the script for the disc startup action

You also might notice that big, Not Applicable section in Figure 7-17. If a particular command cannot use the arguments or comparison areas, you get that lovely note. It actually helps keep the Command Inspector consistent—you always know where to look for the comparison setup on any given command.

Line 2 needs your help next. If you get here, that means you haven't played the intro clip yet. You're about to play it, though, so you want to remember that fact. You'll use the Set GPRM command and the mov operation to store the value 1 in your tracker variable. You'll need that Immediate type again. The resulting command should look like Figure 7-18.

Attaching a script to other things

The last step in making all this work is altering the default action for the disc. Instead of jumping to the menu or the intro clip directly, you want your script to be in charge of what gets played. You can choose a script as a target for the First Play action (and end jumps and menu buttons, for that matter). Go ahead and set up the Menu Lead-in script as the default entry for the disc in the Disc property sheet (see Figure 7-19 for an example).

Save your project and try simulating the disc again. You still should see the introduction clip right when you start the Simulator. But now try playing the main video clip, or going to the Languages menu. When you come back to the main menu this time, your script should run—it will notice that the introduction has been played, and it will pass you along to the main menu without playing the introduction again. Figure 7-20 shows the script's flow in the Graphical view (in DVDSP 3). Note that you don't get to see the links for conditional exits and jumps in the script. But that's OK because you still get a good sense of what's going on.

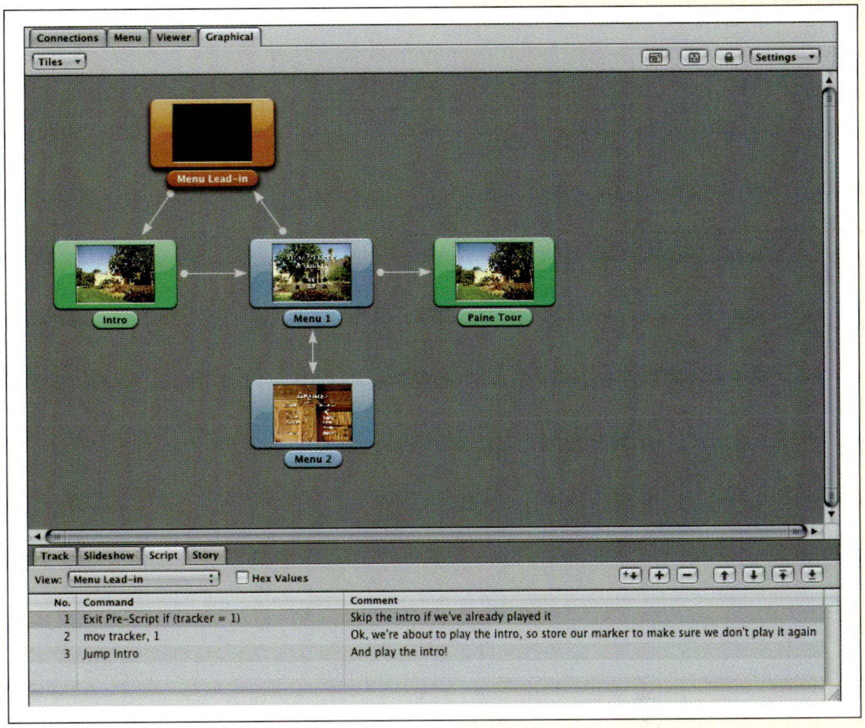

Figure 7-20. The Graphical view of the script project

Good for you! You just made a much more user-friendly version of your projects, while still keeping some of the "oooh" and "aaahhh" factor. You certainly can accomplish this task for the opening menu in other ways, but variables and scripts enable you to apply this technique all over your disc for all kinds of menus and user interactions. It's a powerful technique, so give yourself time to learn it.

Random Access

To help you along your way to mastering DVD scripting, let's look at another popular feature of the DVD scripting language: randomization. You already know that DVDs can jump around to different chapters or tracks at the touch of a button. But what if you don't want to touch that button? Video kiosks that display visitor information, for example, need to run on a continuous loop when no one is around. You could set up all the End Jump properties on the various clips to make a permanent, consistent loop, as shown in Figure 7-21.

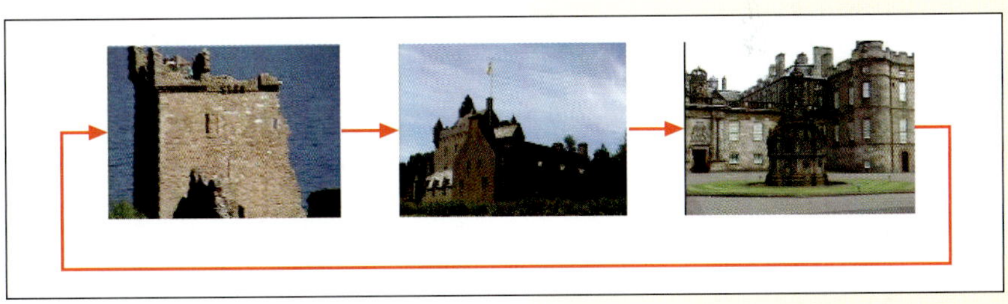

Figure 7-21. One (boring) possibility for playing our castle clips continuously

But that's somewhat boring, and you can do so much more with a DVD! Let's write a script to play the clips in a random order. First, grab the following castle clips again, as well as the new menu background from the *projects/ch07* folder:

- *cawdor.m2v*
- *cawdor.AC3*
- *holyrood.m2v*
- *holyrood.AC3*
- *urquhart.m2v*
- *urquhart.AC3*
- *quick_castle_tours_bkg.png*

Then, perform the following steps:

1. **Create the Menu.** Use the *quick_castle_tours_bkg.png* file to create a new menu. The menu will be your target for the disc (for now). It enables you to "ease into" the kiosk. It has a lone button that starts the random play, as shown in Figure 7-22.

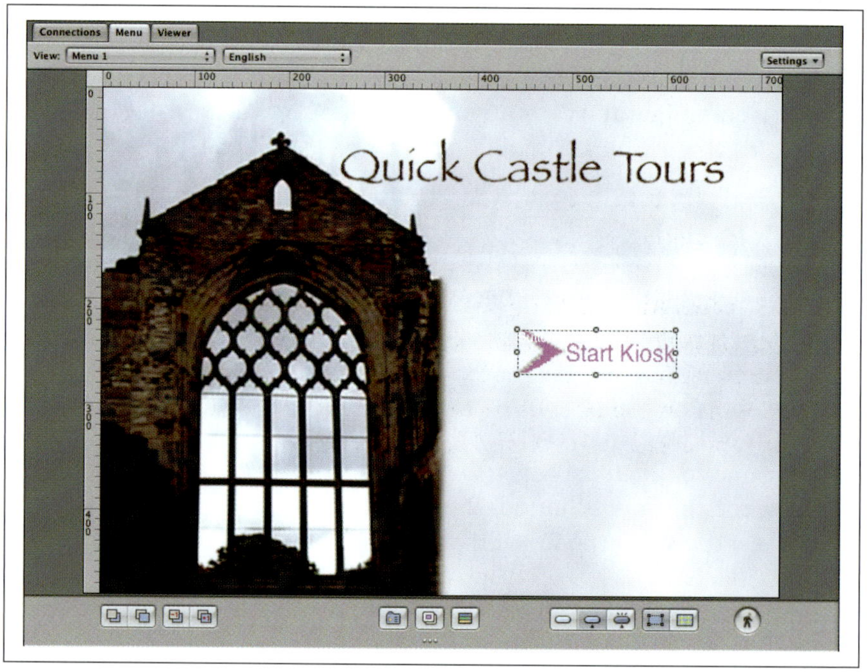

Figure 7-22. Our randomized playback starting point

2. **Create a Single Button.** Go ahead and define the button as shown in Figure 7-23, but don't worry about its target just yet.

3. **Add the Video Clips.** You also need to add the three castle clips. Create new tracks for each clip and name them as shown in Table 7-1 (to make the script more readable).

Table 7-1. Track names for the kiosk application

Track Number	Name in Inspector or Outline Tab
1 (cawdor.m2v)	Cawdor
2 (holyrood.m2v)	Holyrood
3 (urquhart.m2v)	Urquhart

You can name them anything you want, really, you just have to remember their names for use in the script later.

4. **Create a Script That Uses the Random Command.** Speaking of the script, create a new one and give it a name, as shown in Figure 7-24.

Figure 7-25 shows the script that will play a random clip from our small collection.

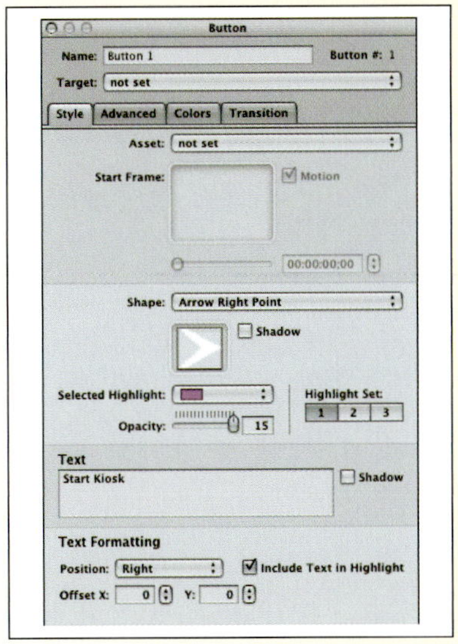

Figure 7-23. The properties for our lone button

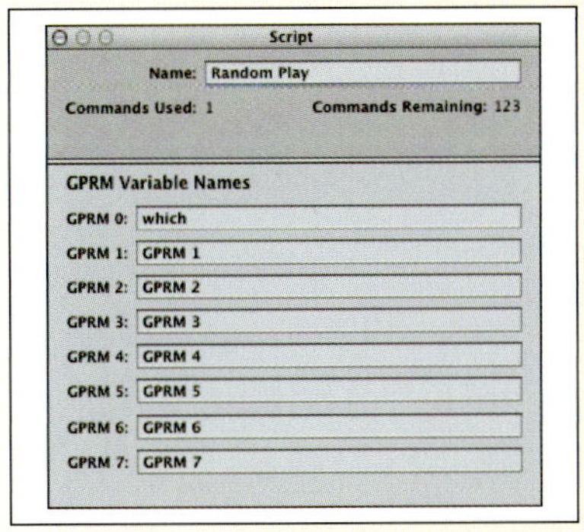

Figure 7-24. Our script and variable names for the random project

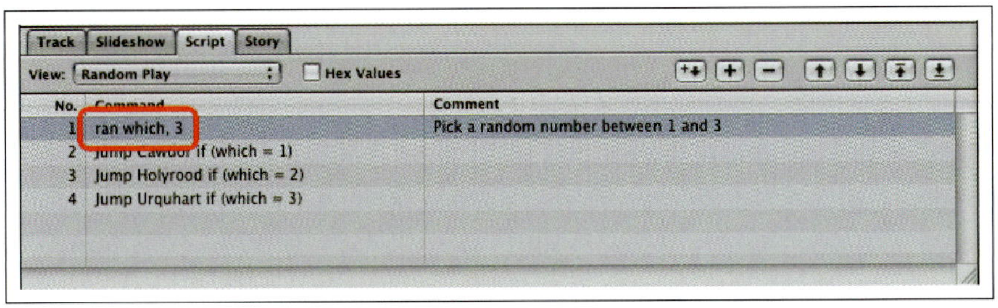

Figure 7-25. The script to play a random clip

The ran operation enables you to store a random number between 1 and some upper limit that you supply into a variable. We have three tracks, so we chose 3 as our limit. You still use the same Set GPRM command as you did in the smart version of the play-once project. Figure 7-26 shows the details on that first line of our script.

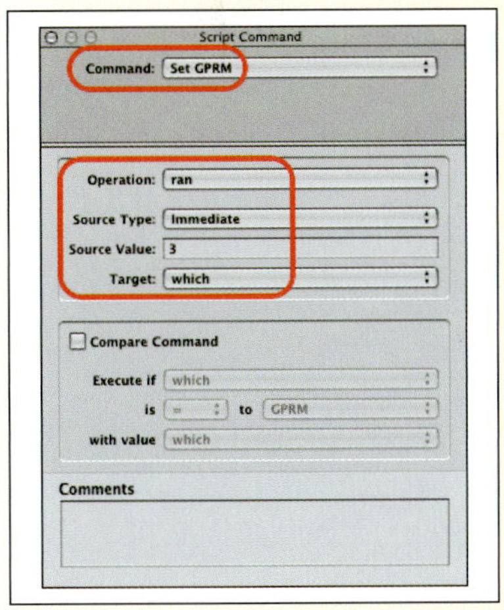

Figure 7-26. Setting up a variable with a random number

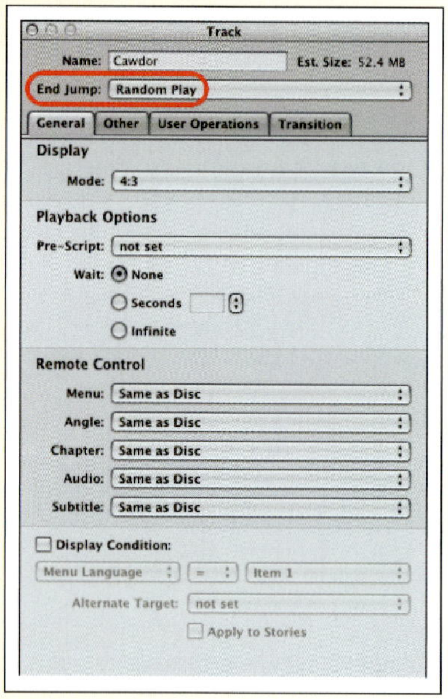

Figure 7-27. The new End Jump property for the Cawdor track

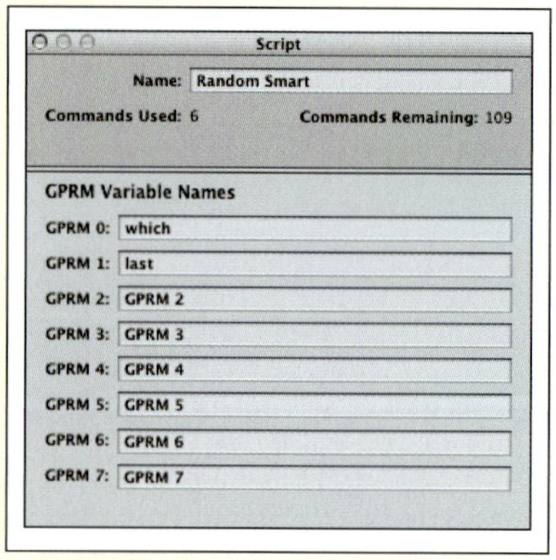

Figure 7-28. Naming another variable for our project

All you have to do to make this script take over playback is set it up as the End Jump action for all three castle clips. That way, when any one clip ends, this script will run and pick a new random clip to start. Figure 7-27 shows the Property Inspector for the Cawdor clip.

No screenshot will do the actual project justice, though, so simulate away. Or if you're a fan of old castles, go ahead and burn the disc to try in your home player. It should run continuously until you press the Menu button on your remote control. More importantly, it should play the clips in a random order—occasionally even repeating one.

Another smarter script

A clip that repeats is another problem you can solve with a smarter script. You can use a variable to track which clip was played most recently and to make sure it doesn't play again back-to-back. This type of setup requires a bit more work, though, because each clip needs to set a variable at some point.

Let's add to our list another variable with a nice name (see Figure 7-28).

Just as you did for the play-once-on-insert project, you can store a bit of information in a variable after it plays for your use later. Figure 7-29 shows one possible scenario for "mostly random" behavior that never will play a given clip twice in a row.

The new lines in this script are 2 and 3. Line 2 says "go back and get another random number if the current one matches the clip we just played." Line 3 copies the value of our variable into our last variable. That way we know what clip was last played.

You can accomplish this task in other ways. Each clip could have a simple pre-script or separate end-jump scripts. As you get more comfortable with scripting you'll find you almost always can handle any given task in multiple ways. The important thing is that your project works the way you want it to. Some methods might be smoother than others, but ultimately, if it works, it's the right way to do it.

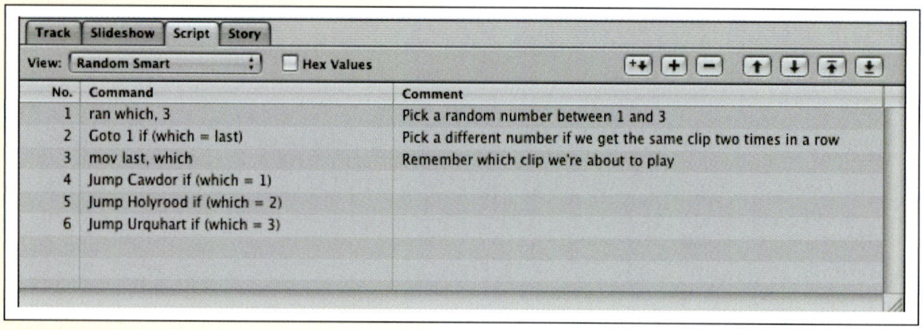

Figure 7-29. A smarter randomizer that won't play the same clip back-to-back

The play-all feature

Another popular feature on discs with lots of episodes or outtakes and such is the Play All button. A menu might have five clips that you can select and play individually, as we have done many times with previous projects. Often, at the bottom (or the top, it doesn't really matter) is a button that enables you to play all the clips, one after the other. You can't do this with simple end jumps—sometimes a clip jumps back to the menu, sometimes it jumps to the next clip. Scripts make easy work of it, though.

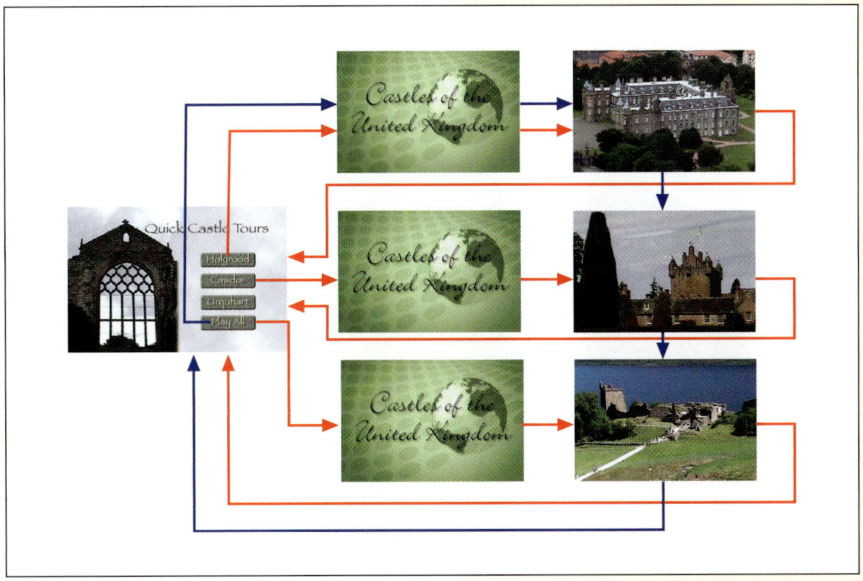

Figure 7-30. The flow for choosing an individual clip (red) or the play-all feature (blue)

For simple play-all capabilities, you'd use a script similar to our mostly random kiosk script: check a variable and jump to the next item in the list. We're going to make our play-all project a bit more interesting. We'll go back to the castle clips, but this time they all have a common introduction that plays at the beginning. If we play one individual clip, we want that introduction to play in its entirety. If we choose the play-all option, however, we want only the first introduction to play. The remaining clips should just jump right into their content. Figure 7-30 highlights the flow we're after.

If you want to just jump in, open the *playall_unfinished.dspproj* file. That has the menu and video clips set up, but leaves out the play-all functionality. Or you can start a new project from scratch and load the following files from the *projects/ch07* directory:

- *cawdor_full.m2v*
- *cawdor_full.AC3*
- *holyrood_full.m2v*
- *holyrood_full.AC3*
- *urquhart_full.m2v*
- *urquhart_full.AC3*
- *castle_playall_scaled.psd*

Here are the steps to follow:

1. **Create the Video Tracks.** Go ahead and make the tracks for each video clip (and attach their respective audio clips). You should notice that each clip comes with a chapter marker that notes when the content of the video actually starts. Those markers will come in handy as you write your script. (They also have a second marker that appears near the end of the clip. That one will be useful when testing the script.)

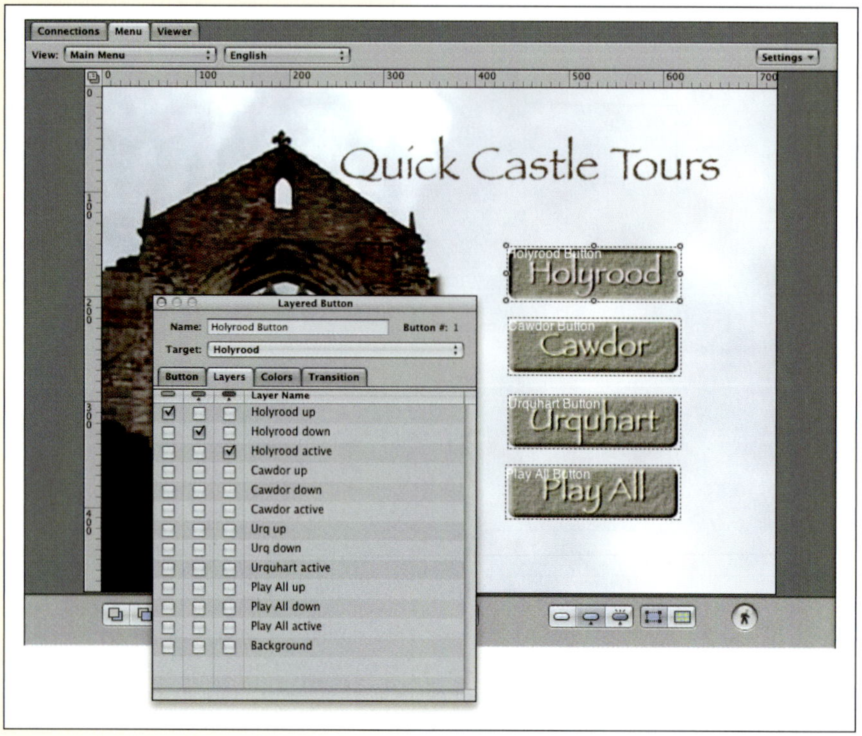

Figure 7-31. The layered menu (and the layered button setup) for the play-all project

2. **Create the Menu and Buttons.** Make a new layered menu with the *castle_playall_scaled.psd* image. You have the standard three buttons for the individual clips and your Play All button. You're using the same layered approach for the buttons as you saw in Chapter 3. Once your buttons are configured, you should see a screen similar to Figure 7-31.

3. **Write the Scripts for the First Three (or Two) Buttons.** The first three buttons have direct targets that point to their respective video tracks (e.g., the Cawdor button's target is Tracks → Cawdor → [Track]). The Play All button, however, will use a script as its target. The video clips will use complementary scripts for their End Jump properties. Those end-jump scripts are pretty easy, so let's tackle them first.

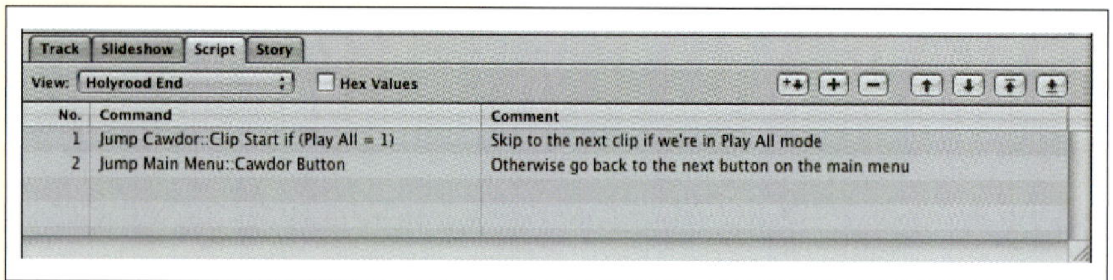

Figure 7-32. An end-jump script for the play-all project

After you finish a particular clip, you have two choices: return to the main menu or play the next clip in the sequence. You've seen scripts like that before—the play-once-on-insert was very similar. You just need to check a variable to see which direction to jump. You'll be using the Play All (renamed GPRM 0) variable for this project. The Holyrood clip end-jump script should look something like Figure 7-32.

You just listed the "next" clip directly in the script. That means this script will work for only the first clip. That's OK. You can have up to 10,000 scripts per DVD! You're in no danger there. You could write a more general (and more complex) script that would work as the end-jump script for all three clips, but that isn't always desirable. You have the space, so separate but simpler scripts probably are easier. Save the big scripts for the big projects—or projects in which your clients change their minds all the time. That's when you want general scripts that you don't have to rewrite just because you swapped the order of two clips.

You'll need a similar script for the Cawdor clip. An easy way to create this script is by selecting it in the Outline tab and pressing ⌘D. That duplicates the script. Rename it and edit the jump targets to match Figure 7-33.

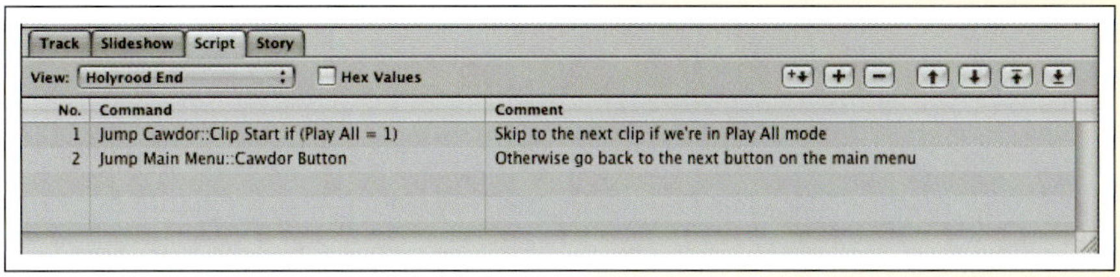

Figure 7-33. The end-jump script for the Cawdor clip

You don't have to write such a script for the Urquhart clip because you always go back to the main menu after that clip completes. You just set the end jump for Urquhart to point to the menu directly. And just in case you forgot, you need to hook up the End Jump properties for the Holyrood and Cawdor clips as well. They should point to their respective end scripts.

4. **Write the Script for the Play All Button.** The big remaining task concerns writing the play-all script. That's actually not too hard because you made separate helper scripts for all the clips. All you have to really do is set the Play All variable to the right value and start the first clip. That script looks like Figure 7-34.

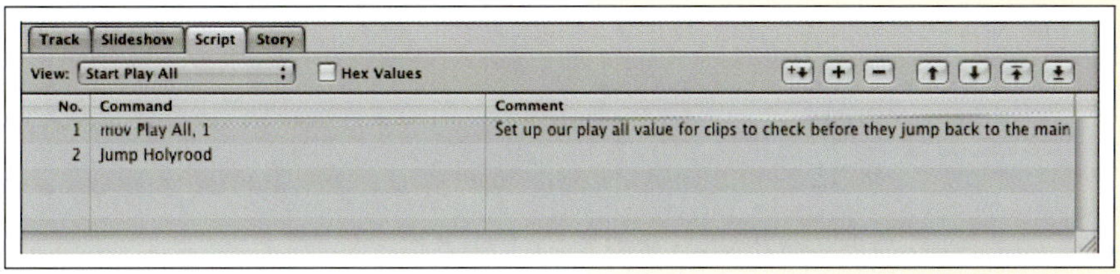

Figure 7-34. The script that launched a thousand clips; well, three, anyway

Three simple scripts give you some pretty impressive functionality. Not a bad start. And it only gets better the more you play with scripts.

Testing and Debugging Scripts

Simulating DVDs with scripts is vital. With all the fun and extra features that scripts allow, you can add a lot of complexity to your projects, so you want to check your scripts thoroughly. The Simulator is the quickest route to verifying (or fixing!) the flow of your scripts. You certainly can burn discs and test them in a player, but that adds both time and money, so it's best to get comfortable with the Simulator.

Most bugs come in the form of forgotten states—for instance, you set the current audio track for one feature and forget to change it back to the default. Fortunately, the depth of DVD scripts is limited, so you can verify that your disc works as intended—it just takes a bit of time and patience.

As an example, our previous project with the Play All button left out a tiny detail. What happens if you choose the play-all option and then come back to the menu to repeat one segment? You never wrote any code to "unset" the Play All variable. After you go through the play-all option once, every clip you play after that will think the disc is in play-all mode. That wasn't obvious (at least to some of us) as we were setting up the project. But it became very obvious once we simulated the disc and played around with the menu.

So, how did we discover the bug? Regrettably, it wasn't via a very scientific method. We played several of the individual clips and they all worked properly. We tried the Play All button and it worked great, too. We were on a roll! But then we went back and tried playing the second clip again just for fun. It started OK, but at the end it went on as if it were part of the play-all sequence. We quickly jumped back to the menu and tried the other buttons. Sure enough, they all were stuck in the play-all mode.

Testing a DVD really just involves a lot of patience. You have to go through all (or at least most) of the combinations of buttons and menus and make sure the disc continues to behave the way you expect. When you find anomalies, note them and keep testing. When you have enough anomalies that it becomes frustrating, use your notes to help you debug the scripts. Then come back and test again. Take more notes, do more debugging, and then test some more. It's not glamorous work, but it does make for a more professional product.

In fact, after you think you've tested the heck out of your DVD, burn a copy and give it to a friend or colleague. Ask her to run through it and try to break things. You'd be amazed what folks in the "real world" will do to a DVD!

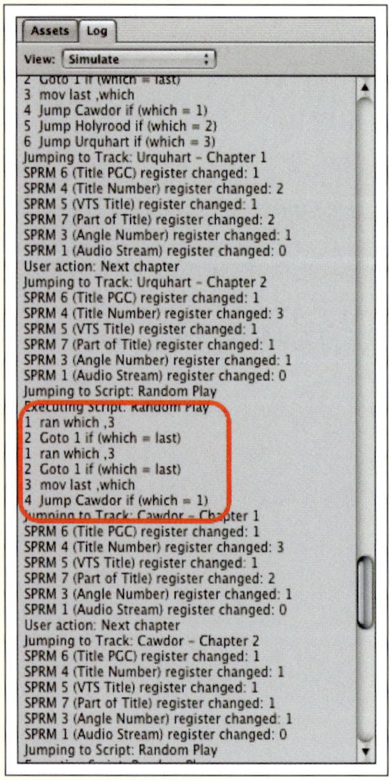

Figure 7-35. Part of the log file produced while simulating our smart randomizer project

Although DVD scripts are not nearly as complex as computer software, they are essentially the same things. You need to test them before shipping a DVD just like software companies do. Or rather, just like software companies should do. We haven't really embraced the idea of "beta" DVDs yet. Make sure your DVD really is ready before sending it off for duplication.

Debugging tools

You do have a few tools available to you for debugging your DVD projects. The first is just you. Make a diagram of the scripts you're going through and see if anything stands out. This is a remarkably powerful approach, if a bit old-fashioned. You can modernize this step somewhat by using products such as OmniGraffle, which ships for free on all new Macs. (Those of you without it can purchase a copy at *www.omnigroup.com*.)

DVDSP also comes with a Log feature that can be very helpful. It's one of the tab-or-window entities. You can Control-click one of your existing windows to load it, or just press ⌘3 to get the log as a standalone window.

You can run through your project in the Simulator and watch the log file. (Here's where the two-headed capability of most Macs comes in very handy if you've got a spare monitor lying around.) Figure 7-35 shows the type of output you're likely to see.

If you look at the highlighted part of the log, you can see one of our scripts running. This log came from simulating the smarter random project. You can see the execution of each line. In this case, the random generator came up with the same number twice, so we went back for another number and then continued, and eventually jumped to the Cawdor track.

You can also watch the state of the player in real time by bringing up the Info panel on the Simulator (click the "i" button on the Simulator's remote control).

SPRMs

You might notice some other entries in the log from Figure 7-35. The System Parameter Register Memory entries (or just "system parameters") can track (and alter) the state of the player. Appendix A lists all these registers and their uses.

> **STUDIO ADVICE**
>
> **Log Burning**
>
> You should remember the Log tab when you're burning new projects. Any unconnected buttons or other warnings/errors will show up here. For complex projects, it's a great idea to "burn" a DVD to disk. You can throw the image out if you want, but you'll have a complete burn log to review before spending the time and money of burning a real disc.

Display Conditions

Before we close out this chapter on scripting, we should mention one other script-like feature available in DVDSP. You can attach a *display condition* to menus and tracks. This is like a simple script that tests some condition before playing the menu or track. If the condition is met, the item will play as usual. If the condition is not met, playback jumps to some other resource you provide. Figure 7-36 highlights the Display Condition area on a track's Inspector page.

For example, you could have written the play-once feature with a very simple script and a display condition. For the script (set as the End Jump target for your intro clip) you just need to record the fact that you played the intro and go on to the main menu, as shown in Figure 7-37.

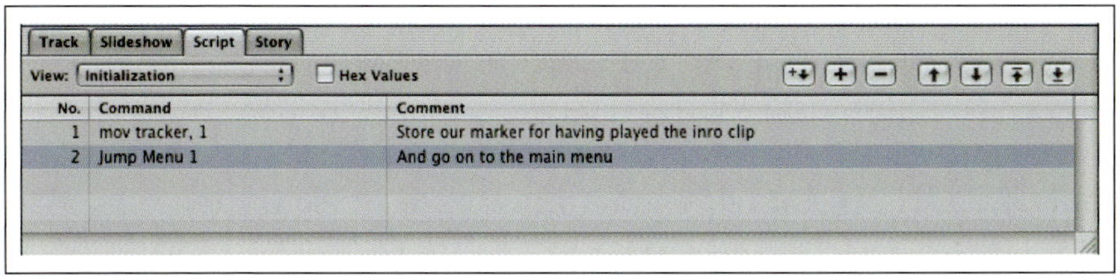

Figure 7-37. A simple script to record the fact that you've seen the intro clip

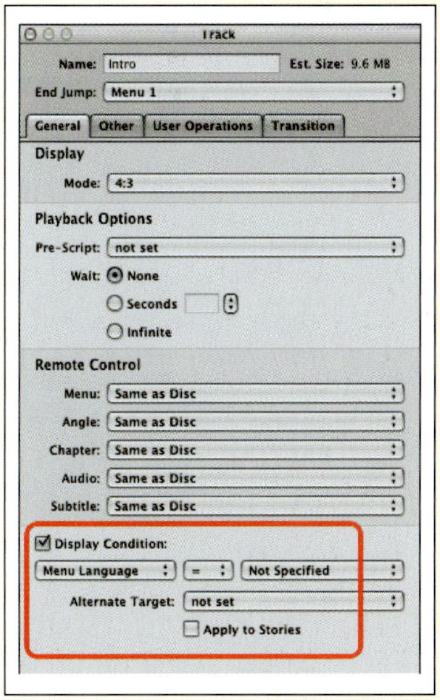

Figure 7-36. The Display Condition section of a track

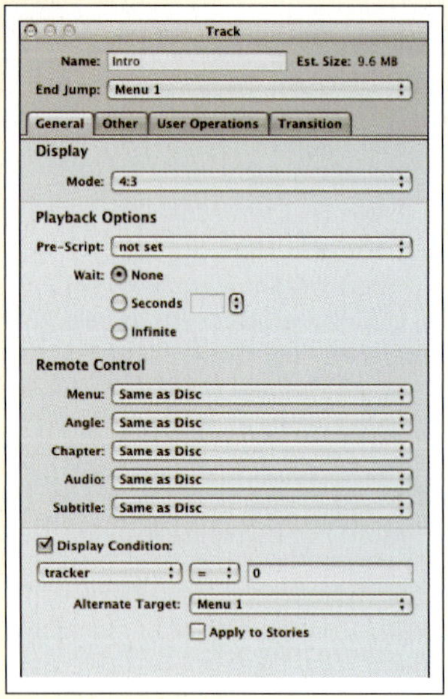

Figure 7-38. The display condition for the new play-once project

Now attach a condition to the intro clip. If the introduction has not been played, go ahead and show it. Otherwise jump to the menu. That configuration looks like Figure 7-38.

Of course, for this to work properly you need to change all the other End Jump properties to point at the intro clip rather than the main menu. And you have to make sure your menu doesn't have a pre-script that points back to the intro clip! (This is an easy mistake to make if you just modify the previous play-once project.) This works, and is just one more example of the fact that multiple solutions exist to almost every scripting task you tackle.

Closing thoughts

There is a lot to DVD scripting that we haven't covered here. For instance, in the next chapter we cover a topic called Easter eggs. Also, Appendix A contains a complete reference of the DVD scripting language. Scripting is an area where your imagination really is the limit. Remember that it's all software, and the Simulator doesn't break. Play around with stuff and don't be afraid to experiment.

The Web has a lot to offer as well. Searching for "DVD scripting" at Google turned up a different batch of sites every time we tried it over the course of writing this book. That's partly an indication of how long it took to get the book done, but mostly it's a reflection of the changing—and increasing amount of—information available on this topic.

If you're a programmer (or have access to one at work or through friends), DVD scripting really is just another language. Programmers will take to this like fish to water. If you're *not* a programmer, you still can get around quite well by snagging other people's projects and tweaking them to work with your content.

In any case, we want to play around with scripts a bit more, so let's get on to the next chapter and start cracking open those Easter eggs!

Games and Easter Eggs

8

This chapter includes some of the lightest themes, but also the heaviest use of DVDs. (Isn't that always true of computer games?) We'll look at adding fun, interactive elements to your DVD projects with some tutorials, Easter eggs, and outright games.

In this chapter

DVD Games

With scripting at your command, you can create some remarkably interactive games on DVDs. You can make simple "choose your path" stories or intricate treasure hunts that track a viewer through complex and interesting menus or video clips.

For the games we'll develop in this chapter, we're going to concentrate on features you've already seen in previous chapters. We want to show you how to use those features in unique ways to accomplish some spiffy projects. In the "Easter Eggs" section, we'll introduce a few new concepts that lend that "next-level" feel to your DVDs.

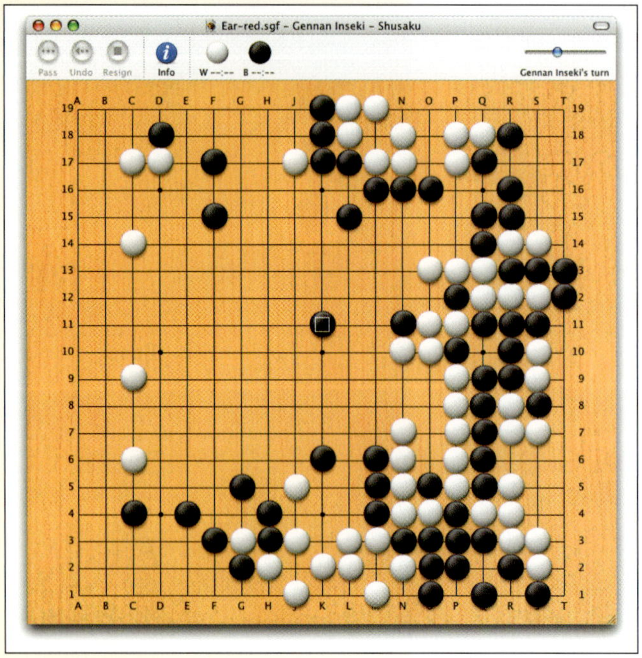

Figure 8-1. A famous game from 1846; black's clever move in the center caused his opponent's ears to turn red

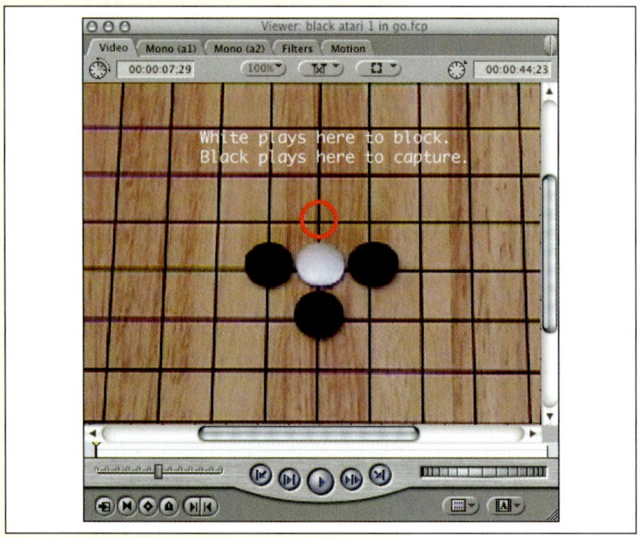

Figure 8-2. A simple atari scenario in Go

Teaching Go

Go is an ancient Chinese game that is enjoying a renaissance throughout Asia. Its rules are remarkably simple, but mastering play takes a lifetime and even then leaves room for learning. Basically you and your opponent take turns placing stones on a board (see Figure 8-1). There's obviously more to it, but that's really the crux of play. If your opponent surrounds your stones with his stones, he can take your stones. You keep going until the game is, well, over. (It's usually obvious when the game is over, but not always obvious who won.) You can find more on Go at *http://senseis.xmp.net/* (among many other great sites on the Internet).

Let's look at a project that will teach viewers of your DVDs about one of the most common Go scenarios: atari. When your opponent can "capture" some of your stones in one move, you are "in atari." Figure 8-2 illustrates the simplest such scenario. Hopefully you can place a stone that will get you out of atari.

The lesson demonstration

The Go lessons are organized as a series of menus and video "results." The lessons start with a quick introduction to the problem (a video clip) and then enable you to choose the next move (a menu). Depending on your choice, you get either the "congratulations" message or the "try again" message (both are video clips).

Grab the following files (in the *projects/ch08/go* folder) to begin a project with two such problems:

- *menu_scaled.psd*
- *problem1_options_scaled.psd*
- *problem2_options_scaled.psd*
- *logo_clip.m2v*
- *problem1.m2v*
- *problem1.AC3*
- *p1m1 wrong.m2v*
- *p1m1 wrong.AC3*
- *p1m2 right.m2v*
- *p1m2 right.AC3*

- *p1m3 wrong.m2v*
- *p1m3 wrong.AC3*
- *problem2.m2v*
- *problem2.AC3*
- *p2m1 wrong.m2v*
- *p2m1 wrong.AC3*
- *p2m2 wrong.m2v*
- *p2m2 wrong.AC3*
- *p2m3 right.m2v*
- *p2m3 right.AC3*

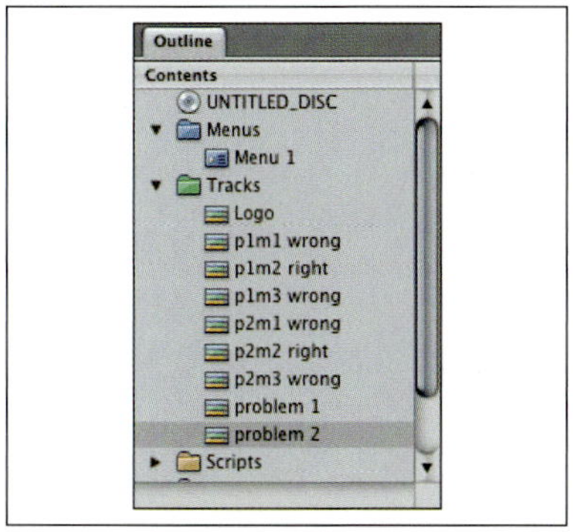

Figure 8-3. A listing of the video tracks in the Go project

Then follow these steps:

1. **Create Tracks for the Video Clips.** All the video clips are standalone clips, so go ahead and put them all in their own tracks. You'll be adjusting the End Jump properties as you build the menus, but for now you just want the video clips out of the way. Be sure to attach their respective audio clips, too. Figure 8-3 shows the outline view of the various tracks so that you can check your project.

2. **Create the Main Menu.** The other initial piece you'll take care of right away is the main menu. You'll present two problems for the viewer to try. Make a new menu from the *menu_scaled.psd* image. You'll just use built-in shapes (the "L Shape" to be exact) to create simple buttons. Figure 8-4 shows the menu and properties for one of our buttons.

   The targets for the buttons are their respective video clips (e.g., *problem1.m2v* for the Problem 1 button). The End Jump properties of those two clips should (eventually) lead to our games menus. We don't have those menus built yet; that's our next topic.

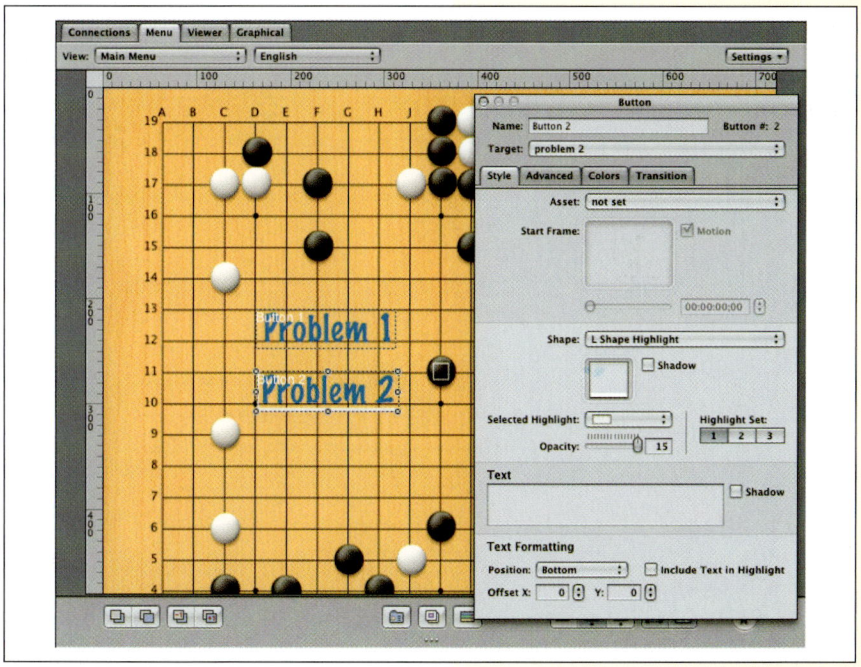

Figure 8-4. Our main Go menu and the properties for the second button

3. **Create the Menus for the First Game Problem.** Make a layered menu (Shift-⌘Y) with the *problem1_options_scaled.psd* layered image. Turn on the Background layer for the menu. You should have a screen

> **NOTE**
>
> *Atari is the Japanese word for "the situation." (The translation, more accurately, is "hit.") Japan has been leading the Go renaissance, so many of the words used to describe various board arrangements are Japanese. At least, that's true in the U.S. and Japan. And yes, the term did predate the game manufacturer, Atari Inc., which presumably took the word as its name when it released Home Pong in the early '70s.*

similar to Figure 8-5. You'll use the other layers to make your buttons become the Go stones.

4. **Create Move Buttons.** In this example, we're going to restrict the possible moves. Generally, you're limited to 36 buttons on a 4:3 aspect ratio menu, but building all the button layers for those 36 possibilities gets beyond cumbersome. We'll enable the viewer to pick from three locations to place the stone. Figure 8-6 shows the button boundaries. Table 8-1 has the exact coordinates for each button as well.

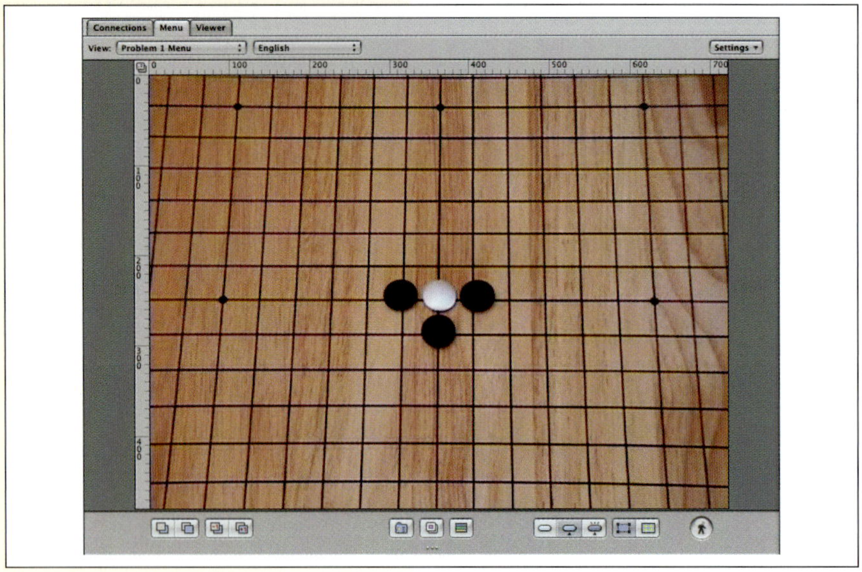

Figure 8-5. Our first game menu

Table 8-1. Coordinates for the buttons in the Problem 1 menu

Button	Top	Left	Bottom	Right
Button 1	195	295	223	332
Button 2	195	342	223	379
Button 3	195	389	223	426

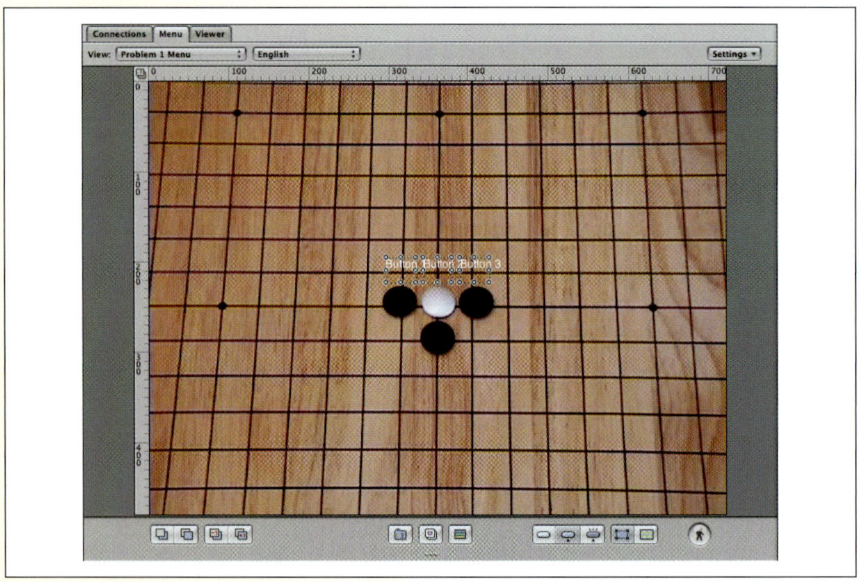

Figure 8-6. The first three moves as buttons in our menu

Recall that layered buttons have three states: normal, selected, and activated. Although the normal state can be completely transparent, we want to show the viewers what choices are available to them. We'll draw the traditional notation outline for the normal state. For the first button, then, choose the move 1 outline layer for the normal state in the button's Inspector, as in Figure 8-7.

Do the same for the remaining buttons. That should give anyone viewing the menu a good idea of where they are allowed to move. Figure 8-8 shows the board with all three buttons in their normal state.

The selected state will be an image of the appropriately colored stone. That layer is called move 1 in our menu. The activated state will be a combination of both the selected stone and the outline. Figure 8-9 shows the layer matrix for the completed first button.

You might want to save your project and simulate what you've got up to this point. Make sure you set your new menu as the End Jump property for the *problem1.m2v* track.

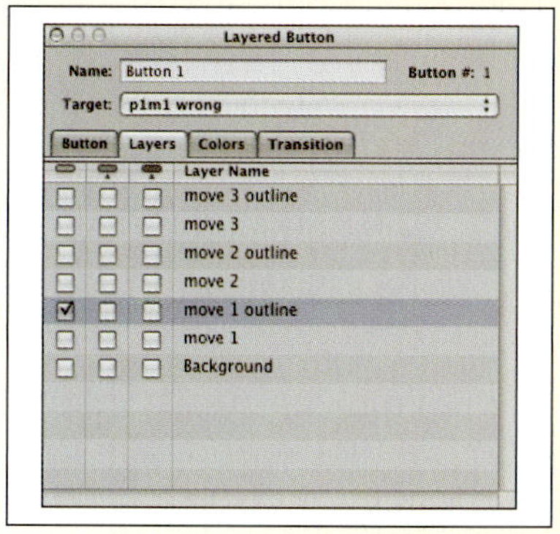

Figure 8-7. The normal state for the first button in the problem menu

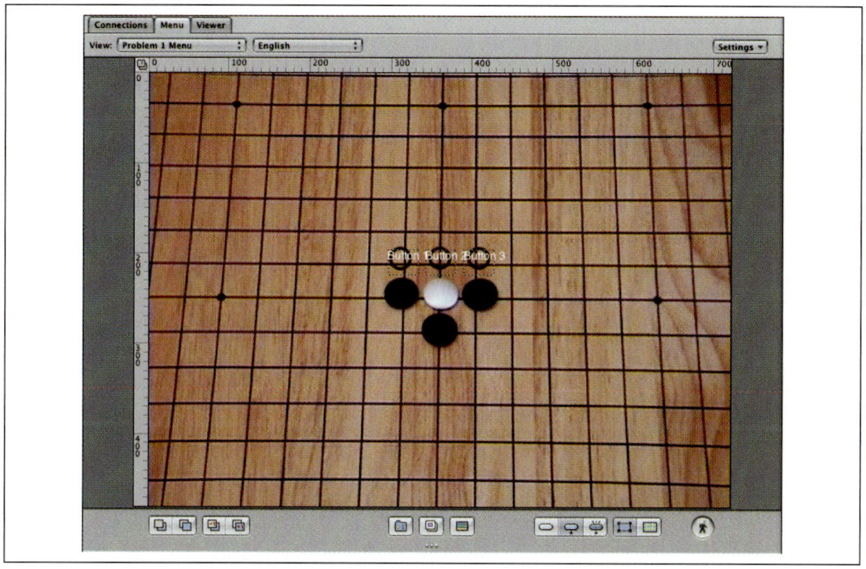

Figure 8-8. The Problem 1 menu with normal outlines defined for all the possible moves

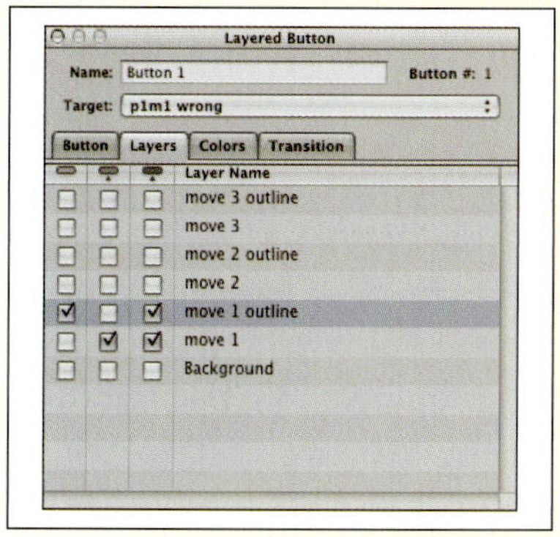

Figure 8-9. The full layer matrix for one of the buttons

5. **Set the Move Button Targets.** Where the buttons lead determines the flow of the game. Two of the buttons (Button 1 and Button 3) are incorrect and lead to individual clips that show the mistake. One of the buttons (Button 2) is correct and leads to a third video clip that highlights the correct move. The incorrect videos should lead back to the same problem menu so that the viewer can try again. The correct-move video should lead back to the main menu so that the viewer can try the next problem.

Go ahead and set the targets for the buttons, as shown in Table 8-2.

Table 8-2. Button targets and End Jump properties for Problem 1

Button	Button Target	Video Track End Jump
Button 1	p1m1 wrong.m2v	Problem 1 Menu
Button 2	p1m2 right.m2v	Main Menu::Button 2
Button 3	p1m3 wrong.m2v	Problem 1 Menu

These end jumps are highlighted in Figure 8-10.

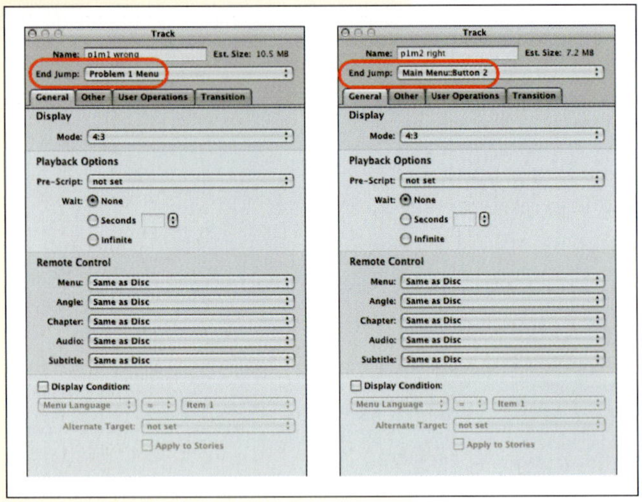

Figure 8-10. The "wrong" jumps (left) and "right" jumps (right)

6. **Create Menus and Buttons for the Second Game Problem.** You'll set up the second problem in a similar way. You'll build the menu from the *problem2_options_scaled.psd* image. Once again you have three possible choices for the viewer, and those choice layers are named the same. In fact, if you duplicate the menu from Problem 1 (highlight that menu in the Outline tab and press ⌘D), all you need to do is drop in the new Photoshop file and reset the background layer. You'll still have to set the button targets and make sure they are in the right positions, but their layers will be turned on already.

   Table 8-3 shows the coordinates of the buttons for this problem. Figure 8-11 shows the end result.

Table 8-3. Coordinates for the buttons in the Problem 2 menu

Button	Top	Left	Bottom	Right
Button 1	117	342	145	379
Button 2	153	389	181	426
Button 3	195	389	223	426

Now that the second menu is in place, go ahead and set the End Jump property for the *problem2.m2v* clip to point here.

Default Menu Buttons and Games

When setting up menus as games, you should remember that the menu will have a default button selected. In the Go game, for example, a stone immediately will appear on the board in the first possible move position (i.e., the first button is in the selected state). If you want to leave all possible moves open, you'll have to create a separate "starter" button that serves as the menu's default. Using the navigation tricks from the treasure hunt in the next section, you can arrange things so that once the viewer leaves the starter button, he can select only valid moves.

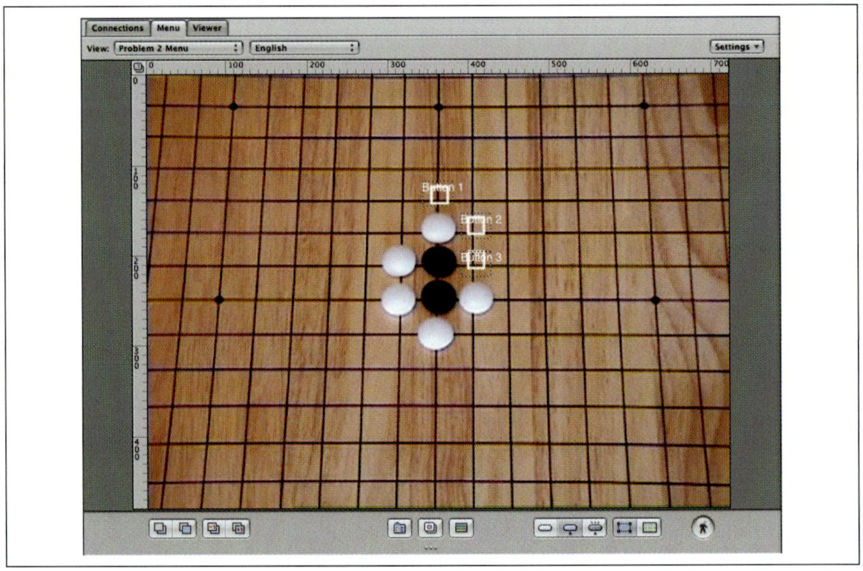

Figure 8-11. The second problem menu in our Go project

The targets for the buttons also are similar to those in the first menu. Here, buttons 1 and 2 are incorrect and should jump to *p2m1 wrong.m2v* and *p2m2 wrong.m2v*, respectively. Button 3 is correct and should jump to the *p2m3 right.m2v* clip.

The end jumps for the first two clips should lead back to the Problem 2 menu, as they did with the first problem. The End Jump property for that congratulations clip should be the main menu. Or, if you want to reward the viewer for getting through both problems, have the end jump lead to the *logo_clip.m2v* track. That track, of course, should set its End Jump property to the main menu. Figure 8-12 shows the final flow for this project in the Graphical view.

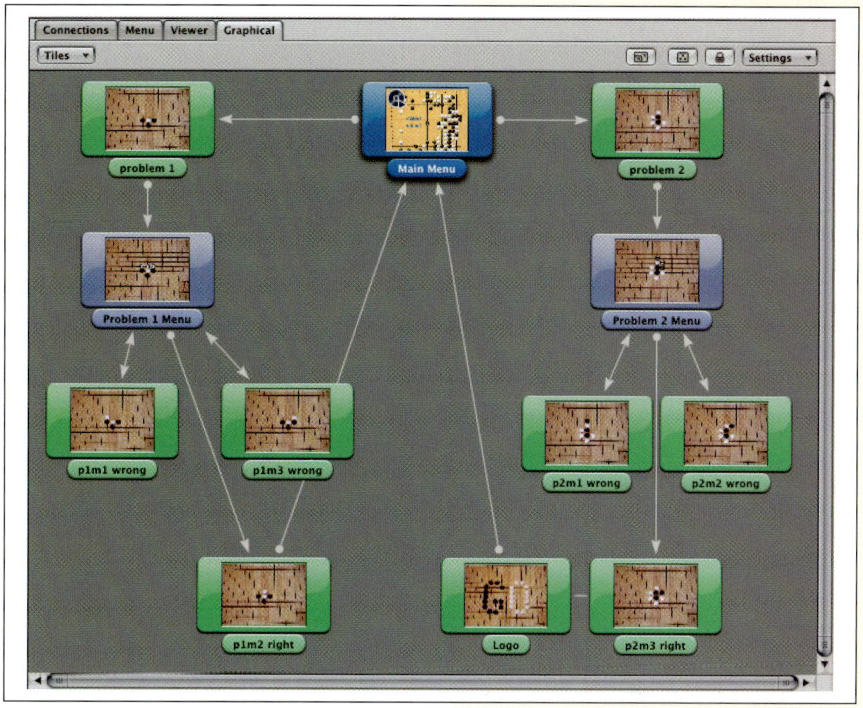

Figure 8-12. The full Graphical view of our Go project

You even can use some of the scripting tricks from Chapter 7 to enable the viewers to complete the problems in any order. (The treasure-hunt game in the next section also explores scripting for that purpose.)

Simulating games

Start the Simulator and test the game. Figure 8-13 shows our project on the first problem. Remember from Chapter 3 that using layered menus in software DVD players often requires that you click the desired button twice: once to select it and once to activate it.

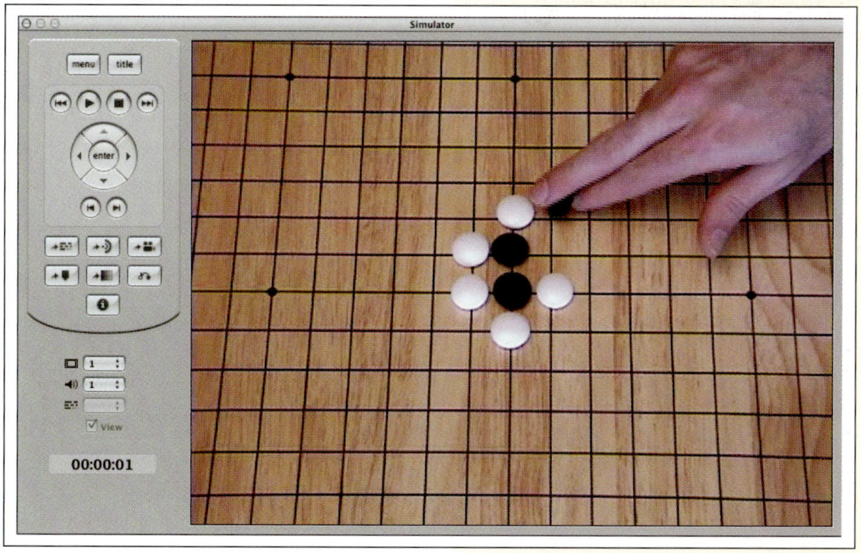

Figure 8-13. Our completed Go lessons, up and running

You can quit the Simulator and start it up again to reset the game's state. Remember that you also can make the Log tab visible and monitor the state of the menus there. Mostly it's back to the patience thing: trying all the combinations by hand is the

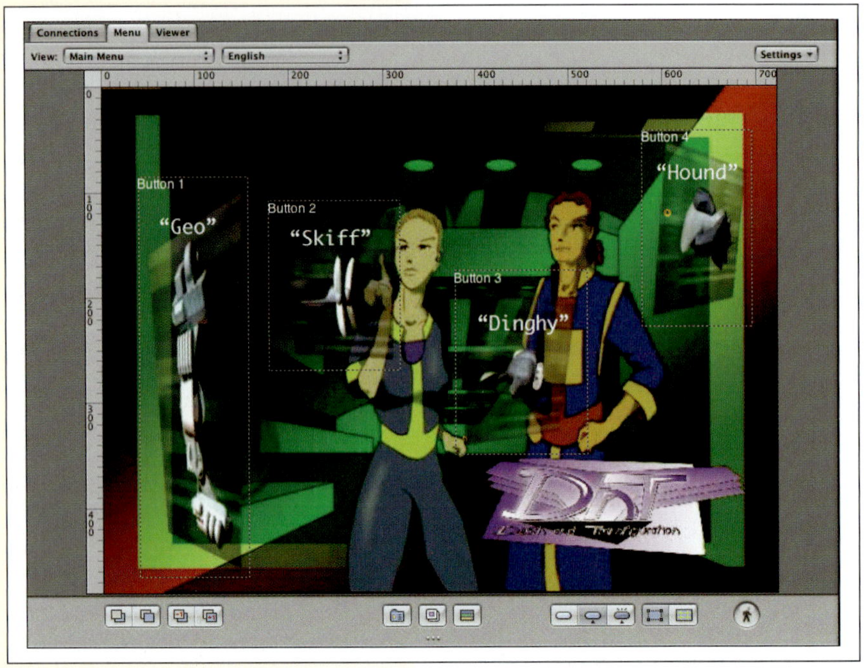

Figure 8-14. The ships from Death and Transfiguration

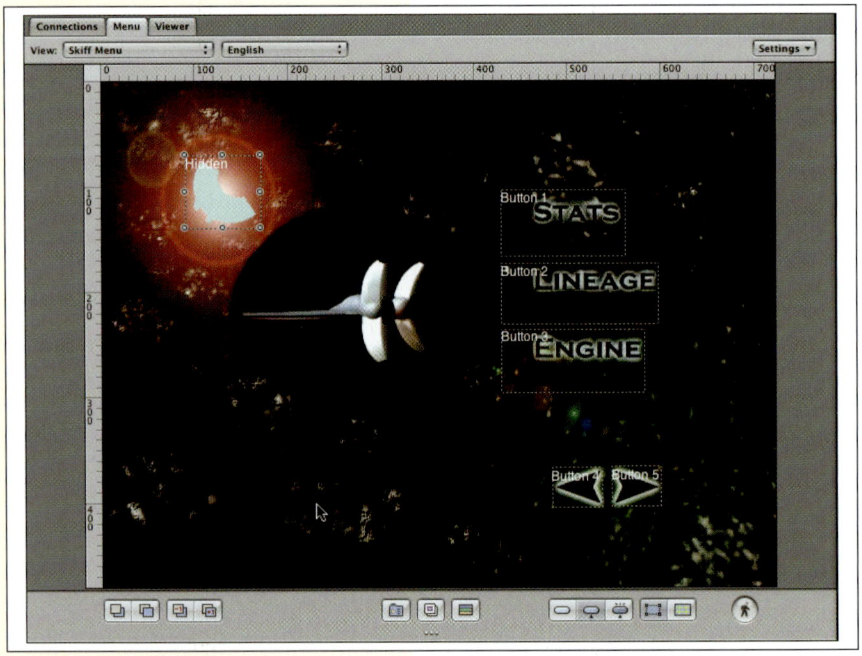

Figure 8-15. The first hidden button (upper-left corner) on the ship menu

best guarantee that everything is working.

Treasure hunts

Another popular game that shows up on DVDs is patterned after the classic treasure hunt. You're given clues to decipher the location of the "treasure." You can extend these games by having chains of clues that lead the viewer toward the ultimate goal, or they can consist of several pieces of treasure to be found throughout the disc.

A great example of such a scattered treasure hunt is the disc of the movie *Practical Magic*. Certain menus have special buttons that are obscured. Finding the buttons becomes the challenge. When you select one of the obscured buttons, a variable is updated. If you find all the special buttons, you are granted access to a previously unavailable video clip.

Let's re-create that type of hunt. We'll be using menus from the anime series *Death and Transfiguration* (for more on that game see Chapter 4, or visit *http://www.smuga-otto.com/dnt/*). The secret video clip will be a collection of original sketches not available anywhere else on the disc.

Obscuring buttons

Our "treasure" will be obscured buttons in the shapes of the various ships from the Death and Transfiguration world. Just for your reference, those ships are recapped in Figure 8-14.

Because we're trying to concentrate on building the game, we'll put each ship on one menu page. Bigger games can scatter the pieces across many more menus. Figure 8-15 shows one such menu with the Skiff as the obscured button.

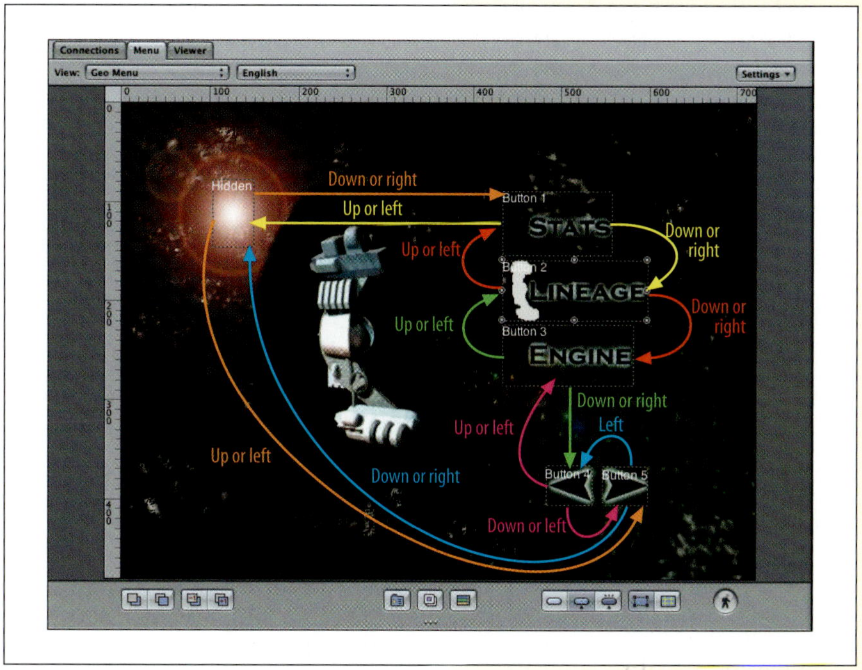

Figure 8-16. The default navigation for remote control buttons in our Geo ship menu

So, how do you obscure the button? Visually you can hide the button outright. If its normal state is completely transparent, the button won't show up at all on the screen. But you need to do a bit more to make sure it's somewhat tricky for the viewer to get to the button when navigating with the remote control.

By default, DVDSP makes each button accessible by simply cycling through the menu using one of the remote control buttons. For example, Figure 8-16 shows the flow of buttons in the menu for the Geo ship.

It's a bit messy, to be sure. And look at all the routes to our supposedly hidden button! This won't do. Fortunately that arrangement is just the default. As with nearly everything else in this project, you can adjust those values to determine which button comes next more explicitly. What

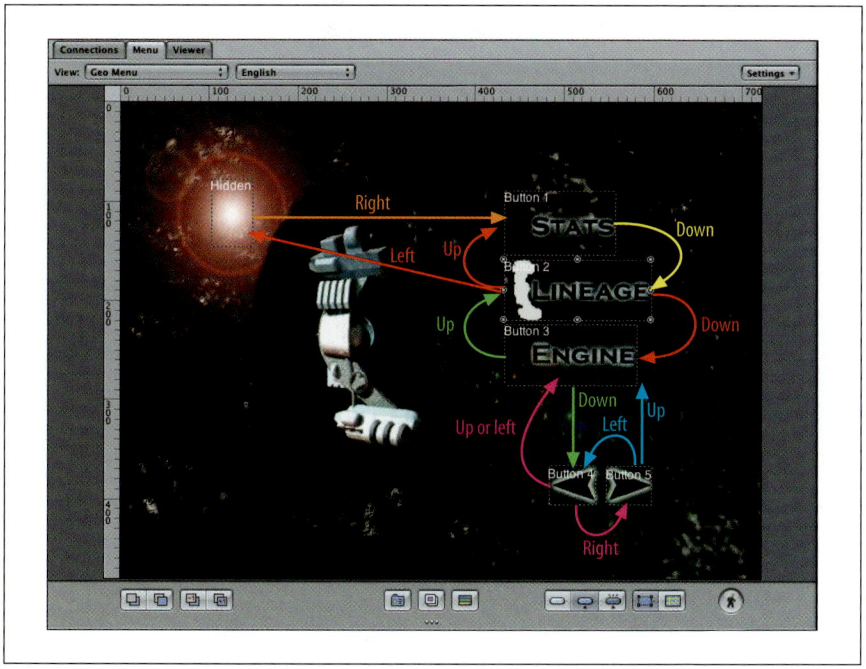

Figure 8-17. A cleaner approach to button navigation

Darn Those Computers

Of course, all this obfuscation does not apply to computers and DVD-playing software. Although most software players (such as DVD Player) provide a remote control with the normal Up, Down, Left, and Right buttons, most viewers simply use their mouse to click the desired button. They still can do that here—there's no real way to stop them.

So, if you're intent on using obfuscation as part of your treasure hunt, you have to do more than just restrict remote control access to the buttons. In larger projects, make a few menus that are just plain menus. That will force treasure seekers to scour the entire menu hierarchy. That's a worthy quest, even on a computer.

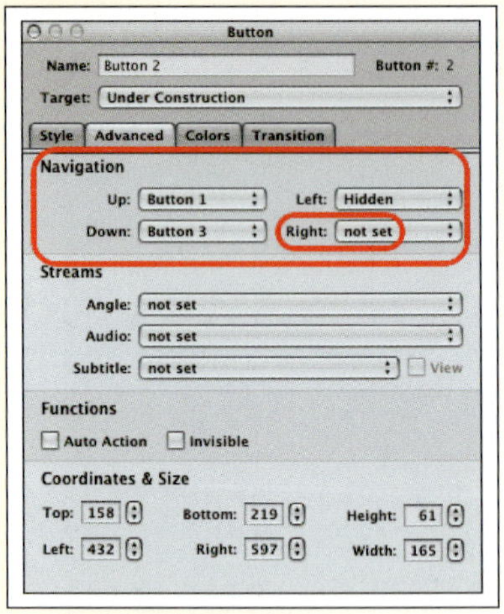

Figure 8-18. Manually setting the navigational properties for a menu button

you really want for your game is a flow similar to that shown in Figure 8-17.

This is much cleaner; well, somewhat cleaner, anyway. Plus, the viewer has to be much more careful before he can reach the obscured button.

To alter a button's navigation properties, use its Inspector. Figure 8-18 highlights the Advanced tab, where you can pick the next and previous buttons for each of the Up, Down, Left, and Right remote control buttons. The button in the figure is the one that serves as the connection to the hidden button.

From those four pop-ups, you can select any other button in the menu. Notice we "disabled" the Right direction by selecting the not set option.

Trying it out

If you want to try this technique, open the *dnt_hunt_unfinished.prj* file from the *projects/ch08/hunt* folder. All the assets have been imported and all the menus—save one—have been completed. You can practice with the navigation controls on the Skiff menu. Go ahead and select that menu in the outline. Your Menu tab should show a picture similar to Figure 8-19.

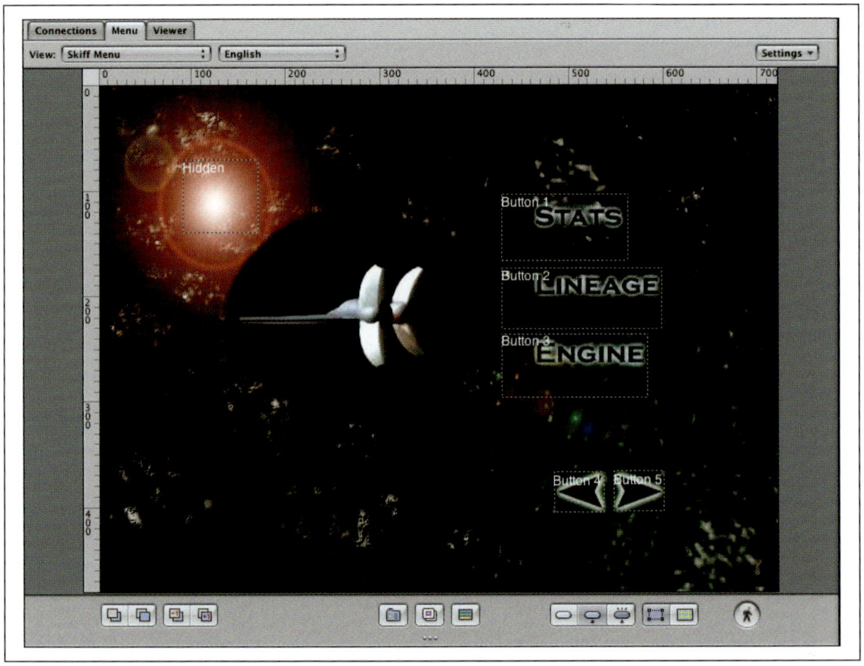

Figure 8-19. The unfinished menu in the treasure hunt project

Define the six buttons, as shown in Figure 8-19. Table 8-4 has their exact coordinates if you prefer not to eyeball the outlines. Don't worry about the navigation just yet. Let DVDSP fill in those values for you.

Table 8-4. Coordinates for the buttons in the Skiff menu

Button	Top	Left	Bottom	Right
Hidden	70	89	139	170
1 (Stats)	103	427	165	561
2 (Lineage)	173	427	230	596
3 (Engine)	235	427	296	581
4 (Previous)	367	482	405	535
5 (Next)	367	546	405	599

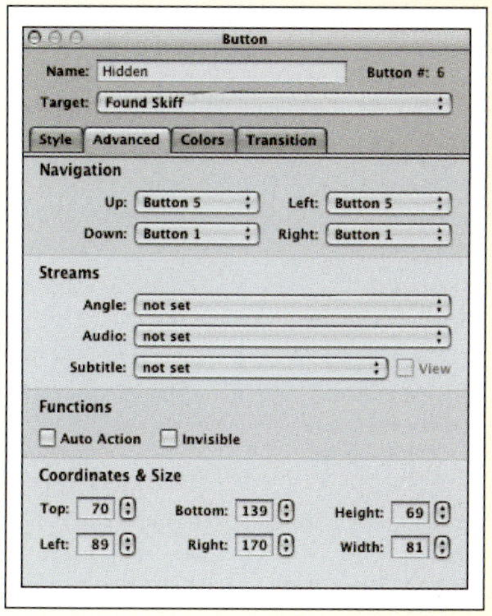

Figure 8-20. The initial state of the Hidden button

Now click the Hidden button and look at its Inspector. It should look like Figure 8-20.

You can reset any of the directions to navigate to a button of your choice—or again, to nothing at all. Figure 8-21 shows the final configuration for the four directions.

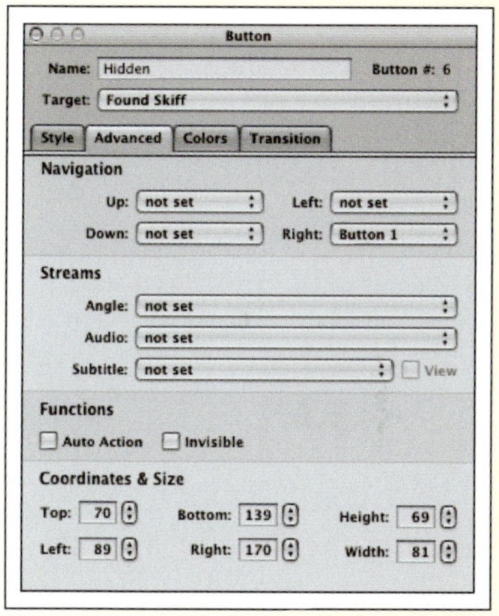

Figure 8-21. The final state of the Hidden button

Try simulating the project. You should be able to get to your Skiff button only from the Previous button at the bottom of the screen, as shown in Figure 8-22. We chose (arbitrarily) to restrict any movement that leaves the Hidden button so that it jumps only to the Right button. Another valid option would be to have every direction lead you to the Stats button.

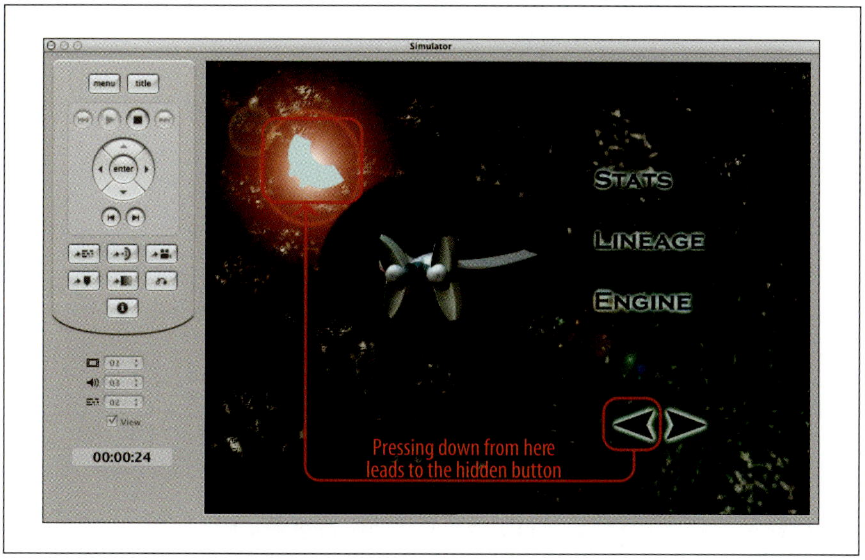

Figure 8-22. A treasure map for the Skiff menu

Treasure-hunt payoff

But how do you "keep score" in this treasure hunt? You effectively need to track four things: whether the viewer has selected each of the four obscured ship buttons. If he has, when he finds the last button, he'll jump to the Hidden video clip.

You have two choices for keeping score. You can use separate variables for each ship and check them all for the winning condition. Alternatively, you can use just one variable and store a particular value in it. This is more of a programmer's trick than you might be ready for, but it can be very flexible—an important trait in the limited world of DVD scripting.

The trick to using a single variable to store four values is using "bit" operations. If you've worked around computers much, you're probably aware that everything is stored (eventually) in *bits* (Binary digITs)—ones and zeros. Even other numbers such as 42 are stored with these ones and zeros.

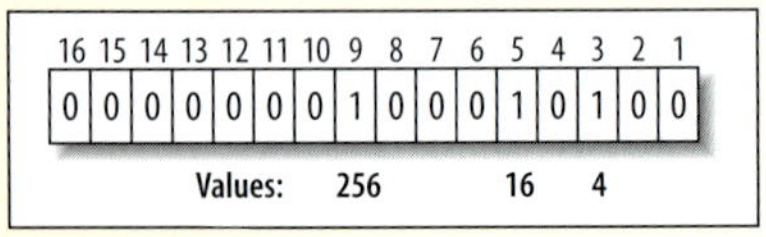

Figure 8-23. The value 276 represented as bits

On DVDs, numbers can range from 0 to 65,535. That translates to 16 bits for each number (see Figure 8-23). The DVD scripting language has several commands (`and`, `or`, `xor`) for manipulating the bits of a variable. If you use the right numbers, you can turn individual bits on and off. Table 8-5 shows the decimal numbers you need to affect each bit in a variable.

Table 8-5. Bit manipulation values for DVDSP

Bit	Test and Set Value	Clear Value (65535—set value)
16	32768	32767
15	16384	49151
14	8192	57343
13	4096	61439
12	2048	63487
11	1024	64511
10	512	65023
9	256	65279
8	128	65407
7	64	65471
6	32	65503
5	16	65519
4	8	65527
3	4	65531
2	2	65533
1	1	65534

Those three operators, `and`, `or`, and `xor`, work on the individual bits between any two numbers. For example, if you `and` the numbers 133 and 27, the computer combines bit 1 from 133 and bit 1 from 27 to get bit 1 of the result. That happens for each remaining bit (2–16). Of course, it happens very quickly on a computer. Table 8-6 shows the results of using each operator.

Table 8-6. Bit operation results

First Bit	Second Bit	and Result	or Result	xor Result
0	0	0	0	0
0	1	0	1	1
1	0	0	1	1
1	1	1	1	0

To manipulate the bits in a variable, you can apply those bit operators from the Operation pop-up in the Set GPRM script command as follows:

- To set the 11 bit in the variable A:
  ```
  A = A or 1024
  ```
- To clear the 8 bit in the variable B:
  ```
  B = B and 65407
  ```
- To figure out if the 5 bit is set in variable A (without changing the value of A):
  ```
  C = A and
  if (C > 0)
  ```

There's certainly a lot more to this if you want to get fancy (such as testing to see if the 5, 6, and 14 bits are set at the same time), but our example does enable you to keep 16 yes-or-no answers in one variable.

We'll use this bit trick to store and test for our four ships. As we find each ship, we'll set a different bit, as shown in Table 8-7.

Table 8-7. The assigned bits for the four ships

Ship Name	Assigned Bit
Geo	1
Dinghy	2
Hound	3
Skiff	4

Figure 8-24 shows the script to execute when the viewer "finds" the Skiff button. We set this script as the target for the button. Activating the button qualifies as "finding" it.

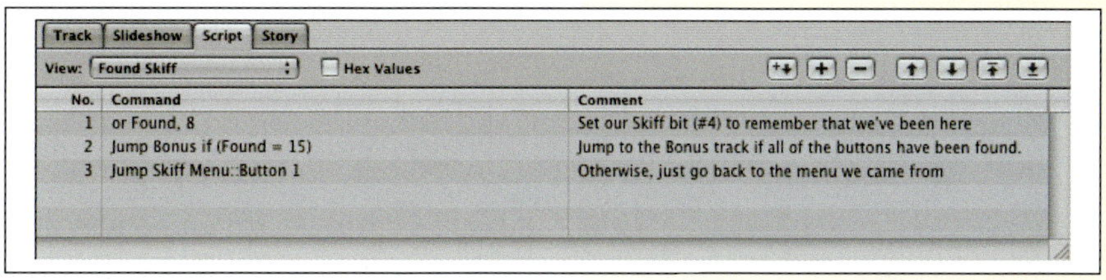

Figure 8-24. The "keep score" script for the hidden Skiff button

Notice that you jump to the bonus track when the Found variable equals 15; 15 is the sum of all the ship bits (8 + 4 + 2 + 1 = 15). Only after you have found all four ships will you be able to see the bonus track.

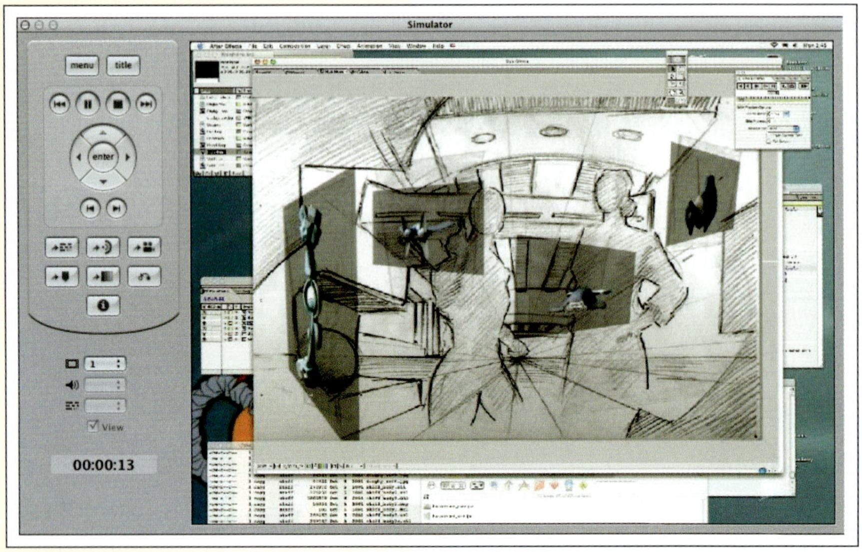

Figure 8-25. The bonus track from our treasure hunt

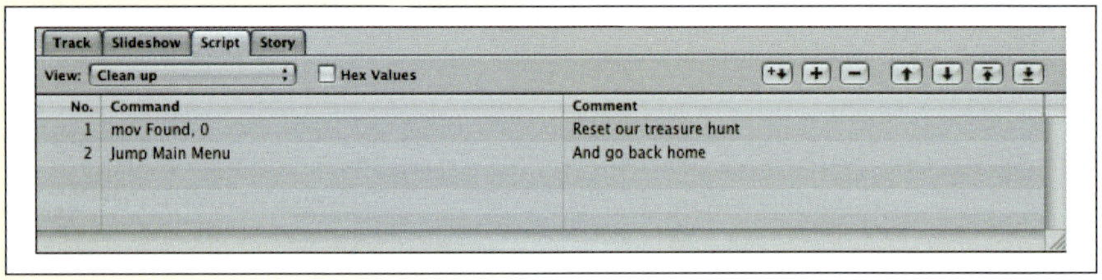

Figure 8-26. Cleaning up the score-keeping variable

Also notice that you can apply the or operation to a bit as often as you want. If the bit is already on, no harm is done. This has two nice side effects: 1) viewers can "find" a ship button multiple times without breaking things; and 2) viewers can't just "find" the Geo ship 15 times to get to the bonus material.

Go ahead and simulate the project (see Figure 8-25 for a glimpse of the bonus material). Try to find the other buttons once before looking at their locations in the project. Can you get to the bonus clip? Great! As you can see, games such as this take a lot of planning, but the work can result in some fun DVDs.

Cleaning up

Just to keep things clean, the End Jump property for the bonus material uses a simple script to clear out the Found variable. Then it just jumps to the main menu. Figure 8-26 shows that Clean up script. Figure 8-27 shows the flow of the entire project in the Graphical view.

Easter Eggs

The popular definition of an Easter egg in the DVD world is any unadvertised feature of a disc. For example, something as simple as a gag reel appended to a movie that plays after the end credits can be an Easter egg. If the term *Easter egg* doesn't give it away, you're supposed to go hunting for these eggs and be thrilled and squeal with joy when you find one. Well, the squealing is optional, but you'd be surprised....

Things such as our treasure hunt can be considered an Easter egg. Usually the kicker is whether you advertise the game. If you have a big bullet item on the packaging that says "Play 'Find the Ship' to see bonus artist sketches!", its not an Easter egg, it's a game. If you don't tell anybody about it and let a rumor start on the Net, you have yourself an Easter egg. In fact, entire web sites are devoted to helping hapless (or lazy) viewers uncover the eggs in commercial DVDs. A good site is *http://www.dvdeastereggs.com*.

Simply obscuring a button can qualify as creating an Easter egg. But we want to try something more interesting. We'll password-protect our next Easter egg.

Passwords and numeric input

One clever trick that falls into the serious Easter egg category is using a "password" to protect a menu or other asset on your disc. There are no actual password entry mechanisms, but you can fake it.

On remote controls, you'll find numeric keys. Depending on the player, you might be able to skip to a particular chapter or time in the movie. They also are used in configuring the DVD player in setup menus and such. In your menus, though, they often still function. It's not discussed frequently, but the numeric keys will activate the onscreen buttons to which they correspond. In English, you say? Pressing the number 5 selects (and possibly activates) the fifth button on the screen. Figure 8-28 shows a menu and calls out the number of

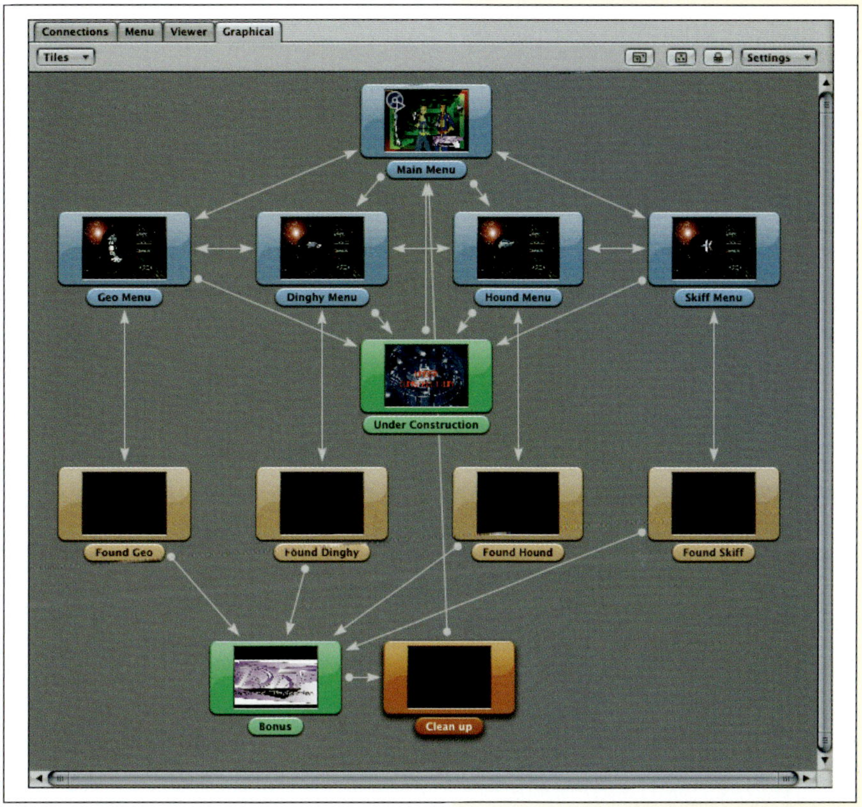

Figure 8-27. The Graphical view of the hunt project flow

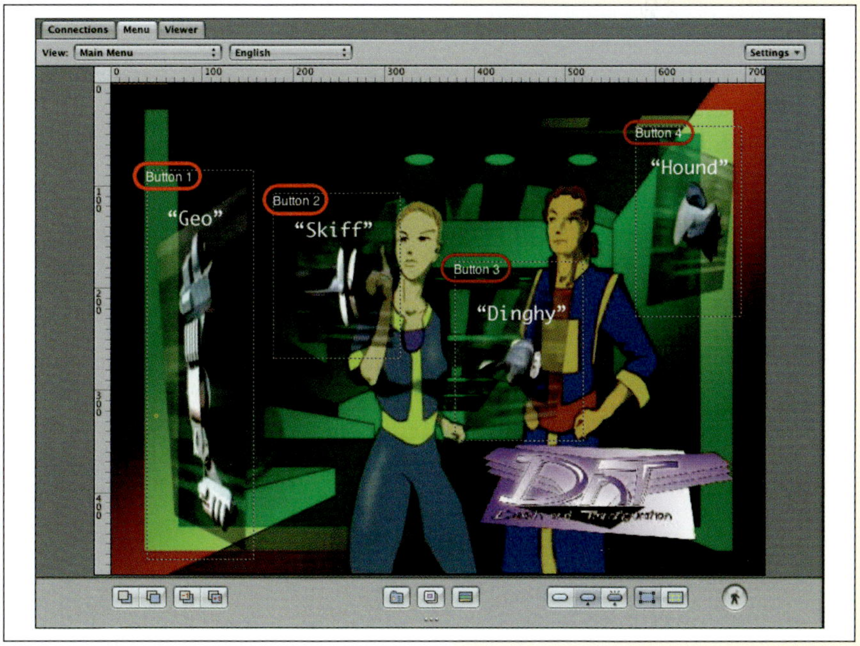

Figure 8-28. The numeric associations for several buttons

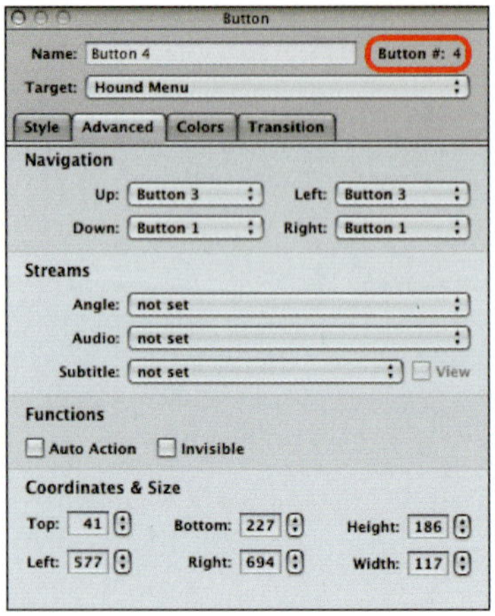

Figure 8-29. Finding the numeric association for a button on its Inspector

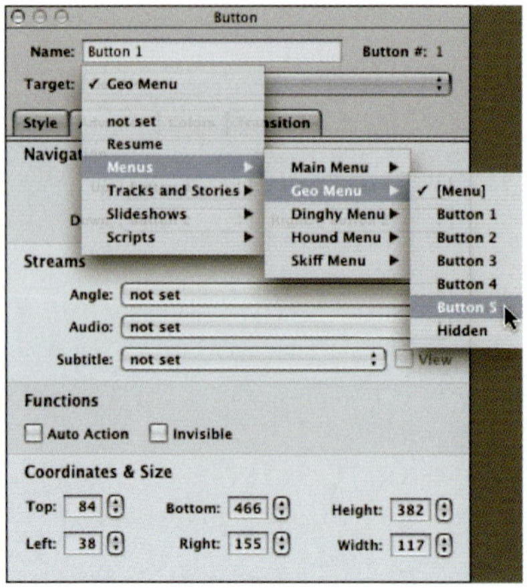

Figure 8-30. Selecting a specific button as a target

each button. Figure 8-29 shows a Button Property Inspector with its number highlighted.

One well-known Easter egg that uses this feature is the Special Edition of *Terminator*. Punching in the date of "the big day" enables you to play a super-special edition with even more deleted footage reintegrated.

But you have to be a little careful here. The 0 button on the remote control usually does not map to any menu buttons. If you have more than nine buttons on the menu, most remotes have a way of entering two-digit numbers. Those will continue to correspond to menu buttons just as the first nine numbers do.

You also have to make sure you don't break the menu's normal functionality. Non–egg seekers shouldn't be punished (at least not severely) for simply navigating your normal menu structure.

Fortunately, DVDSP makes that normal functionality easy to maintain. Instead of simply selecting the menu as the normal target, you can select the menu and one of its buttons. Figure 8-30 highlights the target chain that leads to such a target. We used this feature in previous projects for auto-advancing through a list of episodes or deleted scenes.

You also can use the menu's property sheet to set a "default" button (see Figure 8-31). This has the handy side effect of working wherever you choose the menu as a button or end-jump target. You always can go the menu+button target route, but with a default button, you don't have to remember which menus have hidden buttons.

And that's another thing: if you want to hide your password buttons, just turn up their transparency. Figure 8-32 shows a menu that configures the third button color set for hidden buttons.

Using layered menus makes this process somewhat easier—just leave empty layers for the hidden buttons. Nothing in the DVD specification requires a visual state change—but DVDSP does need at least one layer to be selected for the button to become useable. You mark their boundaries as usual; the size you make them depends on how friendly you want to be to computer users. The size won't matter to viewers using a standard consumer player.

Computers and Numeric Input

One other thing to keep in mind when building numeric input games or eggs is that many software DVD players leave out the numeric input pad. (Some do enable you to use the numeric keypad on your keyboard.) Viewers still can access your hidden content, they just have to find the buttons first and then click them.

Knowing that, you might want to organize your hidden buttons so that computer users have a reasonable shot at deciphering which buttons to click. Or not. It's your Easter egg!

A quick test

If you haven't played around with numeric input and DVD menus before, you might want to load this project and burn it to a disc. It's a simple, single-menu project meant to demonstrate how numeric input works. Figure 8-33 shows the one menu.

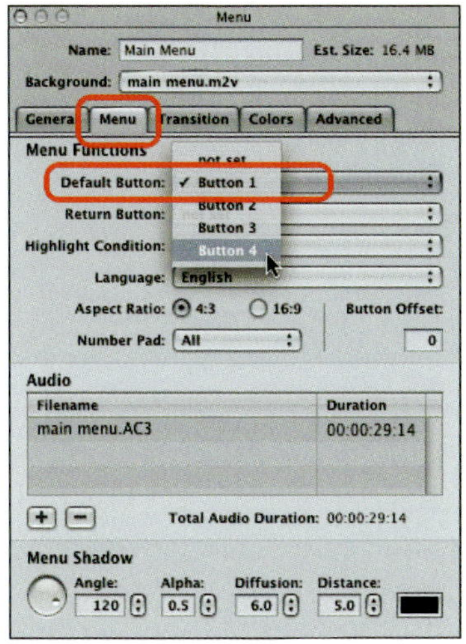

Figure 8-31. Setting the default button for a menu

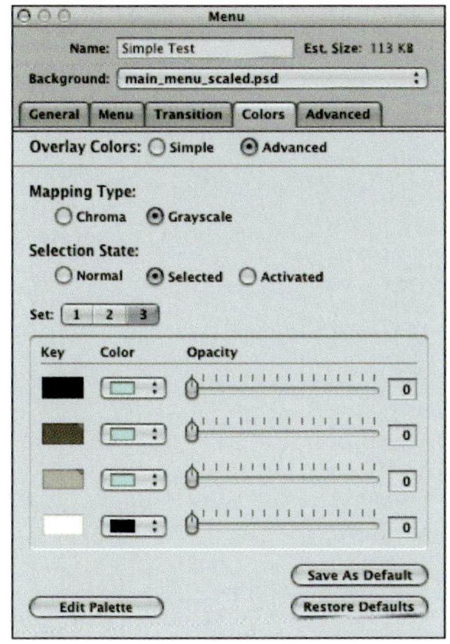

Figure 8-32. Making sure the secret-password buttons stay secret

Each numbered button leads to a simple video segment displaying that number. There's nothing new or exciting to building this project, but it makes the numeric input easy to follow. The tracks you need are in the *projects/ch08/password* folder:

- *simple_menu_scaled.psd*
- *0.m2v–9.m2v*
- *pound.m2v*
- *star.m2v*

Here are the steps to follow:

1. **Create a Simple Menu with the Numbers and Symbols as Buttons.** Just attach the 1 button to the *1.m2v* track, the 2 button to *2.m2v*, and so on. There are no special tricks to making the numbers on a remote control correspond to a given button. Again, the 4 button on your remote will trigger the fourth button on your menu. You'll need to test this with a burned DVD and a consumer player, though, as neither the Simulator in DVDSP nor the DVD Player application comes with numeric pads. (Recall the remote control discussion back at the end of Chapter 1.)

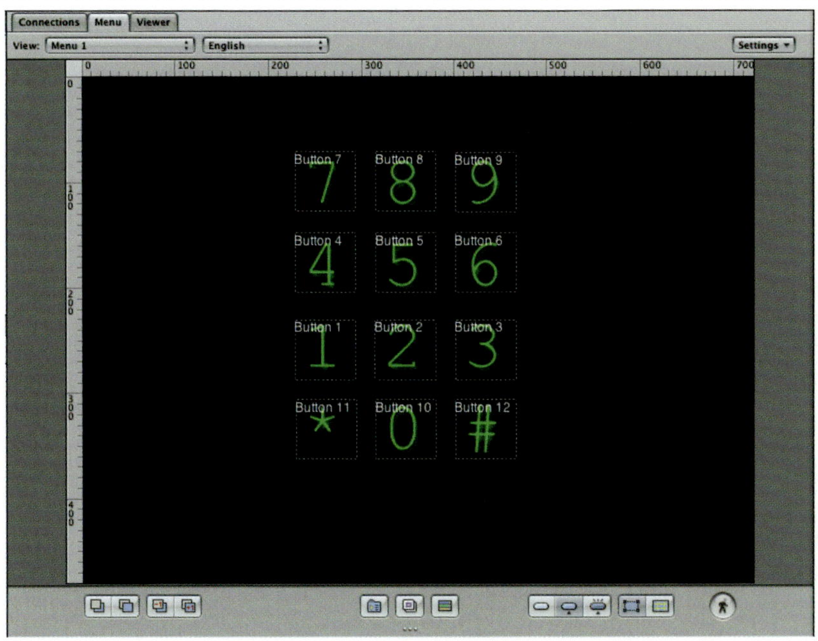

Figure 8-33. A menu for a quick test of numeric input

NOTE

Burning this tiny project all on its own to a DVD probably won't work. Or rather, DVDSP will help you out behind the scenes and you might notice something odd. This tiny 75MB project takes a really long time to burn. If the project is too "small," DVDSP will pad out the disc for you.

Many DVD players need a minimum of data to scan through to make sure everything is working the way it is supposed to work. (It checks things such as the read laser's tracking.) If you have some other DVD burning application that does not enforce this safety feature, try it out. Most likely you'll produce a nice shiny coaster for your coffee table.

2. **Keep Track of the Password.** Of course, getting the buttons on the screen is only half the battle. You need to be able to track what the viewer has entered and then activate the right egg once he's done it correctly. We'll use scripting to accomplish this task.

   Scripting this type of thing can get difficult, depending on how fancy you want your password to be. A one-number password is easy. Two numbers is a bit more difficult but still straightforward. Handling more than that (such as the five-digit date from the *Terminator* disc) requires some patience and skill. So, let's tackle the five-digit version!

NOTE

Remember the various reset scripts we've seen in past projects such as the treasure hunt, as well as in the play-all menu (from Chapter 7). You can make sure a password variable is blank by making one of those reset scripts the prescript for the menu.

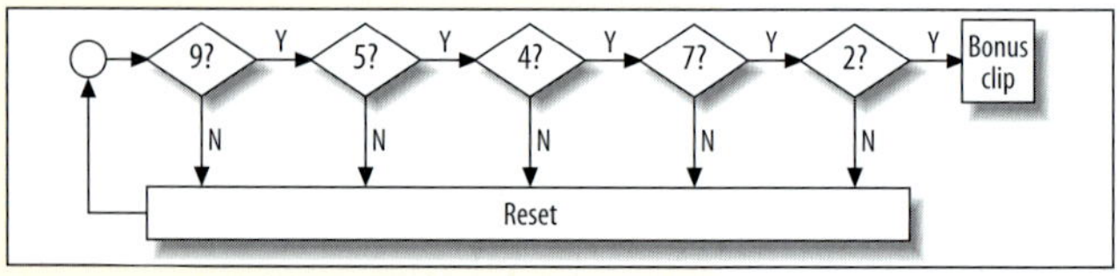

Figure 8-34. The basic flow of the buttons in the secret password

You'll need one variable to keep track of what you're doing and possibly another variable to track your success (or failure). To get going, you need to work out your password and what happens after each key press. For a password, you'll use the ZIP code for O'Reilly Media in Sebastopol: 95472. Figure 8-34 illustrates the flow of our password scripts.

So, let's look at the script for the 9 button. Depending on how mean you are, you can set things up so that if the viewer starts with the wrong number, he has to reset the password to get it correct. We'll enable the viewer to punch in as many wrong numbers as he wants; we'll start tracking the possibility of success as soon as he hits the 9 button. Figure 8-35 shows one way to handle this button.

In your password, several buttons (1, 3, 6, and 8, specifically) always are wrong. Each can use the script in Figure 8-36.

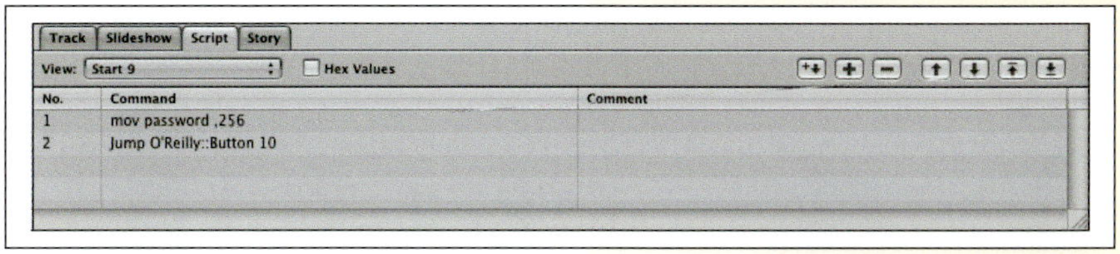

Figure 8-35. The script for setting the first entry in the password

Store the password using bits, just as you did for the treasure hunt. Put the 9 in bit 9, the 4 in bit 4, and so on. You don't have to do it that way—you can use any five bits you want—but this approach is easier to remember.

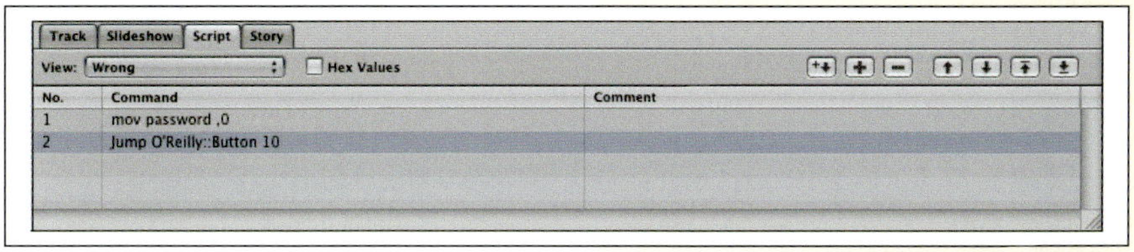

Figure 8-36. The script for all the wrong entries

Now it starts to get a bit more interesting. If you press the 5 button next, you want to continue, but if you press the 5 out of sequence, you want to signal failure. You have to check the variable and make sure the 9 was just pressed. Figure 8-37 shows that script.

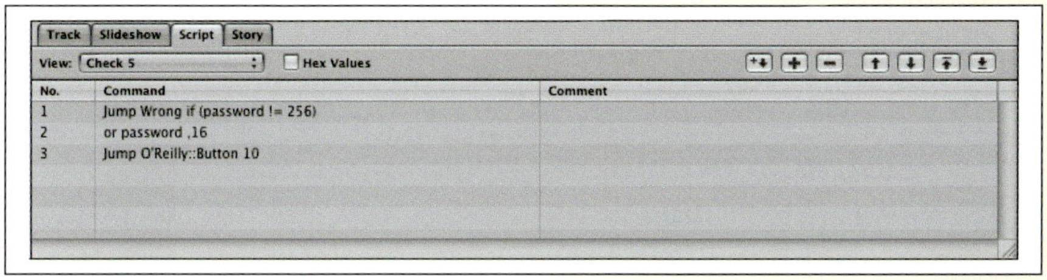

Figure 8-37. Checking the second entry in the password

Check the variable to see if the 9 bit has been set in the first line. If it's on, set the 5 bit. Otherwise, reset the password by jumping to the Wrong script. The 4, 7, and 2 buttons will just use a variation of the 5 script. Those are listed in Figure 8-38.

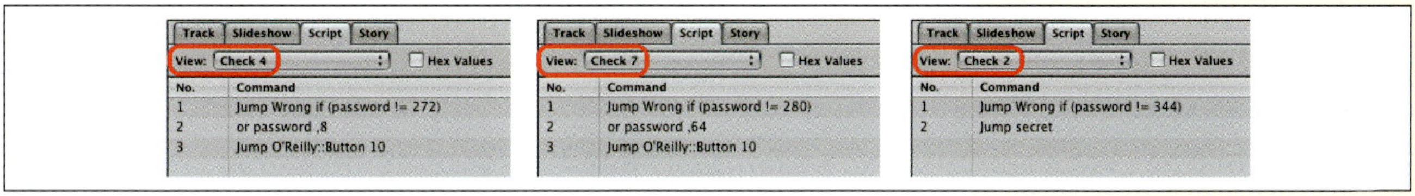

Figure 8-38. The remaining scripts for the password entries

Notice that the 2 script ends by going to a bonus video clip. That way the viewer jumps right into the egg when he gets the password right. You could make this a bit of a treasure hunt as well by "unlocking" some feature of another menu. To do that, you need to set the value of a second variable to make it easy to detect when to activate the other feature.

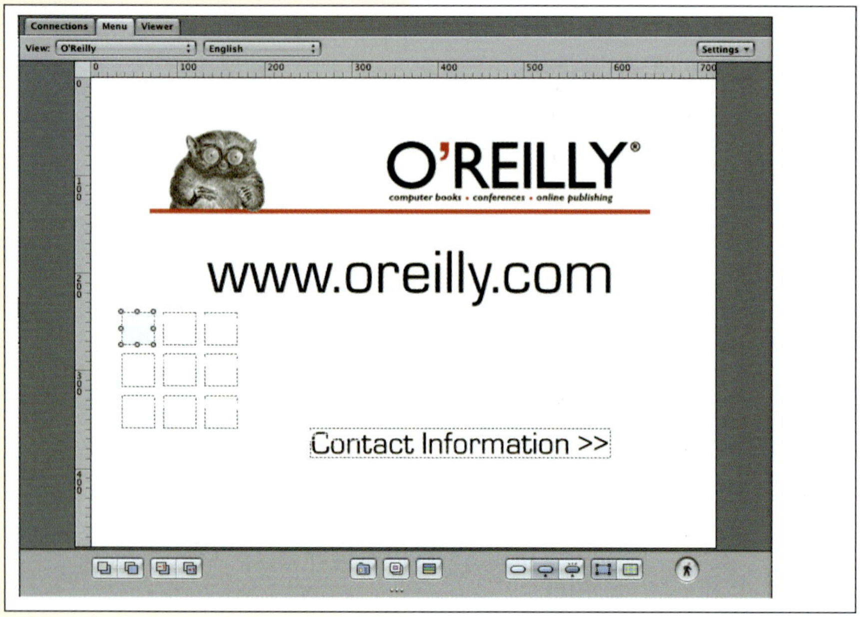

Figure 8-39. The main menu for the password project

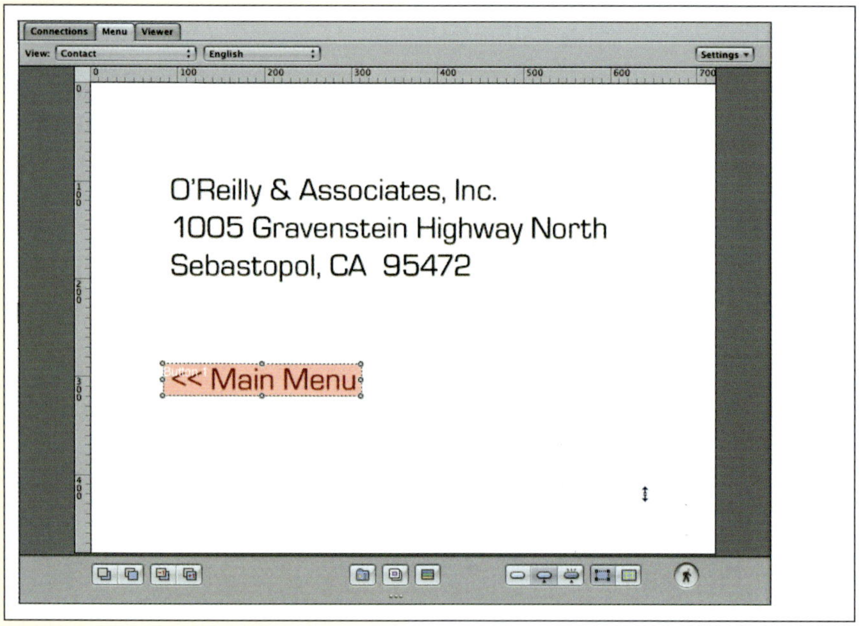

Figure 8-40. The simple Contact menu in the password project

Creating an Easter egg

To use what you've learned in a real project, load the following assets from the *projects/ch08/password* folder:

- *password_menu_scaled.psd*
- *contact_scaled.psd*
- *secret_video.m2v*
- *secret_audio.AC3*

Then follow these steps:

1. **Create the Two Menus.** These are the main menu and a Contact menu. The Contact menu is there mostly to give the main menu something to do besides worry about the password. You'll have one "visible" button on the main menu that leads to the Contact menu. The Contact menu will have a reciprocal button that leads back to the main menu.

   Let's set up the main menu. Use the *password_menu_scaled.psd* file for the background. Off on the left side, define nine buttons. They don't have to be very big. Figure 8-39 shows one possibility.

   Once you've defined those 10 buttons, go ahead and define the Contact Information button. Remember that it's important to define the password buttons first. Otherwise you'll have to use awkward combinations on the remote control to enter the password. Although that might be a good goal for a serious Easter egg, you want to have fairly straightforward access while you're testing!

   The Contact Information button leads to the Contact menu shown in Figure 8-40. That menu also is straightforward. Its only button should lead back to the main menu.

2. **Create the Number Buttons.** The number buttons will use the same scripts we discussed before (see Figures 8-36 through 8-38). So, after typing in the secret password, the special video should show up happily when the viewer hits the 2 button as the last entry in the password. This is the *secret_video.m2v* file from earlier.

Go ahead and try it out. The Simulator shown in Figure 8-41 is still a valid place to test the basic functionality because you can use your mouse to go directly to buttons that would not be accessible otherwise. While you test them, it's worth it to turn up the opacity on those buttons to make it easier to get the right password (as you can see in the figure). Just be sure to make them transparent again once you know everything is hooked up correctly.

This is a pretty slick Easter egg, and you even can use it to keep Mom and Dad out of the part of your family vacation movies that you wanted only your friends and siblings to see. (Did you really want to show them your drunken hula dance?)

Keeping Secret Buttons Secret

You'll want to use the same tricks we looked at in the treasure hunt to keep the "number pad" buttons obscured. You should configure the navigation of your regular menu buttons manually so that viewers don't accidentally end up in your number pad area. Or not. That's another valid egg-like thing to do. Just make sure you're aware of what you're doing!

Buttons over video

Last but not least, you can put little surprises directly into your video streams. *The Matrix* DVD used this next feature to great effect with its "Follow the White Rabbit" option. When you're following the White Rabbit (by choosing that option from a menu), each special effects sequence is accompanied by a little white rabbit. That rabbit is actually a button in a subtitle track that plays over the video. Pressing the Enter or OK button on your remote control activates the Rabbit button, and you get to see a featurette on the making of that scene. Very cool. Too cool, in fact, for many DVD players at the time. Happily, you're not likely to have any such problems these days.

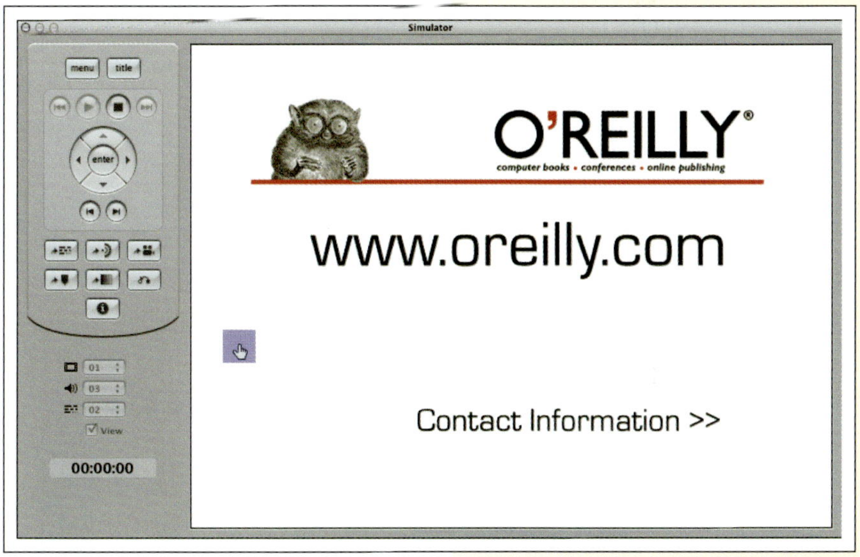

Figure 8-41. The password project running in the Simulator

So, let's look at how to add a button over video.

The button-trick project

The button trick most often starts with putting an image in a subtitle track. You'll just use the same subtitle streams you used back in Chapter 5. This time, however, you'll load a bitmap rather than text. To get started on this project, load the following assets from the *projects/ch08/button* folder:

- *what_is_ripples.m2v*
- *what_is_ripples.AC3*

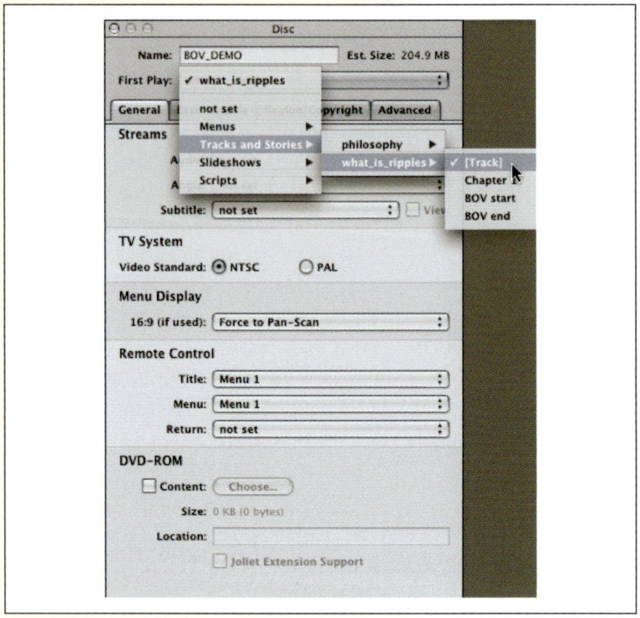

Figure 8-42. Setting the First Play property to make sure you can see the video

- *philosophy.m2v*
- *philosophy.AC3*

You might notice that you're not loading any image assets. This project will jump right into the video portion (no menus). The image you want to use for the button over your video will be loaded dynamically when you're ready for it.

You're going to apply this feature to a promotional clip from Paul Wesselmann and The Ripples Project. The main clip focuses on an introduction to and the overall goal of The Ripples Project. You'll add a button to take viewers to another video clip that explains some of the philosophy behind The Ripples Project. (You'll have to watch the clip to get more information on what a Ripple is!) Here are the steps:

1. **Create the Main Video Track.** First you need the main video track in place. Drag the *what_is_ripples.m2v* file and its audio track to the Track window. You don't have a menu for this project, so make the clip the default action for the disc to take when inserted into a player. Figure 8-42 highlights that property in the Disc Property Inspector for a quick review.

2. **Create the Philosophy Track.** There's nothing special about this one. You just want it handy when you get to the point where you can set the target on the button.

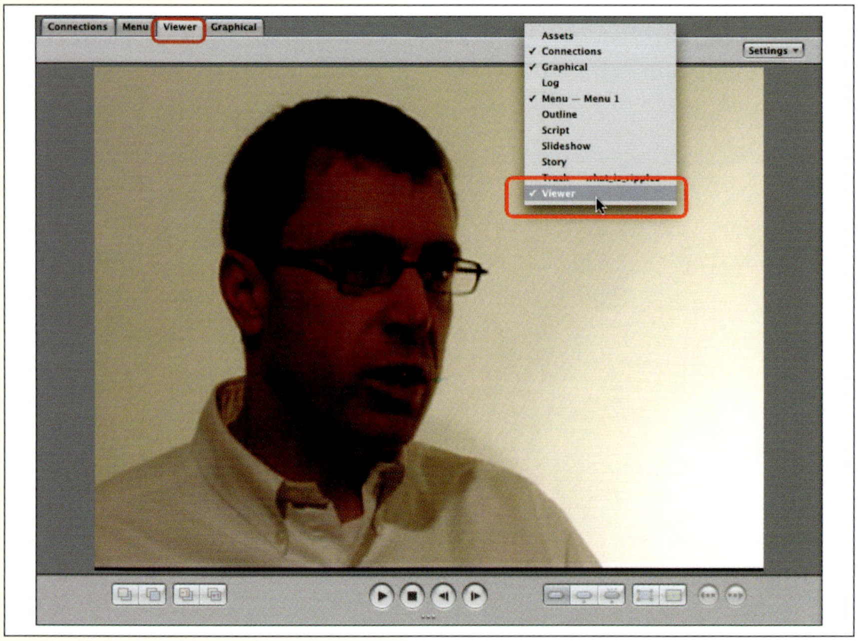

Figure 8-43. Getting to the Viewer tab

3. **Set Up the Video Markers.** First, you'll probably want to make sure you can see the Viewer tab or window so that you can track what's going on in the video clip. Figure 8-43 shows you where to select that window. You also can just press ⌘0 to make it show up.

   As mentioned earlier, you're going to use an image in a subtitle track for the button. (Refer back to Chapter 5 if you need to brush up on subtitle tracks.) Because you're making a live button, though, you have to do a little bit of preparatory work on the video clip. In particular, you need to set two markers:

one at the beginning of the button's appearance, and one at the end. You can create these markers by clicking the marker area of the Track tab, as highlighted in Figure 8-44. By default, the marker will be named Chapter *x*, where *x* is a sequential chapter number. You're free to rename the marker in its Inspector.

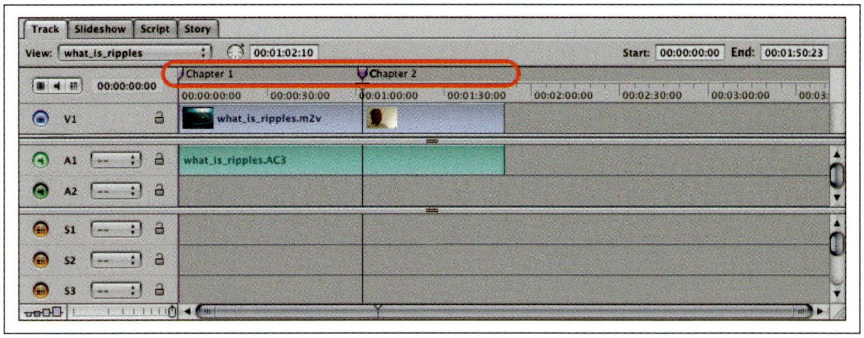

Figure 8-44. The marker area in DVDSP

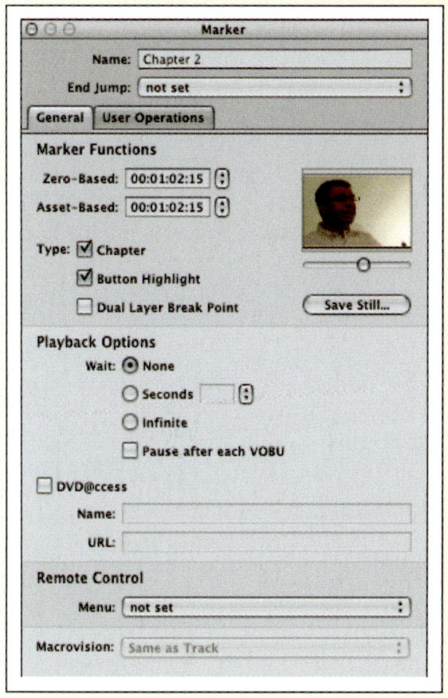

Figure 8-45. The properties for a marker in DVDSP

We placed our markers at 1:02:15 and 1:12:15 to create a 10-second onscreen button. How long you make your button available is up to you. Less than five seconds is really short, but that might be part of the fun, too—catch the buttons while you can!

Next we'll make "button" markers. If you look at the Property Inspector for the marker (you might have to click the marker in the Track tab to make sure the Inspector knows what object to show), you'll notice that there are several possibilities for the type of marker, as shown in Figure 8-45.

You at least want the Button Highlight checkbox checked. You'll probably want the Chapter checkbox checked as well. Set both options for both of the markers. Notice in Figure 8-46 that the marker is both brown and purple, indicating its dual role.

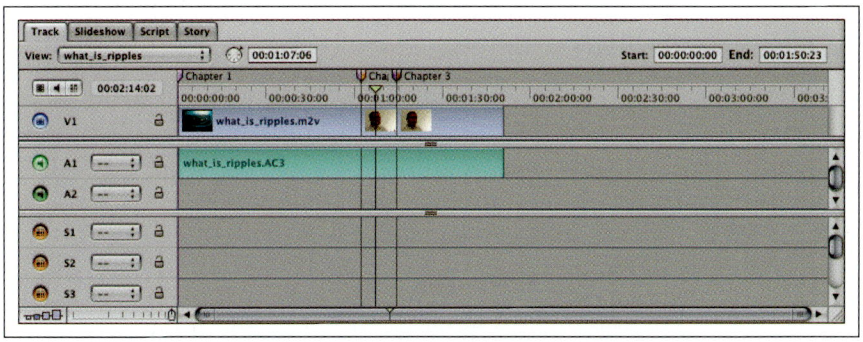

Figure 8-46. The start- and end-point markers for the (eventual) button over video

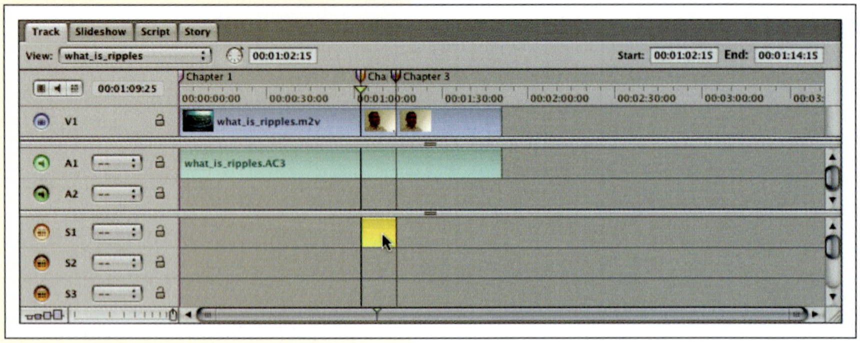

Figure 8-47. Adding a normal subtitle to the video track

4. **Create the Image Subtitle.** To create the subtitle, just double-click in between the two markers, as shown in Figure 8-47. You should end up with a familiar yellow block. The Inspector also should switch to your new subtitle entry. Instead of typing in text for a subtitle, though, you're going to load an image.

Take a look at that Inspector. You're going to load an image using the Choose button highlighted in Figure 8-48.

From the File dialog box that pops up, load the *ripples.gif* image. That image is a four-color logo similar to the menu overlays we looked at in Chapter 4. When you first add the image, it should appear in the upper-left corner, as in Figure 8-49.

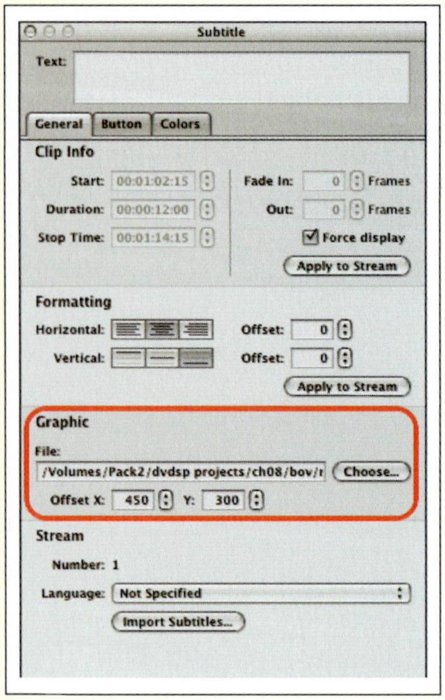

Figure 8-48. The image portion of the Subtitle Inspector

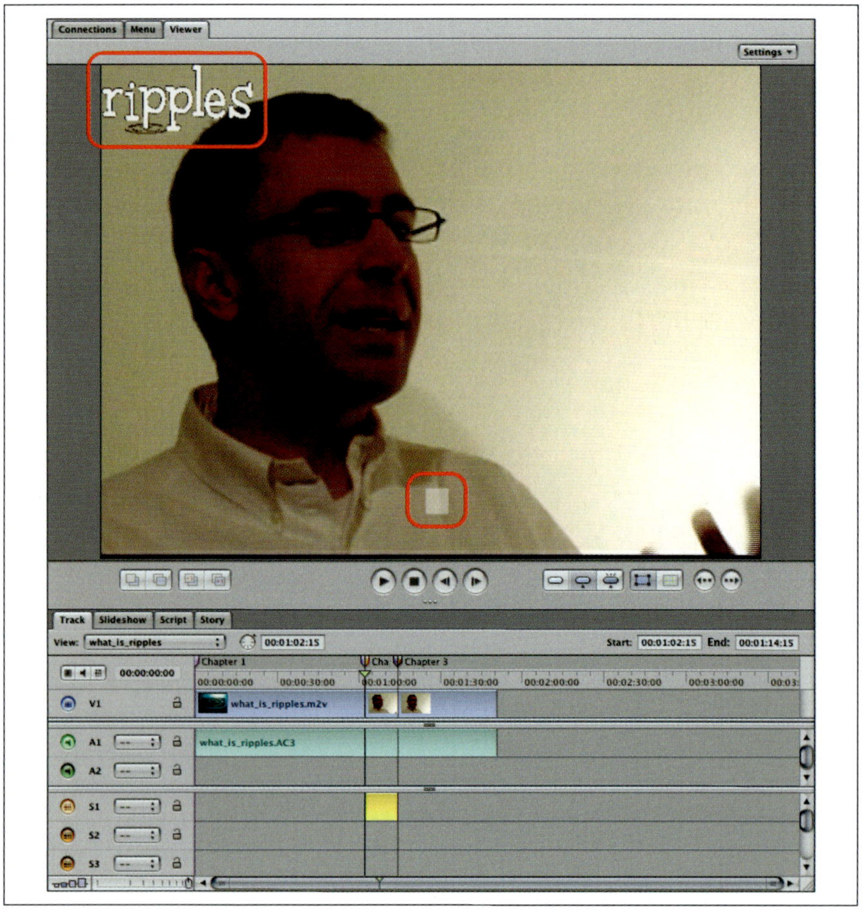

Figure 8-49. The new subtitle with the image loaded (upper-left corner) and the text area (center)

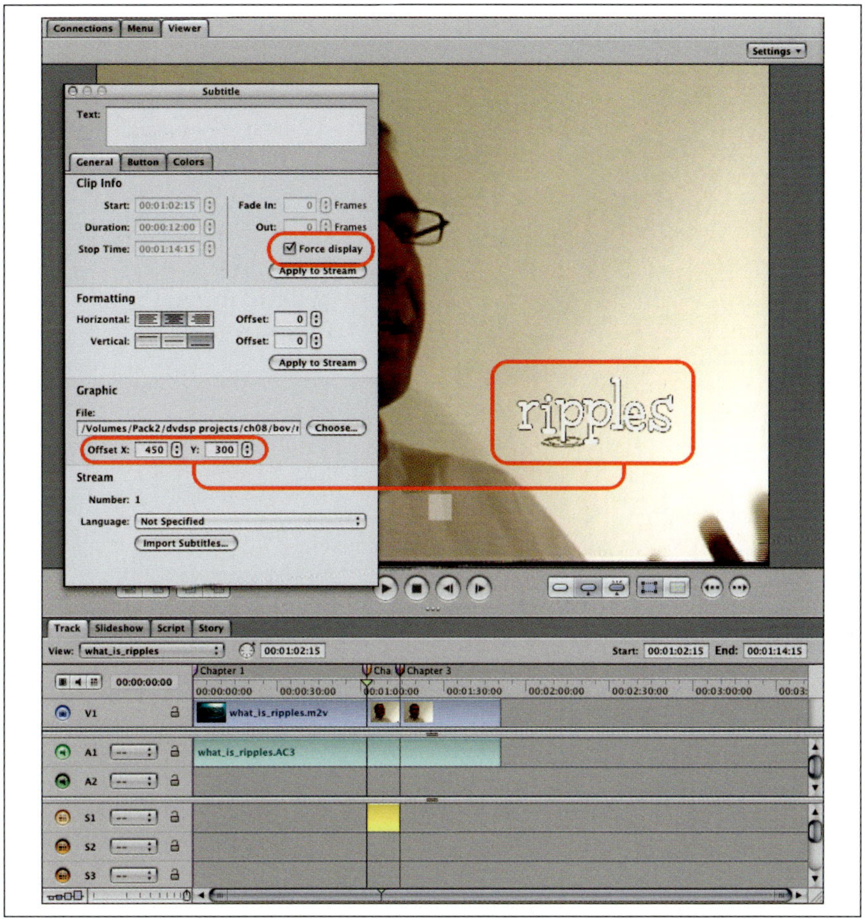

Figure 8-50. Placing the graphic in the Subtitle Inspector

You can set the X and Y offset values in the Inspector to better position the button. Figure 8-50 shows our final X and Y offset values and the results on the video.

Notice we also chose the Force display option. That makes sure the button is visible regardless of the viewer's current subtitle settings.

5. **Create the Subtitle Button.** Now that the image is loaded and placed, you need to turn that into a button. Click once on your new subtitle chunk to make sure it is the active item. In the viewer, drag an outline to create the button just as you would in a menu. Figure 8-51 shows the button creation in progress.

I Can't Drag an Outline!

When using images for buttons over video, the subtitle track occasionally needs a little poke to get going. If you can't drag and create a button outline on your video track, just double-click the subtitle chunk in the Track window. That should reset the focus for the subtitle and cause a cursor to appear over the video, as shown in Figure 8-49. You can ignore the cursor because you're using an image, but you now should be able to create your button outlines.

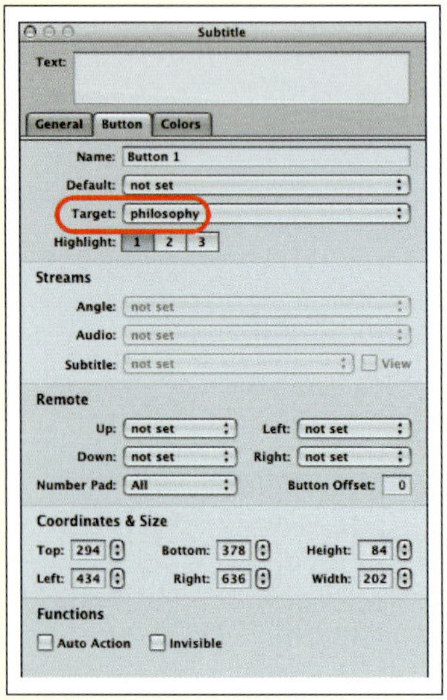

Figure 8-52. Setting the target for the button over video

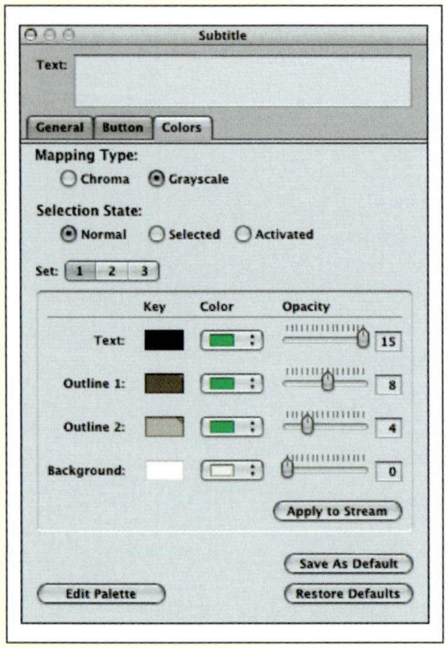

Figure 8-53. The Ripple color settings (designed to assist antialiasing in the image)

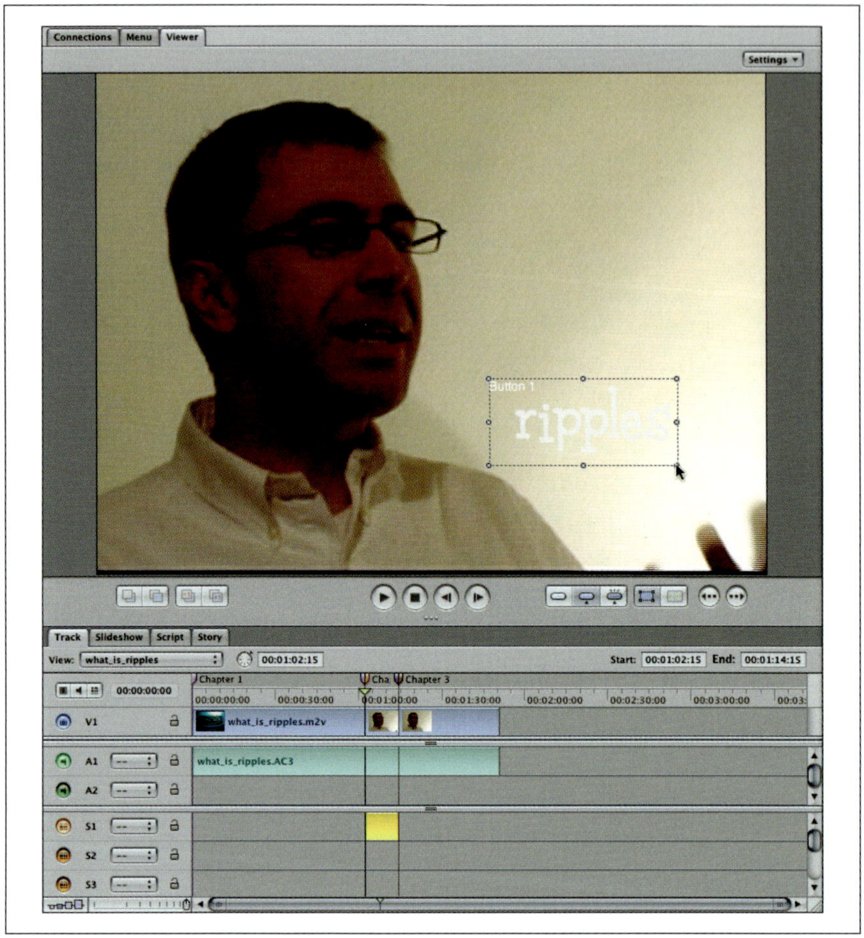

Figure 8-51. Creating a button over the subtitle graphic

6. **Set the Button Properties.** Now you can play with the Button tab for the subtitle in the Inspector window, as shown in Figure 8-52. You should have all the usual button options with the exception of additional text. You can alter the audio stream, alter the video stream, jump to a new target, etc. We're going to jump to the philosophy video clip.

   You also can play with the button's color scheme. As you can see in Figure 8-53, you have exactly the same options as you would for a menu.

7. **Set the Jumpback.** But once you've been on your wonderful detour thanks to the button over video, how do you get back? Unfortunately, the only easy way is to jump to a particular chapter in a known track. That's easy to do; just set the End Jump property on the philosophy track to the second marker you created, as shown in Figure 8-54.

   It would be just as easy to jump back to the beginning chapter marker. If the video material was important, you might want to go ahead and repeat whatever portion the viewer has seen already. The viewer always can fast-forward through things if he gets bored.

   In other scenarios, though, you might need the help of a script to decide where to return. It all depends on how complex your project is. Setting up chapter markers at convenient "resume" points is the most consistent way of ending up in a useful spot.

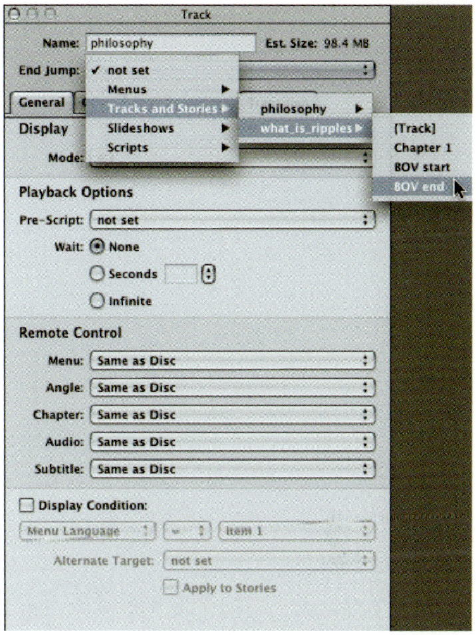

Figure 8-54. Making sure you get back to your regular video clip

Figure 8-55 shows our project playing on a consumer player. You can see the video button in action in the lower-right corner.

Well, that does it for the games and Easter eggs, but there's more fun to be had with DVDs. When you put DVDs in a computer, you can provide links to datafiles and the Internet. The next chapter covers those topics, so keep reading!

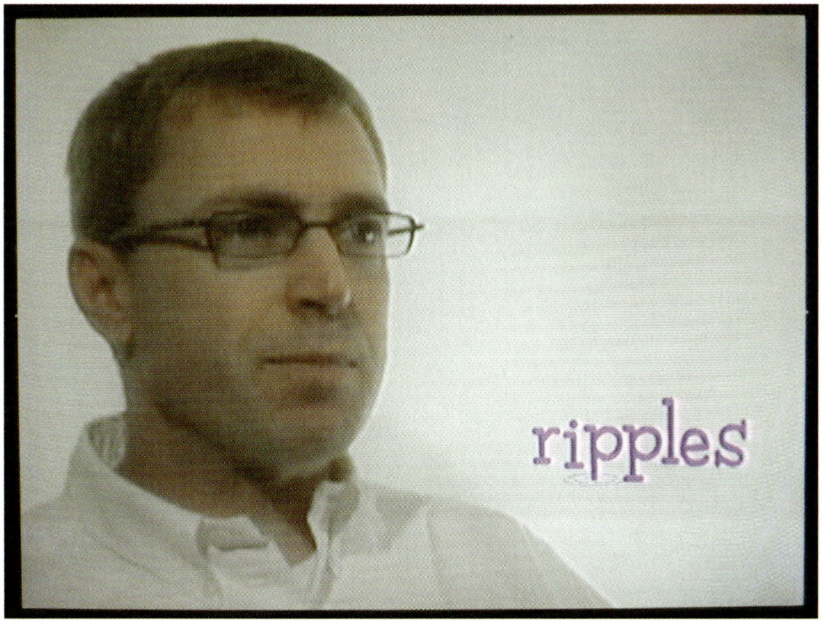

Figure 8-55. Our button-over-video project running on a consumer DVD player

DVDs and Computers

9

If you need to include data with your DVDs, you've come to the right chapter. We'll look at including a data portion on your disc so that computer users can access content beyond the video portion of your DVD. We'll also look at supporting the web-enabled features provided by DVDSP.

In this chapter

Computer Features

From the very beginning, DVDs have had a place in the hearts and minds of computer users. Software outgrew the floppy over the course of a decade and migrated to CDs. Judging by the number of multi-CD software applications out there now, it won't be long before DVDs are the distribution mechanism du jour. One disc (albeit in double-sided, dual-layer mode) can hold 18GB of data. Final Cut Pro and a full array of add-ons almost fill that already!

But DVDs aren't just about space. They're also about entertainment. We've been building projects all along that are meant to entertain and amuse. What's the one thing missing from all this entertainment? That's right: advertising. Well, OK, you might not really miss it, but a DVD playing in a computer also can link to a related web site. That way, anything that is missing from your DVD can be presented. And why not go ahead and add a banner ad for your next theatrical release?

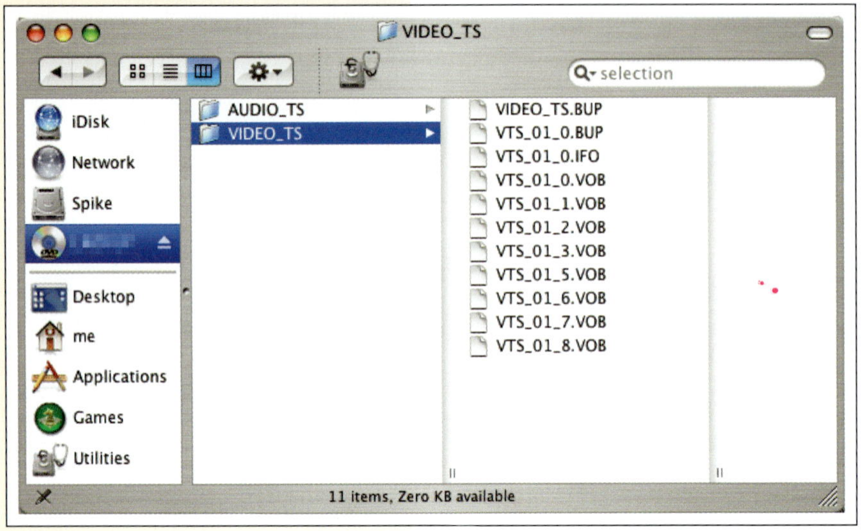

Figure 9-1. The files you typically find in a video-only DVD

DVD disc layout

So, where does the extra stuff go? Let's take a quick look at the files and folders that are stored on a regular DVD (see Figure 9-1).

The main folders are the *VIDEO_TS* and *AUDIO_TS* entries. You might occasionally find a *PROF* folder meant for use on professional players, such as the Marantz PMD930 or the older Philips Professional 170 that have special nonvolatile RAM available for use in scripts. Pop your favorite movie disc into your computer and look at it through the Finder. You should see the same basic hierarchy as the ones shown in the following list—though the exact files in the *VIDEO_TS* folder probably will vary a bit.

VIDEO_TS
: The *VIDEO_TS* folder is where the content is stored for a DVD-Video project. That's the folder consumer players look for to start up a movie.

AUDIO_TS
: The *AUDIO_TS* folder contains the pieces necessary for a DVD-Audio project. It's part of the spec, so DVDSP creates it and leaves it empty.

Beyond that, you are free to use the space on the disc for other folders. Consumer players will blithely ignore them, but computers will understand them and you can treat that portion of the DVD as you would any data disc.

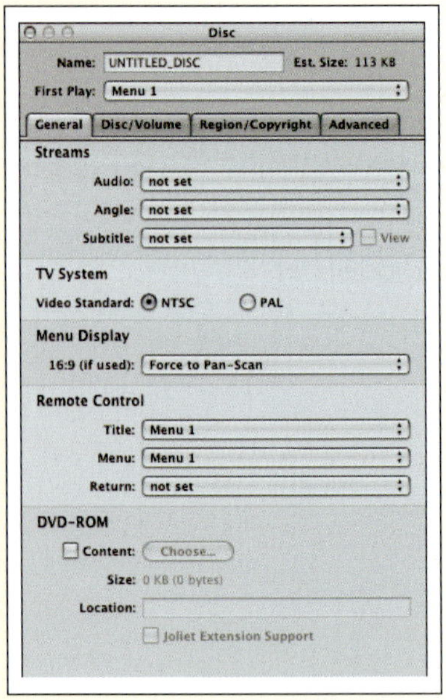

Figure 9-2. The Disc Inspector with DVD-ROM properties shown at the bottom

Video and data setup

So, how would you go about adding a data area to a disc? DVDSP makes that easy. The first step is collecting your datafiles. You should organize into a single folder all the files and folders you want on your disc. That folder's contents—but not the folder itself—will be included on your disc when you burn it. The datafiles and folders will exist alongside the required *VIDEO_TS* and *AUDIO_TS* folders.

To include a folder on your disc, look at the Inspector for the disc. Figure 9-2 highlights the DVD-ROM section of the Inspector.

You select the folder containing your data, and off you go! You have nothing else to worry about until you burn the disc. And even then, you just need to make sure the specified folder has all the right stuff in it. There's nothing else to do in DVDSP.

A video and data project

For our project, we'll put a video presentation on a DVD. We'll be using a presentation built with Apple's KeyNote. We have an MPEG-2 version of the presentation that we'll include as a normal DVD video track, just as we have in all our previous projects. But we'll also include the actual KeyNote presentation file and some of the supporting files mentioned in the presentation.

We'll be using the overview portion of a technical training presentation on Java Swing. Fortunately you don't have to know (or care, really) about Swing to complete the project.* To get started, let's build the simple video portion of the project. Load the following files from the *projects/ch09* folder:

- *overview.m2v*
- *overview.AC3*
- *overview_menu_scaled.psd*

You have one menu with one button, as shown in Figure 9-3.

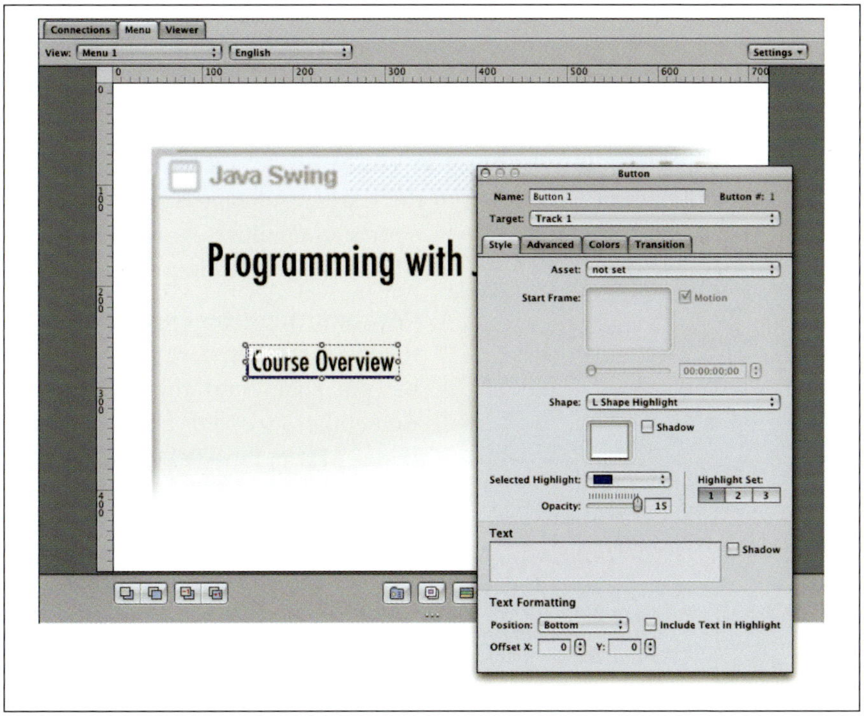

Figure 9-3. Configuring the sole button in the DVD-ROM project

* But if you do know and care, you can check out another fine O'Reilly title: *Java Swing*, Second Edition.

Naming Restrictions

We hate to do this, but we have to. The files and folders you place on a DVD-ROM do have some naming restrictions. On their own, DVDs use a filesystem (UDF) that supports long filenames with Unicode characters. But to be maximally compatible, your filenames should use only uppercase letters, numbers, and the underscore in the name. The name can be 26 characters long, plus a "." and a three-character extension. Directory names should not use an extension. (Yes, that's very similar to the old DOS 8.3 name requirements. Think of it as 26.3.)

You probably noticed the Joliet Extension Support option at the bottom of the Inspector's DVD-ROM section in Figure 9-2. That adds the Joliet naming rules to the mix, but it can upset some older players. (The Joliet extension relaxed some of the restrictions of normal ISO 9660 names. For example, Joliet names can be longer and they support Unicode characters.)

If you can test your video+data disc on your intended playback devices, you have the best guarantee things will work. Barring that, stick to the 26.3 naming conventions if at all possible.

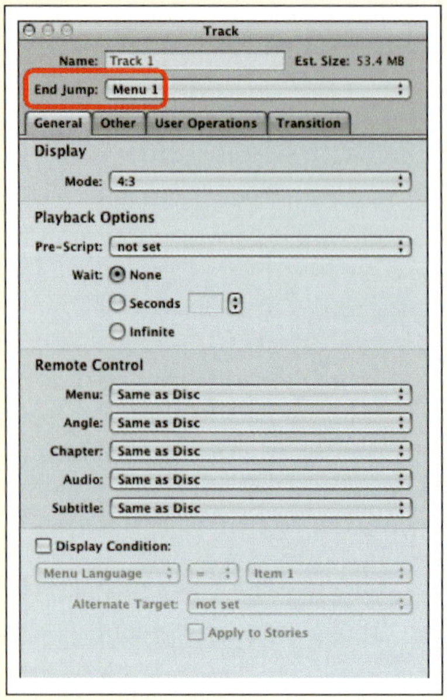

Figure 9-4. The End Jump property setting for the video track

That button should launch the one track. That track, in turn, should set its End Jump property to go back to the menu (see Figure 9-4).

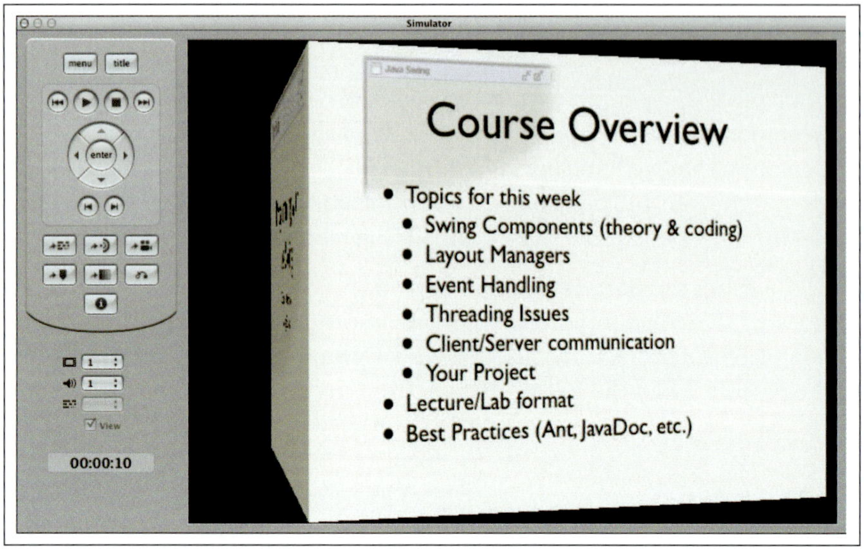

Figure 9-5. Testing the video portion of the DVD-ROM project

Go ahead and simulate the project to make sure the video portion is working. Figure 9-5 shows the presentation up and running in the Simulator.

Collecting the datafiles

For the data portion of this project, you want to include the original KeyNote document, a PDF version of the slides, and some supporting Java source-code files. Organize the datafiles as shown in Figure 9-6.

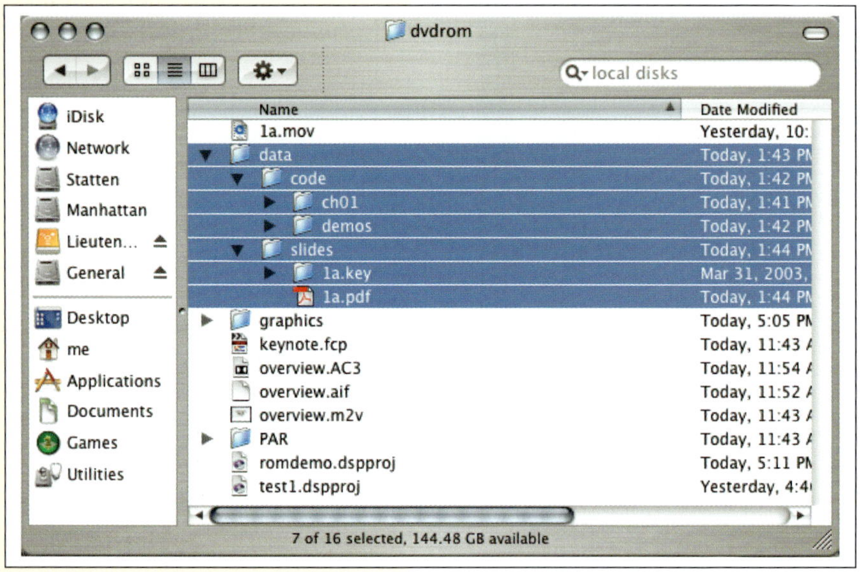

Figure 9-6. The files to include on the DVD-ROM portion

We want the *slides* and *code* folders to go at the top level, so why did we put them into the *data* folder? Remember, we tell DVDSP which folder to use, but DVDSP grabs only that folder's contents. The folder itself is not used at all during the final build and format.

Let's specify that folder for the disc. Select the Disc in the Outline tab and look at the Inspector. Go ahead and check the DVD-ROM Content checkbox and choose the *data* folder. Figure 9-7 shows the results.

That's about all there is to it, really. When you build and format the disc, you'll see verification that you chose a data portion (see Figure 9-8). You also can check the Joliet Extension Support checkbox at that time.

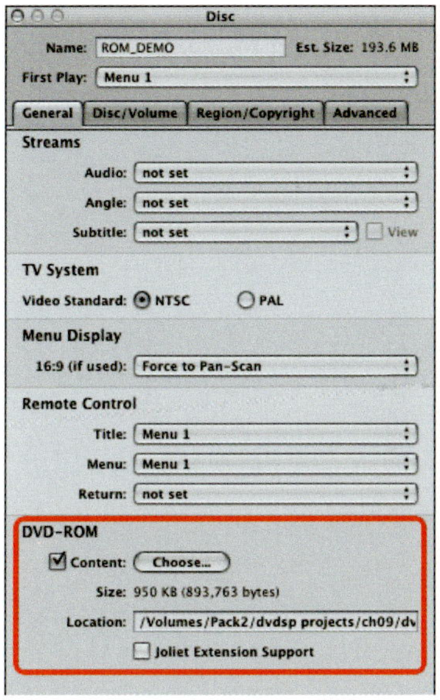

Figure 9-7. The data folder added to the DVD-ROM project

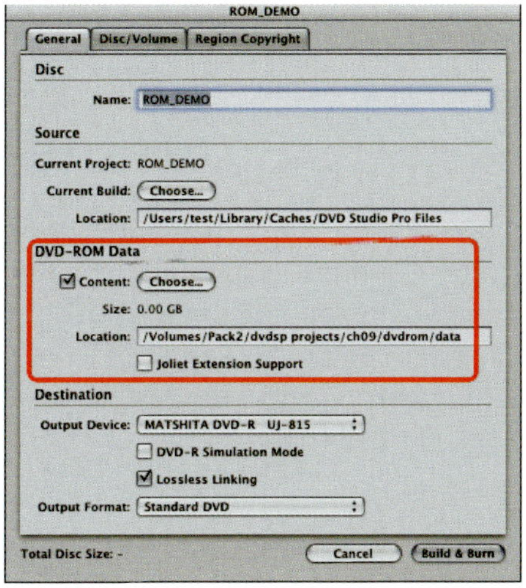

Figure 9-8. The DVD-ROM area of the dialog

Hey wait! You used bad names!

If you look at the Java source code in the *data* folder, you'll notice that we did not follow the 26.3 naming convention. We wanted to, but we couldn't. Java source-code files must end with the "*.java*" extension. No ifs, ands, or buts about it. We're assuming that people developing Java applications will be using a reasonably powerful machine that supports UDF and can use the full names.

But what would happen if they didn't? That's a good question. Some operating systems, such as Linux, for example, actually enable you to control what format the computer applies to the DVD. DVDs have both UDF and ISO 9660 formats available. Joliet would be a third format, if you added that support when you built the disc. Which version your computer uses depends on what your operating system supports. Most Linux systems support all three formats and make a good test bed for this type of work.

When we put the disc in a Linux system and told the computer it was a UDF-formatted disc, we got all the files we wanted and their names were fine. When we tried again but told the computer it was an ISO 9660 disc, all the files were still there, but they were renamed. Table 9-1 highlights a few of the name changes.

Table 9-1. UDF versus ISO 9660 (no Joliet) filenames

UDF	ISO 9660
VIDEO_TS	*VIDEO_TS*
AUDIO_TS	*AUDIO_TS*
slides (folder)	*SLIDES*
1a.key (folder)	*1A*
Calculator.java	*CALCULAT.JAV*

It's important to stress that the files are one and the same. The UDF and ISO 9660 stuff just determines how the computer reads the filenames. Both names end up pointing to the same datafile on the disc. That means anyone on a system that cannot read UDF volumes still can copy the Java files off the disc and rename them once on their hard drive. But that's not always the nicest thing to do.

If we really were producing this disc for mass consumption on all types of platforms, we probably would include a zipped version of the *code* folder. That type of file fits nicely in the 26.3 naming rules. Users can copy the ZIP file off the disc and unpack it. Unpacking the ZIP file restores all the regular filenames, so nothing is lost. We used this trick for distributing Java code on CDs before the Joliet extensions were widely available on non-PC platforms. It's a bit of a hack, but it makes it easier for your end users to get at the data. And that, really, is what this is all about.

The DVD for this book

The data for this book is an example of one of those "known-quantity" situations. We know systems capable of running DVDSP can read the UDF names, so we rely on that fact. The rest of the work in building the data portion of this disc consisted almost entirely of typical file organization tasks in the Finder. Once the folder was ready, we just checked the DVD-ROM checkbox and pointed the Inspector at our top-level folder.

It's great when these things just work!

Hyperlinks

The other key feature DVDs carry when in a computer is the ability to access hyperlinks. It's not quite what you'd expect. For instance, you can't specify a URL as the target for a menu button. But you can have the computer's default browser pop open a web page when a menu is accessed.

DVD@ccess

DVD@ccess is the spiffy marketing name attached to this hyperlink access feature. Enabling DVD@ccess on your disc enables you to include hyperlinks to web sites, local files included on the disc, and email. The software to support DVD@ccess is built right into DVD Player on Macs, and DVDSP includes an installer for adding support to most Windows-based machines.

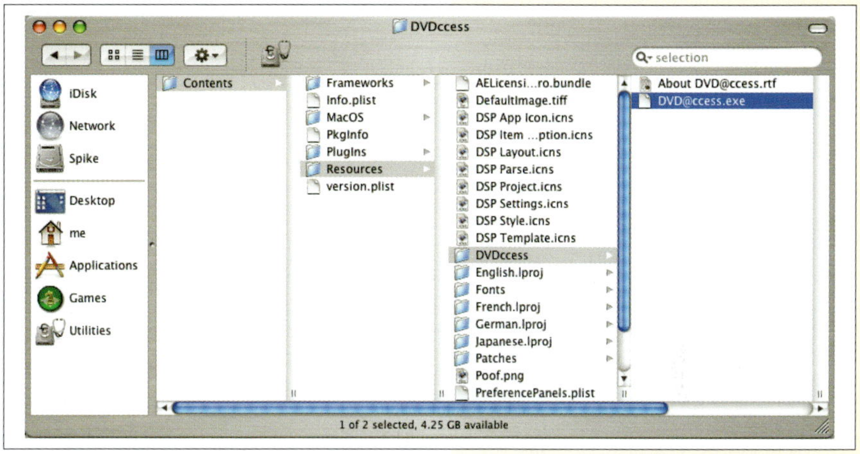

Figure 9-9. Path to the DVD@ccess.exe installation program on Windows-based machines

If you burn your project directly from DVDSP, you don't have to worry about that installer. If you make a project for production by an outside firm, you'll need to make sure you grab the file shown in Figure 9-9 from your DVDSP Applications folder.

To get there, you have to "open up" your DVDSP application. Figure 9-10 shows you how to do this through the Finder.

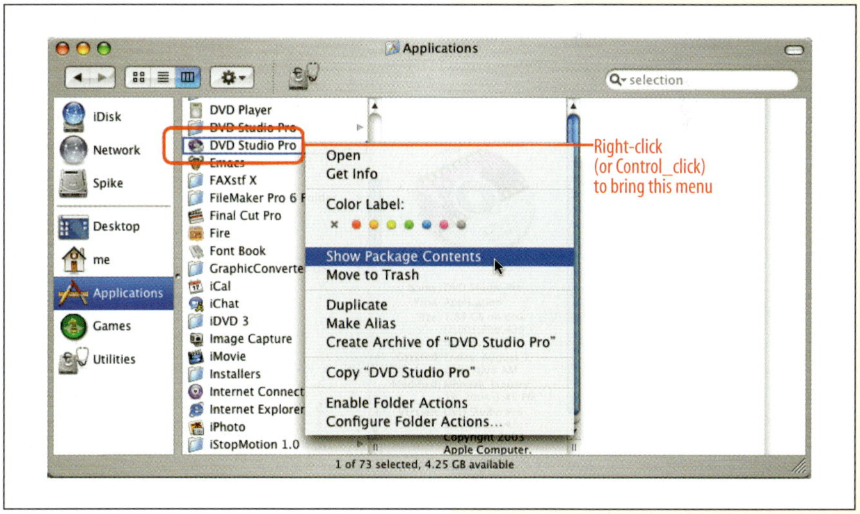

Figure 9-10. Opening up the DVDSP Applications folder in the Finder

Once that's open, you navigate that window just as you would any other Finder window. Grab the *.exe* file and drag a copy to your project's data folder. (Hold down the Option key while dragging to be sure you get a copy—you don't want to remove the original!) Or better yet, drop a copy in a more convenient location for future use.

But before you close that Package Contents window, you should read the RTF that comes along with the installer. It has instructions on how to install DVD@ccess support on Windows-based systems. Again, the support is built into Macs already.

Where can I place a link?

Links can go in three places:

- Menus (when the menu is accessed)
- Markers (when the marker is reached)
- Slides (when a slide is shown)

In each case, you set the URL in the Inspector, as shown in Figure 9-11.

We'll get to the formats for the links in the next section.

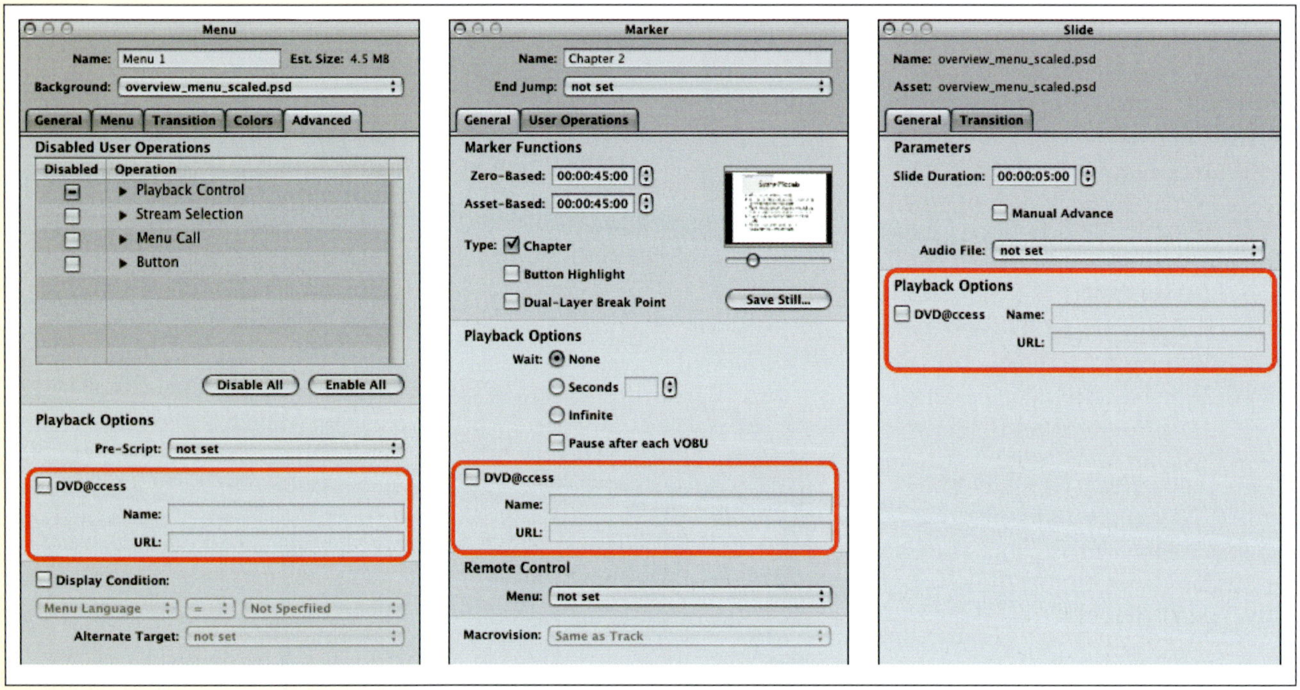

Figure 9-11. The DVD@ccess properties for menus, markers, and slides, respectively

How do I attach a link to a button?

One common request is for menu buttons that launch a web browser. It's a reasonable request, but it's not currently supported. Of course, you can get around that fact fairly quickly. Just have the button go to a new menu with your link on that new menu.

Figure 9-12 shows the General and Advanced tabs for the temporary link menu. We configured this one to use a timeout feature so that you don't even have to place a Return button on the link menu. It shows up for a few seconds and then automatically goes back to the real menu.

We hardcoded the real menu target, but it certainly is possible to create a script that allows the link menu to work with multiple "real" menus. As is often the case, each project will have its own requirements that make it more obvious whether you need hardcoding or scripting.

What links can I place?

So, what types of links are supported? Most of the ones you would think are available. You can use both HTTP and its secure cousin, HTTPS. You can use the "file" protocol to access documents you placed on the DVD. (See the previous project for an example of how to add datafiles to a DVD.) You also can launch the default email client to send an email message.

Web links

Web links are the most common type of link to insert. You format them just like the URL you type into a browser. (They even can carry the same extra information for CGI scripts to use.) Figure 9-13 shows an example of a complex URL—or at least it shows the part that fits in the text box.

HTTPS links really are no different. They rely on the browser to establish a secure connection. The protocol just tells the default application how to get going. Only one restriction is placed on these links: they cannot contain spaces. You can use the "%20" URL encoding instead if you must include a space.

Local data links

If you want to access files that are included on the disc, you can use the *file* protocol, as shown in Figure 9-14.

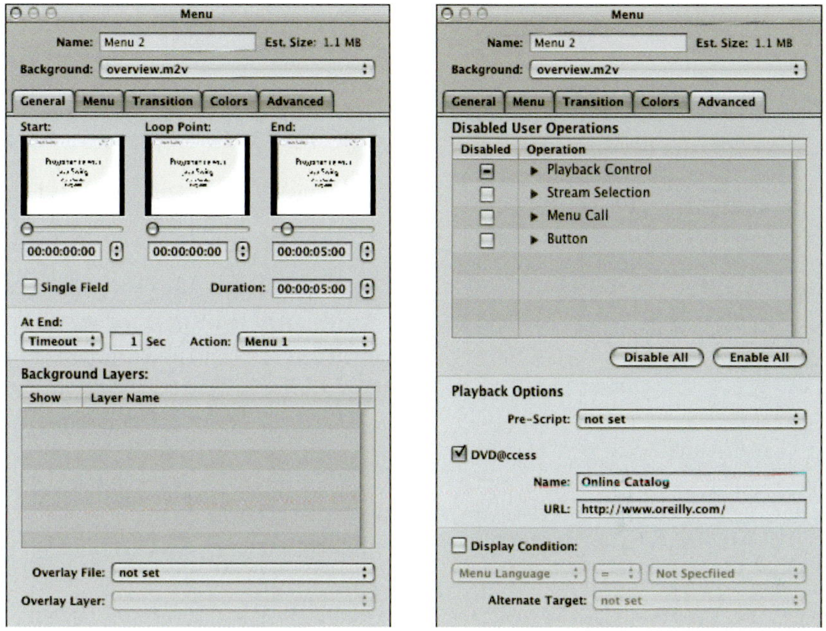

Figure 9-12. Configuring a menu to behave as a "temporary target" for a button (that is really meant to lead to a URL)

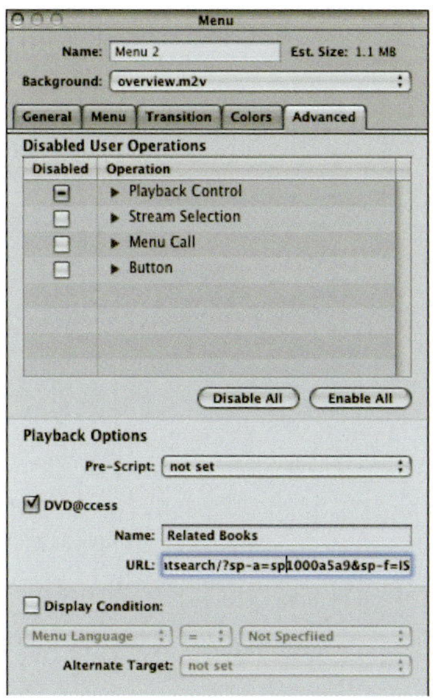

Figure 9-13. A complex URL in the DVD@ccess property

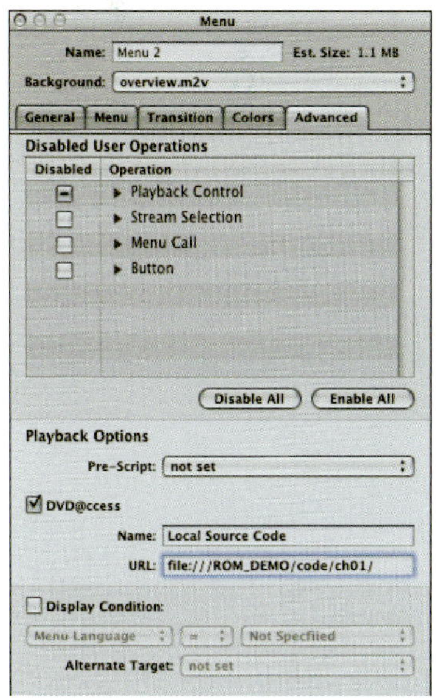

Figure 9-14. A local-file URL in DVD@ccess

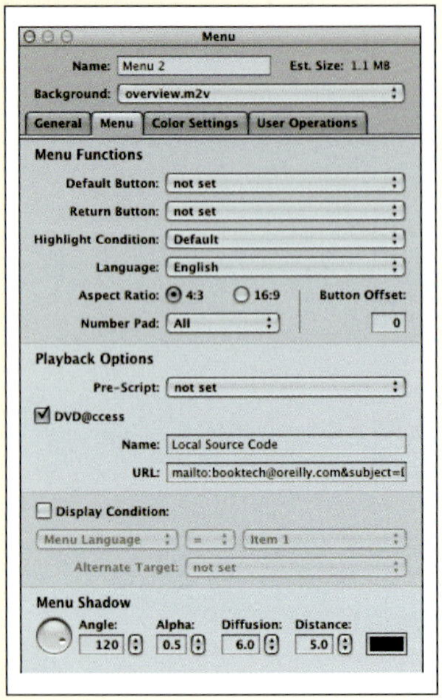

Figure 9-15. A mailto URL in DVD@ccess

The name after *file:///* should be the name you chose for your disc. Your browser will open the file, but most browsers know how to handle a wide variety of files beyond HTML pages. You could use that to launch local multimedia files or word processing documents.

Email

You can use the *mailto* protocol to bring up the local email client with the "To" address filled in. Figure 9-15 shows such a link in the Inspector.

If you're familiar with the *mailto* protocol, you're free to use all the regular features. You can specify the subject (Subject), the body (Body), the carbon-copy list (CC), and the blind carbon-copy list (BCC). For example, to set up an email message to *booktech@oreilly.com* with the subject "DVDSP Book" and the comment "Great Job" in the body, you could use a URL similar to this:

```
mailto:booktech@oreilly.com?subject=DVDSP%20Book&body=Great%20Job
```

Testing DVD@ccess

You can test your DVD@ccess links in the Simulator, but you have to make sure the Simulator is configured properly. Figure 9-16 shows the Simulator properties window. Just make sure the Enable DVD@ccess Web Links checkbox is checked.

If that checkbox is not checked, the Simulator will respond as would a typical consumer player. It is unchecked by default.

Figure 9-16. Enabling DVD@ccess functionality in the Simulator

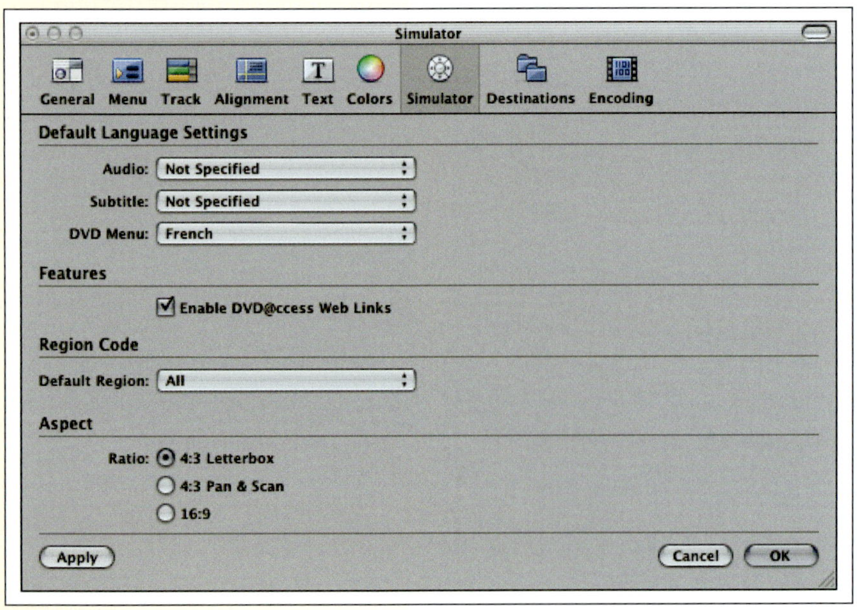

A DVD@ccess project

As we said before, our DVDs have been missing out on great advertising opportunities. Well, we're about to fix that. Let's augment our Java presentation project with some automatic links in the video stream and a link to the O'Reilly web site on the menu.

We'll start with the menu link. Open the *presentation_finished.dspproj* project (or reuse your own copy). Bring up the Inspector for the menu and click the Advanced tab. Select the DVD@ccess option and enter the values shown in Table 9-2 for the Name and URL.

Table 9-2. The property values for the first DVD@ccess link

Property Name	Contents
Name	Online Resource
URL	*http://www.oreilly.com/catalog/jswing2*

Figure 9-17 shows the results.

When you run the project in the Simulator now, you should see the page shown in Figure 9-18 pop up in your web browser.

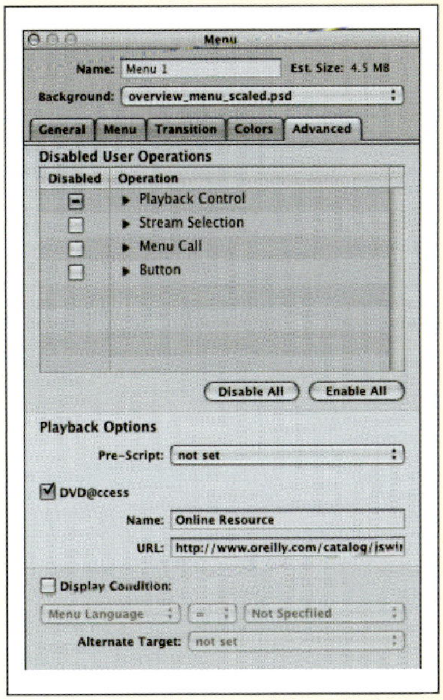

Figure 9-17. Our first DVD@ccess link

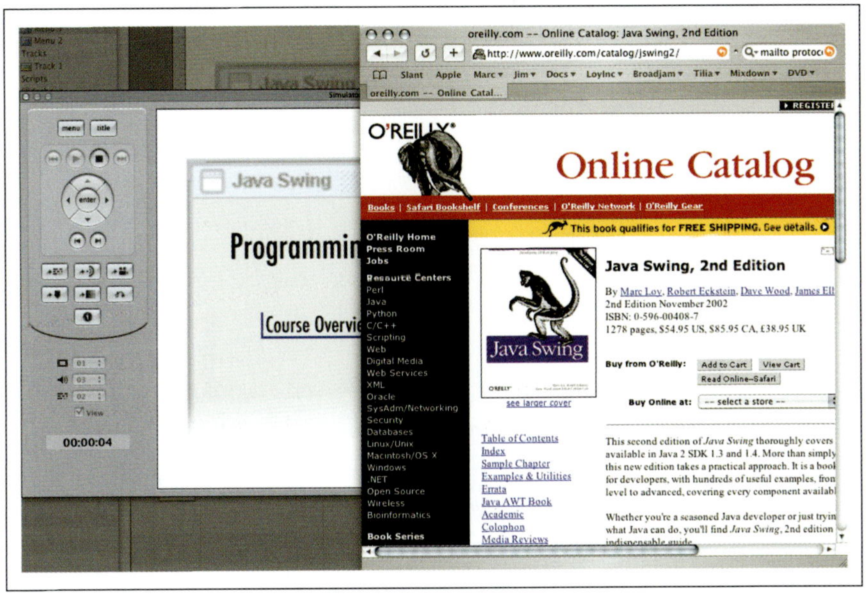

Figure 9-18. The browser page launched when the Simulator starts our DVD@ccess-enabled menu

Links on markers

Well, that was just way too hard. If you're not exhausted from the work, let's add two more links to the video stream. We'll add links to data that is on the disc in the DVD-ROM portion (also from the previous project).

Add two markers to the video track in the Track tab, as shown in Figure 9-19. They don't need to be chapter markers—but they certainly can be.

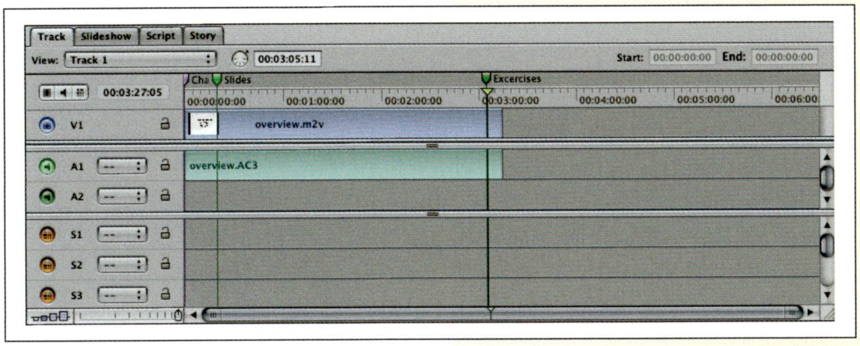

Figure 9-19. Placing two markers for our next DVD@ccess links

Click each marker and look at its Inspector. Set the values for the DVD@ccess Name and URL properties, as shown in Table 9-3.

Table 9-3. The Name and URL properties for our markers

Marker	Property	Contents
1	Name	PDF Material
	URL	*file:///ROM_DEMO/slides/1a.pdf*
2	Name	Code Folder
	URL	*file:///ROM_DEMO/code/ch01/*

Figure 9-20 shows the Inspector with the properties configured.

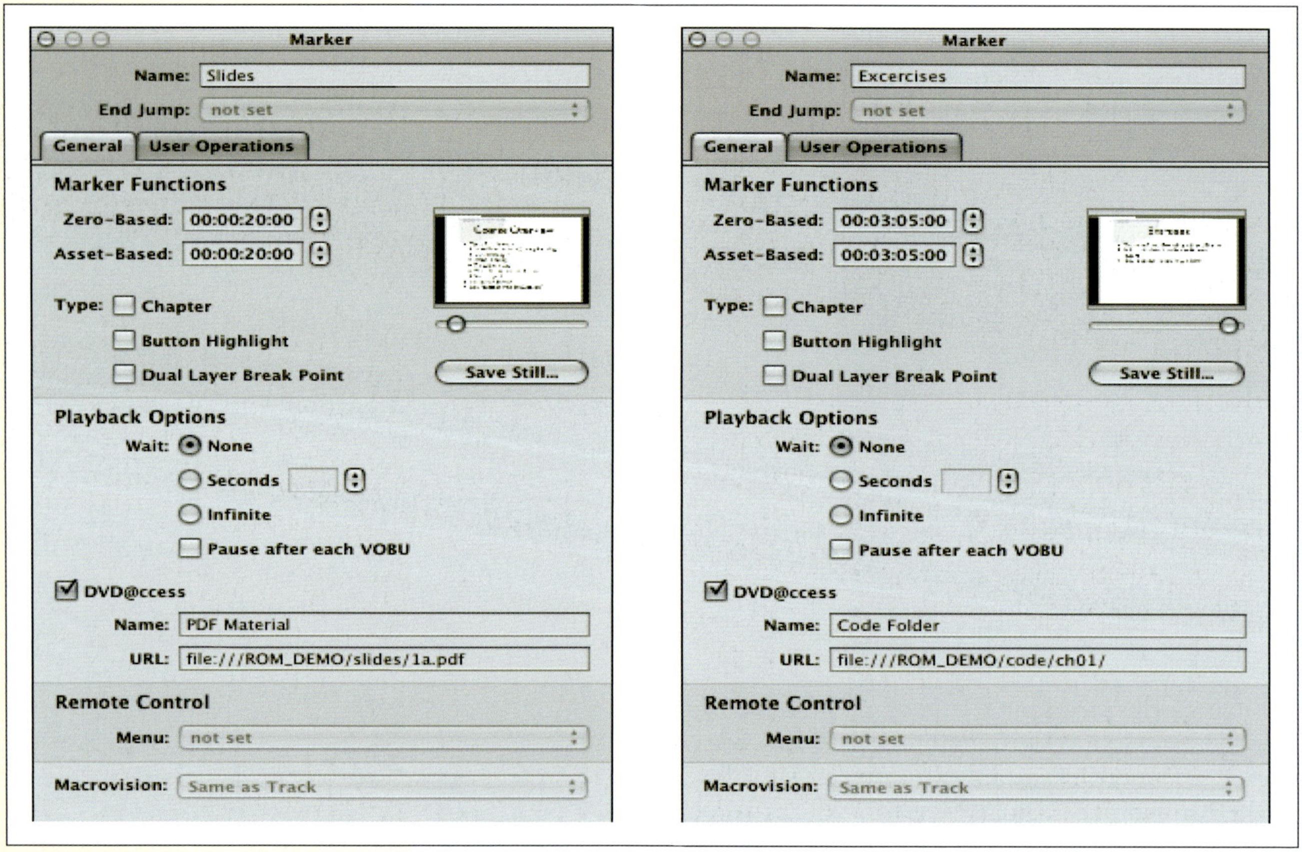

Figure 9-20. Our configured markers

Try it out in the Simulator. Your browser should bring up the PDF document (link 1) and a Finder window (link 2).

Again, another tough assignment. But really, that's the point of software such as DVDSP. You should be spending your time on the content, not the details of making it accessible. We'll take that idea a bit further in the next chapter, where you'll create your own templates that should help future projects go even faster.

10

Templates and Styles

In the first few chapters, you got to see how quickly you can create an aesthetically pleasing DVD project using Apple-supplied templates as a normal part of the DVDSP installation. But what if three sappy wedding themes aren't enough? Well, you've come to the right place. In this chapter we'll look at how you can build your own style and template objects for reuse in your projects. We'll even look at how to bundle a standalone template and make it available for others to use in their projects as well.

In this chapter

Palette Redux

Before we get started building our own styles and templates, let's review the Palette. Once you start building your own menus and buttons, the Palette tends to slip into the background. As we prepare our custom pieces, though, it will be very much in the foreground. Figure 10-1 shows the Palette and its tabs. If you're using DVDSP 2, you'll see slightly larger icons (and slightly fewer of them—DVDSP 3 added a batch of new templates). But don't worry. Everything we discuss in this chapter works in both versions of DVDSP.

Palette tabs

Three of the tabs (Audio, Stills, and Video) merely provide convenient access to your media. They provide some default locations to look for things (the Video tab, for example, automatically loads the *Movies* folder in your home directory), but you can add others. Each tab also has a handy search form (shown in Figure 10-2) that you can use to quickly locate items, even in very large collections.

The other tabs contain items that help you create and manipulate your project. We can add items to these tabs as well, but that takes a bit more preparation.

The Templates tab

Figure 10-3 highlights the Templates tab. Here's where you find complete menu templates. Templates include the background asset, button locations,

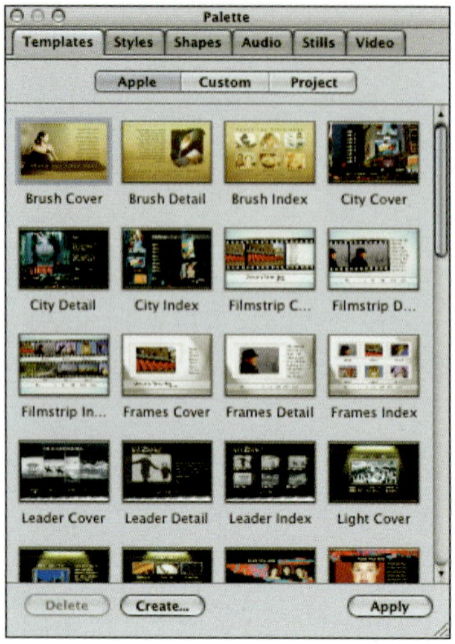

Figure 10-1. The Palette window in DVDSP

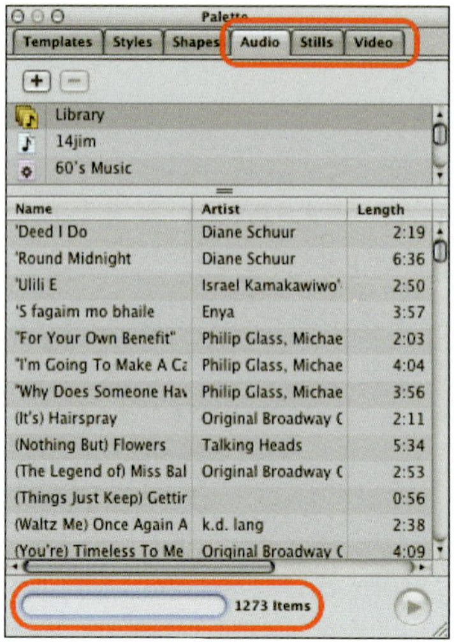

Figure 10-2. The searchable media tabs

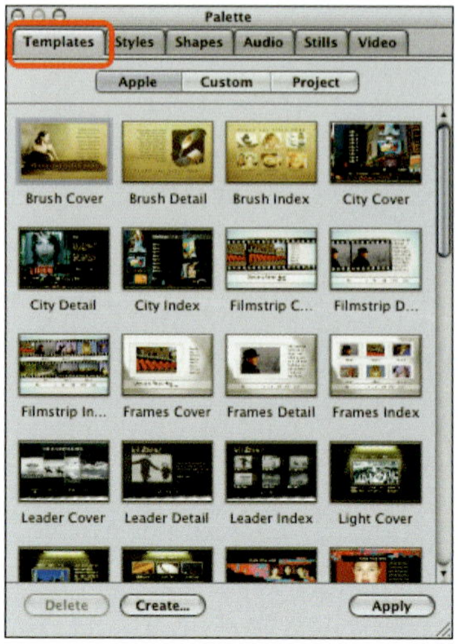

Figure 10-3. The Palette's Templates tab

drop zones for custom images, and any associated audio. Just drag a template icon onto a blank menu and you're almost done. Of course, you need to supply your own images and video and set the targets for the buttons, but the template takes care of the layout and design.

Apple supplies a large set of templates. You can refer back to the list in Figure 3-2 in Chapter 3—or just look at the tab yourself! You can build custom templates, and we will do exactly that later in this chapter.

Projects also can have their own set of templates. That's a great feature for working with clients. As you might recall from Chapter 1, you can change a menu based on templates on the fly. As long as the same buttons and drop zones exist in the new template, everything will stay connected properly. You can build a few variations and switch them on demand as your client reviews the project.

The Styles tab

This tab is a bit more complex. It comprises four categories: Buttons, Text, Drop Zones, and Layouts. The Buttons and Drop Zones categories are pretty straightforward: you pick an item from the list (Figure 10-4), drag it over to your menu (Figure 10-5), and poof! A new button (or drop zone) appears (Figure 10-6).

The Text category of the Styles tab (shown in Figure 10-7) enables you to store two different types of information: a text style (the font face, size, etc.) and an actual chunk of text (which also will have a default style). You can create header and button text styles for use all over your project. You also can create boilerplate text that gets repeated on menus. For example, copyright information might fall into this category.

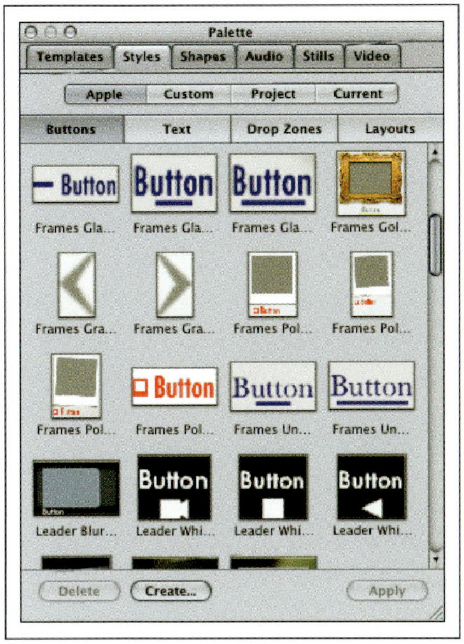

Figure 10-4. Step 1: grabbing a button from the button style portion of the Palette

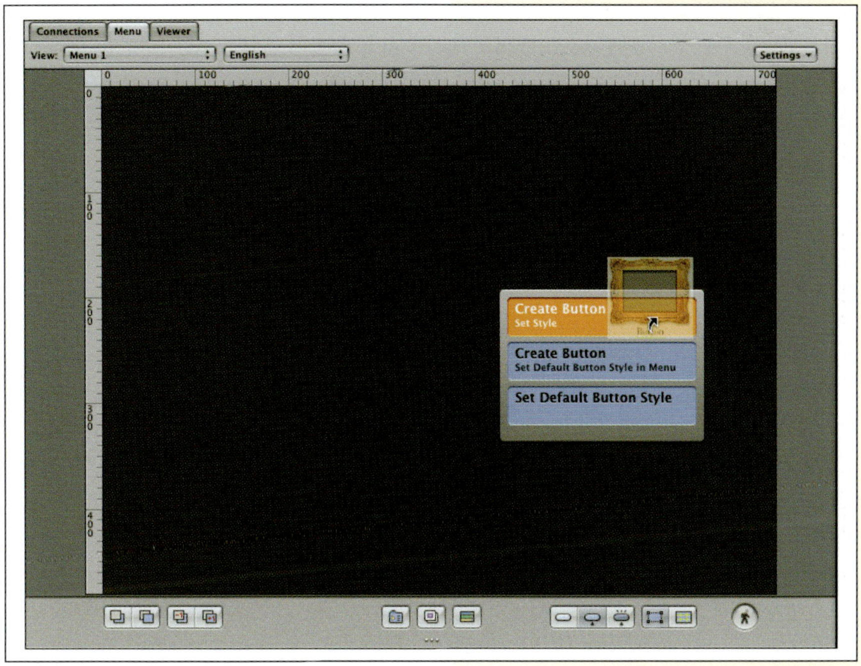

Figure 10-5. Step 2: dragging the button over your menu

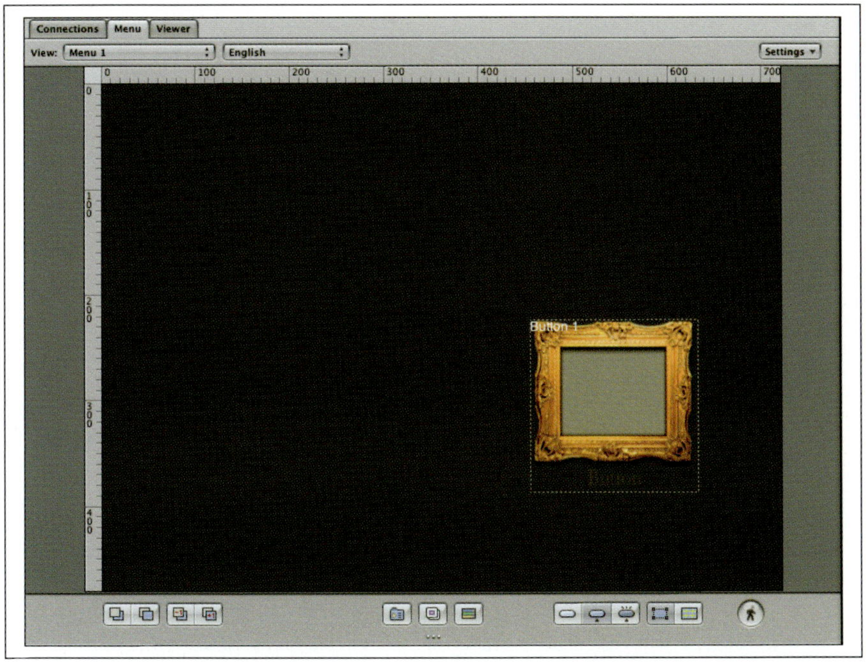

Figure 10-6. Step 3: dropping the button

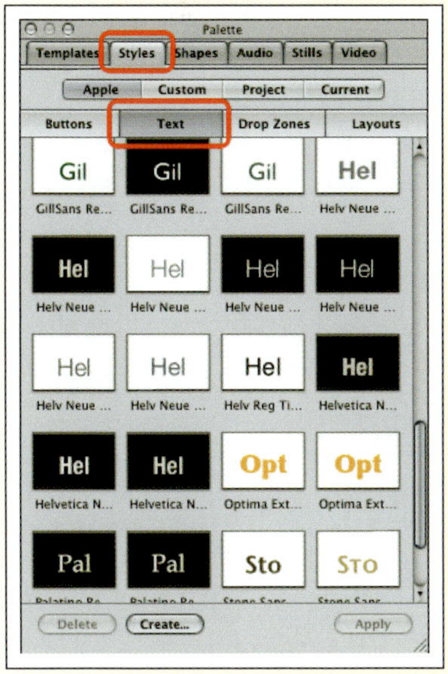

Figure 10-7. The text styles portion of the Palette

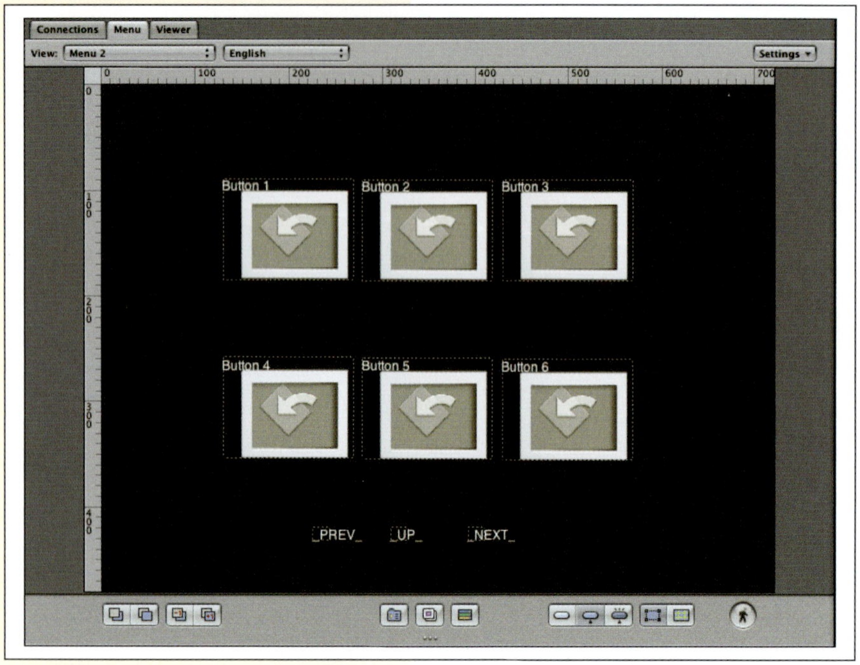

Figure 10-8. The Shelves Index layout style applied to a menu

The Layouts category looks very similar to the Templates tab. However, the Layouts category represents only a subset of the information present in a template. In this category you determine, well, the menu's layout. The number and position of the buttons and drop zones come from a layout. The background and other assets, however, are not carried with a layout. Figure 10-8 shows the result of dragging the Shelves Index layout onto a new menu. You can see the buttons, but no other images are created.

We'll be looking at creating style objects later in this chapter as well.

The Shapes tab

The remaining tab (Figure 10-9) supplies you with shapes that you can use for buttons and drop zones. Figure 10-10 shows the use of a few different shapes on a blank menu. We have created buttons with the shapes, but

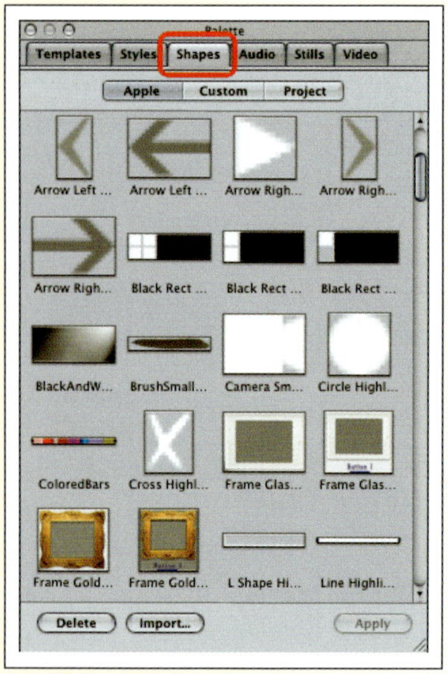

Figure 10-9. The Palette's Shapes tab

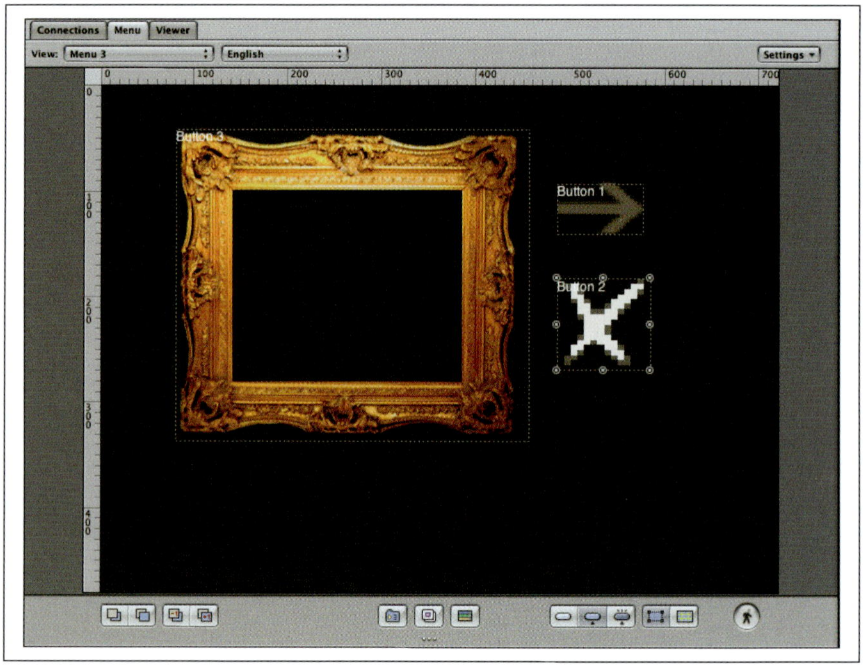

Figure 10-10. Manually created buttons using particular shapes from the Palette

remember that you can convert buttons to drop zones (and vice versa) with the Control-click of a mouse.

Shapes

Shapes are integral to designing your own templates, so let's take time to dig a bit deeper here. We'll look at where you use them as well as how to create one from scratch.

You've seen a simple example (Figure 10-10) of using a shape directly to create a button. Shapes also can leave room for other content. Figure 10-11 shows a frame shape with some of our video footage added as an asset.

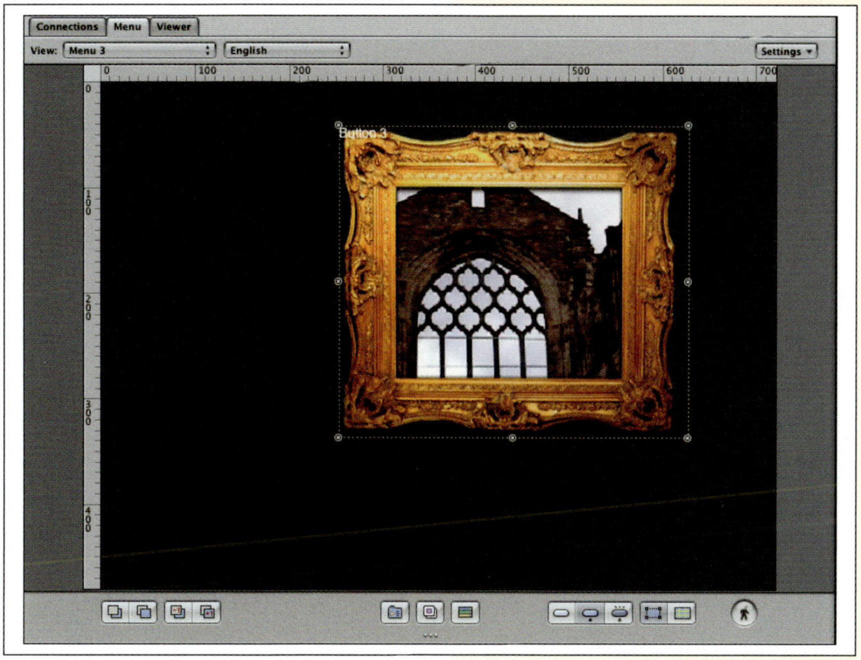

Figure 10-11. A video clip set as the asset for a shape

You can use shapes for drop zones and buttons to help continue the overall look and feel of a menu. You also can make shapes that are interesting in and of themselves. Your corporate logo might be a good candidate here. You just drag-and-drop the logo shape (which can be a full-color image) onto any menu to brand your work.

Creating your own shapes

You can create your own shapes for use in DVDSP, but you need Photoshop. We warned you up front that DVDSP was best used as part of a suite, but this is one of the few places that requires an outside application. The reason for the Photoshop requirement is that a shape, like onions and ogres, has layers: four layers, to be precise. Figure 10-12 highlights those layers.

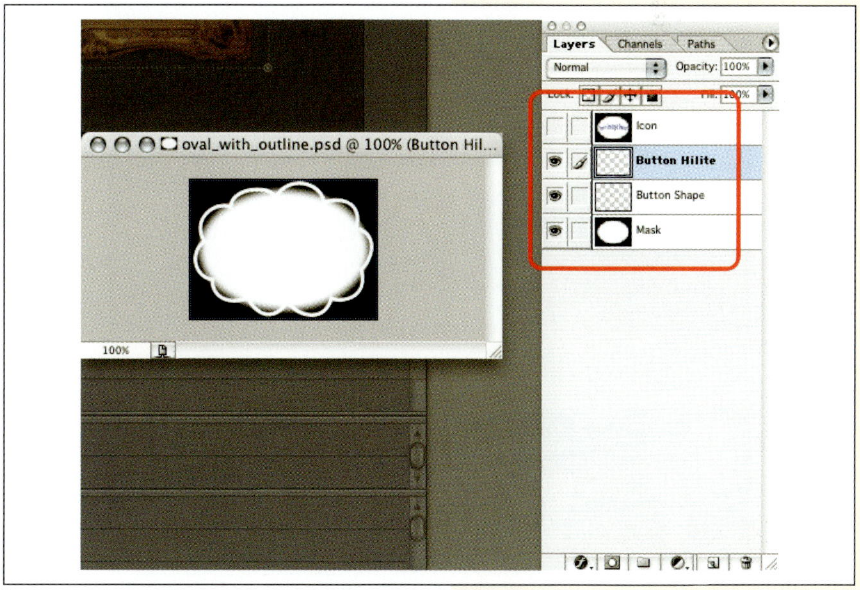

Figure 10-12. The four Photoshop layers used to create a new shape

With the exception of the Icon layer, these layers are stacked together to build a shape that cooperates with the menu around it. You can create transparent and opaque areas. You can fill some parts of the shape with the background menu, other parts with an asset.

So, let's build a shape. In fact, let's build two: one nifty corporate logo and one generic shape that we can use in our Style project.

A logo shape

To start with the corporate logo, it's easiest to create the Button Shape layer. Figure 10-13 shows our logo in its full glory.

Next you have to supply the mask so that DVDSP knows which parts should show up on the screen. This enables you to create logos that are more than just simple rectangles. In our case, we want a mask that matches the block of text and the critter in our logo. Figure 10-14 shows that mask.

Any black or gray appearing outside the button shape will be transparent (i.e., the menu will show through). The button shape always will show up. The white areas show any content we've assigned to the button (as an asset). If you don't assign an asset, the white areas will be transparent as well. Figure 10-15 shows an example of how these pieces get combined.

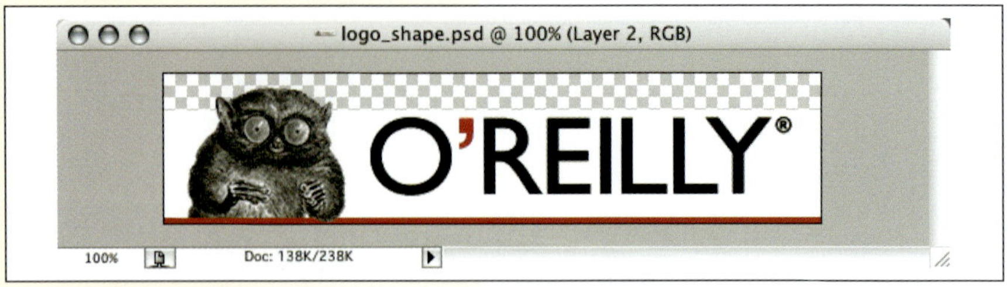

Figure 10-13. The Button Shape layer of our new logo shape

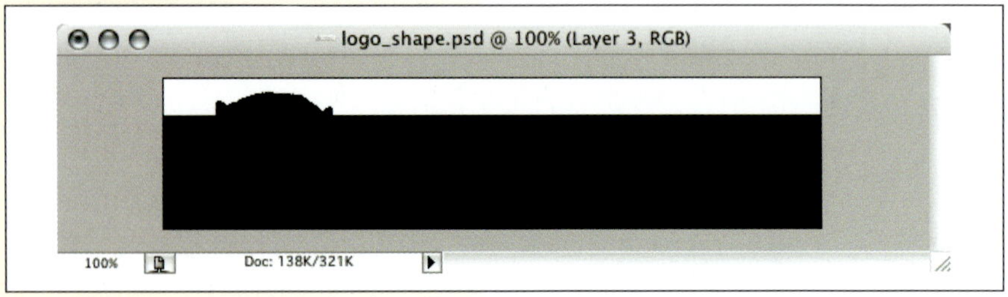

Figure 10-14. The Mask layer of our new logo shape

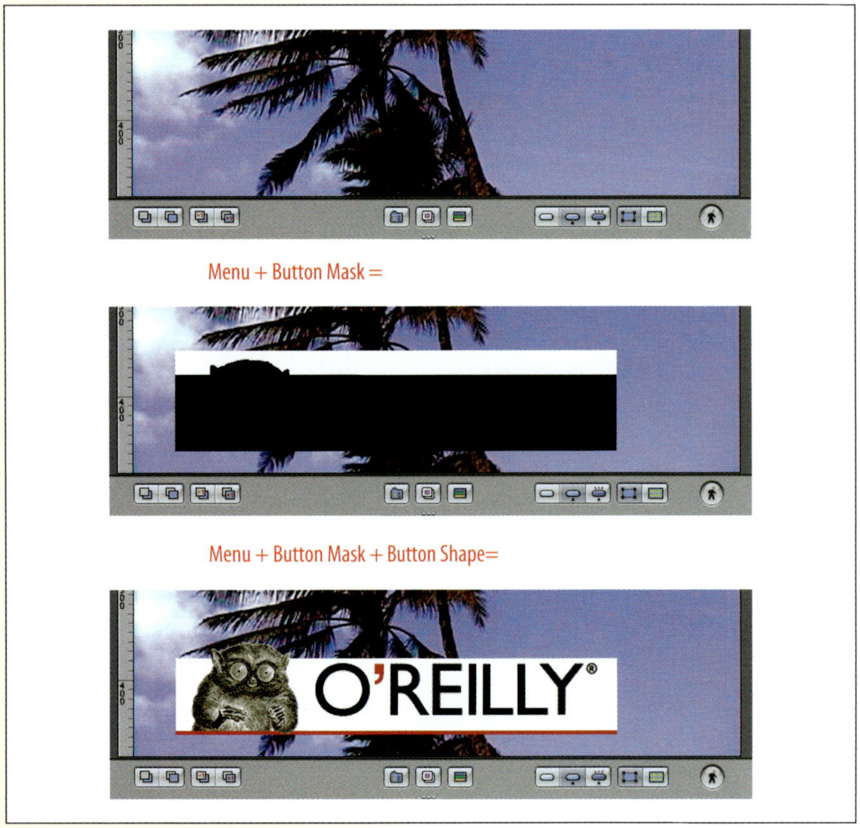

Figure 10-15. A sample menu (top) with a shape mask (middle) and a shape (bottom)

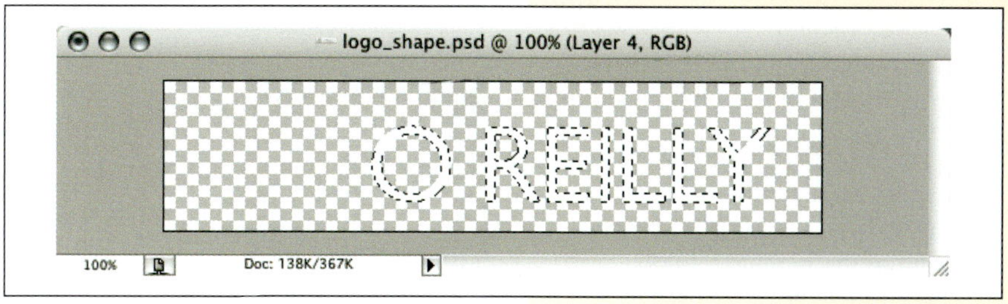

Figure 10-16. The Button Hilite layer of our logo shape

The Button Hilite layer enables you to specify the shape that would appear when the button is selected or activated. If your shape is used only as a drop zone, you can leave this layer empty—but don't leave it out! DVDSP counts the layers. That means you are free to name them as you want. Just make sure you have them in the right order.

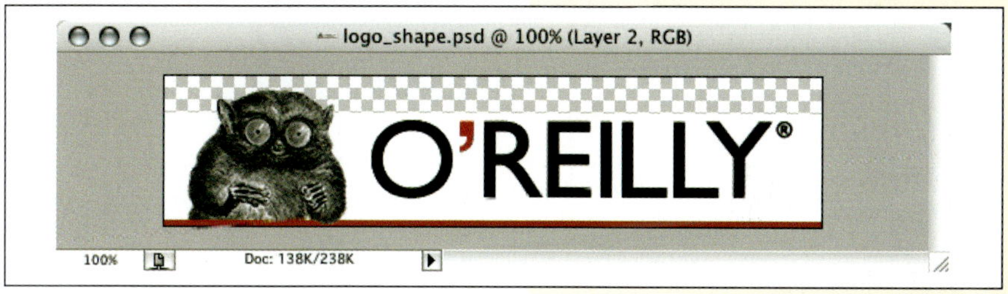

Figure 10-17. The Icon layer of our logo shape

Just in case you want to add a link to the logo (maybe to jump to a production credits page or some such), we'll go ahead and outline the text to make it the active part of the button. Figure 10-16 shows that layer for our logo.

Note that we used a transparent background for the Button Hilite layer. We also should mention that if you use the 2-bit grayscale approach for your color schemes, your Button Hilite layer can use 2-bit colors, too.

The last layer is the Icon layer. This layer is optional but very useful. It's the layer that contains the image DVDSP uses in the Palette. If you don't supply this fourth layer, DVDSP defaults to using the Button Shape layer (Layer 2). Figure 10-17 shows our Icon layer for the logo. It's really just a copy of the logo, but we want to be explicit and create all four layers for this example.

That's it! If you're building your own shape, be sure to save the file somewhere so that you can get to it easily. We'll look at how to import a shape in the next section.

A more generic shape

The next shape we'll build is more general and meant to be used with an asset the user supplies. That means we need to leave a "hole" for the asset to show through. We'll build a soft-edged oval with a cloudlike highlight. The layers are broken down in Table 10-1.

> **STUDIO ADVICE**
>
> ## Alpha-Channel Transparency
>
> A shape's Button Hilite layer uses its alpha channel to manage the highlight colors. The multicolor buttons we built back in Chapter 4 used levels of gray: black (100%), dark gray (66%), light gray (33%), and white (0%) could be mapped to any four colors in your DVDSP project. You have the same four "colors" in the Button Hilite layer, but they are transparency levels, not shades of gray.
>
> Because DVDSP relies on the alpha channel and not the color, you actually can use black or white for drawing your outline in this layer; 100% is one color, 67% (not 66%—it's very picky about that) is the second color, 33% is the third, and 0% usually is left as the completely transparent color. When painting a particular color in this layer, you just need to adjust the transparency of your paint tool to get the desired color.

Table 10-1. Our Soft Oval shape layers

Photoshop Image	Layer	Description
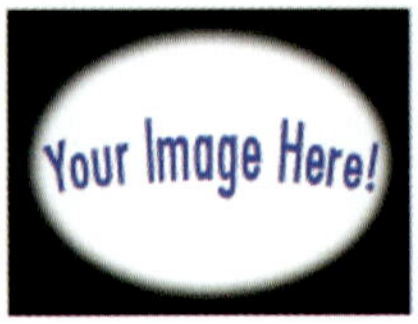	Icon (4)	We just want a nice, simple icon to show so that users can supply their own image inside the oval.
	Button Hilite (3)	Here we draw our outline shape. The outline is similar to the multicolor overlay buttons from Chapter 4. Notice that we leave most of the area empty so that the user-supplied asset will show through, even when selected. (The cyan background is for clarity—it is not in the actual image.)
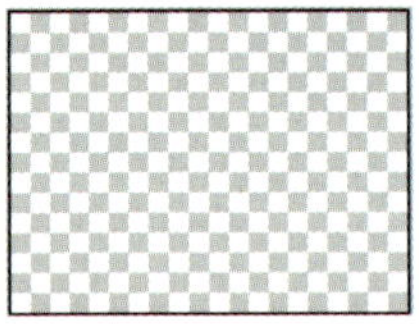	Button Shape (2)	For our shape, we leave this layer empty. We don't want any highlights or marks on the user's asset if the button isn't selected. An empty Button Shape layer is fine as long as we supply the Icon layer so that DVDSP has something to display in the Palette.
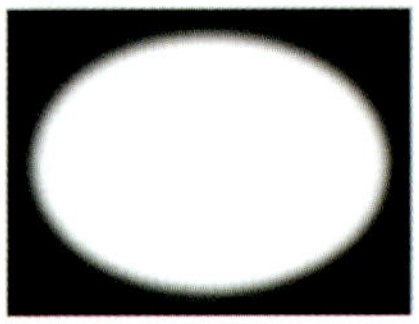	Mask (1)	We want the asset to show through as an oval. We use a grayscale gradient around the edges to make the oval softer. Black is 100% transparent. White is 100% opaque (i.e., 100% the user's asset). Grays are simply in between. This layer is a full 8-bit grayscale layer. You can use any gray in Photoshop.

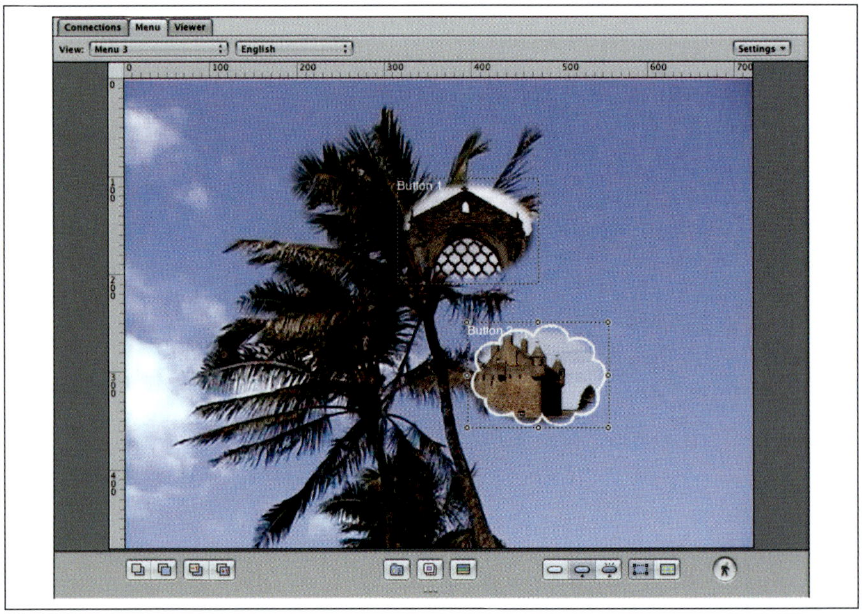

Figure 10-18. Two buttons created with our new Soft Oval shape

Figure 10-18 shows what happens with this shape and a user asset when set up as a button in a menu.

Although it doesn't show in this simple screenshot, both buttons are using video assets. Mastering how to create custom shapes definitely is worth your effort!

Importing shapes

To import your newly built shapes for use in future projects, head to the Palette and click the Shapes tab. Once there, click the Custom category. Your screen should be blank, similar to Figure 10-19.

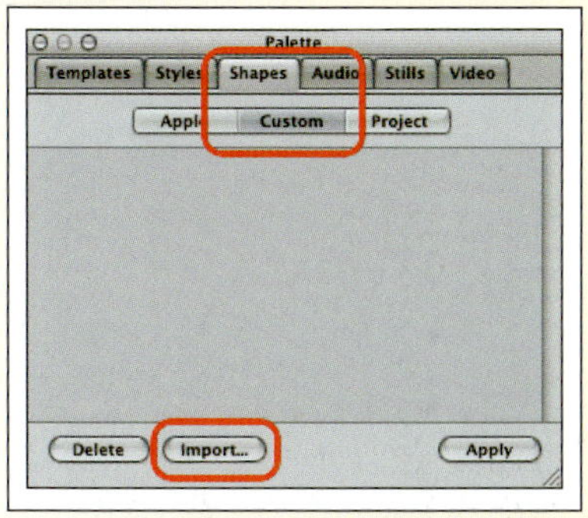

Figure 10-19. Importing custom shapes

Notice the Import... button at the bottom of the Palette. Click that button to launch a file selection dialog. Find your Photoshop file and click Import, as shown in Figure 10-20.

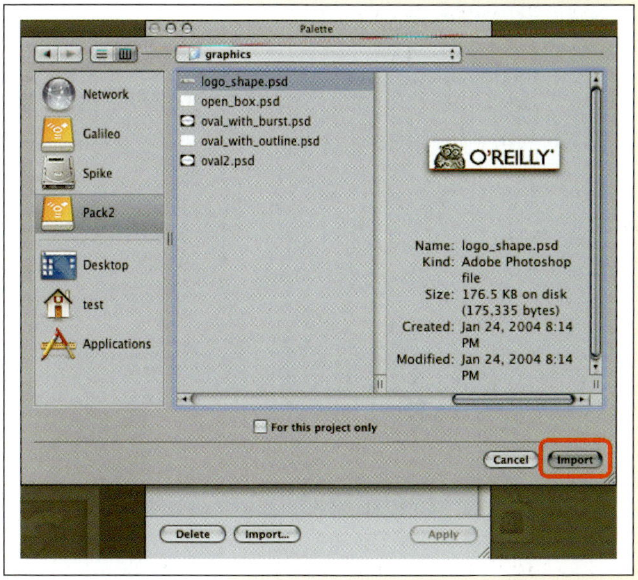

Figure 10-20. Selecting the Photoshop document for import

You also should take note of the "For this project only" checkbox near the Import button. You can check that checkbox if you want to place this shape in the current project *only*. The imported shape will be placed in the Project category on the Shapes tab. Leaving that checkbox unchecked will make your shape available to all projects and will place the imported shape in the Custom category.

If you want to grab the prebuilt versions of these shapes, you can find them in the *projects/ch10* folder.

At this point, the Custom category should look like Figure 10-21.

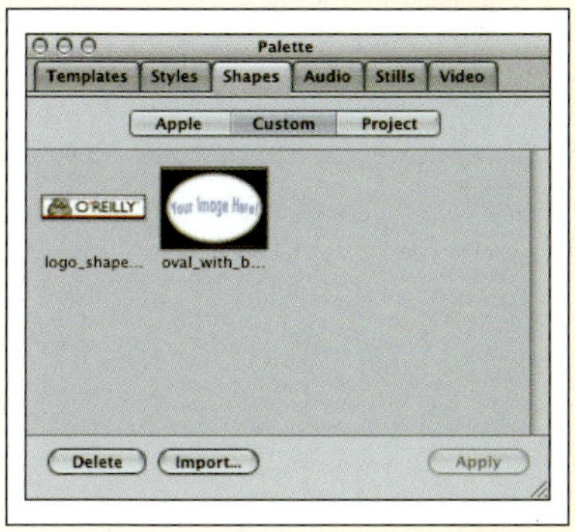

Figure 10-21. Our freshly imported shape objects

Now those shapes are available for you to use in any other projects you start. If you want to delete a custom shape, just select it in the Shapes tab and click the Delete button that you see at the bottom of Figure 10-21. Note that the deletion is permanent and cannot be undone. You can, of course, reimport the shape if you have a copy of the file somewhere else.

You also can update a shape by reimporting it. Just import it as usual and use the same name. You'll be prompted to make sure it's OK to replace the existing shape. Buttons in the current project that use the shape should be updated immediately. Previous projects that used the old shape also will be updated to the new shape the next time you open them.

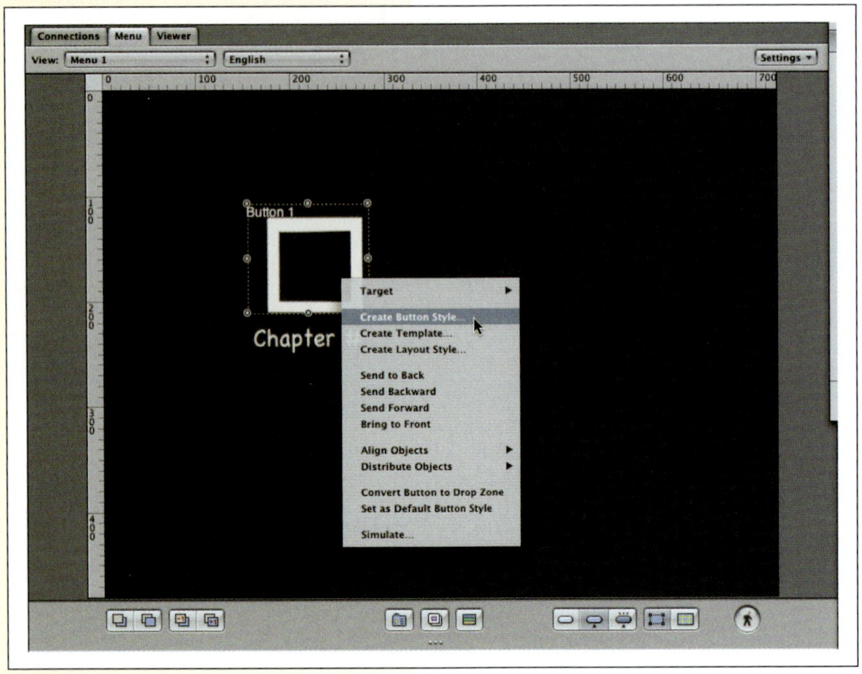

Figure 10-22. Turning an existing button into a button style

Styles

As we noted before, styles are part of what makes up a template. You have four different types of styles: Buttons, Drop Zones, Layouts, and Text. Let's look a bit deeper at how to create and use styles in your project.

Creating styles

Creating a new style is fairly straightforward. Mostly you just start designing a new menu and then "save off" a piece as a style. As a quick example, take a look at Figure 10-22. We have a simple menu with a button defined on it. The button has a particular font and highlight shape. We can make that a style by using the context menu (Control-click the button as shown in Figure 10-22). Select the Create Button Style... option and off you go!

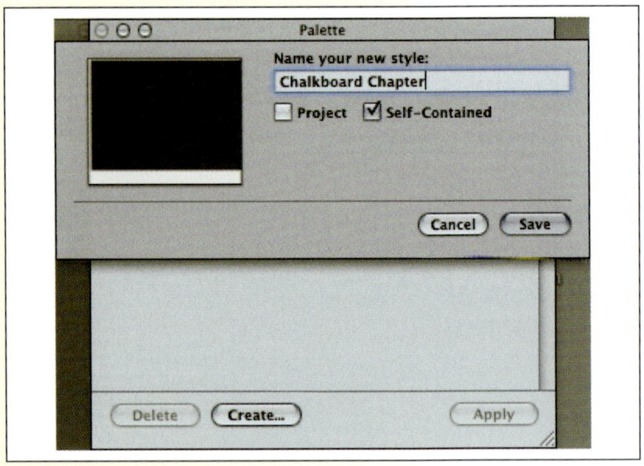

Figure 10-23. The New Style dialog

You'll be prompted for three pieces of information:

- The name of the style
- Where to store the new style (Custom or Project)
- Whether it should be self-contained

Figure 10-23 shows the dialog box in question.

Button styles

As you can see in the previous short example, creating a button style really involves creating a button. A button's font, shape, text position, size, default asset, and default text are stored in a button style. You can use that style to create new buttons or to update existing buttons.

You also can create a button style by selecting an example button on your menu and clicking the Create button at the bottom of the Palette.

To create a new button from a button style, just drag the style onto your menu and drop it there. Figure 10-24 shows a drop in progress.

To update an existing button (or buttons, actually, which can be *very* handy), select the button(s) in the menu, select the desired button style in the Styles tab, and click the Apply button (highlighted in Figure 10-25). All the selected buttons should be updated to use the selected style. Any asset or text changes you made should be left alone.

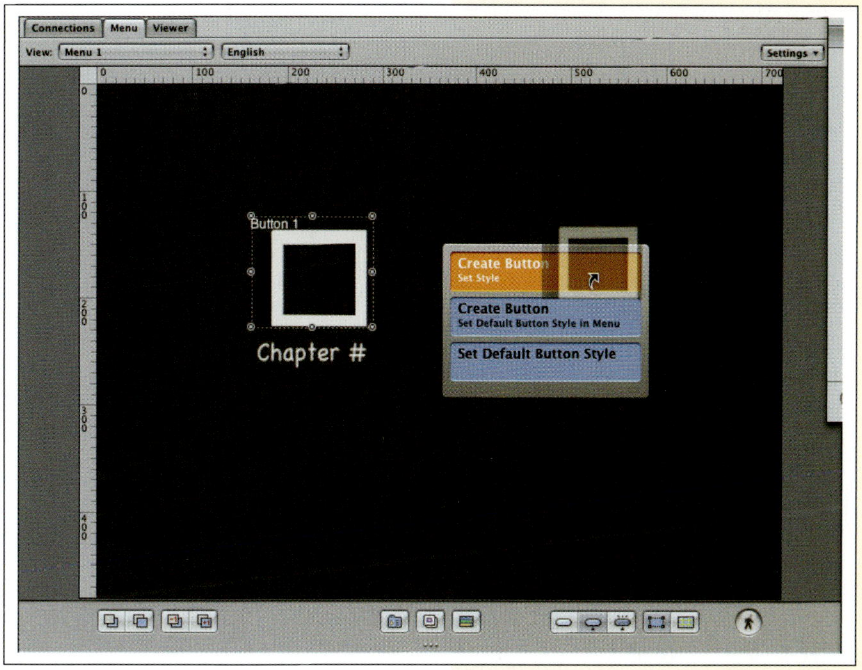

Figure 10-24. Creating new buttons with style objects

Drop zone styles

Drop zone styles are very much like button styles. The main difference is that drop zones don't have highlights, as they aren't active parts of menus. You can create a new drop zone style just as you do a new button style. The same characteristics are stored.

Text styles

As we mentioned before, a text style carries two kinds of information: the look of the text and the text itself. You can apply text styles to text objects or to buttons that have associated text.

To create a text style you do pretty much the same thing you do for buttons and drop zones. Create some text on any menu (double-clicking the background should get you a text box). Apply any font (⌘T) or color (Shift-⌘C) changes you want. Then, Control-click the text and select the Create Text Style... option from the menu, as shown in Figure 10-26.

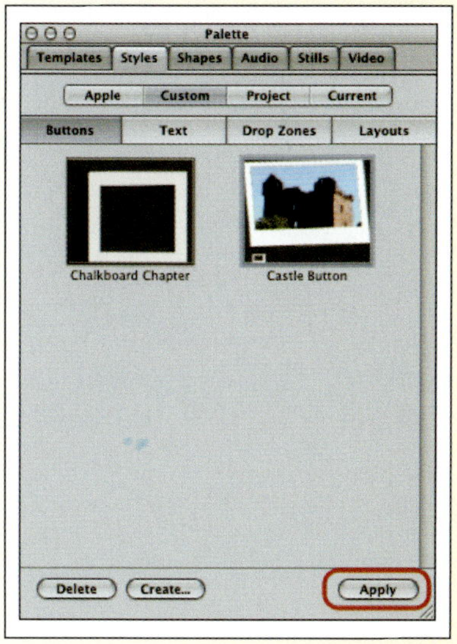

Figure 10-25. Applying a style to existing buttons

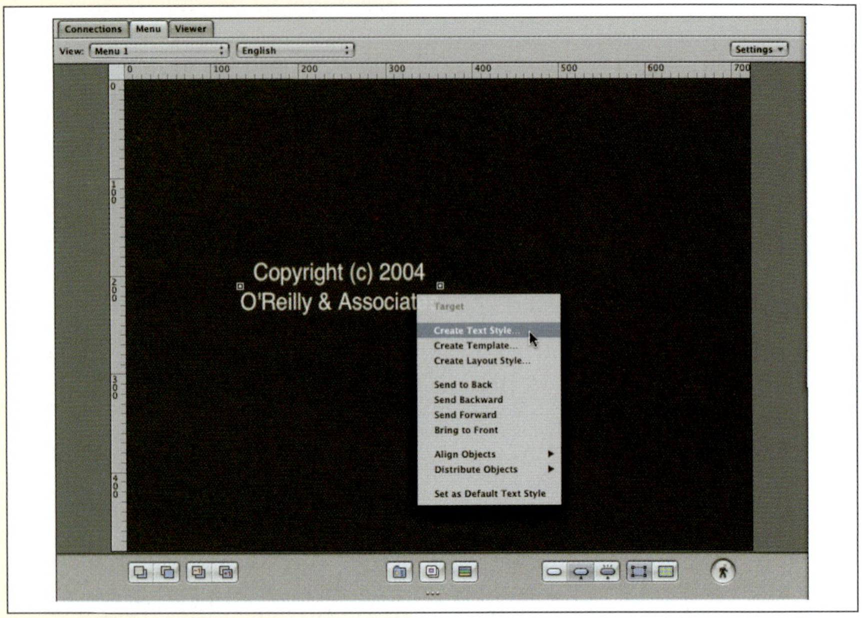

Figure 10-26. Creating a new text style

You should end up with a new text style that shows your font and some of the text. Figure 10-27 shows our copyright example in the Palette.

You can create a new text object by dragging a text style onto your menu. You also can select one or more text objects on the menu and click the Apply button in the Palette.

Layout styles

Creating a layout style can save you a great deal of time on future projects. You can set up a layout for different types of menus. You don't even have to know the specific content. Your library of layouts will be ready to spring into action on any new menu.

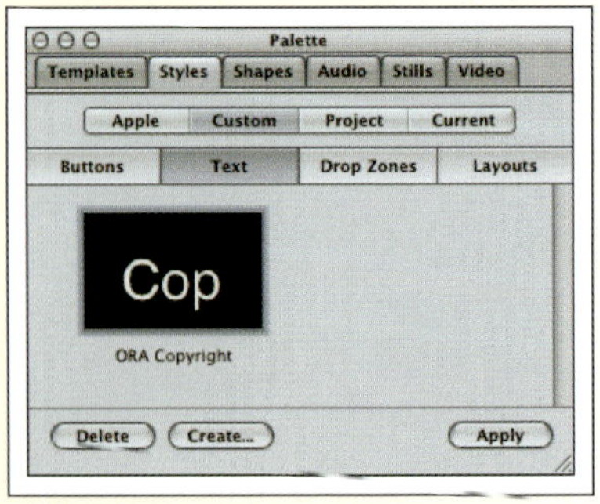

Figure 10-27. Our newly created text style in the Palette

You can create a layout style much like other styles. Lay out a menu with the buttons you want, set their styles, and save the styles by Control-clicking the menu background and selecting the Create Layout Style... option. At this point you can apply that layout to any menu. For example, start a new menu and place four buttons on the compass points, as shown in Figure 10-28.

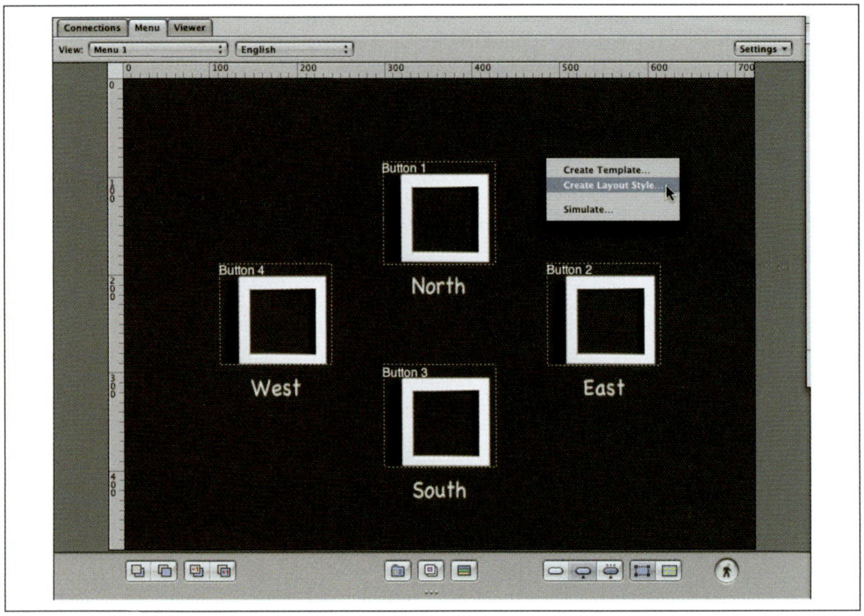

Figure 10-28. Creating a layout style

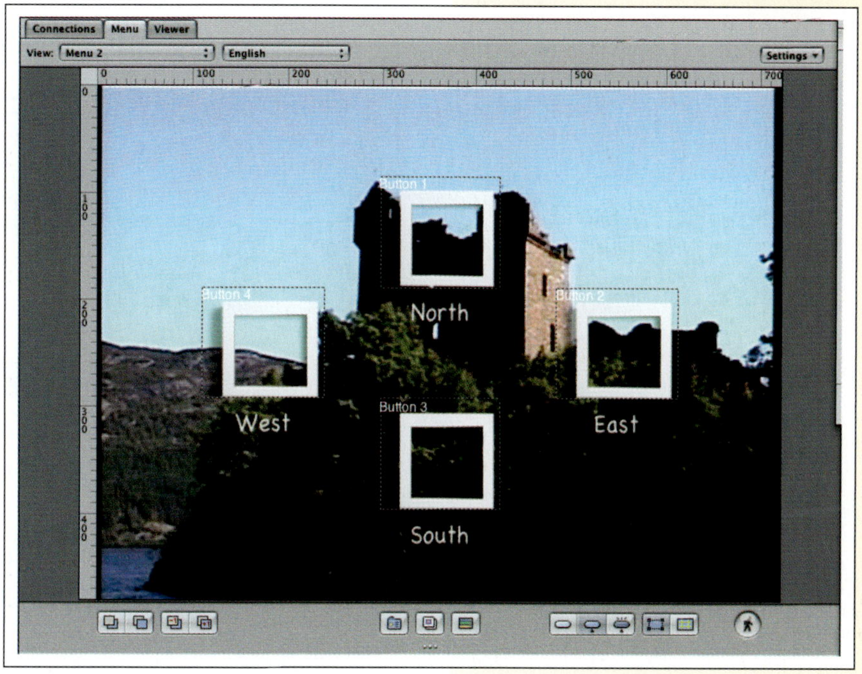

Figure 10-29. A new menu with the Compass layout style applied

Save this layout. (We named ours Compass.) Now start another new menu and pick a nice background for it. Drag your new style onto the new menu and poof! The compass buttons should appear as shown in Figure 10-29.

Of course, now you're free to update the styles of each button or rearrange them, but at least you have a good starting point.

Using styles

That point about having someplace to begin is really at the heart of styles in DVDSP. Over time you should build up a library of styles you like (or that your clients like!) and use them as interior designers use swatches. Try them on for size. Take your project for a test spin in the Simulator.

Buttons and other items on a menu remember that they've been styled. Applying a new style (for buttons, text, drop zones, or even layouts) simply shuffles things around to match the new style. Where possible, none of the content is altered. Styles aren't perfect, though—they don't know when you change a button name. Buttons that don't match the names in a layout, for example, simply are left alone.

Styles are very powerful tools for appraising a menu's appeal. Try using them in a project and see if you get hooked. It makes later alterations so much easier, too. If we update the style (we changed the font in this example), it's easy to propagate that change.

Button names

Speaking of button names in layout styles, you should know about three button names in particular. You can use some layouts (and templates, which we'll see later) to build automatic index pages for the chapters in your video streams. We'll be making both styles and templates that you can use with this very spiffy feature.

But most larger projects have more than four or five chapters. Even our simple alphabet example from Chapter 6 had 26 chapters! In that case, DVDSP creates multiple menus. Each menu uses all the available buttons. For example, using a six-button layout creates five new menus: four with six buttons (A-F, G-L, M-R, and S-X) and one with two buttons (Y and Z).

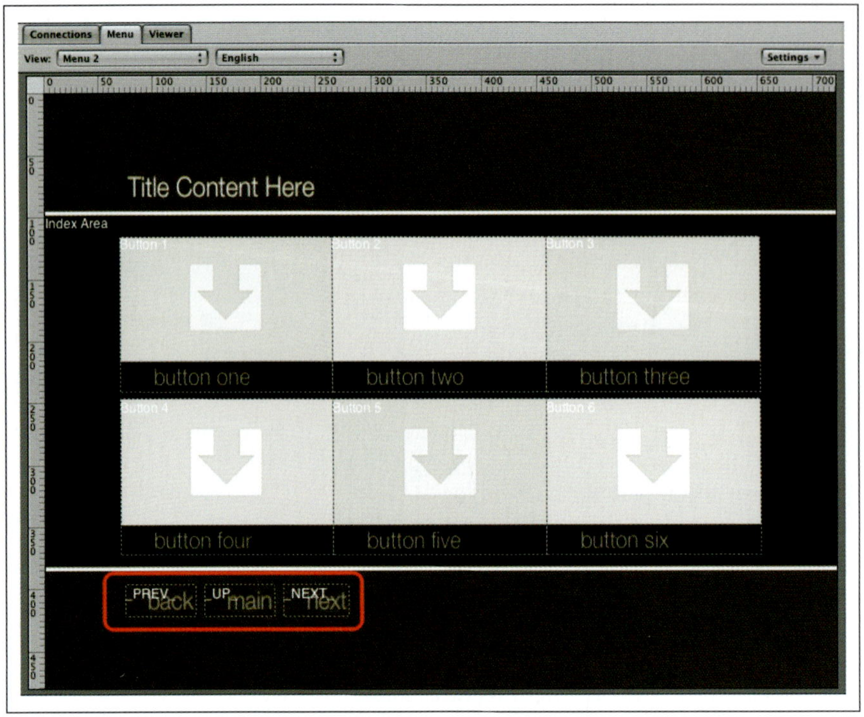

Figure 10-30. The three navigation buttons available with Index templates

If you want to produce a template for chapter indices, you should add three buttons for the intermenu navigation. Figure 10-30 highlights these buttons in one of the Apple-supplied templates. Table 10-2 lists their names and functions.

Table 10-2. Template and layout style navigation buttons

Button Name	Button Function
_UP_	Returns to the menu containing the button that launched this series of chapter index menus
NEXT	Takes you to the next chapter index menu in the series
PREV	Takes you to the previous chapter index menu in the series

Following these naming rules enables you to do some pretty fantastic stuff in very short order. For example, try out this quick test. Start a new project and import the *alphabet.m2v* video asset from the *projects/ch10* folder. Create a new menu using the Filmstrip Cover template. Now drag the *alphabet.m2v* asset over the new menu. You should see a pop-up similar to the one in Figure 10-31.

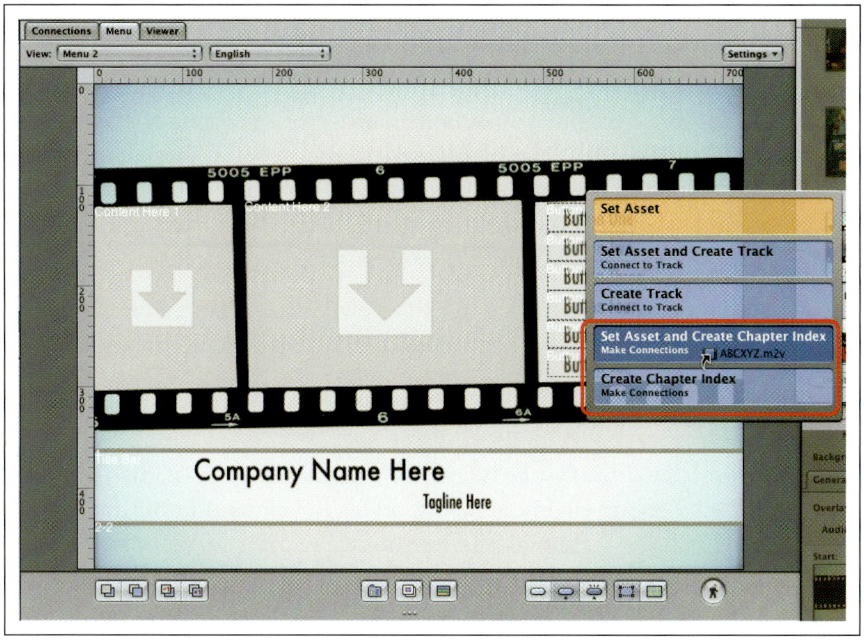

Figure 10-31. Automatically creating chapter index menus with templates and styles

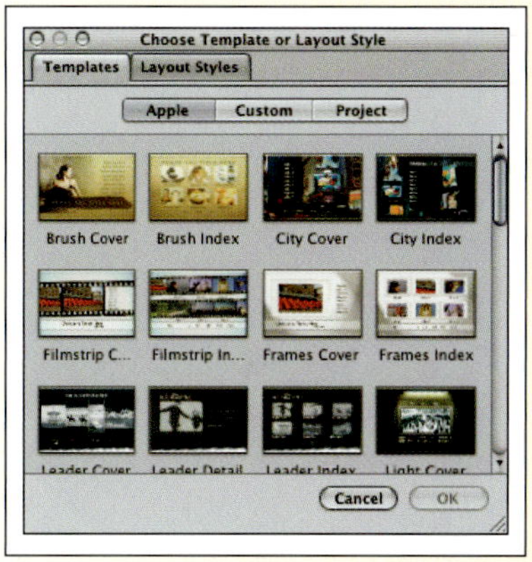

Figure 10-32. The chapter index choice prompt

Choose Set Asset and Create Chapter Index, or choose Create Chapter Index. You'll be prompted to pick a template or layout style, as shown in Figure 10-32. Select the Filmstrip Index template. Now stand back and watch the magic.

You should have a new track with the *alphabet.m2v* file loaded. You also should have a new button on your menu that leads to the chapter indices. And—best of all—you should have five new chapter index menus with their content filled and their targets set. Try simulating the project. You should be able to see the new menus, navigate through them, and select a chapter. Figure 10-33 shows one such index menu in action. Talk about quick work! If you named your chapter markers appropriately, your whole project is done. That's power.

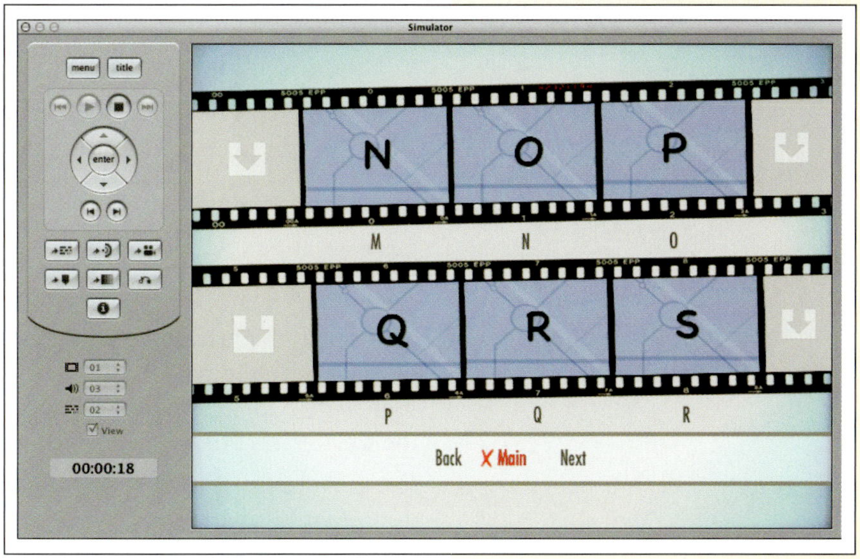

Figure 10-33. An autogenerated chapter index menu

Updating styles

Unfortunately, updating an existing style is not easy. However, it is easy enough to create a new button or layout using an existing style, to modify it, and to save the modified version. That's pretty much your best bet at the moment.

Be careful about deleting "old" styles, though! If you update a custom style and delete the old version, you might confuse old projects that used the style. It's a good idea to "move" old styles into specific projects that use them. Just create a new object with your style (Buttons, Drop Zone, Text, or Layout) and save it again, but check that Project checkbox in the creation dialog. Now you can delete your old custom style without fear.

Of course, you can just keep saving the new versions with new names or maybe version numbers. No harm there, but it does take up disk space.

A style project

Let's start on that library of styles. We'll create a new "cameo" button style. We'll also create two new layout styles for managing buttons: 3-Up and 6-Up. Figure 10-34 shows the layouts we're after (each button in the layouts uses our new button style).

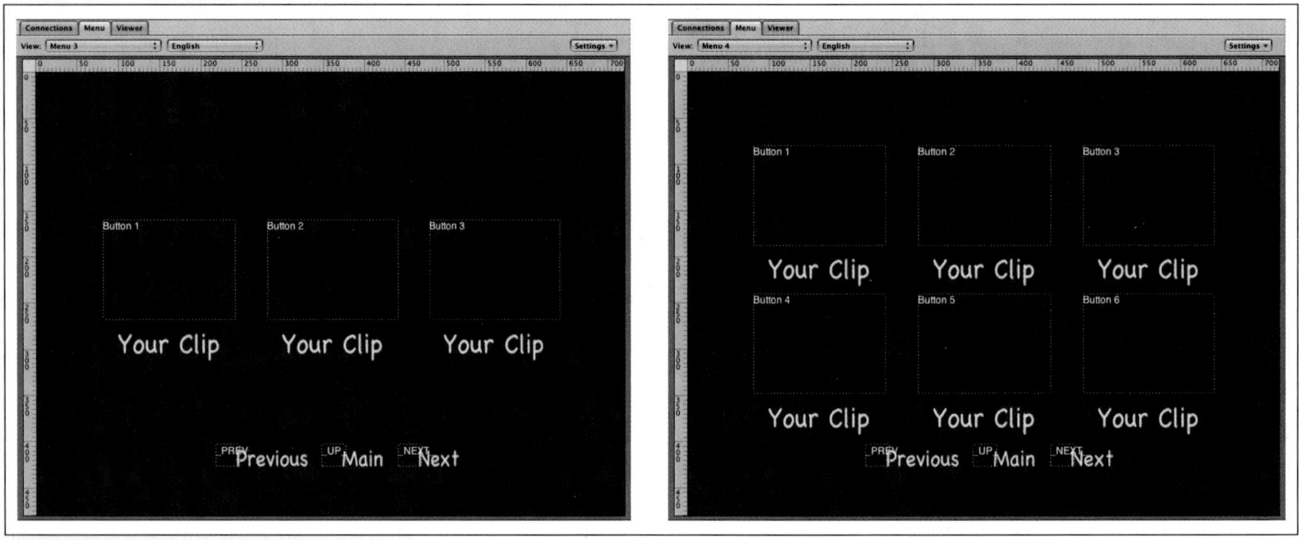

Figure 10-34. The 3-Up and 6-Up styles we'll be creating

A button style

We'll build a button style based on our generic oval shape. (If you didn't go through the "Shapes" section earlier, you quickly can import that shape now. Click the Import button in the Shapes tab of the Palette and load the *oval_with_outline.psd* file from the *projects/ch10* folder. It should appear in the Custom section of the Shapes tab now.)

Now, follow these steps:

1. **Create a Button with the Shape.** Drag the *oval_with_outline.psd* shape from the Palette onto a new menu. Select the Create Button option, as shown in Figure 10-35.

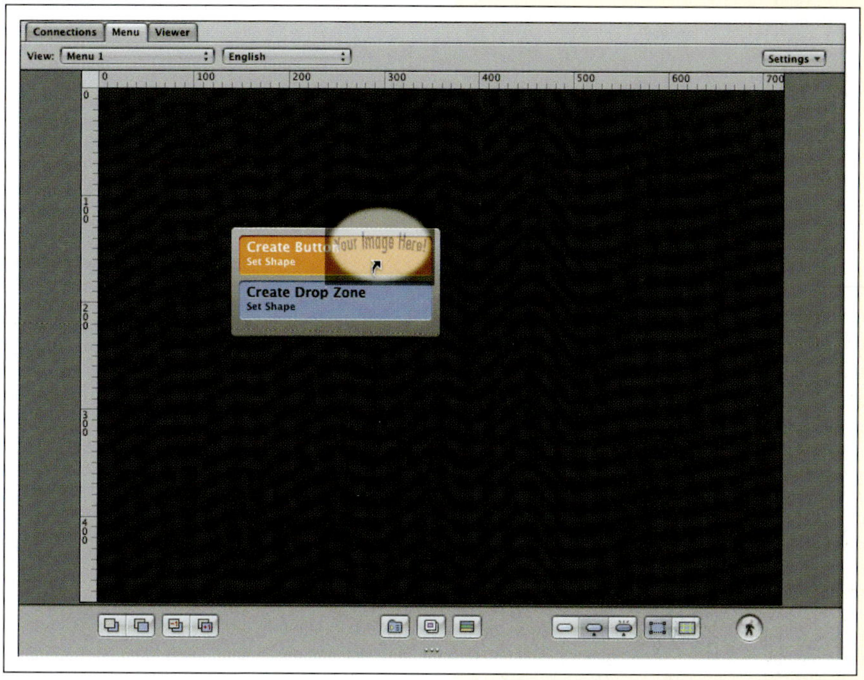

Figure 10-35. Creating a button with the oval shape

2. **Enter the Button Text and Position It.** Type "Your Clip" as the text for the button and choose a font for it. We went with Chalkboard Regular at 30 points. We also placed the text at the bottom of the button. Go ahead and save the button style as we discussed back in Figure 10-23. Figure 10-36 shows the options we chose for this style. As you can see, we named our style Soft Oval, made it available to all projects, and made it self-contained.

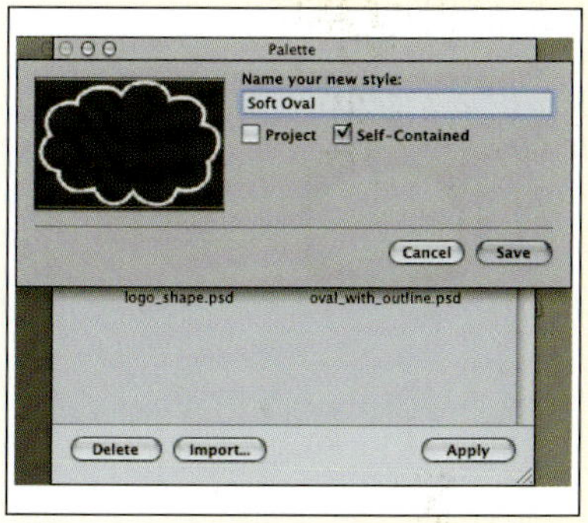

Figure 10-36. The settings we chose for the Soft Oval button style

Move the button to the center-left position, as shown in Figure 10-37.

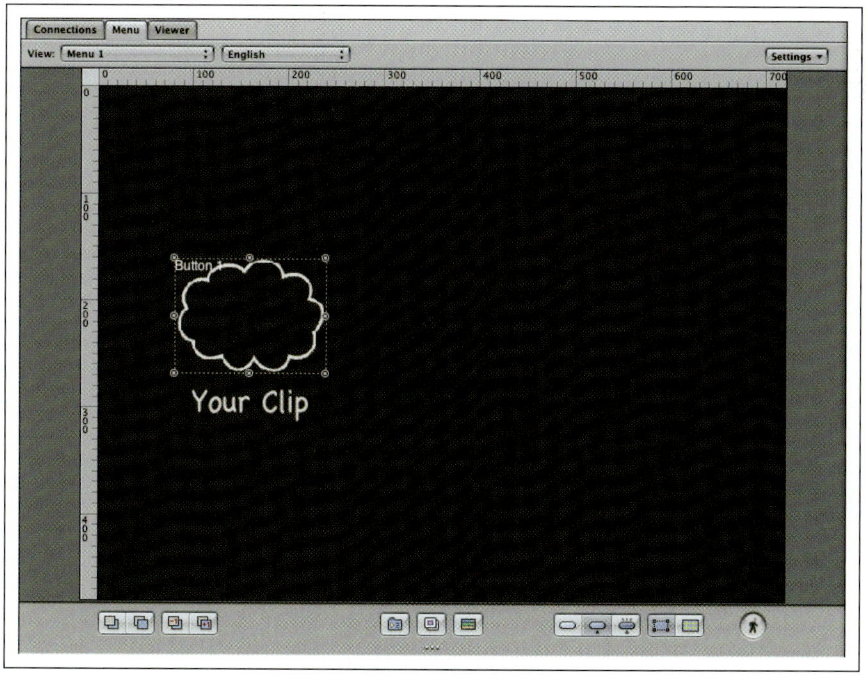

Figure 10-37. Building a layout style with a styled button

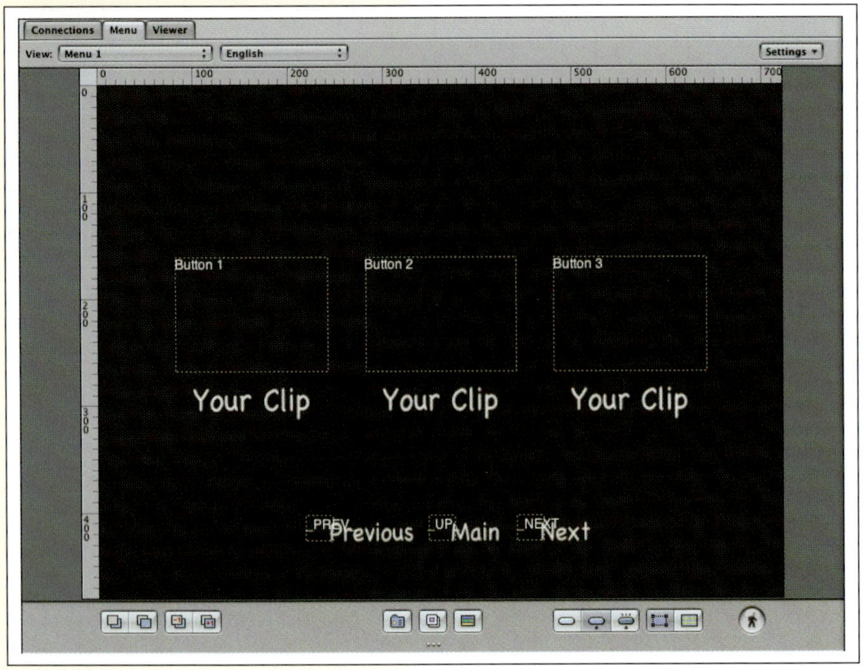

Figure 10-38. Our completed 3-Up layout

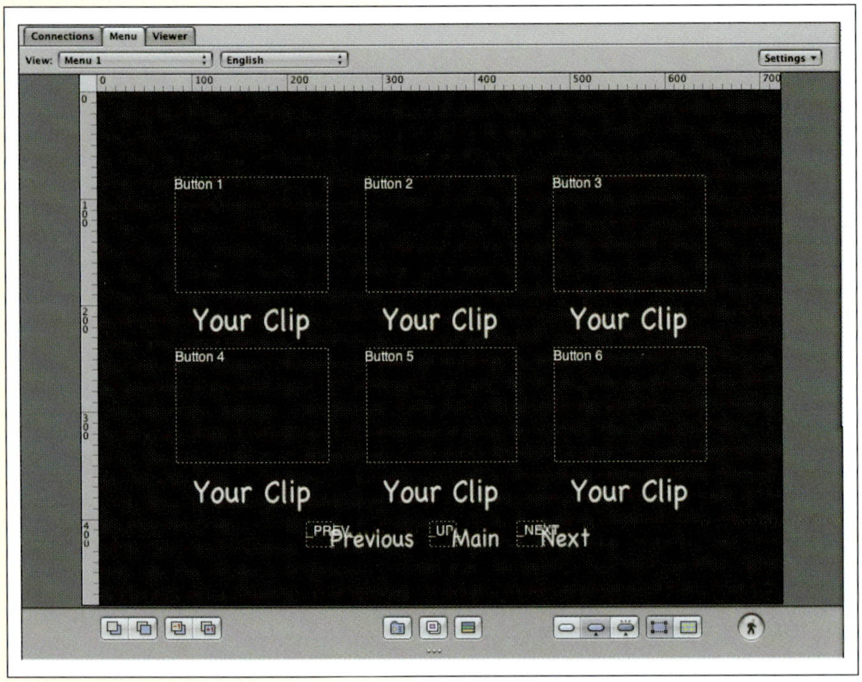

Figure 10-39. The 6-Up layout style

3. **Create Two More Buttons.** Create two more buttons to fill the center and right positions. We also want to add the navigation buttons in case our layout gets used for a chapter index page. (It's best to add these now. Deleting unwanted buttons later is much easier.) We put three cross-highlight buttons along the bottom with the fairly common Previous, Main, and Next labels. Remember that to work nicely within DVDSP, you should name these buttons _PREV_, _UP_, and _NEXT_, respectively. You should end up with a screen similar to Figure 10-38.

   That will round out our 3-Up style, so go ahead and create the style. We named ours 3-Up, but you can choose anything that suits you.

4. **Create the Next Style.** While we have the row of three buttons handy, let's make our 6-Up style. Select all three buttons and press ⌘D. That should give you three more buttons slightly offset from the first three. Go ahead and reposition the buttons so that they appear in two rows, as shown in Figure 10-39. Save that layout style.

   That should leave you with several new layout styles available for use later. Figure 10-40 shows our Custom Layouts tab in the Palette.

Templates

Templates bring all your style pieces together and fill in a few more of the initial details for you. Chief among those missing details are the assets for a menu (its background and audio, for example). Templates give you one more leg up on the get-started-quickly ladder and also make switching among radically different designs almost painless!

Template elements

Templates can have all kinds of elements in them, including assets, buttons, drop zones, pieces of text, and even layout styles. In fact, if you already have a nice library of styles, creating a new template based on a combination of those styles will be straightforward.

Creating a template

Creating a new template is (mostly) a simple task. The upshot is you create a menu exactly as you want to see it, and then you save it. The *mostly* qualifier refers to the creative work involved in designing the menu. The logistics are simple, but that's only a tiny part of the whole process. Figure 10-41 shows the first logistical step (choosing an item from a pop up) for saving a completed menu as a template.

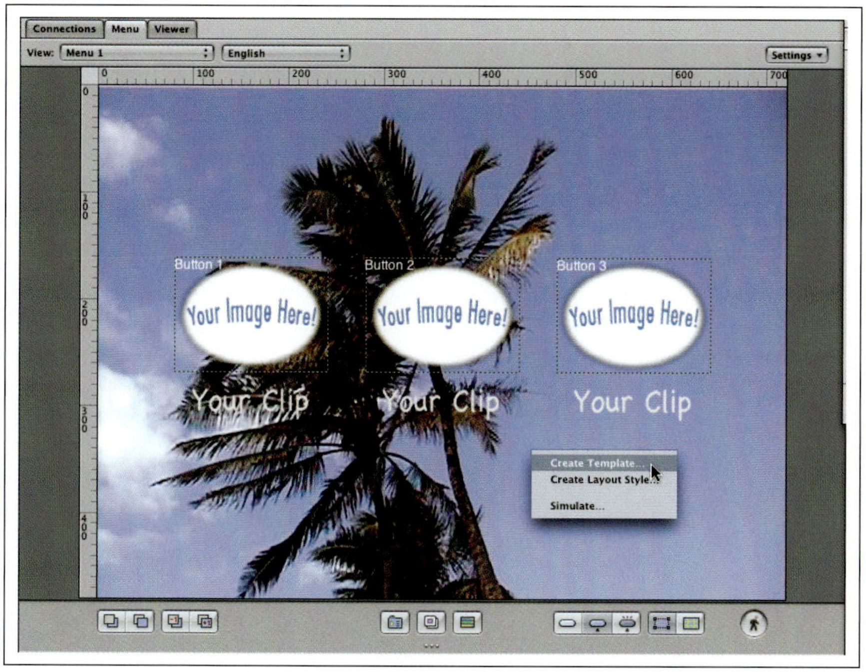

Figure 10-41. Saving a completed menu as a template

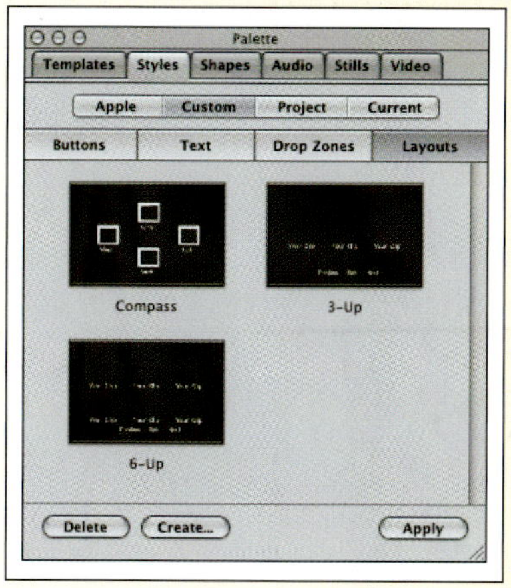

Figure 10-40. The new 3-Up and 6-Up layouts in the Palette

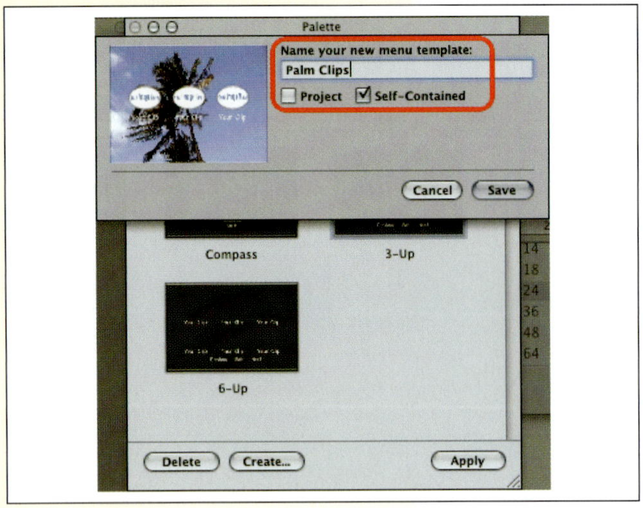

Figure 10-42. The new template dialog

You immediately should see the dialog shown in Figure 10-42. You have to fill out this dialog to save the template.

It's quite similar to the other dialogs we saw before when saving styles. First you supply the name for the template. You can opt to place it in this project only, or make it available for all projects. You also can make the template self-contained.

That last option is a bit more interesting with templates. With a big menu comprising a motion background or an audio file, a self-contained template can start to take up real disk space.

Of course, "dependent" templates have consequences, too. If a template is not self-contained, all its resources must be available when you use the template (or when you open a previous project that uses the template). This can be a big deal if you use network storage or removable media—including external hard drives.

As always, it's a trade-off. If you have the space, go the self-contained route. It usually means fewer headaches later on.

Using templates effectively

Once you've created your template, you'll want to use it in your projects. If you create a library of templates, be sure to use the same names where possible for your various buttons and drop zones. Keeping names the same enables you to switch templates on the fly without losing any of the nontemplate content (button targets, or a video asset for a drop zone, for example) that you already supplied.

Something else to keep in mind is that the Project category can be really useful. It's a handy spot for leaving elements you might reuse. The need to make project templates self-contained is ameliorated by the fact that the assets probably are part of your project and you know they'll be available. That means you pay almost no penalty for creating button and text styles and menu templates. It can speed up those last-minute changes that inevitably crop up.

A standalone template project

So, let's build a template of our own. We'll start off with a single standalone template just to get the taste of it. (We'll also build a series of templates for doing things such as the chapter index trick from the previous section.)

Let's make a Vacation Clips menu. We'll add three simple buttons and a motion background and save it as a custom template. To get started, create a new project and import the *palm.m2v* file from the *projects/ch10* folder.

Go ahead and drop the video clip onto a new menu and set it as the background. Use the cameo buttons you built in the Style project for your new template. (If you didn't complete the Style project, you can use a regular button or one of the existing button styles from Apple.) Drag the button style onto your menu to create a new button. You can do that three times or just duplicate the first button. Place the three buttons into positions similar to those shown in Figure 10-43, or use the 3-Up layout style from the previous section.

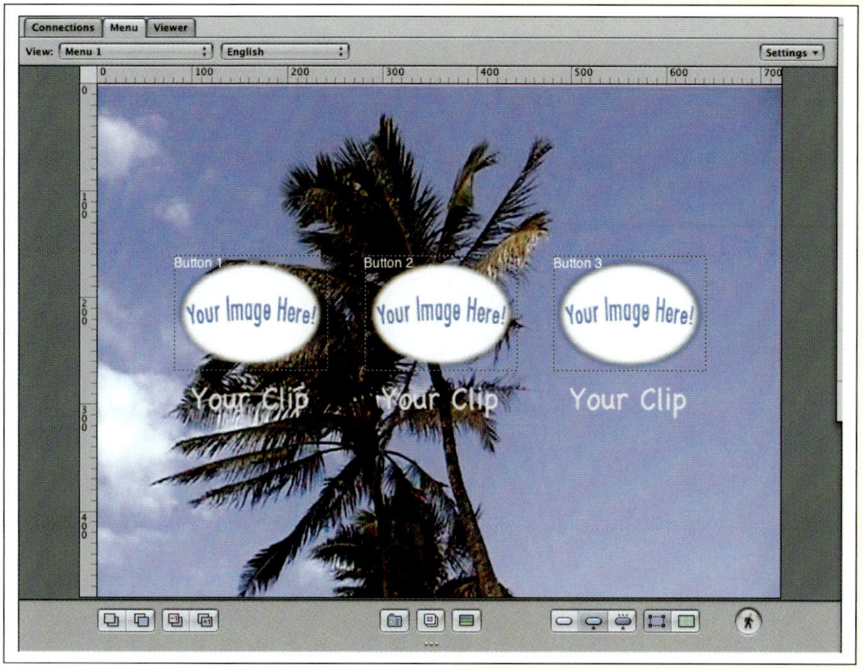

Figure 10-43. The first stages of creating a DVDSP template

Great! You're almost done. You just want to add a text title for your menu. Double-click the menu in the upper-left corner and start typing. We chose the oh-so-creative title "Vacation Clips"—but feel free to add your own flair here. You should end up with a menu similar to Figure 10-44.

OK, now just Control-click any of the unoccupied areas of the menu and select the Create Template option. Fill out the dialog box and *voilà*! The new template is done! Good job! If you're up for more, try the next project to create a themed set of menus similar to the themes supplied with DVDSP.

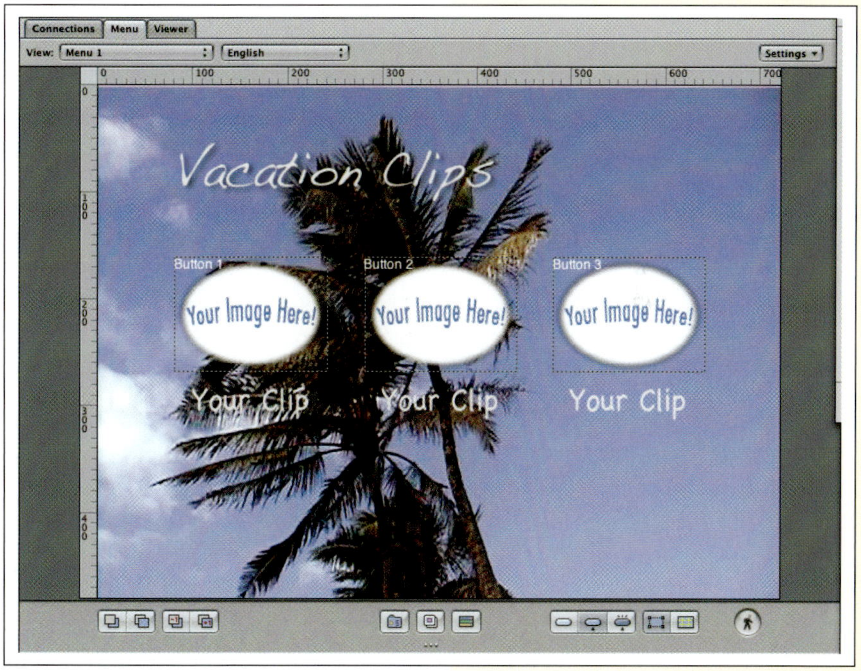

Figure 10-44. The finished standalone template

A template series project

This project is quite similar to the previous project. We'll be making three menus in a style comparable to the templates Apple supplies. The only really key thing to note as we go through this project is the names of the objects. We want to use the same names Apple did so that our templates are interchangeable.

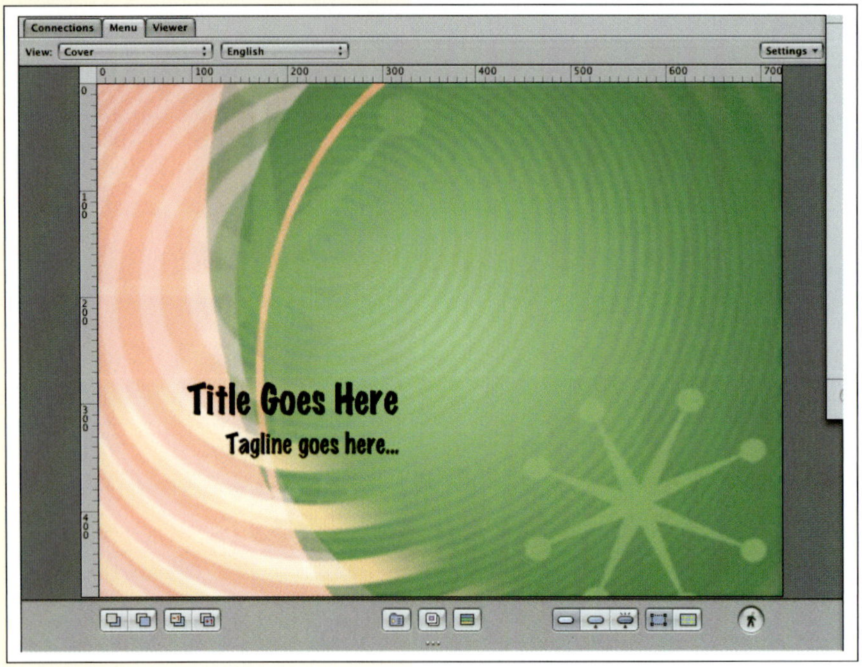

Figure 10-45. The first step in our Wobble Cover template

Grab the following still images for your menus (you'll add a few shapes and do everything else with styles for this project):

- *wobble_cover.psd*
- *wobble_detail.psd*
- *wobble_index.psd*
- *add_image.psd*

For the shapes, import the following Photoshop documents into the Shapes tab:

- *wobble_big.psd*
- *wobble_tiny.psd*

You can use either the Custom area or the Project area; just remember which you chose. See Figure 10-19 for a refresher on how to import shapes.

Now follow these steps:

1. **Create a Cover Template.** Let's start with the Cover template. You have one drop zone and two text objects to deal with. You'll also create four buttons that you can use however you see fit. For this template, grab the *wobble_cover.psd* graphic for the background. You want title and tagline text objects as shown in Figure 10-45.

2. **Add the Drop Zone and the Four Buttons.** Now throw in the drop zone and the four buttons (see Figure 10-46). Drag the *wobble_big.psd* shape over the menu and choose the Create Drop Zone option. Set an asset to the *add_image.psd* file so that users know to drop in their own content.

   The four general buttons use the *wobble_tiny.psd* shape and have text on the right. Create the first button and duplicate it to make the others. That should leave you with a menu similar to Figure 10-46.

   Select all four buttons and set their text style using the Font window. (It's easiest to grab that window by pressing ⌘T.) We chose Marker Felt for the font, as you can see in Figure 10-46. We also used the Color Window (Shift-⌘C) to make the text black.

   Go ahead and save that menu as a template. (Control-click the menu background and select Create Template.) We named ours Wobble Cover and made it self-contained.

3. **Add an Index Template.** For the index page, we'll have six buttons big enough for still images or video clips, and the three navigation buttons. Start off with the *wobble_index.psd* image for the background.

   Use your 6-Up style to place the six main buttons and the three navigation buttons. Reset the shape of the first button to use the recently imported *wobble_big.psd* shape. You'll probably need to resize it to something more reasonable. You'll also want to set its asset to *add_image.psd* so that it has some default content, and you'll want to set the font to match what you used on the Cover template. You should have a menu similar to Figure 10-47.

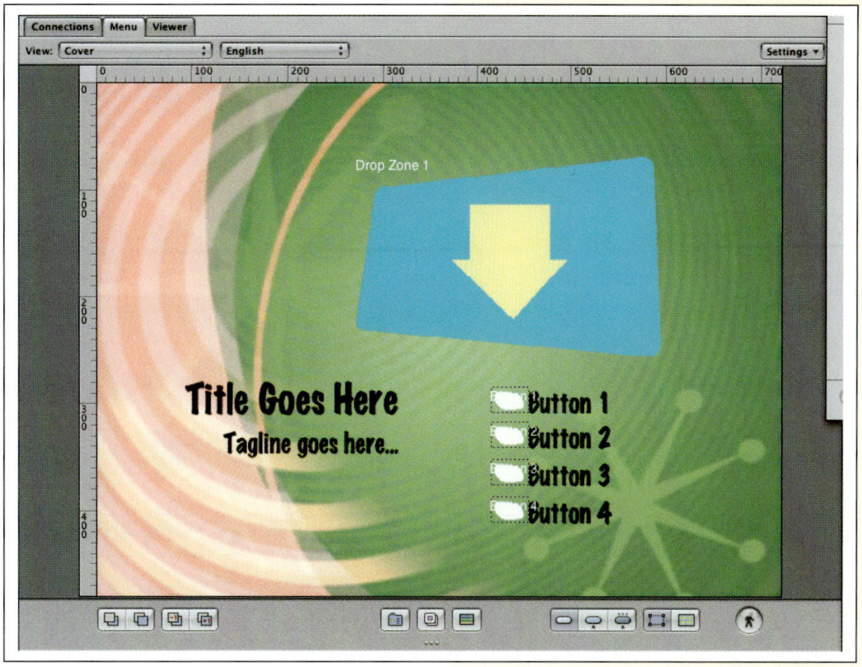

Figure 10-46. The Wobble Cover template with its drop zone and buttons added

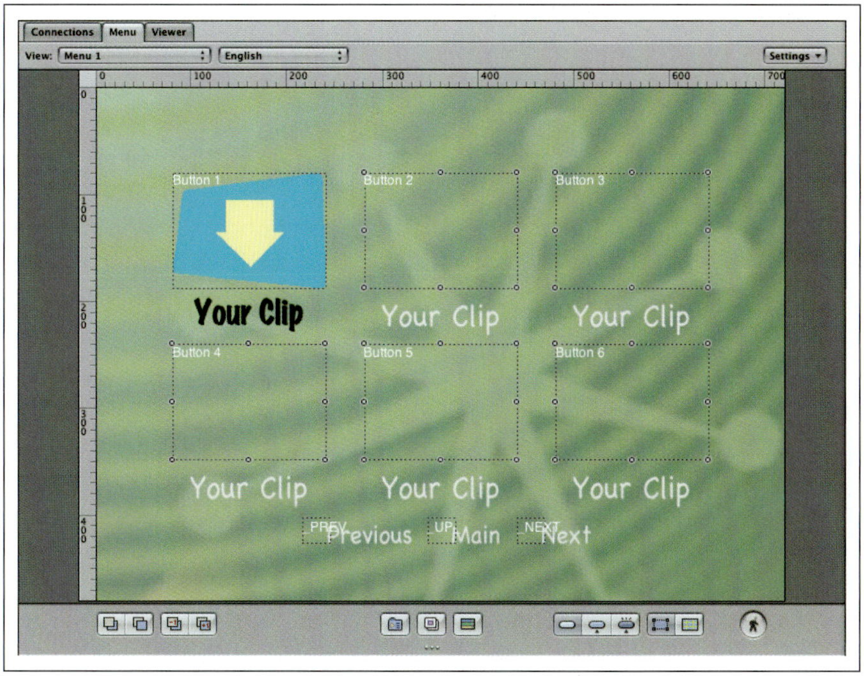

Figure 10-47. Our first steps toward creating an Index template

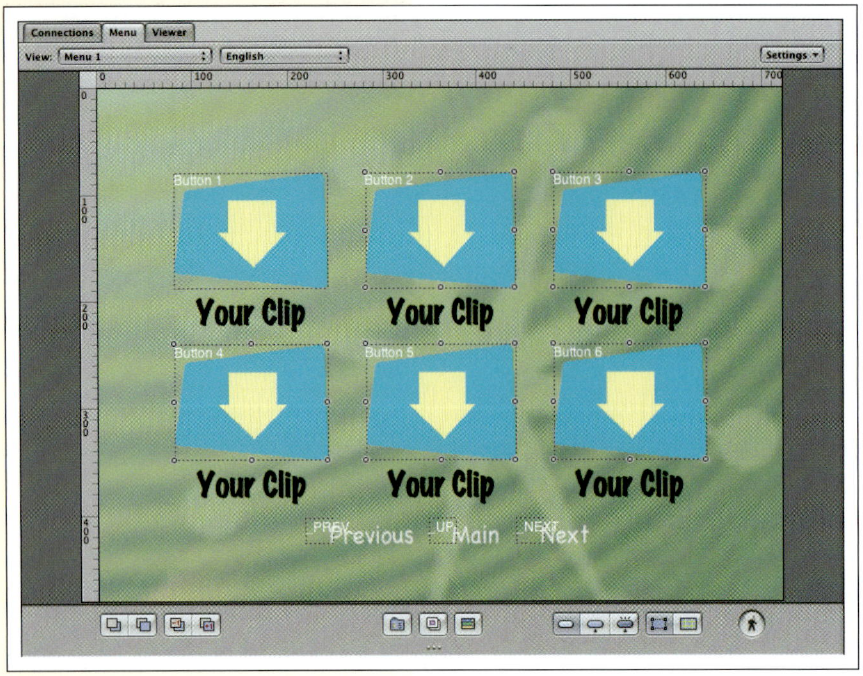

Figure 10-48. The main buttons for the Index template

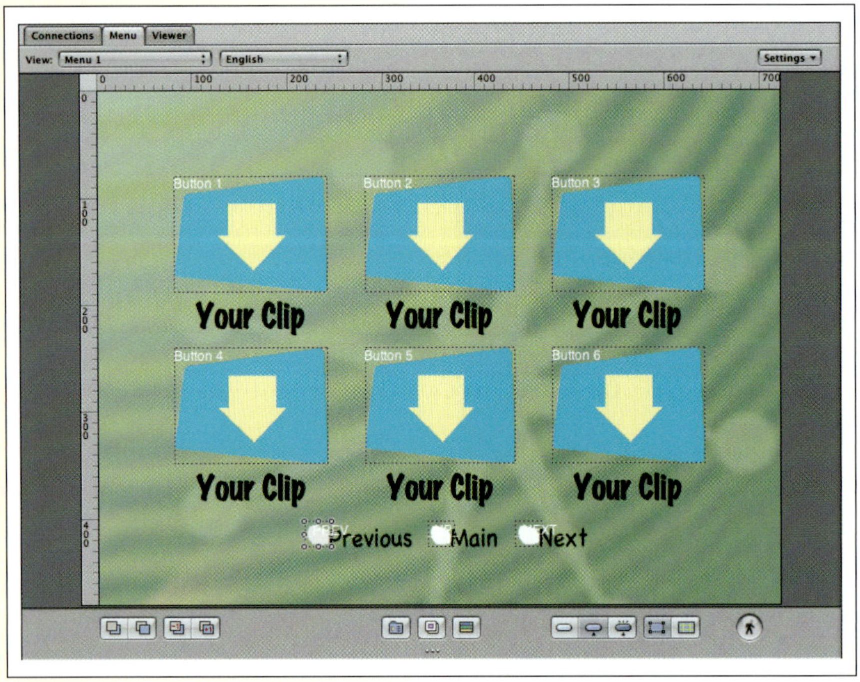

Figure 10-49. The completed Wobble Index template

Now save that first button as a button style for this project. Next, select the five remaining main buttons and apply that new button style. All of them should change to look like the first button, as shown in Figure 10-48.

At this point, you need to style the navigation buttons. Select all three buttons and apply the font and color of your choice (although you probably should match the Cover template). Change the shape of each button to use the *wobble_tiny.psd* custom shape. You can change the shape one button at a time, or make another project button style if you want.

Your final index menu should look similar to Figure 10-49. Go ahead and save this menu as a template. We named ours Wobble Index.

4. **Create a Details Menu.** Last but not least, you want a Details page. This page will have one big drop zone, one piece of text (a description area), a Play button, and a Back button.

   Use *wobble_details.psd* for the background. You can lay out the text object and the drop zone as shown in Figure 10-50, or you can improvise. We turned on the title-safe borders (Shift-⌘E) to make sure the text was within the proper limits.

You'll place the buttons next. We used the same style we used for the buttons on the Cover menu and placed them below the drop zone. Figure 10-51 shows the results (with the title-safe borders turned off).

Save this template as Wobble Details.

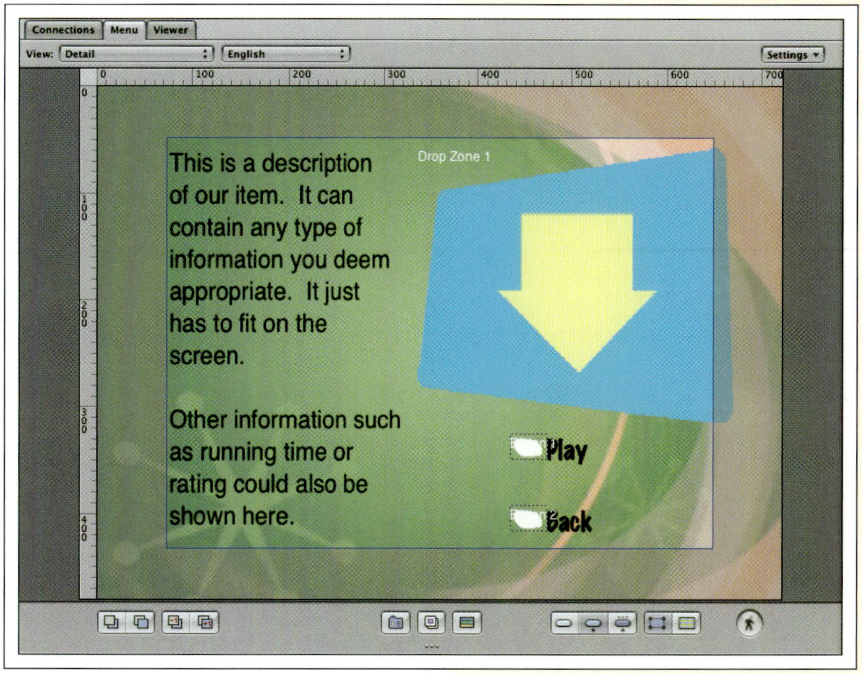

Figure 10-50. The first pieces of our Details template

Testing the templates

Congratulations! You've completed all the big projects for this chapter. But just to be safe, you probably should test your templates. Try the same test we discussed back in Figure 10-31 with the alphabet video sequence. This time, however, use your own Cover and Index templates.

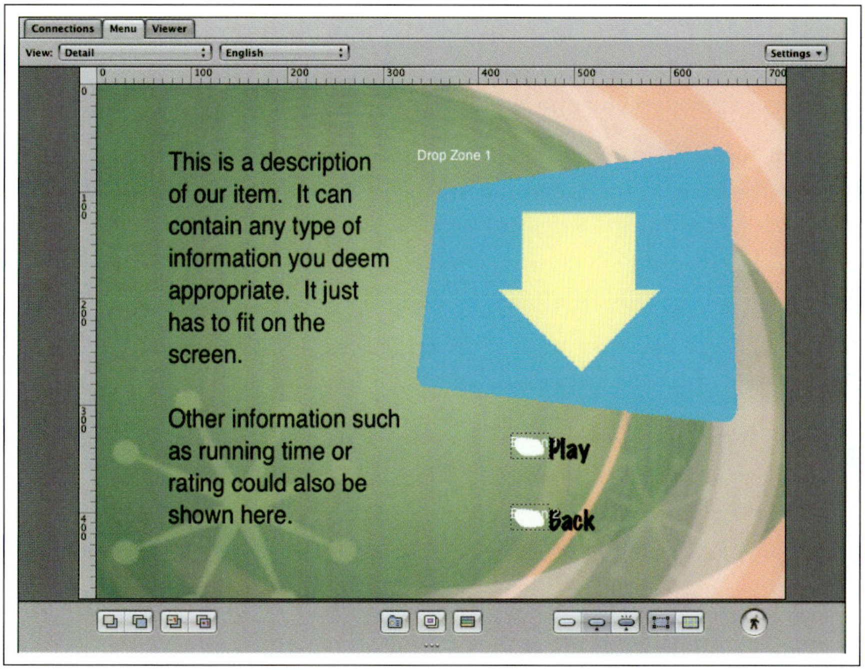

Figure 10-51. The completed Details template

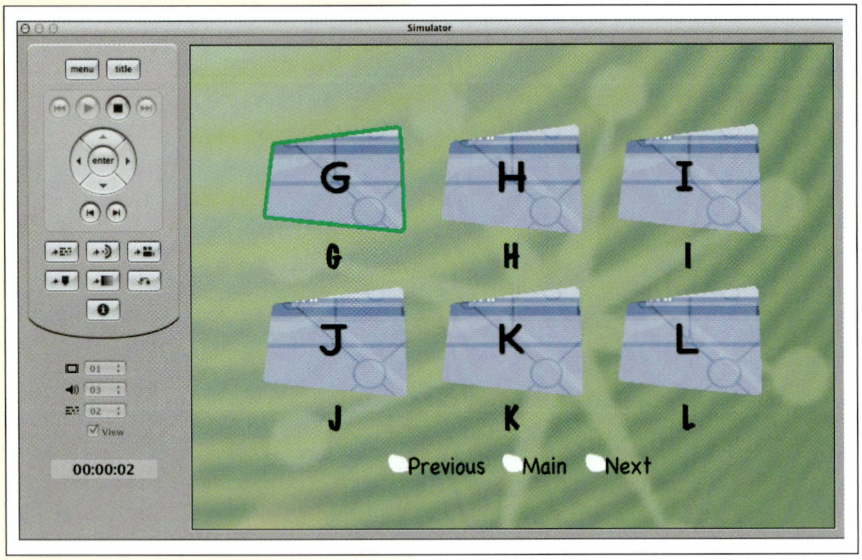

Figure 10-52. The Wobble Index template in action

Figure 10-52 shows our test up and running in the Simulator. We navigated to one of the middle index pages to make sure all the navigation buttons were connected properly.

Distributing Styles and Templates

Whew! Almost done. The last thing we want to cover before releasing you into the wild on your own projects is how to distribute templates. If you've got colleagues working on similar projects, this is a great way to share these time-savers. If you're just a hobbyist, this is a great way to share your templates with other hobbyists. If you're a hobbyist with an entrepreneurial bent, this is a great way to make money off your friends and colleagues.

In short, you want to be able to get these into someone else's hands.

Custom elements

If you develop a set of styles that you want to give to friends or colleagues (or sell on eBay), you just need to know where to look.

You can find styles, templates, or shapes that are not part of a particular project (the ones in the Custom category in the Styles tab on the Palette) in the */Library/Application Support/DVD Studio Pro* folder. Or rather, you'll find three fairly obvious folders there: *Shapes*, *Styles*, and *Templates*. Figure 10-53 shows the Finder window on one of our development systems.

Figure 10-53. The folders containing custom elements

These elements are easy to distribute if you made them "self-contained." (See Figure 10-23 for a reminder of where to do that.) Just grab a copy of

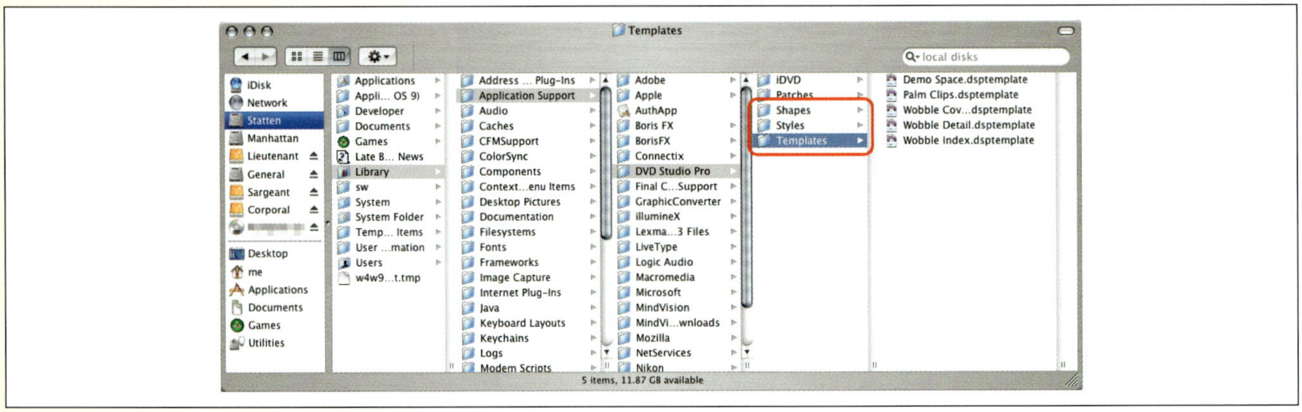

your style or template entry (it's actually a folder similar to the way Apple bundles its applications) and put it somewhere others can get to it. That might mean on a network storage server or on a CD or DVD. If you have several styles and templates, put them all on the same DVD for a collection. You even can build a video tour of the templates that plays in consumer players and includes the actual templates in the DVD-ROM portion. (Remember Chapter 9?)

If you're in a group that makes new templates fairly regularly, your best bet is to save a copy of each template in a common folder. Your teammates can grab a copy from the common folder and drop it into their DVDSP folder for later use.

That's really all there is to it. Apple tried to make this distribution process accessible, if not completely easy.

Project elements

If you designated your styles and templates as part of a particular project, you still can distribute them; you just have to dig a bit deeper. They are located in similar *Shapes*, *Styles*, and *Templates* folders, but those folders are part of the DVDSP project file.

That project file is really another of those Apple packages. You can open the package in the Finder by Control-clicking your project file and selecting the Show Package Contents option.

If you did not make the style or template self-contained, be sure to include any graphics files necessary to complete the element.

One other (very mild) word of warning: DVDSP does not differentiate between the various style types. Make sure you name your styles so that you know what you're getting when you grab them for distribution. The User's Manual even recommends putting the type of style in the name (e.g., Soft Oval Button).

More Customization to Come!

After all this custom template work, you might be wondering where the custom transitions are. Not to fear. This chapter was long enough without them, so we promoted the transitions to their own chapter. Chapter 11 has a starter project for those of you still in DVDSP 2 to make sure you have access to video transitions between your menus. If you build a few custom templates, you easily can augment them with such video transitions.

The rest of Chapter 11 will tackle building custom transitions for use specifically with DVDSP 3. Although this definitely is a topic for the advanced user, if you've got a spare afternoon and some bottled-up creativity just dying to get out, forge ahead!

11 Transitions

In this chapter

Smooth transitions from one menu to another (or from a menu to a video clip) are common in DVDs. With DVDSP 3, they're also easy to add. But DVDSP 2 didn't include transitions. And sometimes, even in DVDSP 3, you want more than the basics. This chapter includes two projects: one to create a transition completely from scratch that works with any version of DVDSP, and one to create a custom transition that works with the DVDSP 3 transition framework.

Custom Video Transitions

If you're still using DVDSP 2, you don't have access to the spiffy transitions we looked at back in Chapter 4. But fear not! We're going to look at using the End Jump property of video tracks a bit creatively to accomplish the same effect. For those of you using DVDSP 3, if the built-in transitions are enough, stick with those. They're much easier and allow flexibility that building your own transitions just doesn't afford. But if you need something special in the transitions, read on. This technique works just as well in DVDSP 3—all you need is time!

Getting started

The most difficult part about providing video transitions is building the actual transition. DVDSP 2 doesn't provide any animation tricks for you, so you have to create these transitions in a separate application. You could use iMovie and attach transitions, just as we did at the end of Chapter 3. We'll detour into Adobe After Effects for most of this work, but any video editor or animation tool, such as Apple's Motion, will do the trick.

For any particular project, you can create simple, themed transitions that you apply to all menus. Smaller projects often can use customized transitions because the user has only one real navigation path.

We'll start by re-creating the transitions we used in the expanded Castle Tours project from Chapter 4. For those of you with DVDSP 3, the online project includes a few extra tricks that you can't perform in DVDSP directly, just so you have something interesting to play with as well!

With three tours, that means we need six transitions: from the main menu to each historical menu (three), and back to the main menu again when the historical menu is completed (three more). We'll use a quick but elegant animation transition for each transition. As a little detour to show you some of the other tools that are available, we'll build these transitions in Adobe After Effects.

And if you don't have After Effects, don't worry. Macromedia Flash or any other animation tool that can output using QuickTime will work. Plus, we put the finished transitions in MPEG-2 form in the projects folder for this chapter, so you still can play with the DVDSP aspects.

Start up After Effects and load the stills you'll use to build the transitions. You can use the File → Import → Multiple Files... menu option to grab the following files:

- *main_menu.pct*
- *cawdor_menu.pct*
- *holyrood_menu.pct*
- *urquhart_menu.pct*

Go ahead and save your project. We called ours *transitions.aep*. Your Project window should look something like Figure 11-1.

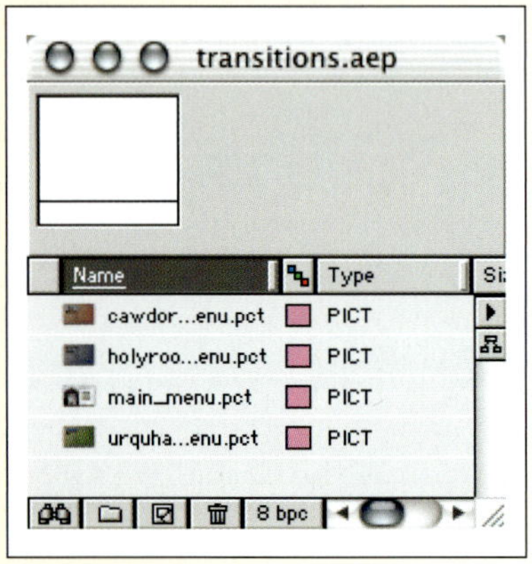

Figure 11-1. The files necessary to build the video transitions

Now you just have to settle in and build the transitions. One quick thing to remember is that video transitions should be fairly quick. Transitions that take a full minute might look really cool the first few times the user sees them, but before long they will become slow and very annoying. If you're not going to shut them off after the first time using scripting (refer back to Chapter 7), keep 'em short!

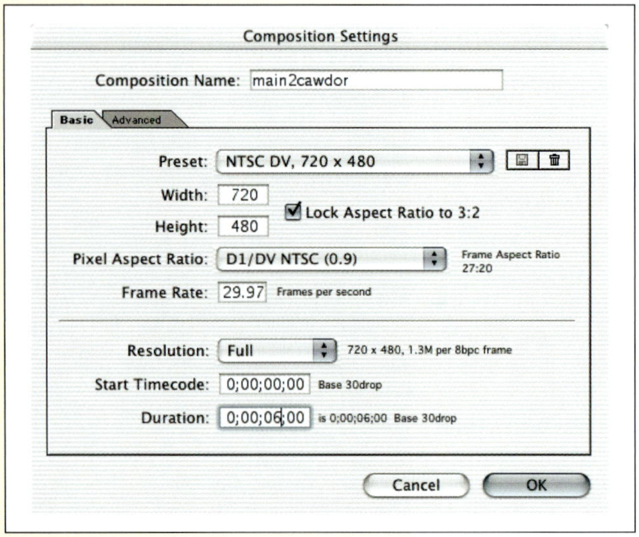

Figure 11-2. Setting up the basic properties for the After Effects project

1. **Create a New Composition.** Let's start with the transition from the main menu to the Cawdor menu. Create a new composition (select Composition → New Composition, or simply press ⌘N). Most of the defaults should be fine, but you want to make sure you're building a project with 720x480 as its dimension and 29.97 frames per second. We've chosen to go for six seconds as the transition duration, as shown in Figure 11-2.

   Drag the *main_menu.pct* and *cawdor_menu.pct* resources from your Project window down to the Timeline window. You can use the little triangle to the left of the resource name in the timeline to expose that resource's properties. We're particularly interested in the images' position, which is grouped under the Transform heading (twirl that triangle down as well). Set the position of the *cawdor_menu.pct* resource to (1080, 240), as you see in Figure 11-3. That should leave the Cawdor menu off to the right of the screen. Our transition will slide that menu in as it slides the main menu off to the left.

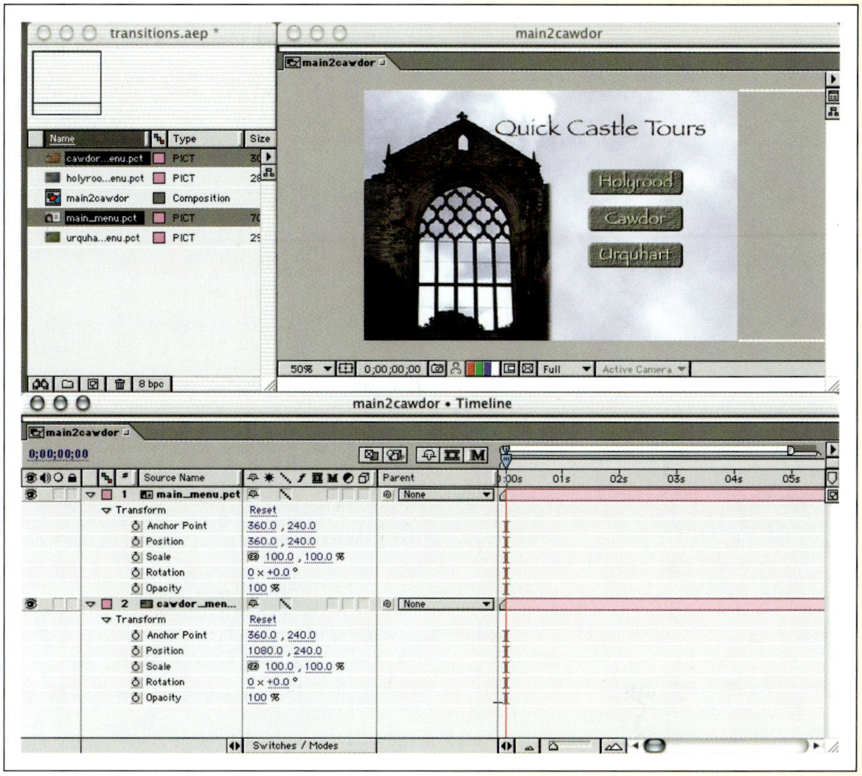

Figure 11-3. The initial settings for the first transition

Go ahead and close the triangles, as you're done looking at the details of those resources for the moment.

2. **Create an Adjustment Layer.** After Effects provides a nifty feature known as an Adjustment Layer. You can use this layer to uniformly "adjust" many other things. Let's create a new adjustment layer in our project by Control-clicking the timeline and selecting New → Adjustment Layer from the menu that pops up (see Figure 11-4). You can name it anything you want; we just left the name alone.

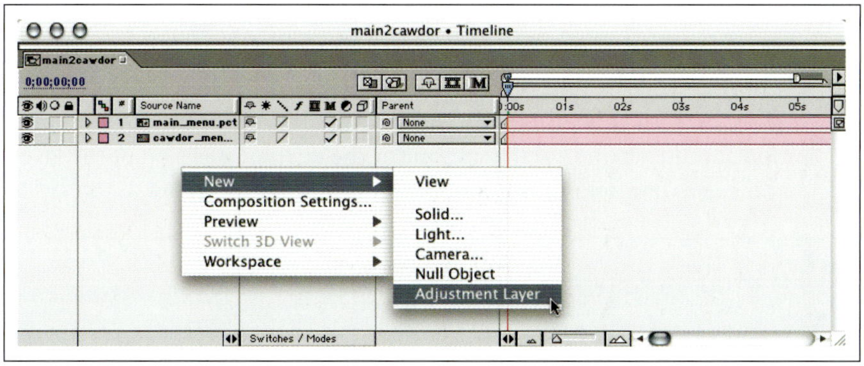

Figure 11-4. Creating an adjustment layer in After Effects

Once the adjustment layer is in your menu, use it as the "parent" for your other two layers. An entire column is devoted to showing a layer's parent. Just use that pop-up button to select your adjustment layer, as in Figure 11-5.

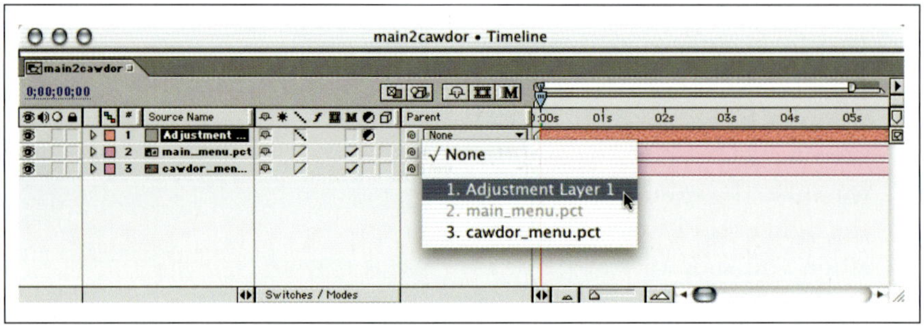

Figure 11-5. Using the adjustment layer as a "parent"

With the parent set, you can change the position of the adjustment layer and everyone else will follow suit. This makes it a lot easier to modify timing and other such things.

3. **Set Marker Keyframes.** After Effects enables you to animate resources using the tried-and-true technique of *keyframing*. Keyframes are single snapshots of a full frame at a specific point in time. If you set two or more keyframes with different property values (e.g., one has a position of [0, 240] and the next has a position of [1080, 240]), the computer can interpolate and create the frames that go in between. This creates the illusion that the object is smoothly shifting its position. This technique also goes by the name *tweening*.

   You need two keyframes for your animation. Open up the details on the Adjustment layer and make sure the timeline marker is at the beginning of the composition. Click the icon that looks like a stopwatch to the left of the Position line, which tells After Effects that you want to use keyframes with this property. Once you click the stopwatch, the application should set an initial keyframe at the current position of the layer. You'll see a diamond in the timeline area similar to Figure 11-6.

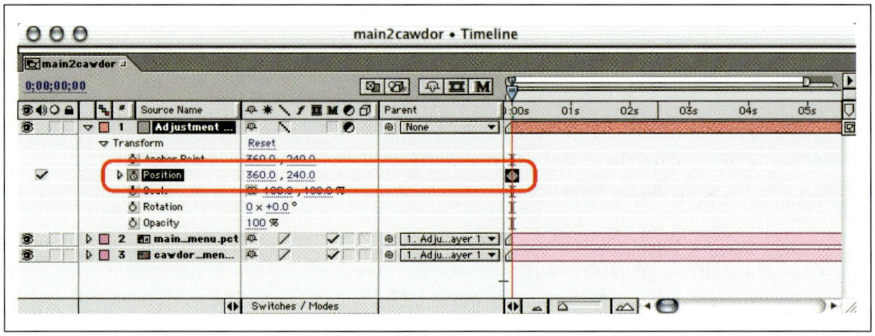

Figure 11-6. Our first keyframe in After Effects

Now comes the hard part. You have to set the second keyframe. First, move the timeline marker to the end of the composition. On the far left side of the Timeline window should be a checkbox. Go ahead and check that checkbox to set the second keyframe. Figure 11-7 shows the results.

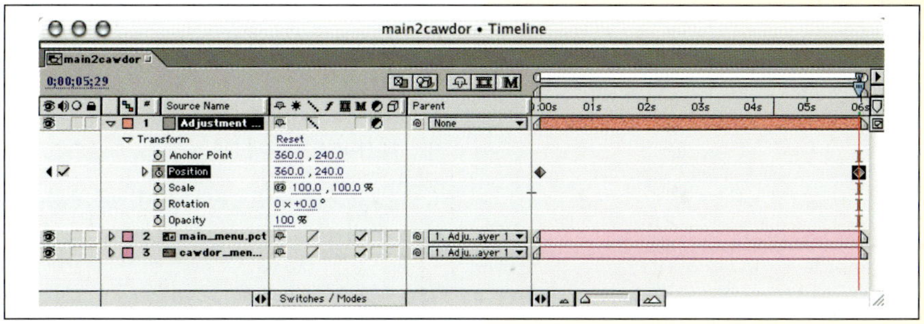

Figure 11-7. Setting the second keyframe for the transition

While the timeline marker is still at the end of the composition, change the position of the adjustment layer to (–360, 240). You should see the Cawdor menu in your Preview window (see Figure 11-8) along with a new line that's tracking the motion you just added.

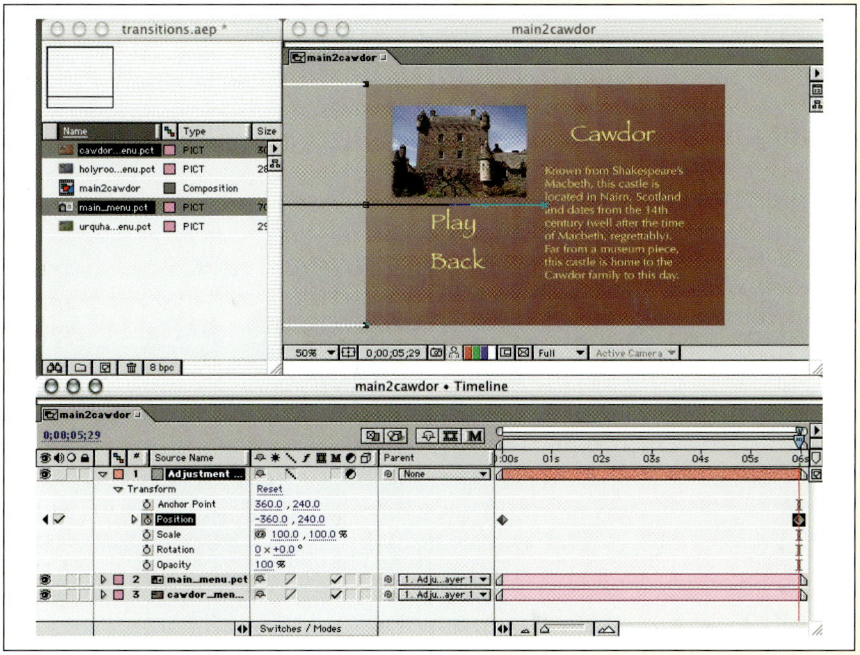

Figure 11-8. The end position of the transition

4. **Smooth the Motion.** Technically, you have enough for your transition. The pictures will slide at an even pace from left to right. To make the motion look a little less clunky and computerish, you can play with the "handles" for the motion by revealing the position details (click that triangle to the left of the Position field). Dragging the start and end handles down to the bottom of the Position row will create a curve similar to Figure 11-9. That curve means the transition will start out slow, speed up, and then slow down toward the end.

   You can verify it looks the way you want by previewing the composition. Choose the Composition → Preview → RAM Preview menu option. It'll take a minute to render all the frames, but then it should provide a close-to-real-time preview.

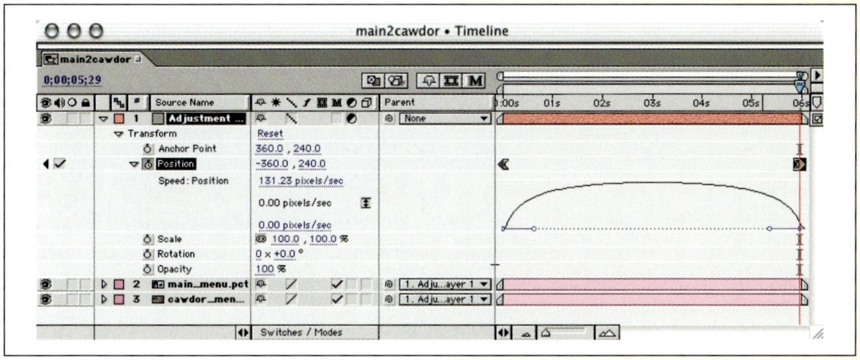

Figure 11-9. Setting up a more natural acceleration and deceleration curve for the transition

5. **Export the After Effects Project into DVDSP.** Now you just have to export the transition to a format DVDSP can use. Unfortunately for us, After Effects 5.5 has no direct MPEG-2 export. But that's not a big deal, just a minor bump in the road. You *can* use After Effects to export to full-quality DV, so that's exactly what you'll do.

   Make sure your timeline is the active window and select File → Export → DV Stream, as shown in Figure 11-10.

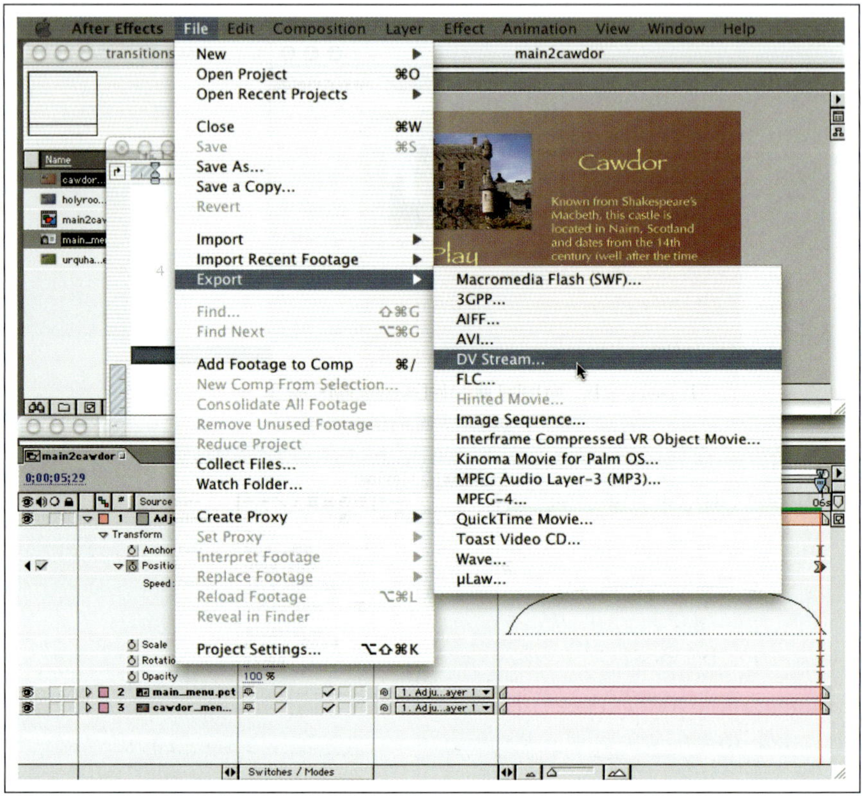

Figure 11-10. Exporting the transition to DV

   You'll be prompted to pick a folder to save the file. Save the file and then use any of the regular tools (QuickTime Pro, Compressor, Final Cut Pro) to convert it to MPEG-2. Or just let DVDSP do it for you in the background. Again, you get more control if you do it yourself, but for such simple things, the DVDSP defaults will be fine. If you try anything fancy, though, you probably should do the conversion in a tool such as Final Cut Pro so that you can run a good NTSC-safe filter on the resulting video as well.

Lather, rinse, and repeat for the other five transitions. Or, if you're in a hurry, just grab the *.m2v* files from the DVD included with this book and dive on into the DVDSP portion of this project.

Preparing the project

To get to the fun part of this project, you'll need to go ahead and set up the same basic Castle Tour menu you had before. In fact, just grab that project as a starting point and save it with a new name. Then import the following new assets:

- *cawdor_menu.pct*
- *holyrood_menu.pct*
- *urquhart_menu.pct*
- *main2cawdor.m2v*
- *main2holyrood.m2v*
- *main2urquhart.m2v*
- *cawdor2main.m2v*
- *holyrood2main.m2v*
- *urquhart2main.m2v*

Now follow these steps:

1. **Set Up the Menus.** Go ahead and set up the three historical menus for the castles. Just use simple shapes, as you did with the quick menus in Chapter 1. You don't need anything fancy here. Hook up the buttons so that you can play your video clips and make sure you can get from the main menu to each historical menu and then back again. Figure 11-11 shows the Graphical view of the main menu. Figure 11-12 shows the connections for the Cawdor menu. (The other historical menus are similar.)

2. **Add Transitions to the Menus.** Now for the fun part. Instead of having the buttons on the main menu go directly to the historical menus, you'll point them at your transition videos. You can

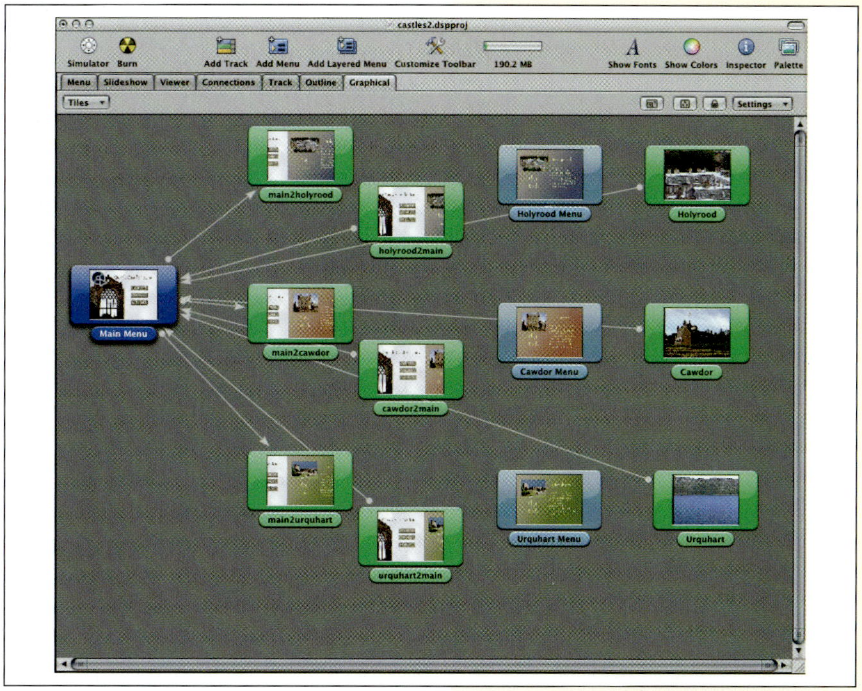

Figure 11-11. The connections set for the main menu

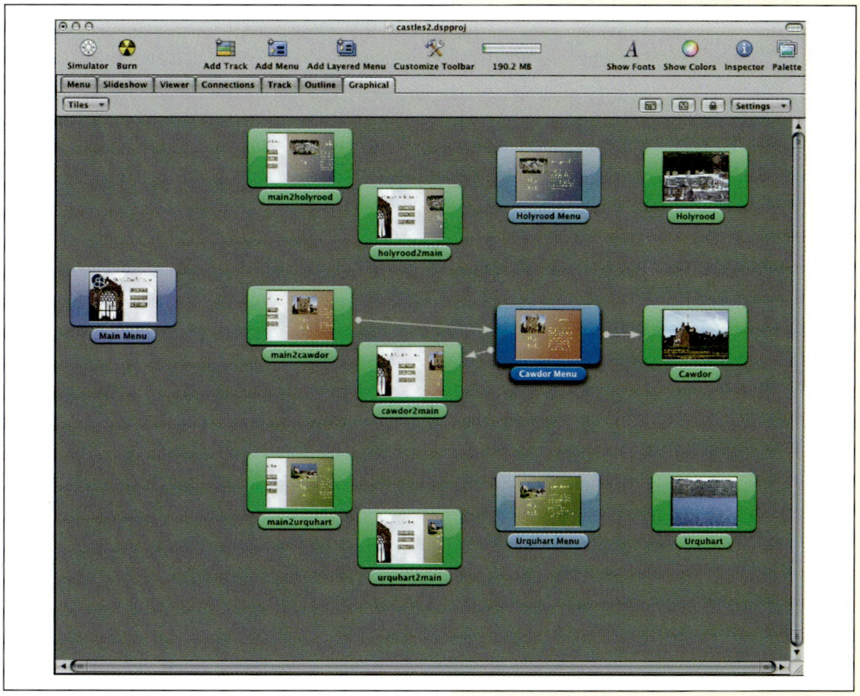

Figure 11-12. The connections set for the Cawdor menu

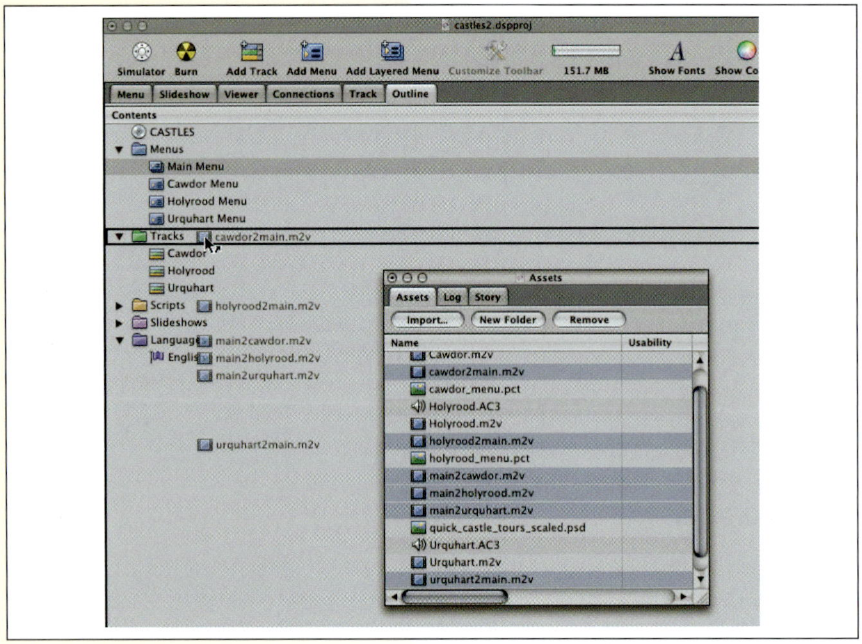

Figure 11-13. Making the transition available as a normal track

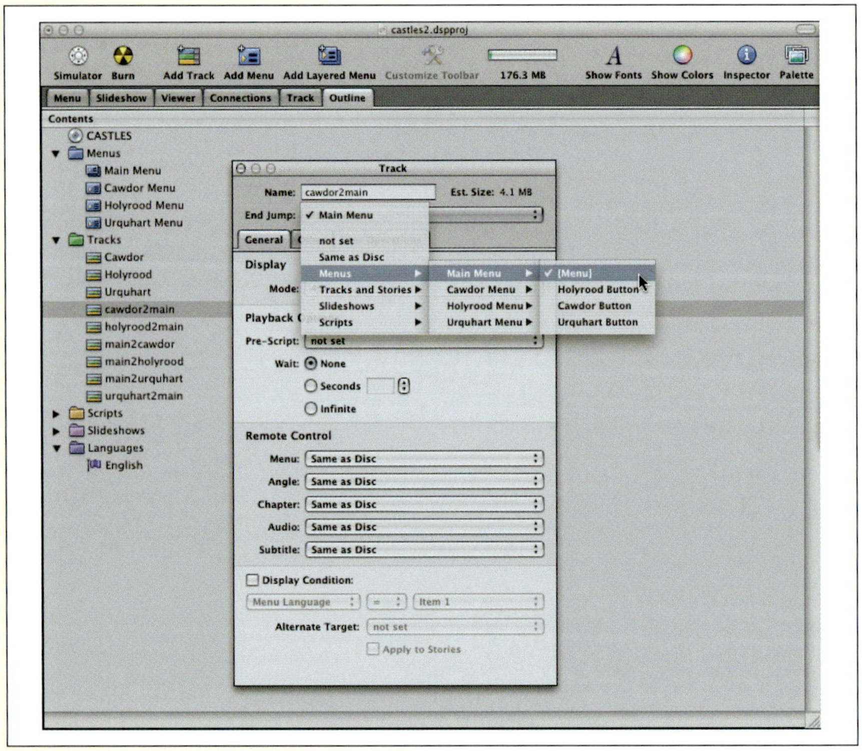

Figure 11-14. Making sure your transition ends by going to the correct menu

use the End Jump property to make sure you still end up at the historical menus. The *main2...* clips will end at their appropriate historical menus and the *...2main* clips will end back at the main menu.

Start by making tracks for each of the six transitions. The quickest way to do that is by marking all six tracks in the Asset window and dragging them to the Tracks folder in the Outline window (see Figure 11-13). Note that you might need to put the Outline view in one of your other windows temporarily. It all depends on what window layout scheme you're using.

3. **Set the End Jump.** With the Outline view still active, you easily can set up the End Jump properties for all your transitions. Bring up the Property Inspector (Option-⌘I) and highlight the first transition (*cawdor2main.2mv*, in our case). You should see something similar to Figure 11-14. In the Inspector, set the End Jump property to be the main menu.

Figure 11-15 shows a quick graph of the full set of connections between the main menu and the three historical menus.

4. **Rearrange the Button Targets.** Now go back and alter the button targets. In the main menu, each button should point at the transition—not at the historical menu directly. Figure 11-16 shows the switch for the Urquhart button.

   For each historical menu, the Back buttons should point at their respective transition clips. Figure 11-17 shows the reciprocal button for Urquhart.

And you're done! You'll definitely want to simulate projects such as this before burning them. You have so many more connections to check that it's vital you get in the habit of running through all possible navigation options.

You might notice video flashes as the transitions start and stop. These flashes won't be nearly as bad on the real thing. Sometimes there's also a slight—but noticeable—shift in color as the video transition kicks in. Again, this is stuff you should test on a consumer player once you know the navigation works correctly.

One parting note to this approach: if you want audio associated with your transitions, this is the only way to go. The transitions are just short video clips—there's nothing stopping you from adding an audio clip as well.

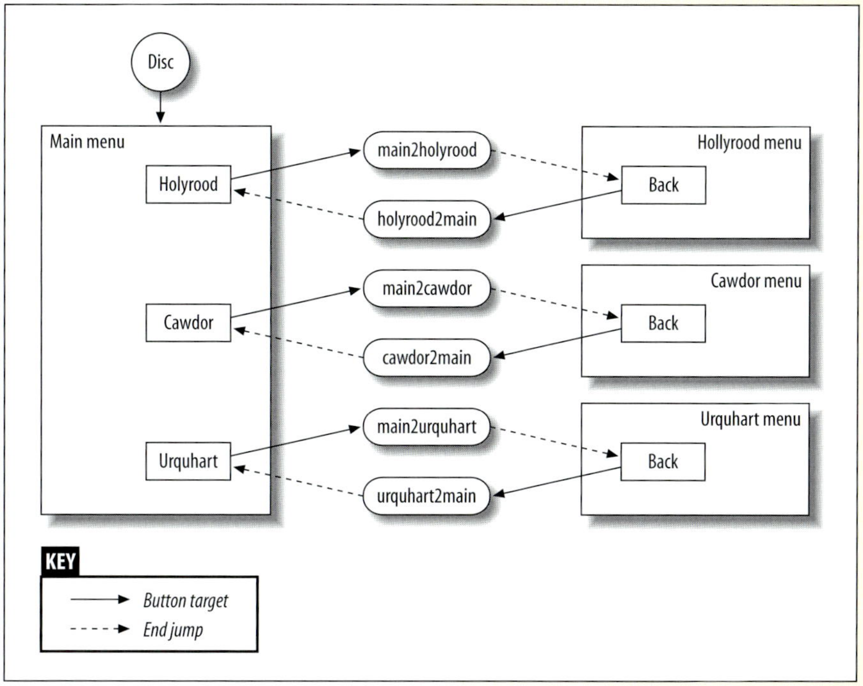

Figure 11-15. The full set of transitions for this project

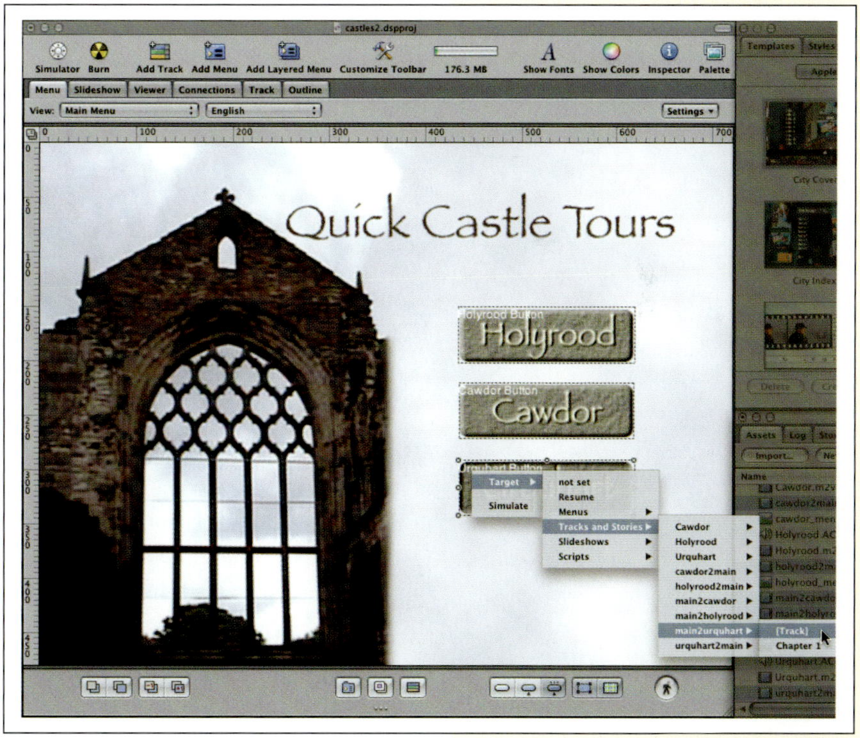

Figure 11-16. Changing the button targets to use the new transitions

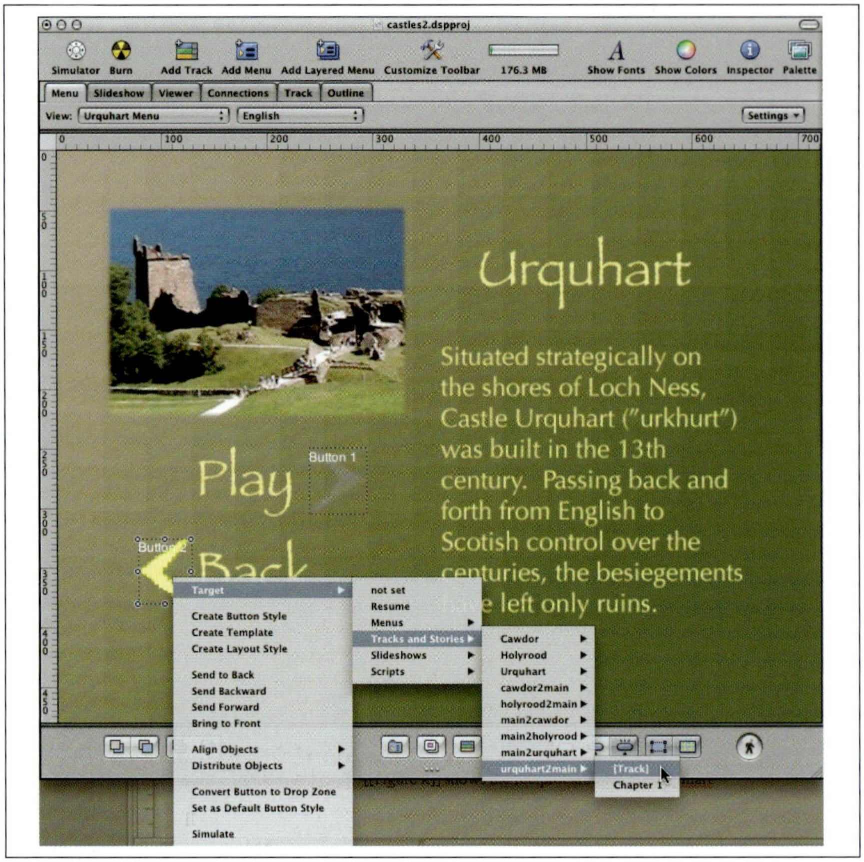

Figure 11-17. Setting up the reciprocal transition target

Alpha Transitions

So, now that you've put all that energy into creating those specific transitions, they are good for exactly one project. Sigh. Wouldn't it be great if you could spend your time making a generic transition and then be able to reuse it over and over? Well, with DVDSP 3, you can. You can build custom transitions and then add them to your Application Support library and they'll appear in the pop-up lists, along with the built-in transitions.

But how, you ask, can one make a "generic" transition? That's why you see "Alpha" in the heading for this section. DVDSP can use the alpha channel of a movie to decide what parts of a transition should show which assets. Actually, DVDSP allows fairly complex transitions, so it takes more than just one movie with an alpha channel, but we're getting ahead of ourselves.

Before we get any further, we need to set down a few ground rules. Well, not so much "ground rules" as vocabulary terms, but they are important vocabulary terms! First off, when talking about custom transitions, the word *asset* gets a bit overused. So far we've talked about assets in terms of images, video clips, and audio clips that you use to build a DVD. Anything you can put in the Asset view is, well, an asset. However, Apple also uses *asset* in transitions to refer to any external video involved in a transition. Because the names of things are important in building custom transitions, we're going to rely on that second definition of asset. So:

- *A* will refer to the first slide or menu you are transitioning from.
- *B* will refer to the slide, menu, or other target you are transitioning to.
- *Asset* will refer to any video used in your transition that's not A or B.

The other term you'll see over and over is *matte*. In movies, a matte is used to block out part of a scene that will be filled in with some other footage later on. That's exactly what you'll be building for your transitions—things that get filled in with the A and B images later on.

> **NOTE**
>
> *We're using After Effect to build these custom transitions, but After Effects is not the only option. We could use any video editing software. It all depends on how fancy we want to get. iMovie can help with simple transitions. Final Cut Pro or Premiere can create fairly advanced transitions. Motion and After Effects can create wildly complex transitions. We'll point out the key features of what we're doing so that you can translate them to your own editing or animation software. Fortunately it's not difficult stuff, but it can be time-consuming.*

Basic A → B transitions

If you need a transition that relies only on the content of your A and B images in DVDSP, you can create a fairly simple matte without worrying about the asset stuff. Let's build a quick example of such a transition. We'll create a "card" transition that simulates those big crowds in stadiums where fans hold parts of a picture to form a gigantic image.

To do this, you need to create a *background matte*. This matte is a movie that tells DVDSP where to draw A and where to draw B. Figure 11-18 shows a still frame from our matte to help illustrate the point. Everything that's white shows A; everything that's black shows B. Everything that's gray shows A at various levels of transparency. The closer you get to black, the more transparent that portion of A will be.

To get started, you'll need three images for the project in After Effects:

- *white.psd*
- *black.psd*
- *burst.psd*

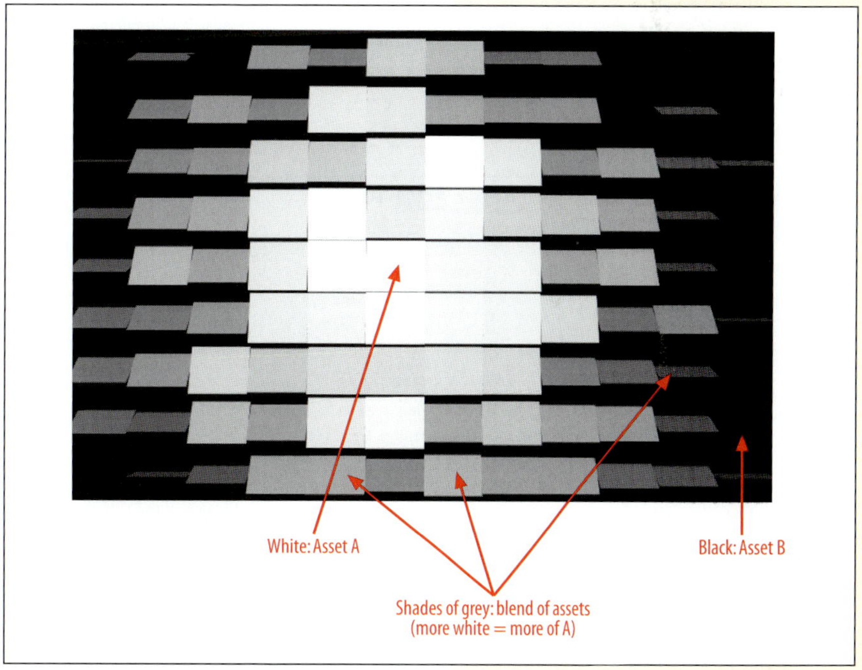

Figure 11-18. Background mattes using grayscale movies to transition between assets

Having white and black images around generally is useful for building these mattes. You can do it with blocks of color, too, but sometimes it's just easier to have an image to drag-and-drop.

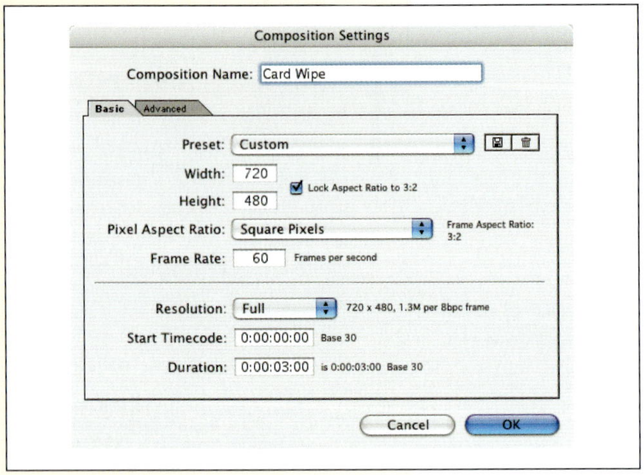

Figure 11-19. Starting a simple transition project in After Effects

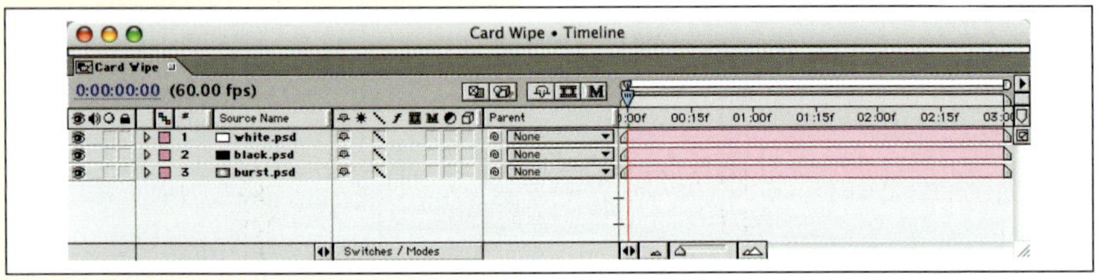

Figure 11-20. The three layers you need to make the Card Wipe transition

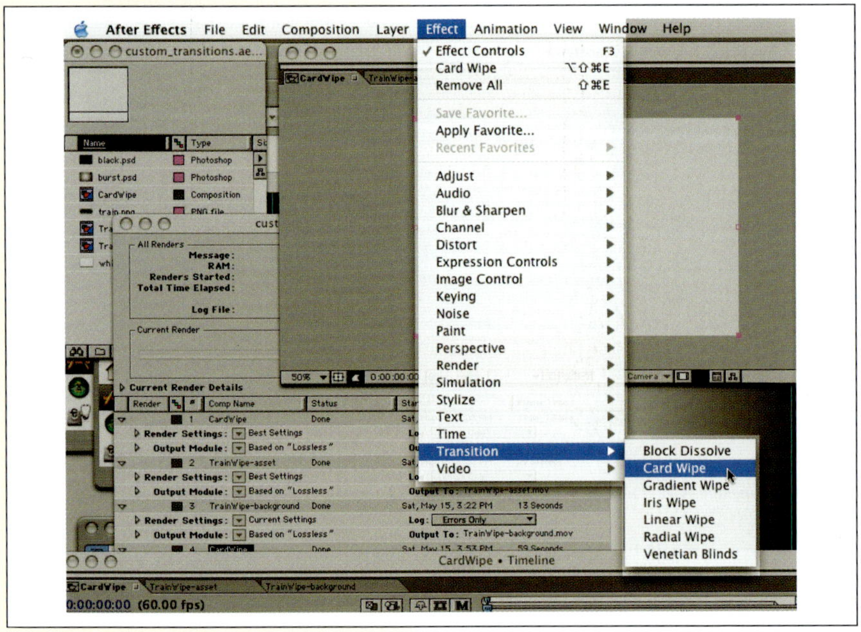

Figure 11-21. Adding the Card Wipe transition to the white layer

Once you've obtained the aforementioned three images, follow these steps:

1. **Create a New Composition.** Create a new composition and make it last for three seconds. (You'll be clipping that a bit as you go.) Apple recommends creating your transitions at 60 frames per second for NTSC video and 50fps for PAL. This creates fantastically smooth transitions, but it's not required. If you have the luxury of building the transitions in an animation tool, you probably can use 60fps. If you're building things in iMovie, you'll have to stick to 29.97. No real harm there, but use 60fps if you can. Figure 11-19 shows our initial settings when we created the composition.

2. **Add the Images in the Timeline.** Drag all three images to the timeline. They should be in the same order as the earlier bullet points: white is layer 1, black is layer 2, and burst is layer 3 (Figure 11-20).

3. **Add a Transition to the White Layer.** Figure 11-21 shows the menu choice to pull up the card effect you're after. (That card effect is one of the free plug-ins you get from Adobe for registering After Effects 6 ... clever ploy on Adobe's part!)

   Adding that transition should leave you with a timeline similar to Figure 11-22. You need to set keyframes on the transition effect at the beginning and end of your movie. You'll tackle the details of configuring the Card Wipe transition next.

4. **Configure the Card Wipe Transition.** By default, you can set the cards to turn over, starting from any side and going to the opposite side. You want

to get a more circular pattern going to help your transition stand out from the ones that are available in DVDSP already. To accomplish that, you can use a third gradient image; that's what the *burst.psd* file is for (see Figure 11-23). The cards start flipping over wherever it's black and work their way up to the white areas. You get an outside-in pattern from this gradient.

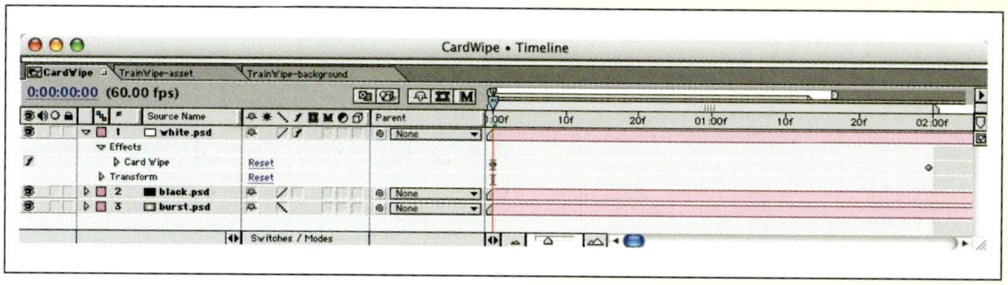

Figure 11-22. The transition in place in the project timeline

You can tweak the percentage of completion to get the transition to fill as much or as little of your timeline as you want. We set our working area to be just two seconds and then altered the Card Wipe effects as shown in Figure 11-24. We used our black layer as the back layer and our burst layer as the gradient that drives the card flipping. We also bumped the randomness up a bit to give the transition more of an authentic crowd feel.

Figure 11-23. The gradient pattern that drives the card animation (darker areas will flip their cards first)

5. **Render It as a Matte.** To save this movie, you just add it to the render queue as usual. For background mattes such as the one you're building, you can use the Animation Compressor and output 256 grays for the color palette. Figure 11-25 highlights the settings we used for this matte. If you do this very often, you'll definitely want to make a template out of this so that it's easy to render future mattes.

That's it! At least, that's all you have to do in After Effects for this transition. You still need to get the transition into DVDSP, but let's build our other transition first.

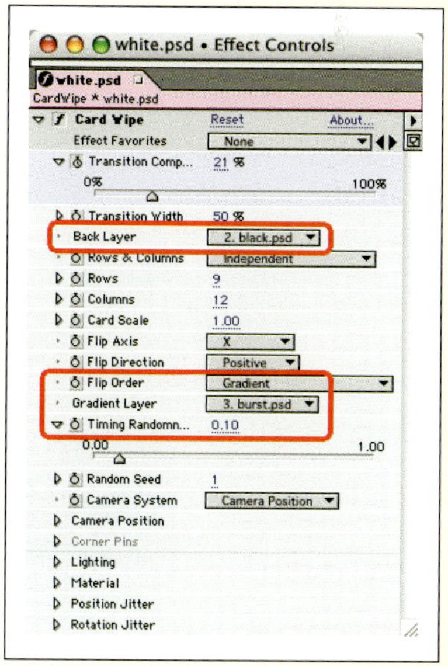

Figure 11-24. Possible options for the card transition

Transitions with assets

What if you want a transition in which you actually supply more images than just A and B? That's entirely possible; you just have a bit more work to do.

In addition to the background matte, you can add extra video information to the transition. In fact, let's tackle that right now.

1. **Start a New Composition in After Effects.** You can use the same settings as the previous project. The only file you need to import is *train.png*.

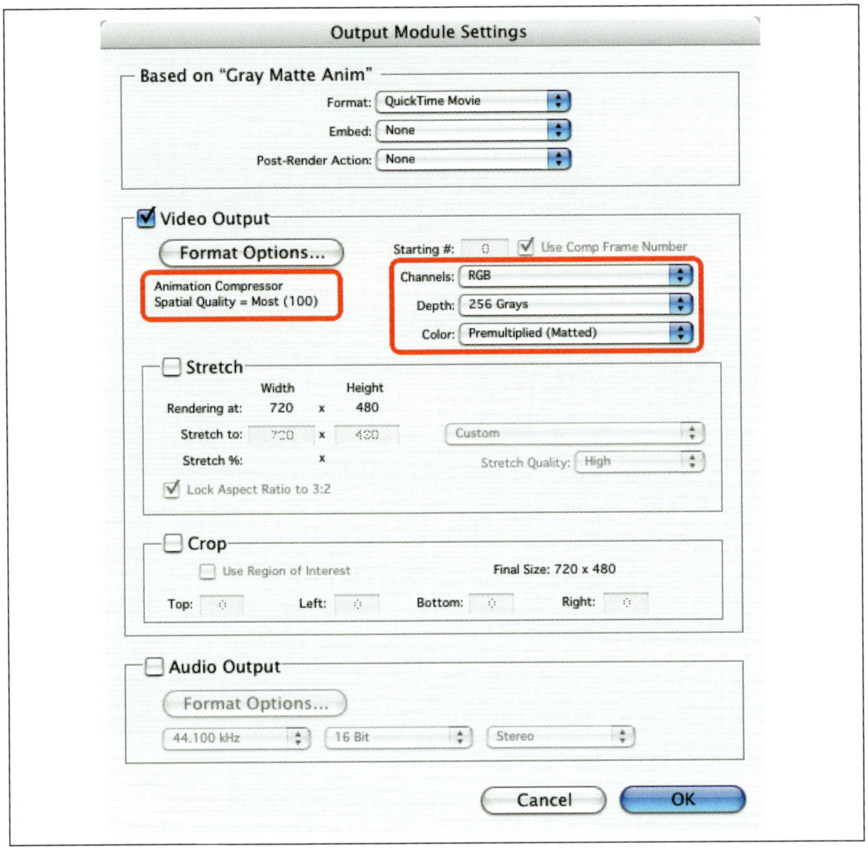

Figure 11-25. The output settings for a matte movie

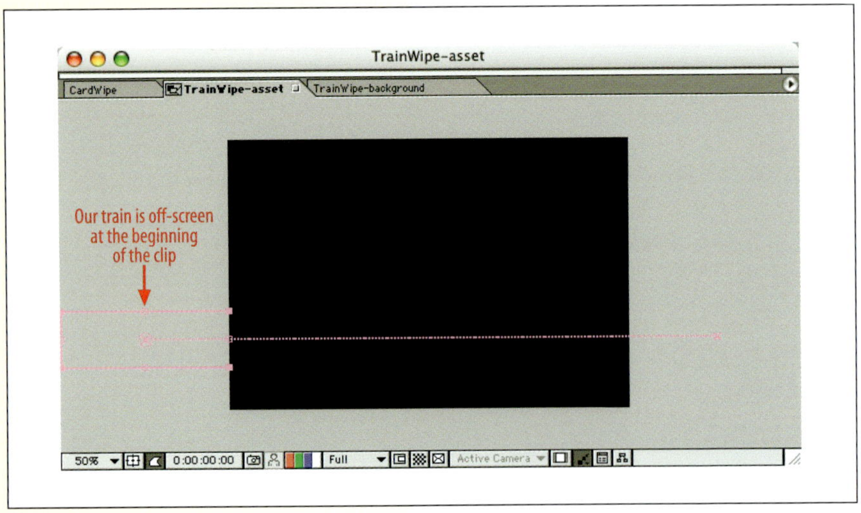

Figure 11-26. The starting position of our train animation

2. **Create Starting and Ending Keyframes for the Train.** Go ahead and drag the train image to the timeline and create starting and ending Position keyframes. You'll start this transition with your train completely off-screen so that it doesn't just appear out of nowhere. Figure 11-26 shows the starting position we chose.

We end with the train off-screen on the opposite side. Figure 11-27 shows that position. Figure 11-28 shows the train in one of its intermediary frames.

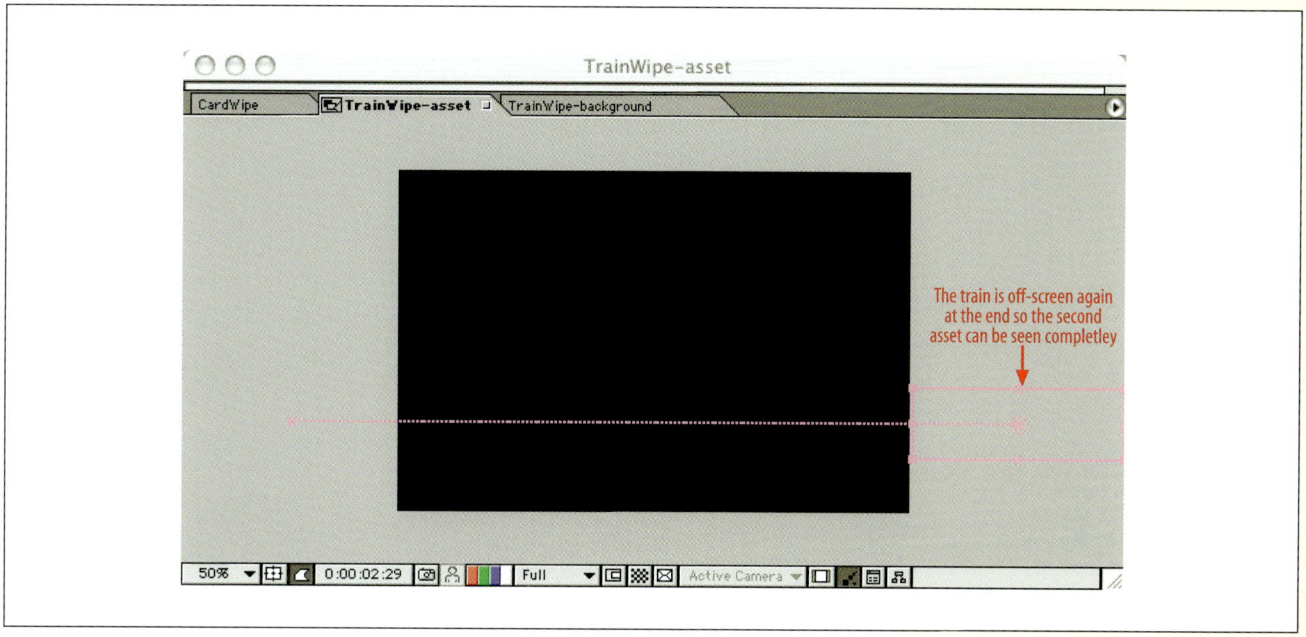

Figure 11-27. The end position of our train

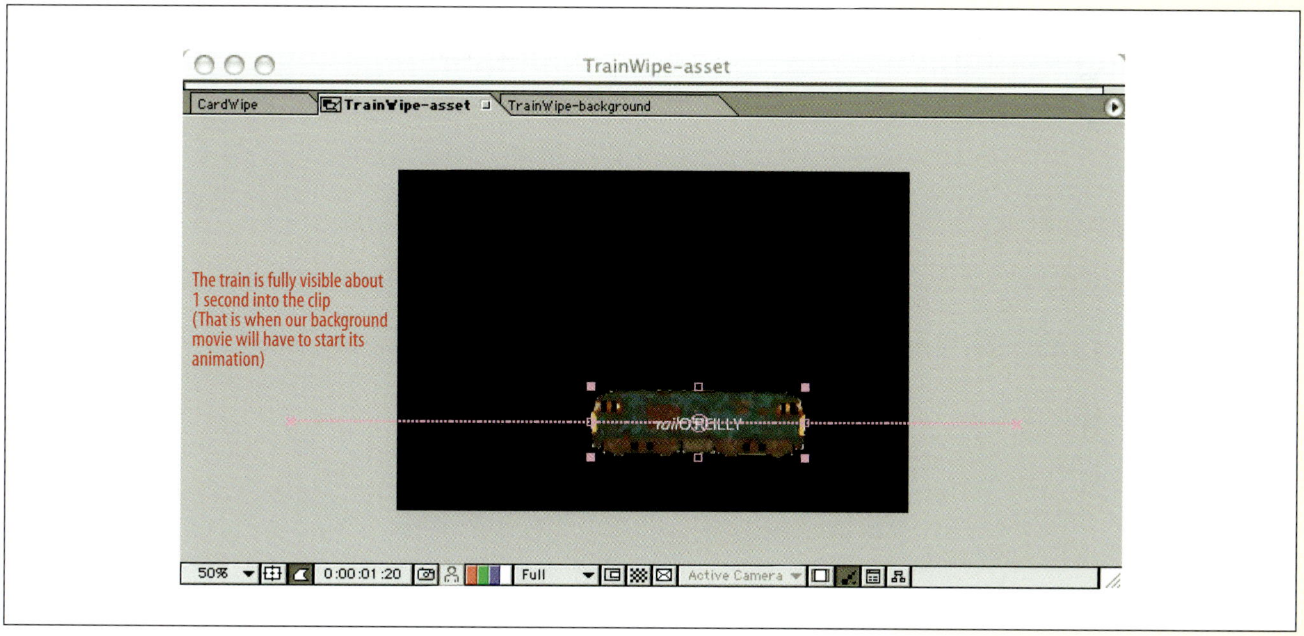

Figure 11-28. Our train, fully visible while in midtransition

3. **Create a Background Matte.** The function and creation of the background mattes for asset transitions is basically the same as it is for creating standalone background mattes. The only real difference is that you have to pay attention to your asset movie. In our asset movie, for example, the train starts off the edge of the screen. Until the train is entirely visible, you don't want your background changing at all. You have to take that into account when building your background matte.

   Otherwise, this background matte is pretty boring. You just want the A image to follow the train off to the right side of the screen. You'll create one more composition in After Effects and reuse the *white.psd* and *black.psd* images. (Put white in layer 1 and black in layer 2.)

   When you start out, the white card should occupy the entire frame, as shown in Figure 11-29.

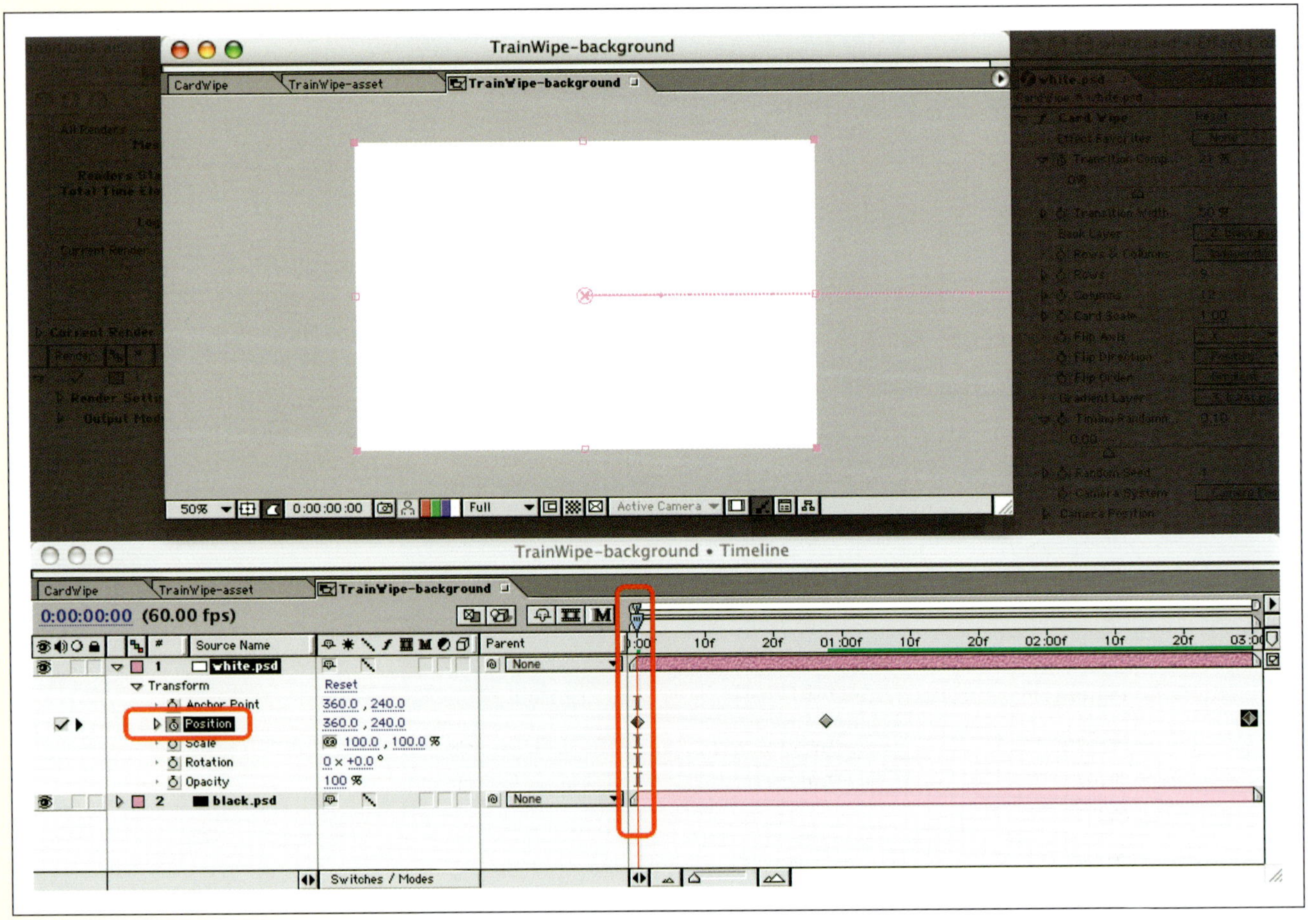

Figure 11-29. The starting position of the white image in the background movie

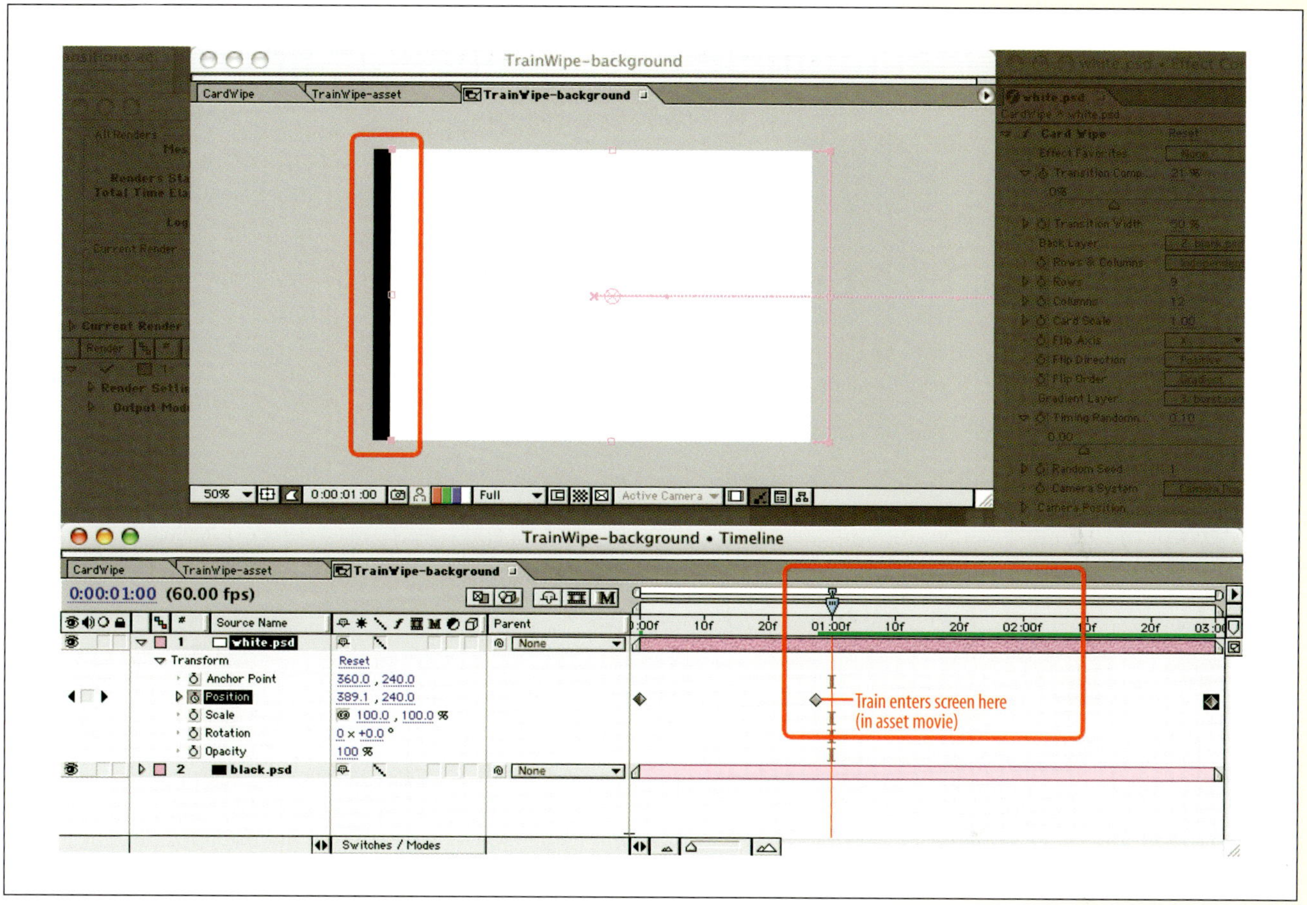

Figure 11-30. A midpoint in the background movie

But you need to make sure it stays put for about one second (to give the train time to get into the frame). To do that, just position the cursor in the middle of the timeline and click the keyframe box in the timeline to create a new keyframe. Don't modify anything about this keyframe. Nothing changes, so the white card will stay put. Figure 11-30 highlights this new keyframe.

Notice in Figure 11-30 that you can see the black portion of the background starting to peek out. That's the animation kicking in. In the completed transition that black part will provide the first glimpse of B.

For the end of the movie, you want the white card off-screen, so create a new keyframe and then drag the image in the viewer off to the right side, as shown in Figure 11-31.

4. **Render the Results.** You can (and indeed, should) save the background matte with the same settings you used for the standalone background matte in the previous project. The asset movie, however, needs its alpha channel rendered. You might recall the alpha-channel discussion from Chapter 4 when you built your video menus. The idea is that you can store the transparency of parts of the image along with the colors you usually store.

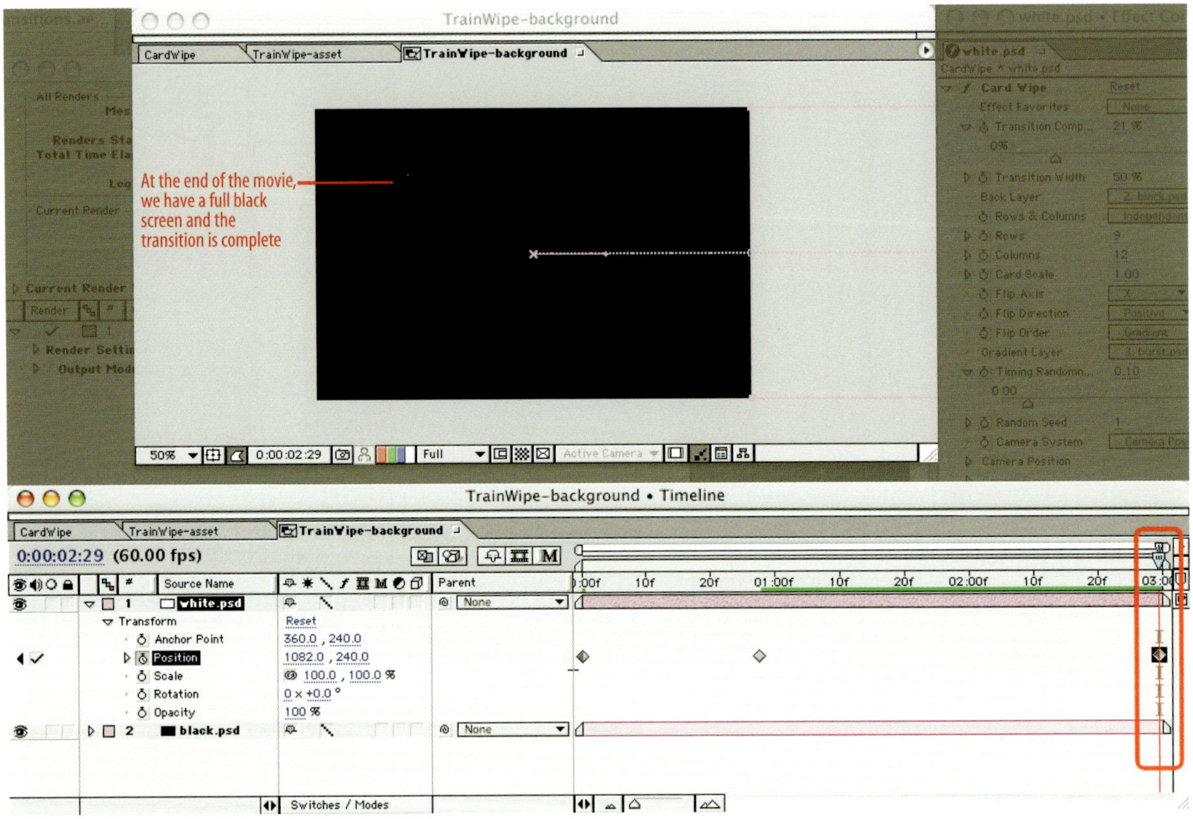

Figure 11-31. The final position of the white image

To save the alpha-channel information in After Effects, you use the Output Module Settings. Figure 11-32 shows the settings we used for the train asset movie.

Notice that the Channels setting now lists RGB + Alpha. The other pop-up lists default to the right settings: Millions of Colors + and Premultiplied (Matted) for Depth and Color, respectively. The premultiplied part is required for DVDSP to do its work properly. (Fancy compositing software—such as After Effects—can take advantage of unmultiplied clips to accomplish very intricate and smooth layering effects.)

You can verify your settings before rendering your movie, as shown in Figure 11-33. Fortunately these transition movies usually are only a few seconds long, so re-rendering doesn't pose much of a hassle.

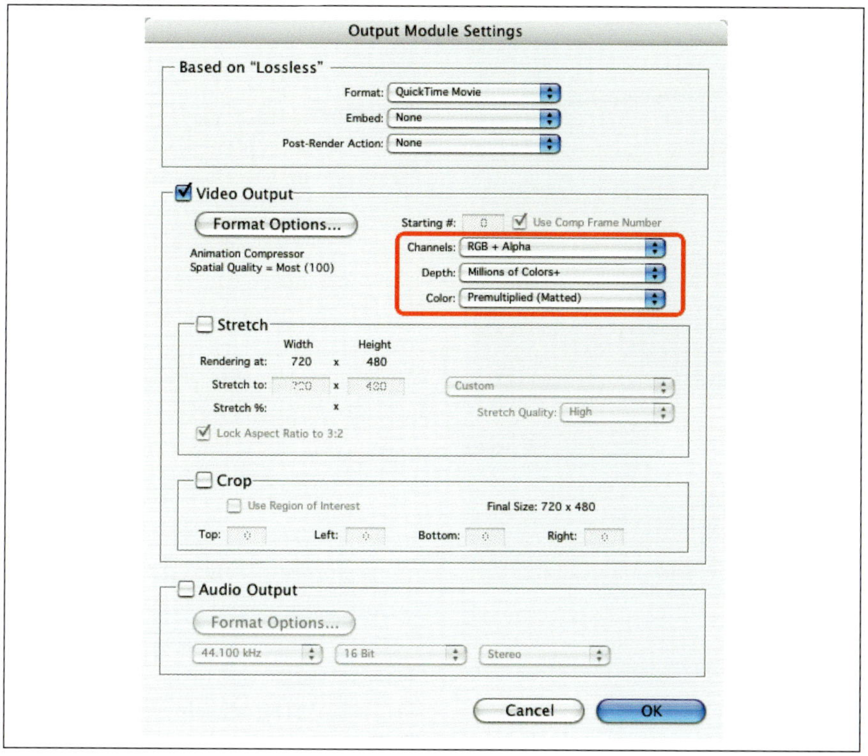

Figure 11-32. The output settings for an asset movie

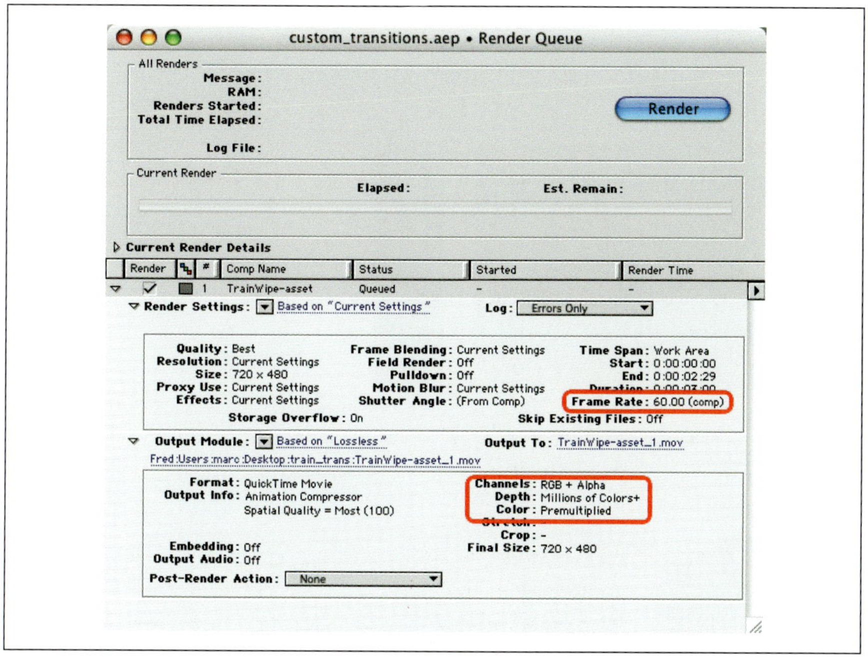

Figure 11-33. Verifying our settings before we render the asset movie

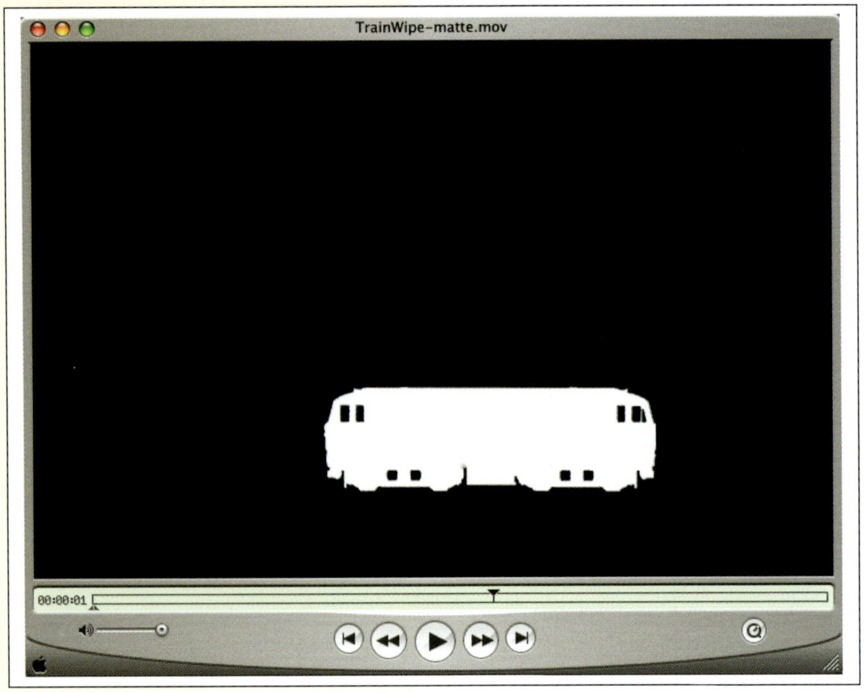

Figure 11-34. An example of an asset matte movie

What if I don't have After Effects?

Don't worry! You still can accomplish some fun transitions with other tools you do have. In fact, you can use iMovie for the train transition if you really want to. You'd have to settle for 29.97fps video and you'd have to fake the alpha channel. How do you fake the alpha channel? DVDSP gave us a great alternative to premultiplied alpha video. You can supply a third video clip that uses grayscale information instead of the alpha channel.

In the third movie (known as the *asset matte movie*), white represents fully opaque and black is fully transparent. You put white where you want your asset to show and black where the A-to-B transition should show through. Figure 11-34 shows a snapshot from such a matte for our train clip.

Shades of gray usually serve to make smooth edges around your white areas, but you can use them for fancy effects such as a bubbling lava transition. Figure 11-35 illustrates how the three movies get put together in DVDSP.

Figure 11-35 Compositing the three transition movies with the menus to create the final transition

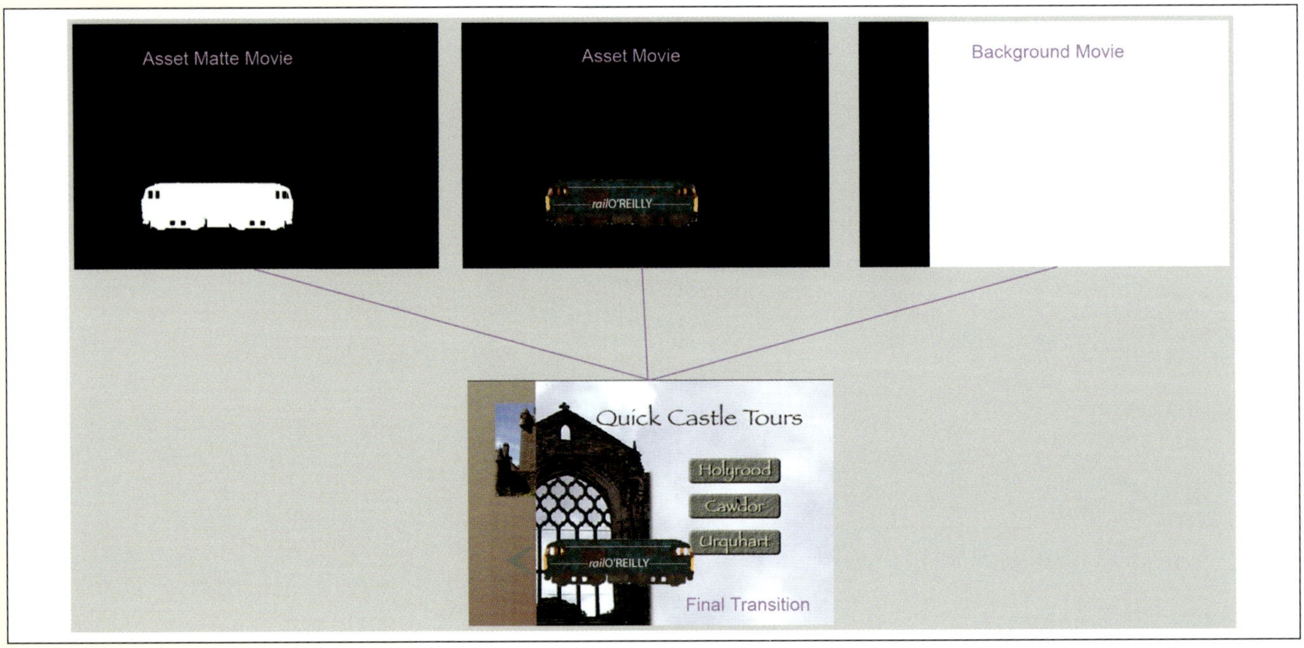

If your asset movie carries an alpha channel (as ours does), the third movie isn't necessary.

Adding transitions to DVDSP

Now that you've got your various transition movies, how do you get them into DVDSP? This part is remarkably straightforward. You just put them in the right folder and name them correctly. One quick note before you get started: you must do this before you start DVDSP. If you add transitions while you're using DVDSP, they will be ignored until you restart.

The "right folder" depends on who needs to use the transition. You can place transitions in your own library for your personal use. Or you can place them in the DVDSP library for everyone to use. (You have to be an administrator on your system to add transitions for everyone.)

For example, to add our Card Wipe transition, we would create a folder named Card Wipe in the *Library/Application Support/DVD Studio Pro/ Transitions* folder. You might have to create the *Transitions* folder if this is your first transition. You can create another folder for the Train Wipe transition. Figure 11-36 highlights the folder path for placing the transitions in your personal library. To add the transition to the application for everyone, you would start at the top-level *Library* folder, but the rest of the path would be the same.

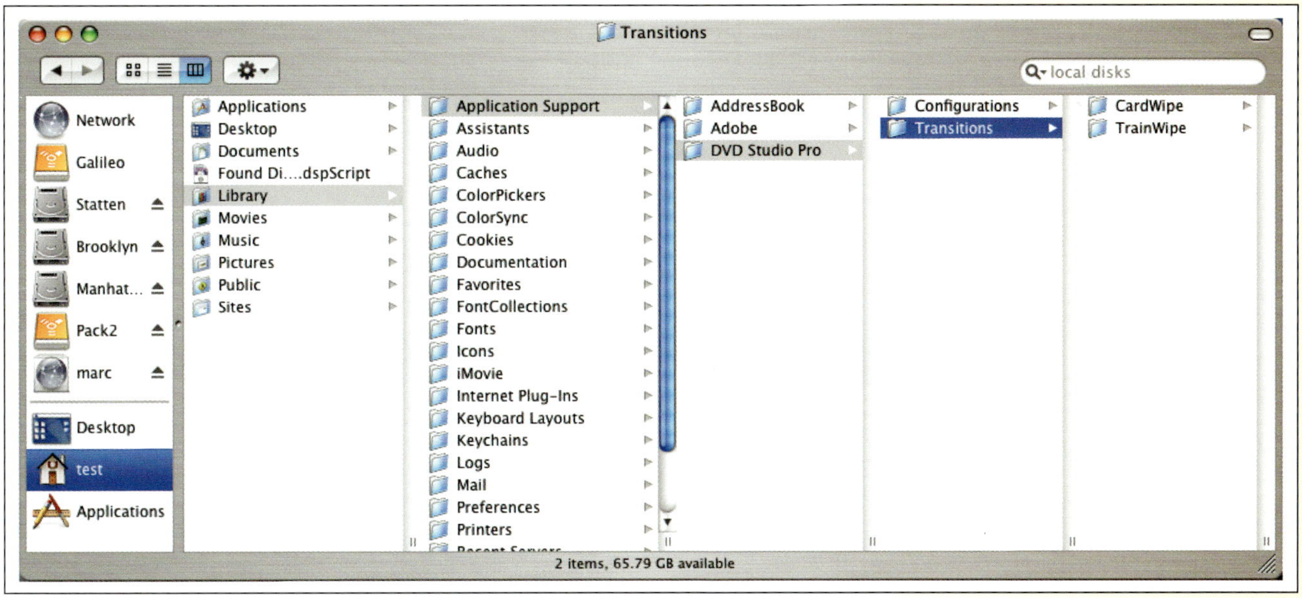

Figure 11-36. The folder path for personal custom transitions

Inside the folders for your transitions, the files have to follow certain naming rules.

Asset movie

Name this the same as your folder. You can have any extension you want on the name—for example, *TrainWipe.mov*.

Asset matte movie

Name this file the same as your folder and then append "-matte" to the name. Again, you can have any extension you want. If we had used the hypothetical train asset matte from Figures 11-34 and 11-35, that movie would be named TrainWipe-matte.mov.

Background matte movie

Name this file the same as your folder and then append "-background-Matte" to the name. (Notice the uppercase "M" in the suffix.) As with the other movies, you can have any extension—for example, *CardWipe-backgroundMatte.mov* or *TrainWipe-backgroundMatte.mov*.

Figure 11-37 shows the contents of the *TrainWipe* folder for illustration.

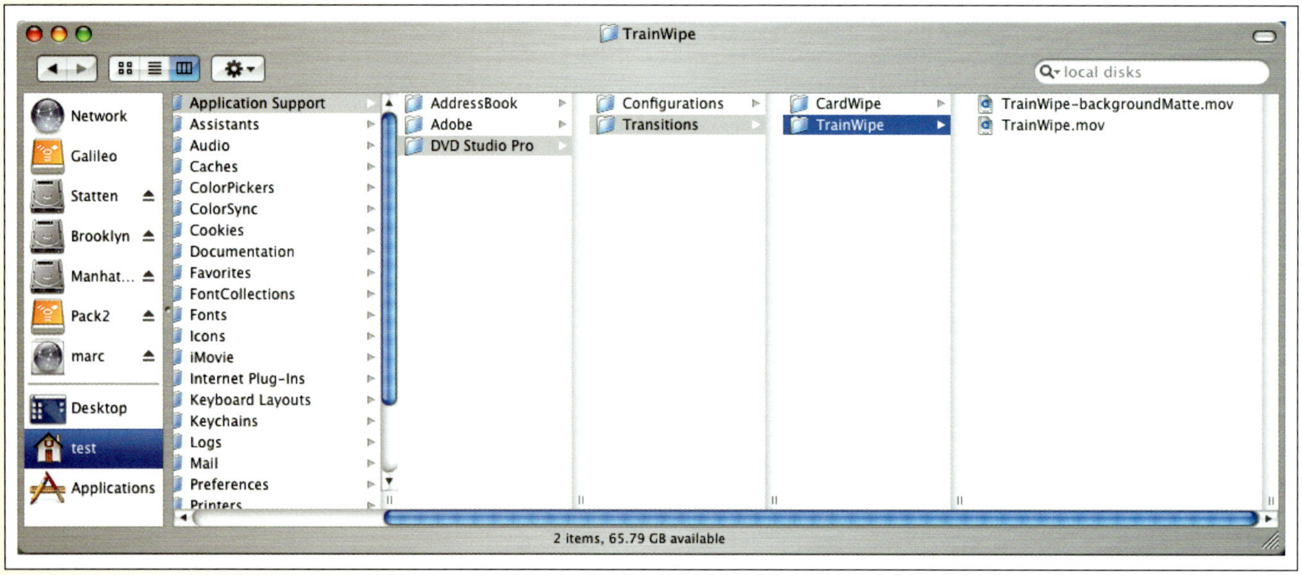

Figure 11-37. The contents of the TrainWipe transition folder

Getting back to DVDSP

Go ahead and start DVDSP. You should see your custom transitions in the various Transition pop-up lists. Figure 11-38 illustrates one such list.

Just for fun we applied the transitions to our Castle Tour project. We use the train transition when moving from the main menu to the individual castles (Figure 11-39) and the card transition when moving back to the main menu (Figure 11-40).

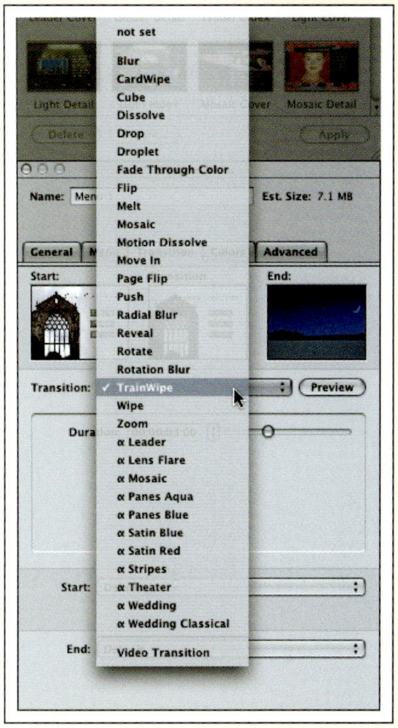

Figure 11-38. Our transitions showing up in the pop-up list

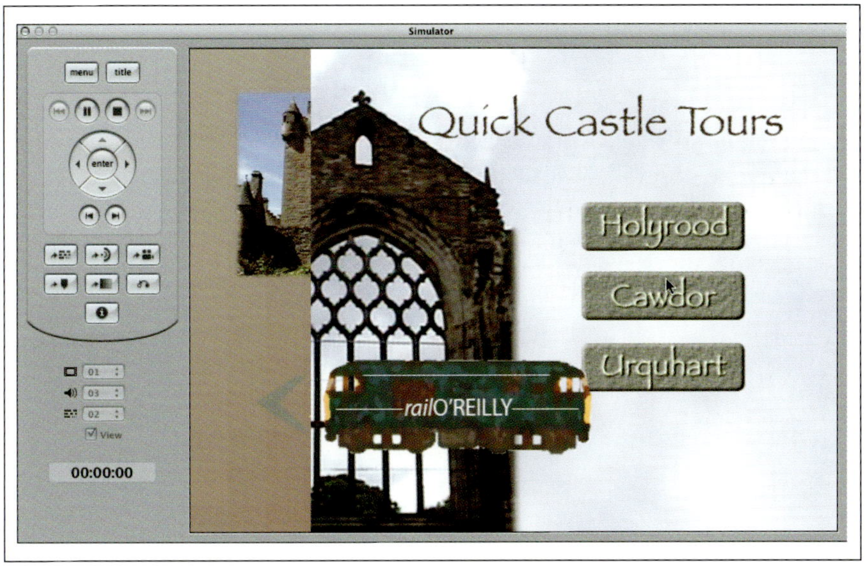

Figure 11-39. Our train transition in action

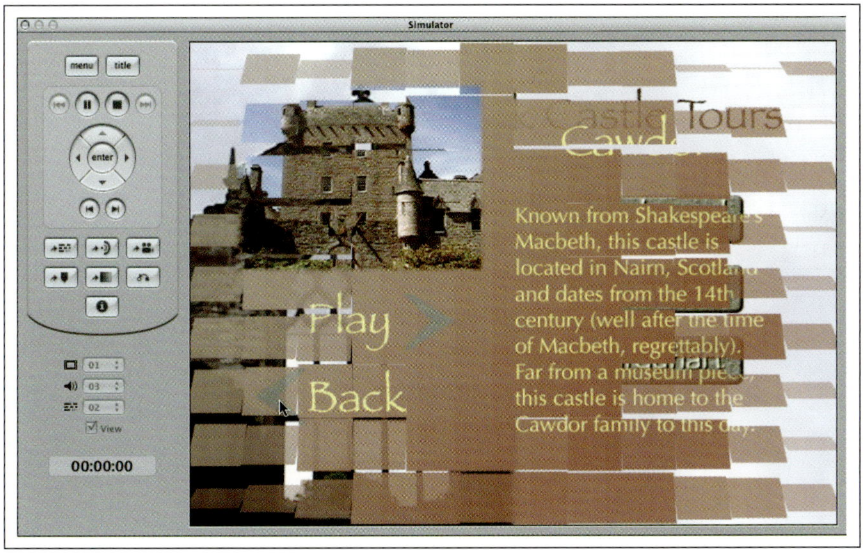

Figure 11-40. Our card transition in action

What's Next?

So, what's next? Well, lots and lots of practice. We didn't have the space to cover 100% of the details and features in DVDSP, but we hit the biggies. Hopefully you've got a sense of confidence in your ability to work in DVDSP. You should go play with it now. Back up all your files and then go to town! Try things out. As we said at the beginning of the book, don't be afraid to burn a few test discs to see if things really are working the way you expect. Now that you've gone through the book, we hope you won't be burning as many coasters as before, but even so, experimenting is OK.

Actually, we encourage it! Have fun with the projects and be sure to check out the "best practices" recommendations in Appendix B. Most of those recommendations also are in the video portion of the DVD if you want to see them in action. The introduction to Appendix B has more details on the various tests you can watch.

So, get going already! Thanks for playing with us, and good luck!

Script Command Reference

A

This appendix takes an in-depth look at the scripting commands available to you in DVDSP. You can find more information on scripting in the DVDSP 3 User's Manual.

Argument Types

Before we list the commands, we need to enumerate the types of arguments the commands can use. Figure A-1 highlights the argument sections of the Script Command Inspector.

From the pop-ups highlighted in Figure A-1, you can select one of five types for storing or comparing:

GPRM
: General Purpose Register Memory (global variables)

SPRM
: System Parameter Register Memory (global status information)

Immediate
: Numeric values such as 1 or 255

Jump Target
: The same options available to menu buttons and end jumps

Special
: Other jump targets such as Current Item, Last Item, and Last Track

Type reference

Let's go through each type in a bit more detail.

GPRM

GPRM types represent the variables of DVD scripting. DVDSP gives you access to eight such variables. You can store a variety of information in a GPRM. Typically they hold immediate values or jump targets for use throughout a script.

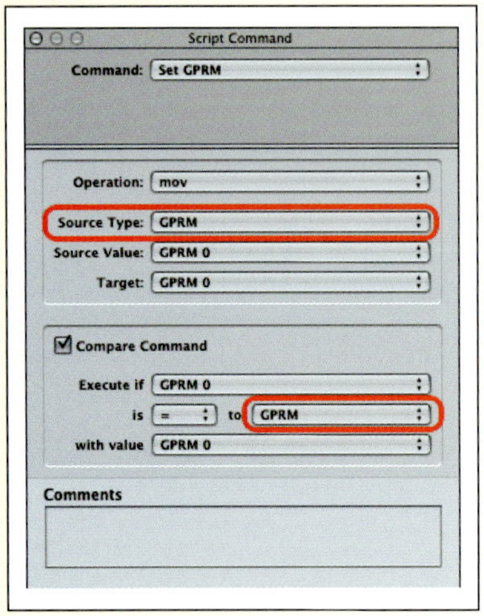

Figure A-1. The sections that use arguments in the Script Command Inspector

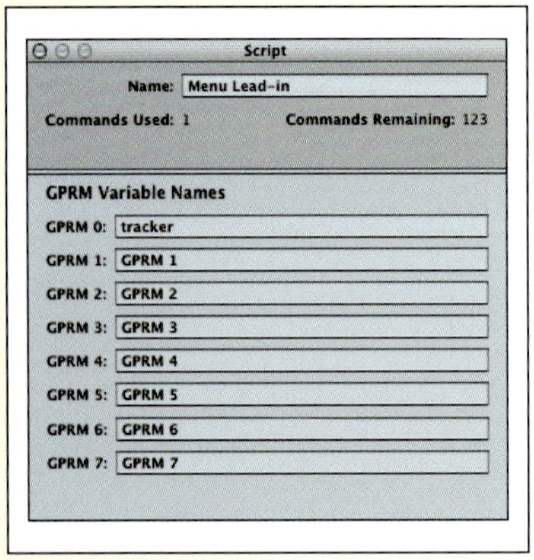

Figure A-2. Setting the GPRM entry name via the Inspector

By default, these variables are referred to as GPRM 0 through GPRM 7. You can rename them for each project, though. The Inspector window for both the disc and any script will give the list of variables, as shown in Figure A-2. The eight names are global—you can't have different names for each script.

The names are meant to make the scripts you write easier to read. They are purely a convenience—but one you should get into the habit of using.

SPRM

The system parameters reflect the state of the player. You can use a total of 24 accessible parameters in your scripts. Table A-1 lists these parameters.

Table A-1. SPRM parameter map

#	R/W?	Description
0		Preferred menu language (two-letter code; see Tables A-4 and A-5)
1	Y	Current audio stream number
2	Y	Current subtitle stream number
3	Y	Current angle number
4	Y	Current title number
5	Y	Current Video Title Set (VTS) number
6	Y	Current Program Chain (PGC) in title
7	Y	Current chapter in title
8	Y	Current highlighted button number
9		Navigation countdown timer
10		Navigation timer target
11	Y	Karaoke mixing mode
12		Parental management country code
13	Y	Parental management level (1-8 or 15 for "none")
14		Video configuration (16:9, letterbox, etc.)
15		Audio configuration (DTS, PCM, etc.)
16		Initial audio language code
17		Initial audio language code extension
18		Initial subtitle language code
19		Initial subtitle language code extension
20		Region code
21		Reserved
22		Reserved
23		Reserved

They always are referred to by number in the Inspector, so you might want to keep this list handy. The R/W? column indicates whether this parameter is both readable and writable. If it's not read/write, you only can query the value in a comparison or copy its value to some other variable. You cannot change it using the *Set* command.

Immediate

Immediate values sometimes are referred to as literals in other languages. These values usually are numeric in nature. You would use an Immediate type to count things or to set a flag. Many of the Easter eggs from Chapter 8 use Immediate types in this manner.

Many programmers who hang out around bit-level operations (which you have to use in DVD scripting) become fond of hexadecimal format (base 16 as opposed to base 10). One hexadecimal digit represents exactly 4 bits. It becomes easier to read and understand exactly which bits are being used in a given operation. You can use hexadecimal values (the regular numbers 0–9 plus the first six letters of the alphabet, A–F) in your Immediate-value entries if you check the Hex Values checkbox in the Script editor (see Figure A-3).

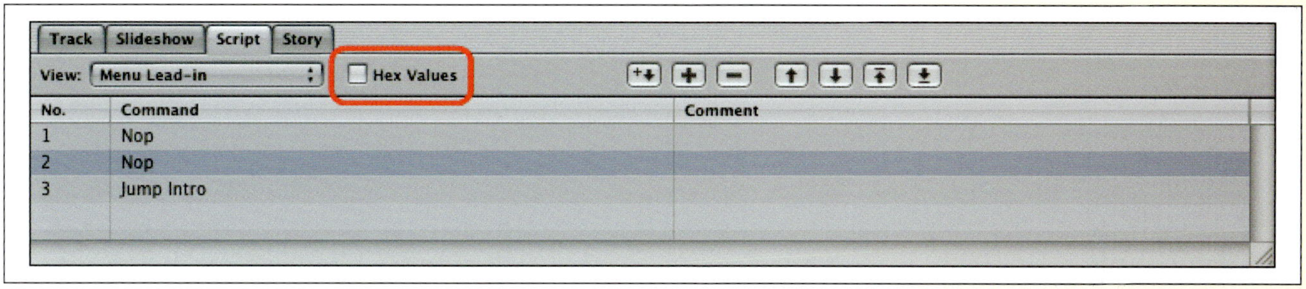

Figure A-3. Using hexadecimal values in a script

Using the hexadecimal format does not affect the function of or reasons for using an Immediate value. It just changes the presentation.

Jump Target

You also can store a Jump Target value in a variable. This can be useful for complex branching applications that keep track of where the user was (or should go next). This becomes important for generalized scripts that get used in many places throughout the project. (This is a common alternative to having numerous scripts that are hardcoded with one specific Jump Target value.)

Special

There are three Special values:

Current Item
: This script, unless this is a pre-script (in which case this is the item to which the script is attached)

Last Item
: The item that started this script (menu, track, etc.)

Last Track
: The last track that was played

You can store these values in variables for later retrieval. You also can use them in comparisons for executing commands such as "Jump to Menu B if the last track was Track 5."

Command Conditions

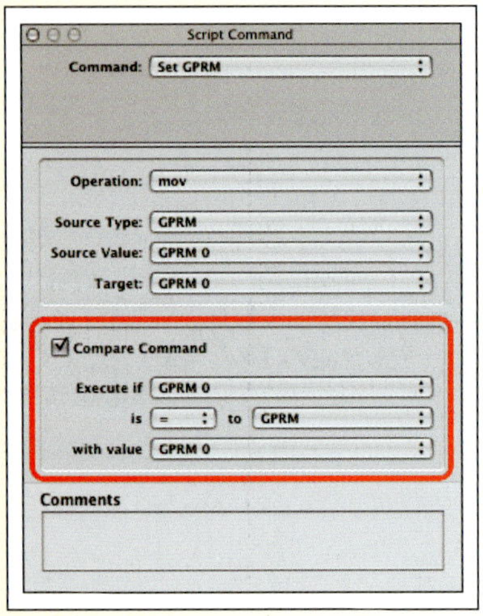

Figure A-4. The conditional area for script commands

Another bit of background before we tackle the reference section concerns conditional execution. In many languages you have some type of "if" statement. If some condition is true or exists, you perform one command. If that condition is false or does not exist, you perform a separate command—or often no command at all. DVDSP goes the one-command-or-none-at-all route. Figure A-4 highlights the condition section of the Script Command Inspector.

The basic setup of a comparison is to look at one of the GPRM or SPRM values and compare it to something else (another variable or maybe an Immediate value).

With the exception of the No Operation (Nop) command, you can apply a condition to any script command. You just check that condition checkbox and enter your condition.

You can compare items using the same types as those listed in the previous section (namely GPRM, SPRM, Immediate, Jump Target, and Special). If the condition you set up is true, your command will execute. Otherwise your command is skipped and execution picks up on the next line of the script.

Comparison types

You can use seven comparisons in the Compare Command area. Table A-2 lists those tests with a brief description of their purpose.

Table A-2. Compare Command tests

Option	Test	Description
=	Equals	Returns true if the variable and the value are the same
!=	Does not equal	Returns true if the variable and the value are different
>=	Greater than or equals	Returns true if the variable is bigger than or the same as the value
>	Greater than	Returns true if the variable is bigger than the value
<=	Less than or equals	Returns true if the variable is less than or the same as the value
<	Less than	Returns true if the variable is smaller than the value
&	Bitwise AND	Returns true if the variable has at least the same bits turned on as the value (i.e., the variable can have more bits turned on, but cannot have fewer bits turned on)

Script Commands

You choose script commands from a pop-up list in the Inspector, as shown in Figure A-5. You configure one line of your script at a time through this interface.

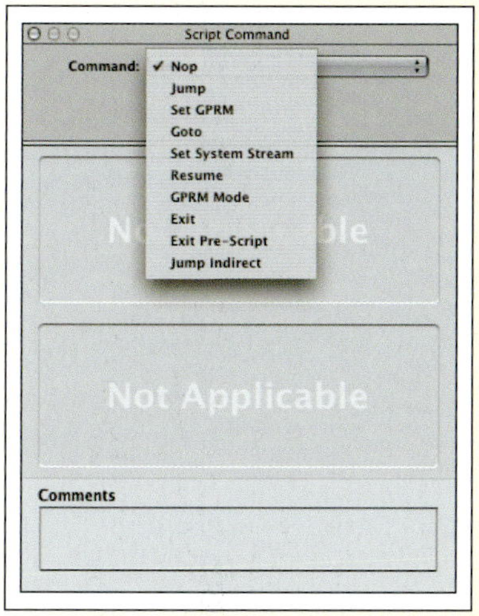

Figure A-5. The various commands available for your DVD scripts

As you can see, all commands can have comments. Where appropriate, commands can take arguments. Most commands also enable you to place a condition on them. Each command listed in the following sections shows a full screenshot of its Inspector window and enumerates the valid arguments and whether conditions are applicable.

For quicker reference, we listed the commands here in alphabetical order.

Exit

The `Exit` command shown in Figure A-6 leaves a script.

Note that exiting a script—rather than jumping to another item—will halt playback of the title. This can be useful for things such as password-protected titles or mismatched region codes.

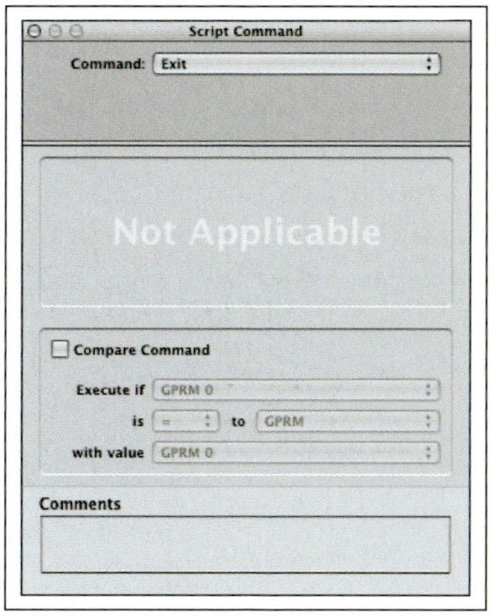

Figure A-6. The Exit command

Exit Pre-Script

You can use the `Exit Pre-Script` command (Figure A-7) only in a pre-script. It jumps directly to the item using the pre-script.

It's important to use this operation to leave your pre-script and get on with playing the item associated with the pre-script. Using the *Exit* command will halt playback altogether.

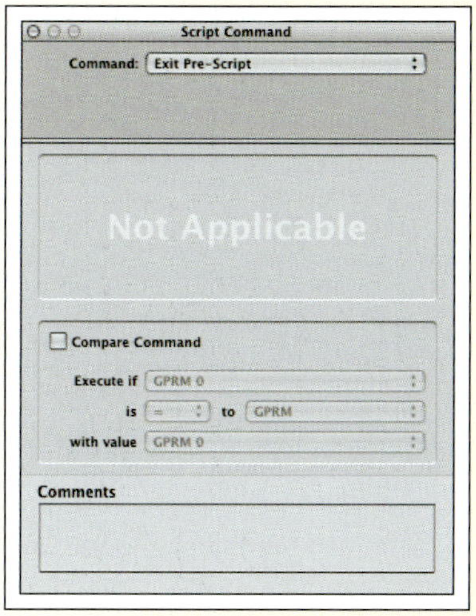

Figure A-7. The Exit Pre-Script command

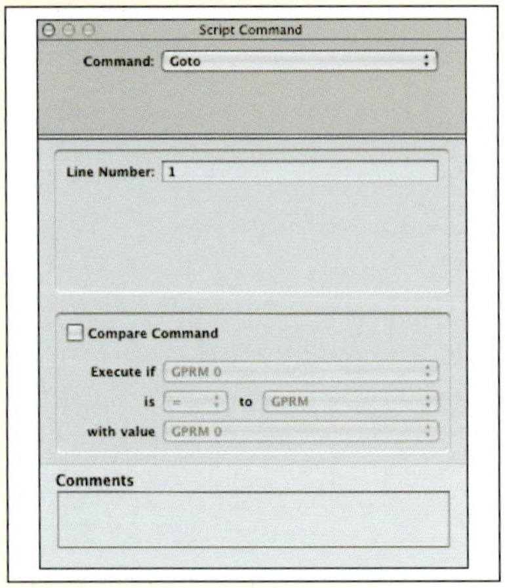

Figure A-8. The Goto command

Goto

The Goto command (Figure A-8) enables you to jump to a different line in your script. This is the sole means of creating complex flow or loops.

To build a loop with the Goto command you can use a portion of script similar to the script shown in Figure A-9.

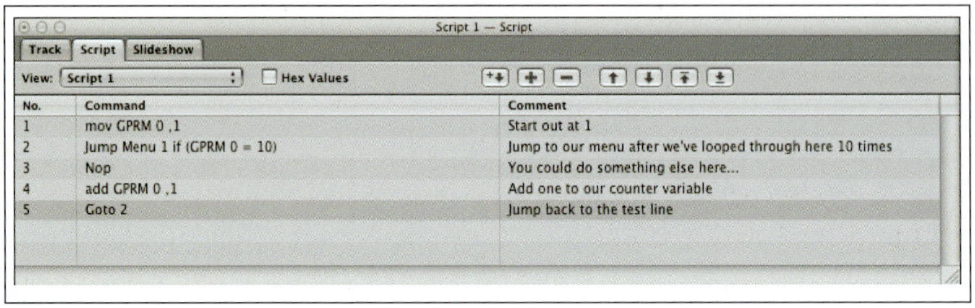

Figure A-9. A simple loop using the Goto command

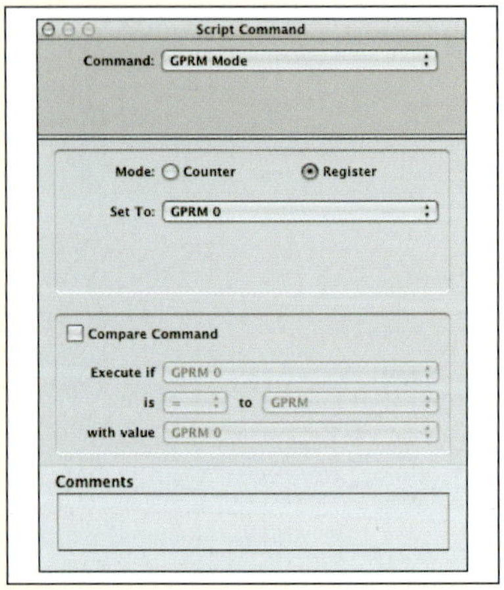

Figure A-10. The GPRM Mode command

GPRM Mode

The GPRM Mode command (Figure A-10) enables you to switch between treating a GPRM entry as a normal variable and as a countdown timer.

If you put the GPRM into countdown mode, placing a value in it will start the "timer." It will count down by one every second until it reaches zero. Nothing in particular happens when a countdown entry reaches zero, but you can test for that value in other scripts (or later in the same script, too).

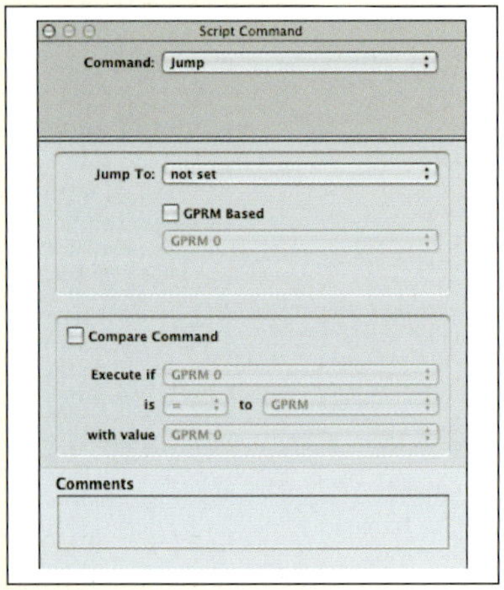

Figure A-11. The Jump command

Jump

You use Jump to, well, jump to another asset. The other asset (Figure A-11) can be a track, menu, slideshow, or another script. This is the same set of options you have when configuring menu buttons or End Jump properties.

The one interesting addition to the usual list of targets is the GPRM Based option shown in the figure. Depending on the type of target (track, menu, slideshow, etc.), that option will enable you to pick a subelement based on the value in a GPRM variable. For example, if you jump to a menu, that option turns into a "GPRM Based button" that enables you to select a button based on the value in one of the variables. If you were tracking where the user had been, you could use this option to automatically highlight the button for the next track the user should view.

Jump Indirect

You can use `Jump Indirect` (Figure A-12) to jump to another asset, but that asset must be stored in a GPRM variable.

For example, you could store a jump target in a variable and then use the `Jump Indirect` command to use that variable as the jump target. This is a valuable command for Easter eggs, and it opens the doors for some clever programming tricks.

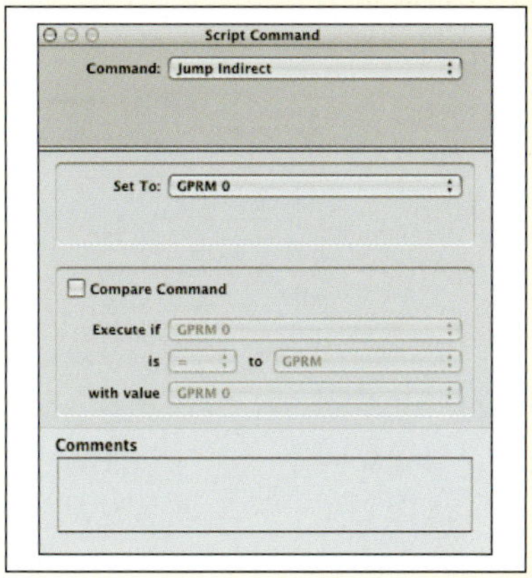

Figure A-12. The Jump Indirect command

Nop

This simple no-operation command (Figure A-13) serves as a placeholder for future work.

If the `Nop` command is the last command in a script, however, it acts like the Exit command and halts playback. If the `Nop` command is the last command in a pre-script, it acts like the `Exit Pre-Script` command and jumps to the item associated with the pre-script.

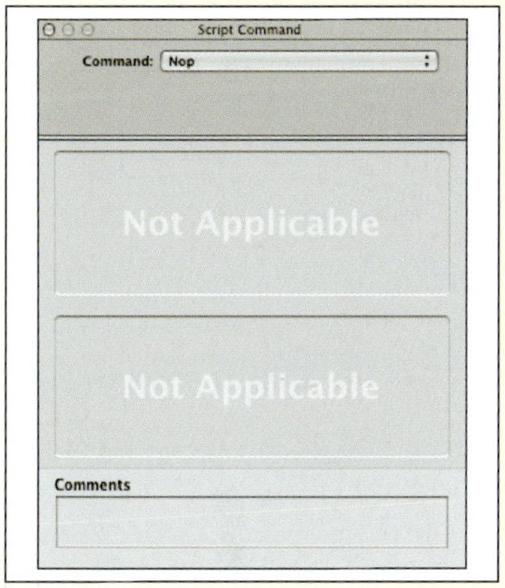

Figure A-13. The Nop (no operation) command

Resume

The `Resume` command (Figure A-14) enables you to exit a script and return to the point you left off. Usually that means resuming playback of a video track from the point you pressed the Menu button on your remote control or after hitting a marker that executed a script.

One notable exception to the "where you left off" rule applies to pre-scripts. If you issue a `Resume` command inside a pre-script, it behaves like the `Exit Pre-Script` command and starts playing the item associated with the pre-script.

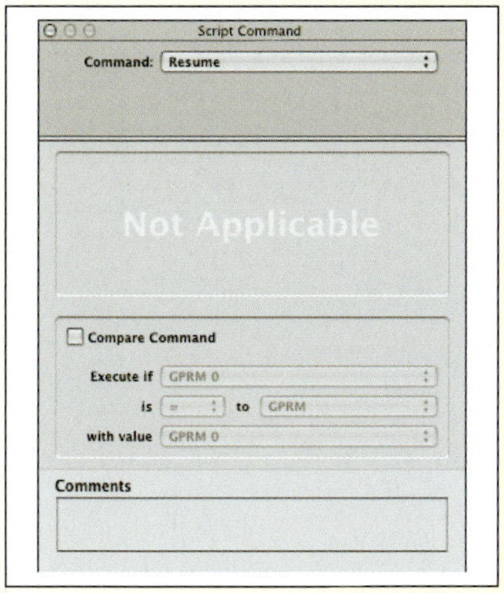

Figure A-14. The Resume command

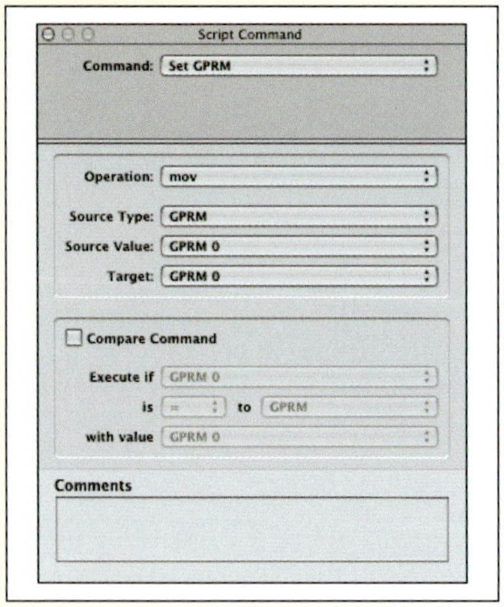

Figure A-15. The Set GPRM command

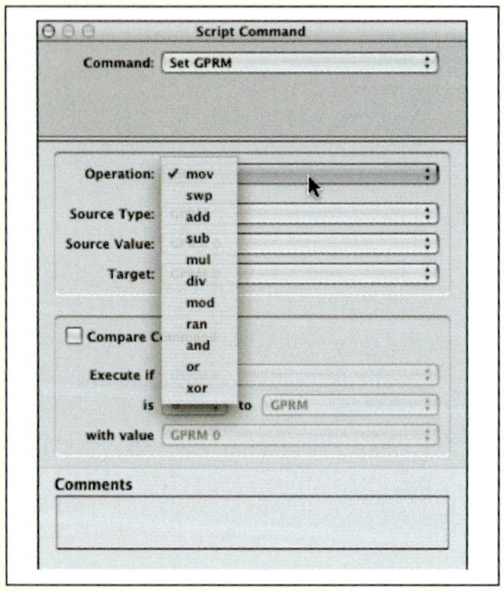

Figure A-16. The Set command operations

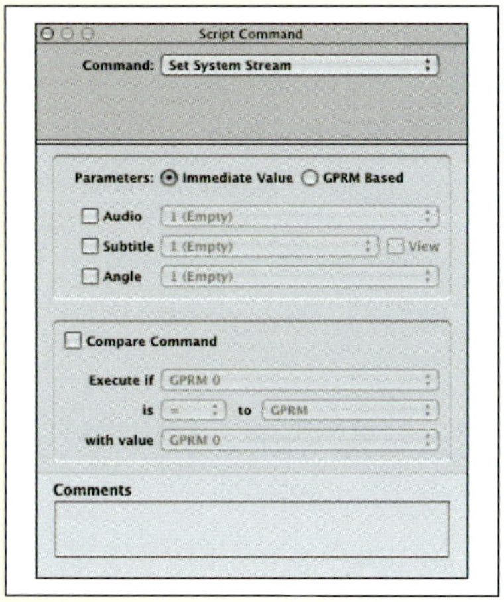

Figure A-17. Switching system streams

Set GPRM

You use `Set GPRM` (Figure A-15) to store a value in a variable or to perform a calculation.

This command shows up in almost every script. You use it to initialize variables, alter their contents, and store jumps for future use. You can store the values of the SPRMs for safe keeping as well. Note that this is the `Set GPRM` command. Although it's tempting to use this command to alter the writable SPRM entries, other commands (such as `Set System Stream`) are responsible for those operations.

Set operations

You have 11 operations to choose from when setting a variable's value. The pop-up list of these operations is shown in Figure A-16.

Table A-3 details those operations.

Set System Stream

You can switch system streams (audio, video, or subtitles) with this command (see Figure A-17).

This is one of the commands that directly affect the writable SPRM entries. For example, switching to the second audio stream would set SPRM 1 to 2. You can set each different stream type with one command. You can use Immediate values (the default) or GPRM variables. If you use GPRM variables, they should contain the number of the stream you want to use.

Also note that you can turn on the forced subtitles with the View checkbox. Leaving that box unchecked will cause the player to follow the user's current configuration for displaying subtitles (i.e., if subtitles are on, the new stream will be shown, but if subtitles are shut off, this stream change will have no effect on the playback).

Table A-3. The Set GPRM operations

Operation	Name	Description
mov	Move or assign	Use this command to store a value in a variable. You can use all five value types with this command.
swp	Swap	This command swaps the values in two variables. This can be useful in programming more complex logic. This command uses only the GPRM source type.
add	Add to	Adds the source value to the target. This command can use GPRM and Immediate source types.
sub	Subtract from	Subtracts the source value from the target. This command can use GPRM and Immediate source types.
mul	Multiply by	Multiplies the source value and the target (and stores the results in the target). This command can use GPRM and Immediate source types.
div	Divide by	Divides the target by the source and stores the results in the target. This command can use GPRM and Immediate source types.
mod	Modulus of	This command divides the target by the source and stores *the remainder* in the target. This command can use GPRM and Immediate source types.
ran	Random	Generates a random number and stores it in the variable. The Source Value argument is used as the upper range. The random number will be between 1 and that value. The source type must be GPRM or Immediate.
and	Bitwise AND	ANDs the source bits with the target bits and stores the results in the target. (Every bit that is on in *both* parts remains on in the result.) This command can use GPRM, SPRM, and Immediate source types. Usually used to test the value of the target for a particular bit.
or	Bitwise OR	ORs the source bits with the target bits and stores the results in the target. (Every bit that is on in *either* part remains on in the result.) This command can use GPRM, SPRM, and Immediate source types. Usually used to set a particular bit in the target.
xor	Bitwise XOR	ORs the source bits with the target bits and stores the results in the target. (Every bit that is on in *either but not both* parts remains on in the result.) This command can use GPRM, SPRM, and Immediate source types. Usually used to toggle a particular bit in the target.

Language Codes

Table A-4 contains the various language codes that pop up in the values of SPRM 0 (menu language), SPRM (audio language), and SPRM 18 (subtitle language). This table lists the 136 possible choices alphabetically by language. Table A-5 lists the same 136 choices, but they are arranged alphabetically by the two-letter language code.

Table A-4. Language codes by language

Language	Code	Decimal	Hex
Abkhazian	ab	24930	6162
Afar	aa	24929	6161
Afrikaans	af	24934	6166
Albanian	sq	29553	7371
Amharic	am	24941	616D
Arabic	ar	24946	6172
Armenian	hy	26745	6879
Assamese	as	24947	6173
Aymara	ay	24953	6179
Azerbaijani	az	24954	617A
Bashkir	ba	25185	6261
Basque	eu	25973	6575
Bengali; Bangla	bn	25198	626E
Bhutani	dz	25722	647A
Bihari	bh	25192	6268
Bislama	bi	25193	6269
Breton	br	25202	6272
Bulgarian	bg	25191	6267
Burmese	my	28025	6D79
Byelorussian	be	25189	6265
Calician	gl	26476	676C
Cambodian	km	27501	6B6D
Catalan	ca	25441	6361
Chinese	zh	31336	7A68
Corsican	co	25455	636F
Croatian	hr	26738	6872
Czech	cs	25459	6373
Danish	da	25697	6461
Dutch	nl	28268	6E6C
English	en	25966	656E
Esperanto	eo	25967	656F

Table A-4. Language codes by language

Language	Code	Decimal	Hex
Estonian	et	25972	6574
Faeroese	fo	26223	666F
Fiji	fj	26218	666A
Finnish	fi	26217	6669
French	fr	26226	6672
Frisian	fy	26233	6679
Georgian	ka	27489	6B61
German	de	25701	6465
Greek	el	25964	656C
Greenlandic	kl	27500	6B6C
Guarani	gn	26478	676E
Gujarati	gu	26485	6775
Hausa	ha	26721	6861
Hebrew	iw	26999	6977
Hindi	hi	26729	6869
Hungarian	hu	26741	6875
Icelandic	is	26995	6973
Indonesian	in	26990	696E
Interlingua	ia	26977	6961
Interlingue	ie	26981	6965
Inupiak	ik	26987	696B
Irish	ga	26465	6761
Italian	it	26996	6974
Japanese	ja	27233	6A61
Javanese	jw	27255	6A77
Kannada	kn	27502	6B6E
Kashmiri	ks	27507	6B73
Kazakh	kk	27499	6B6B
Kinyarwanda	rw	29303	7277
Kirghiz	ky	27513	6B79
Kirundi	rn	29294	726E
Korean	ko	27503	6B6F
Kurdish	ku	27509	6B75
Laothian	lo	27759	6C6F
Latin	la	27745	6C61
Latvian	lv	27766	6C76
Lettish	lv	27766	6C76
Lingala	ln	27758	6C6E

Table A-4. Language codes by language

Language	Code	Decimal	Hex
Lithuanian	lt	27764	6C74
Macedonian	mk	28011	6D6B
Malagasy	mg	28007	6D67
Malay	ms	28019	6D73
Malayalam	ml	28012	6D6C
Maltese	mt	28020	6D74
Maori	mi	28009	6D69
Marathi	mr	28018	6D72
Moldavian	mo	28015	6D6F
Mongolian	mn	28014	6D6E
Nauru	na	28257	6E61
Nepali	ne	28261	6E65
Norwegian	no	28271	6E6F
Occitan	oc	28515	6F63
Oriya	or	28530	6F72
(Afan) Oromo	om	28525	6F6D
Pashto	ps	28787	7073
Persian	fa	26209	6661
Polish	pl	28780	706C
Portuguese	pt	28788	7074
Punjabi	pa	28769	7061
Pushto	ps	28787	7073
Quechua	qu	29045	7175
Rhaeto-Romance	rm	29293	726D
Romanian	ro	29295	726F
Russian	ru	29301	7275
Samoan	sm	29549	736D
Sangro	sg	29543	7367
Sanskrit	sa	29537	7361
Scots Gaelic	gd	26468	6764
Serbian	sr	29554	7372
Serbo-Croatian	sh	29544	7368
Sesotho	st	29556	7374
Setswana	tn	29806	746E
Shona	sn	29550	736E
Sindhi	sd	29540	7364
Singhalese	si	29545	7369
Siswati	ss	29555	7373

Table A-4. Language codes by language

Language	Code	Decimal	Hex
Slovak	sk	29547	736B
Slovenian	sl	29548	736C
Somali	so	29551	736F
Spanish	es	25971	6573
Sudanese	su	29557	7375
Swahili	sw	29559	7377
Swedish	sv	29558	7376
Tagalog	tl	29804	746C
Tajik	tg	29799	7467
Tamil	ta	29793	7461
Tatar	tt	29812	7474
Tegulu	te	29797	7465
Thai	th	29800	7468
Tibetan	bo	25199	626F
Tigrinya	ti	29801	7469
Tonga	to	29807	746F
Tsonga	ts	29811	7473
Turkish	tr	29810	7472
Turkmen	tk	29803	746B
Twi	tw	29815	7477
Ukrainian	uk	30059	756B
Urdu	ur	30066	7572
Uzbek	uz	30074	757A
Vietnamese	vi	30313	7669
Volapuk	vo	30319	766F
Welsh	cy	25465	6379
Wolof	wo	30575	776F
Xhosa	xh	30824	7868
Yiddish	ji	27241	6A69
Yoruba	yo	31087	796F
Zulu	zu	31349	7A75

Table A-5. Language codes by code

Code	Language	Decimal	Hex
aa	Afar	24929	6161
ab	Abkhazian	24930	6162
af	Afrikaans	24934	6166
am	Amharic	24941	616D
ar	Arabic	24946	6172
as	Assamese	24947	6173
ay	Aymara	24953	6179
az	Azerbaijani	24954	617A
ba	Bashkir	25185	6261
be	Byelorussian	25189	6265
bg	Bulgarian	25191	6267
bh	Bihari	25192	6268
bi	Bislama	25193	6269
bn	Bengali; Bangla	25198	626E
bo	Tibetan	25199	626F
br	Breton	25202	6272
ca	Catalan	25441	6361
co	Corsican	25455	636F
cs	Czech	25459	6373
cy	Welsh	25465	6379
da	Danish	25697	6461
de	German	25701	6465
dz	Bhutani	25722	647A
el	Greek	25964	656C
en	English	25966	656E
eo	Esperanto	25967	656F
es	Spanish	25971	6573
et	Estonian	25972	6574
eu	Basque	25973	6575
fa	Persian	26209	6661
fi	Finnish	26217	6669
fj	Fiji	26218	666A
fo	Faeroese	26223	666F
fr	French	26226	6672
fy	Frisian	26233	6679
ga	Irish	26465	6761
gd	Scots Gaelic	26468	6764
gl	Calician	26476	676C

Table A-5. Language codes by code

Code	Language	Decimal	Hex
gn	Guarani	26478	676E
gu	Gujarati	26485	6775
ha	Hausa	26721	6861
hi	Hindi	26729	6869
hr	Croatian	26738	6872
hu	Hungarian	26741	6875
hy	Armenian	26745	6879
ia	Interlingua	26977	6961
ie	Interlingue	26981	6965
ik	Inupiak	26987	696B
in	Indonesian	26990	696E
is	Icelandic	26995	6973
it	Italian	26996	6974
iw	Hebrew	26999	6977
ja	Japanese	27233	6A61
ji	Yiddish	27241	6A69
jw	Javanese	27255	6A77
ka	Georgian	27489	6B61
kk	Kazakh	27499	6B6B
kl	Greenlandic	27500	6B6C
km	Cambodian	27501	6B6D
kn	Kannada	27502	6B6E
ko	Korean	27503	6B6F
ks	Kashmiri	27507	6B73
ku	Kurdish	27509	6B75
ky	Kirghiz	27513	6B79
la	Latin	27745	6C61
ln	Lingala	27758	6C6E
lo	Laothian	27759	6C6F
lt	Lithuanian	27764	6C74
lv	Latvian, Lettish	27766	6C76
mg	Malagasy	28007	6D67
mi	Maori	28009	6D69
mk	Macedonian	28011	6D6B
ml	Malayalam	28012	6D6C
mn	Mongolian	28014	6D6E
mo	Moldavian	28015	6D6F
mr	Marathi	28018	6D72

Table A-5. Language codes by code

Code	Language	Decimal	Hex
ms	Malay	28019	6D73
mt	Maltese	28020	6D74
my	Burmese	28025	6D79
na	Nauru	28257	6E61
ne	Nepali	28261	6E65
nl	Dutch	28268	6E6C
no	Norwegian	28271	6E6F
oc	Occitan	28515	6F63
om	(Afan) Oromo	28525	6F6D
or	Oriya	28530	6F72
pa	Punjabi	28769	7061
pl	Polish	28780	706C
ps	Pashto, Pushto	28787	7073
pt	Portuguese	28788	7074
qu	Quechua	29045	7175
rm	Rhaeto-Romance	29293	726D
rn	Kirundi	29294	726E
ro	Romanian	29295	726F
ru	Russian	29301	7275
rw	Kinyarwanda	29303	7277
sa	Sanskrit	29537	7361
sd	Sindhi	29540	7364
sg	Sangro	29543	7367
sh	Serbo-Croatian	29544	7368
si	Singhalese	29545	7369
sk	Slovak	29547	736B
sl	Slovenian	29548	736C
sm	Samoan	29549	736D
sn	Shona	29550	736E
so	Somali	29551	736F
sq	Albanian	29553	7371
sr	Serbian	29554	7372
ss	Siswati	29555	7373
st	Sesotho	29556	7374
su	Sudanese	29557	7375
sv	Swedish	29558	7376
sw	Swahili	29559	7377
ta	Tamil	29793	7461

Table A-5. Language codes by code

Code	Language	Decimal	Hex
te	Tegulu	29797	7465
tg	Tajik	29799	7467
th	Thai	29800	7468
ti	Tigrinya	29801	7469
tk	Turkmen	29803	746B
tl	Tagalog	29804	746C
tn	Setswana	29806	746E
to	Tonga	29807	746F
tr	Turkish	29810	7472
ts	Tsonga	29811	7473
tt	Tatar	29812	7474
tw	Twi	29815	7477
uk	Ukrainian	30059	756B
ur	Urdu	30066	7572
uz	Uzbek	30074	757A
vi	Vietnamese	30313	7669
vo	Volapuk	30319	766F
wo	Wolof	30575	776F
xh	Xhosa	30824	7868
yo	Yoruba	31087	796F
zh	Chinese	31336	7A68
zu	Zulu	31349	7A75

DVD Style Guide

B

Although none of the recommendations that follow is a hard and fast rule, you should consider each of them for your projects. Over time the recommendations you like and appreciate will become second nature. In the meantime, you can use this appendix as a checklist.

Many of the items listed here have counterparts throughout the book that provide more details or information on implementing the recommendations. Where appropriate, we've tried to include cross references to make it easy to go back and refresh your memory.

You also can take a look at the Style Guide section of the DVD included with this book. It contains several examples (and nonexamples!) of the recommendations listed here. You can see for yourself that tiny fonts in bad colors make for poor menus.

In this chapter

Asset Organization

You should get into the habit of organizing your assets in a consistent manner. You can use folders to organize by function (graphics, video, audio, etc.). You can duplicate this structure in the Assets tab in DVDSP and other applications such as Final Cut Pro and After Effects. It makes locating and editing the different parts of your projects much simpler.

You also should work on planning out your project before just diving in. Many of the projects in this book are tiny one-menu tasks that illustrate a particular feature. Larger projects always will benefit from an outline of where you're going. This is especially true when getting into scripting and Easter eggs.

Menus

The first thing a viewer sees on a DVD is your menu. (Well, maybe the FBI warning, but still....) Making sure your menus function well is just as important as making them look good. Here are a few points to keep in mind when designing your menus:

- Use legible fonts. Different fonts have different thresholds for legibility. Sans-serif fonts such as Arial or Gill Sans can go much smaller than something fancy such as handwriting or script fonts.

  Three pixels are a good threshold for the thickness of any part of your text. Less than three might be fine, but three or more always should be safe.

- Select a reasonable contrast between the foreground and background colors. Some colors don't mix and match very well. It's important to remember that many of your DVDs actually will be played in consumer players and on consumer television sets. That equipment won't always look as nice as your 23-inch Cinema Display!

- If you use motion menus (audio, video, or both), make sure the menu is at least 20 seconds long. When players reach the end of a motion menu, it usually takes a second or two to jump back to the beginning of the menu. During that pause, the user cannot interact with the menu. A one-second loop followed by a one-second pause can make for a very frustrating menu!

- Every submenu should have a direct link back to its parent menu. Most DVD players don't have a "back" button, so you need to provide it. Although most remote controls will have a button to the top-level menu, it's a good idea to make explicit buttons for basic navigation whenever possible.

  But be careful here! This is an area where you can have too much of a good thing. You can have too many buttons on a menu. You want to strike a balance between functionality and appearance.

Color Schemes

Colors have a large effect on the usability of your DVD content. High-contrast combinations with sharp edges can end up being difficult to read because television sets can't cope with the crisp edges. You'll see a shimmer around the letters.

Of course, the biggest rule here is to use NTSC-safe colors if you're headed for display on a television set. Regardless of the image (photo, menu, text, etc.), the colors shouldn't be outside the NTSC range.

Something else to remember is that NTSC displays augment the contrast in any image. You can plan ahead and keep your image a bit below whatever levels you usually use and it should come out nicely on a TV set.

With this in mind, here are a few points to keep in mind when designing color schemes:

- Stay away from bright colors. White and red in particular cause problems. NTSC white actually is closer to the RGB values of (234, 234, 234), not the (255, 255, 255) of computers. In fact, that 234 number is a good cap to use for all your colors.
- At the other end of the spectrum, too much black also is a bad thing. The low end should be (16,16,16), not (0, 0, 0).
- For text, light text on a dark background seems to be the most readable. You still can achieve the "paper feel" with dark text on a light background if you need it for your design; just use a gray or beige rather than a hard white.

Also, be sure to test everything on a television set. And use a cheap one!

Subtitles

When building subtitle streams, you should consider the content of the subtitles as well as how DVDSP works with subtitles. We'll look briefly at both topics.

Subtitle content

When considering the content of subtitles, it's important to remember that subtitles are not closed captions. They are meant to be a translation (or at least an interpretation) of a film's dialog. Subtitles are aimed at viewers who can hear the music and distinguish between the actors who are speaking. Closed captions provide a transcript of the audio portion of a film for viewers who are hearing impaired.

Many DVDs blur the lines, though. The English subtitle track often is used for both subtitles and closed caption information. Some discs use subtitles as an alternative to the more costly process of closed captioning. You certainly can do the captioning work on your computer, although DVDSP can import only closed caption files—it has no built-in creation or editing tools.

With all that said, it is clearly a matter of choice whether your subtitles are merely dialog or function more like closed captions. Many TV shows released on DVD create their subtitle tracks from the closed captions that were built for the original broadcast. Many films have no such track and produce the more traditional dialog translation subtitles.

If you're planning to go with caption-like subtitles, here are a few standard practices to keep in mind:

- Action sounds and music cues are described in square brackets [].
- Off-screen dialog or sounds often are shown in italic.
- When two separate actors have speech in a single caption, each actor's lines often begin with a dash ("-") to provide visual separation.
- Dialog also can be prefaced with the name of the character who is speaking (e.g., "PETER: Be there in a minute!"). This can be useful when the character is not onscreen or is in a crowd.
- When a sentence is spread out over multiple captions, ellipses are used as connectors. You often see one set at the end of the first caption and then another set at the beginning of the second caption.
- Singing or music often is described with a musical note (one or two eighth notes, usually) bracketing the text.

Subtitles in DVDSP

In DVDSP, you should take advantage of the features available to make subtitles as simple and efficient as possible. You want to spend your subtitling time on the content, not on getting the subtitles into your DVD project.

Here are a few practices to keep in mind when creating subtitles in DVDSP:

- Get into the habit of building STL files, even for small projects. DVDSP 2 has no export mechanism for subtitle tracks, so editing or rebuilding video can mean retyping all the subtitles. Deleting a track and reimporting an edited text file is easy.
- Configure the defaults for subtitles (font, duration, etc.) in the DVDSP preferences so that you don't have to worry about them. (Refer back to Figure 5-52 in Chapter 5.)
- Remember that you can place subtitles anywhere on the screen. (Captions often rely on this feature.)
  - If you are creating caption-like subtitles, place them at the top of the screen whenever interesting action is occurring at the bottom of the screen.
  - Traditional subtitles do not take advantage of this ability. They stay at the bottom to provide consistency.

Compression

It's not a coincidence that the words *compression* and *compromise* look a lot alike. Compression is all about compromising. You should ask yourself a series of questions. The answers will help determine your compression rate. Of course, waiting for terabyte discs to come out is a fine way to avoid the compromise, but in the meantime, consider these questions.

How much space do you have?
: Are you building a DVD-5, -9, -10, or -18 project? If you have space, use it! You also need to remember to account for any DVD-ROM content that might be associated with the project.

How long is your video/audio content?
: A 20-minute corporate video can use a hefty 8Mbps constant bit rate encoding with no problem. If you're encoding a feature-length film, you need to combine this question with the first one. A DVD-9 (like many commercial DVDs) easily can hold 100 minutes of high-quality video. A DVD-5 (for those of us on home-movie projects) will have to go with a lower-quality encoding.

What's in the video?
: Is it a talking-head video or a sports game? Here the variable bit rate encoding can do wonders. You can have an average of 4Mbps, while peaking at 7Mbps for the action sequences.

: The two-pass VBR encoding that's available turns out some nice MPEG-2 files. On any big project, you should spend the time it takes to make the second pass.

What's in the audio?
: Is it a single speaker in a lecture hall or is it an orchestral score with laser blasts and angst-ridden cries of lifelong commitment? You always should head for A.Pack (or your own encoding application), but think about the purpose of the audio. You can encode speaking at lower rates without adverse affects. If you've got a large collection of corporate presentations that are going onto a single disc, steal a bit of space from the audio tracks to give the video tracks a bit more room.

You also should get into the habit of logging your decisions. If you come back to a project, make a tweak, and then go to re-encode it, you'll know exactly what settings to use. It also will help you start building empirical knowledge on what works, what is merely acceptable, and what is unacceptable.

Tools such as Compressor can help in that regard by enabling you to store compression settings directly in the application. If you're stuck with QuickTime export (not a bad thing!), you can just use a plain-text file or maybe a spreadsheet.

DVD-ROM Content

The big thing to remember when supplying DVD-ROM content is that you have to consider that different computer systems have different defaults for showing data. Most computers these days can handle the UDF filesystem, so you should be in good shape, but if you want maximum compatibility, go for the 26.3 naming conventions discussed in Chapter 9. To recap those rules:

- Use only letters, numbers, and the underscore character in the name (no spaces!).
- Files can have 26 characters in the name plus a dot (".") and a three-character extension.
- Directories cannot have an extension.

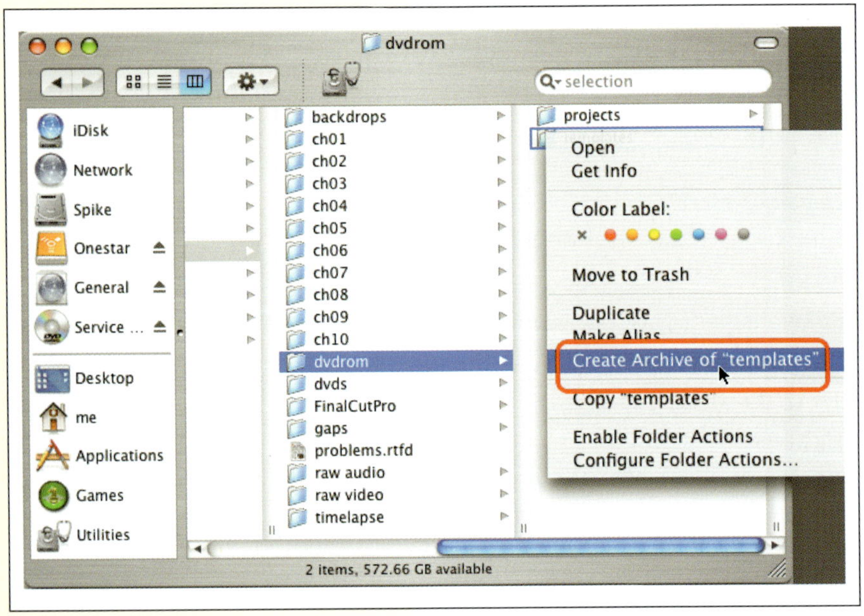

Figure B-1. Archiving a folder to create a ZIP file for inclusion on a DVD

For data headed to both Macs and PCs, you can use the Joliet extensions to help support longer filenames.

You also can use the Create Archive of ... feature in the Finder to build a single ZIP file that will preserve all filenames and structure for delivery to any machine. To use this feature, just Control-click any folder and choose the archive option, as shown in Figure B-1.

Note that you'll still need to place the ZIP file into its own directory, as DVDSP requires you to specify a directory to include. You can't supply a single file.

Templates

There's not a lot to say about the best thing to do with templates (and styles, for that matter) except this: *use them!* Even if you know your buttons will be used in only one project, make a project-only style out of them. You'll be pleasantly surprised at how easy it is to add a last-minute menu for your clients after they OK the "final" copy of the disc.

And we did lie a bit. There's another thing to say about templates and styles. When possible, make them self-contained. This does take up space, but space is cheap and time is not. One round of "spring cleaning" can render an entire project useless. DVDSP politely copies things out of the reach of mortal users, and you always will have the necessary assets available for future projects.

If you must keep the assets linked, just make sure you use a common location so that it's easy to maintain the links. A shared network folder can be a great place to store such things; just make sure no one else can delete them. (Making shared assets read-only also is a good idea.)

Another clever thing about templates is that they can serve as a rudimentary form of copy-and-paste in DVDSP 2.x. There is no facility (yet!) for copying a button and duplicating it in a different menu. Trying to move things to a different project is even more difficult. Making custom styles and templates can be an efficient way to copy pieces of one project and include them in another. When you're done, you always can delete the template or style if you really don't need it.

C

DVD-9 Projects

Back in Chapter 2, we mentioned that dual-layer burners were imminent—so imminent, in fact, that they became available in July 2004. Not only did they become available, but also, they launched at a street price of about $240! Quite the introduction, don't you think?

In this chapter

If you rushed out to buy one of the new models, you might have been a bit disappointed that DVDSP wouldn't work with it. Or more precisely, that DVDSP could not use the DL media. But fear not, you do have options.

Toast 6 from Roxio can burn to dual-layer media. The media itself still was expensive (around $12/disc) when we went to press, but that price should drop as manufacturers catch up. Obviously, it's still cheaper to use two single-layer discs when you can, but it sure is convenient to put long movies (or several short ones) on one disc. This chapter goes through the steps required to build a DVD-9 project in DVDSP. We'll also show you how to burn an image to a DL disc via Toast.

Dual-Layer Details

First, let's take a quick look at what makes dual-layer media work. It uses essentially the same setup as single-layer media, but using two layers does involve a few quirks. Figure C-1 recaps the physical construction of single-sided, dual-layer media that we looked at briefly in Chapter 2.

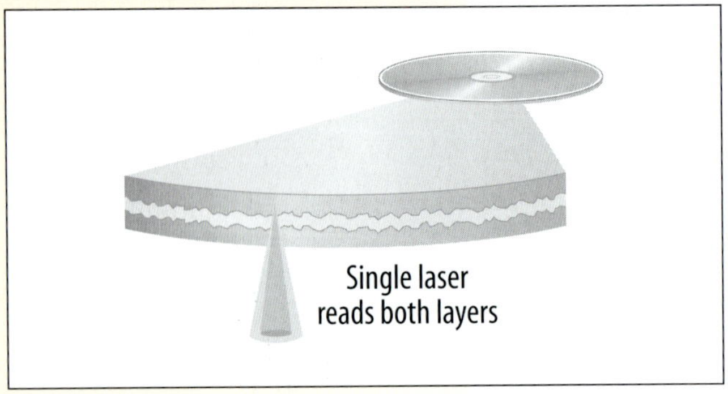

Figure C-1. Dual-layer media recap

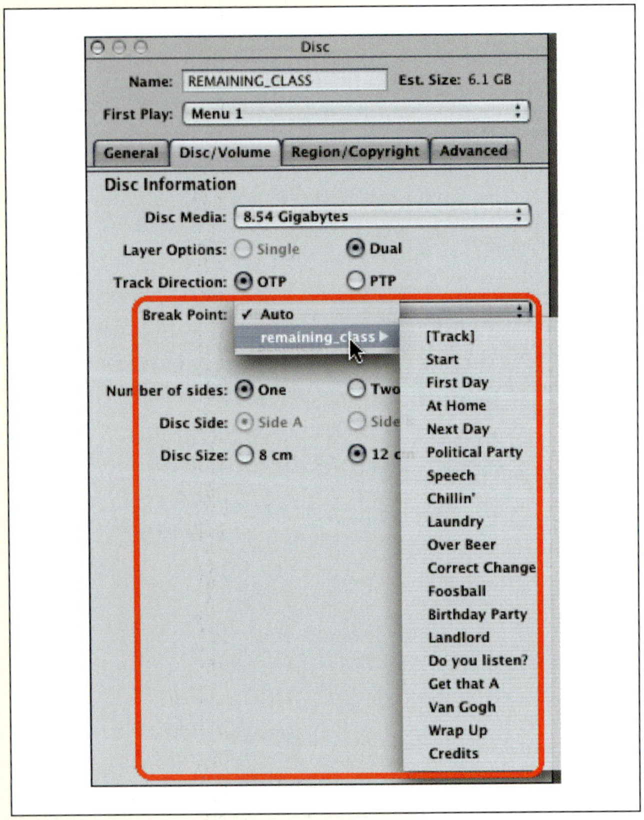

Figure C-2. Break point options in DVDSP for dual-layer discs

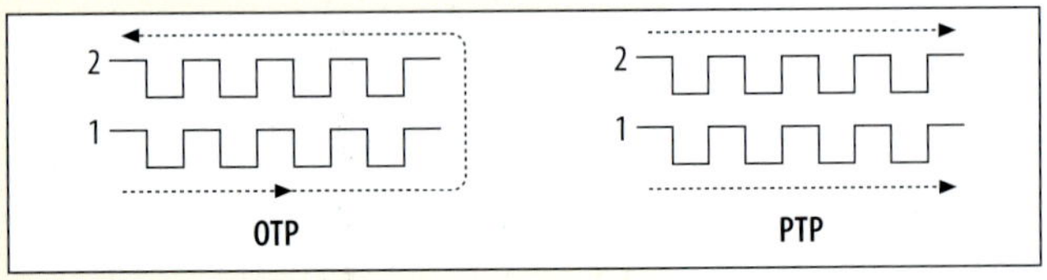

Figure C-3. Opposite versus. parallel track path layouts for dual layers

The two layers use slightly different specs to help keep the tracks distinguishable. The pit length goes up from 400 nanometers on a single-layer disc to 440nm on both layers of a dual-layer disc. That's why a dual-layer disc has only 8.54GB instead of 9.4GB of space. Still, 8.54GB is more than enough to hold a big movie. Even the 194-minute film *Titanic* fit on a dual-layer disc with room left for three audio tracks and a few extras.

Layer breaks

If your movie spans both layers, at some point the DVD player switches over from the first layer to the second. You've probably noticed that switch point while watching commercial DVDs. There's often a slight pause in the movie about halfway through. That's the layer change.

DVDSP offers two options for setting the layer break point. You can let DVDSP do it for you automatically (which is great if you have several small, separate clips), or you can specify a particular chapter marker to serve as the break point. Figure C-2 shows the Disc Inspector window and highlights the controls for this option.

If you pick your own marker, you need to make sure it's a valid marker in the middle. In other words, you can't have 5.6GB of your movie on the first layer. That said, if you have a perfect fade-to-black break point after only 3.5GB, don't hesitate to waste a bit of space on the first layer.

Track path directions

The other decision you must make in the dual-layer world is which direction the second layer should use to read/write its data. The most common option is known as *opposite track path* (see Figure C-3). This option allows the single reader head to stay put when switching layers. For long movies with a break in the middle, you'll definitely want this option. The parallel track path option causes the second layer to start back at the "beginning" of the disc. This option is required if you have more data in the second layer than in the first.

Figure C-4 shows the Disc Inspector window with the Track Direction controls called out. The default is opposite track path (OTP).

Of course, you'll be able to choose a track direction only if you're using dual-layer media. (The Single/Dual Layer Options choice was made for us when we selected the 8.54GB option in the Disc Media list.)

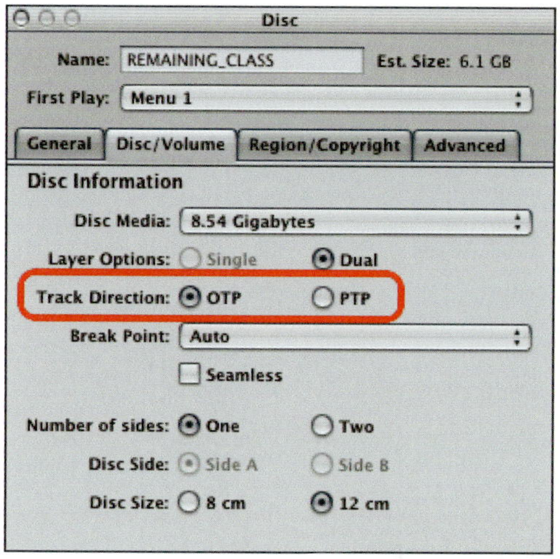

Figure C-4. Selecting OTP or PTP in DVDSP

Building a DVD-9 Project

To prepare your movie for burning to a dual-layer DVD, you still encode the video in MPEG-2 and the audio in AC3. The only thing you really need to do differently is to think about that break point. Of course, now you can encode longer movies at much higher quality. You can get nearly four hours of video at VHS quality, or two hours of video at top-notch quality.

The rest of the project will proceed exactly like a traditional project with menus and tracks (and scripts and stories). We put together an example project using the independent film *Remaining Class*, which we covered in Chapter 6. At just less than two hours, it's a perfect candidate for high-quality encoding and burning to a dual-layer disc.

We encoded the film at a constant bit rate of 7Mbps. The file weighed in at 5.7GB. (The audio was a standard AC3 file and took up about 375MB of space.) We just dropped the clips into a standard track via the handy auto-chaptering mechanism we looked at back in Chapter 4. Figure C-5 shows the Outline view of our big project. The estimated size is highlighted in the toolbar.

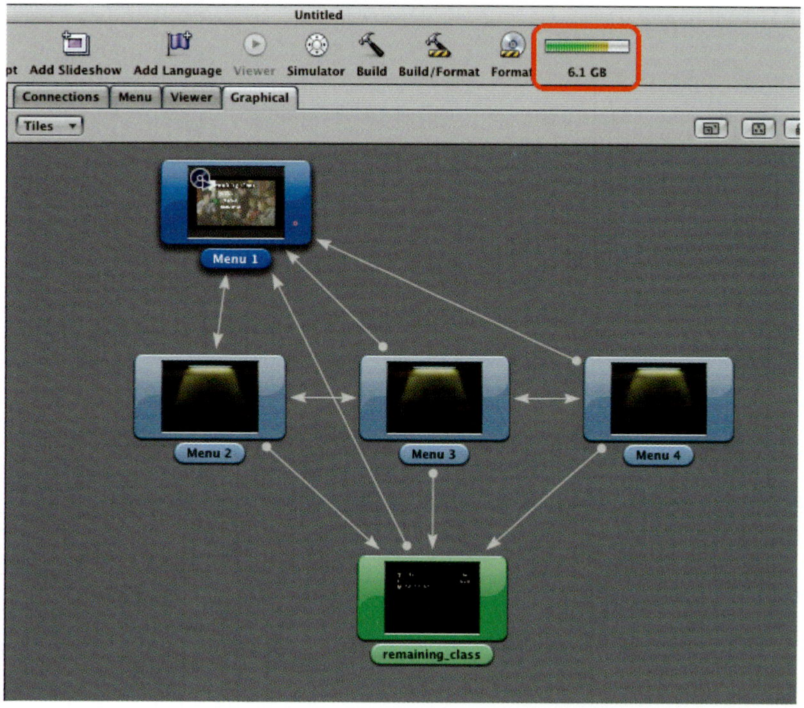

Figure C-5. Our long (but simple) film project

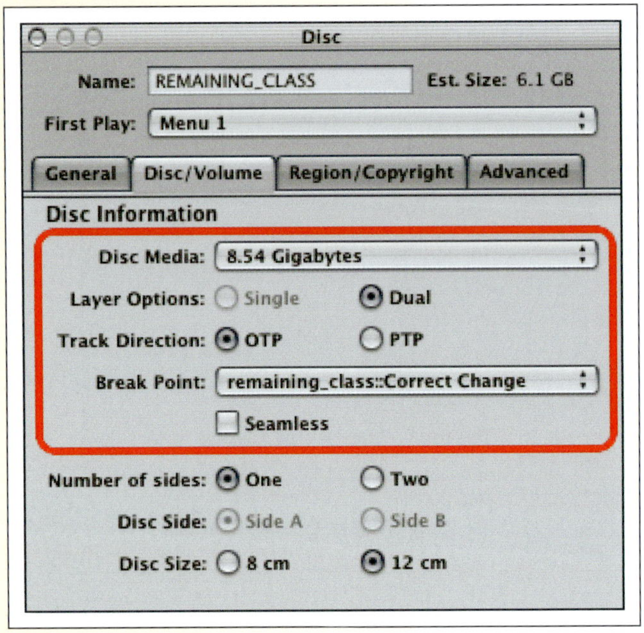

Figure C-6. Setting up the dual-layer options

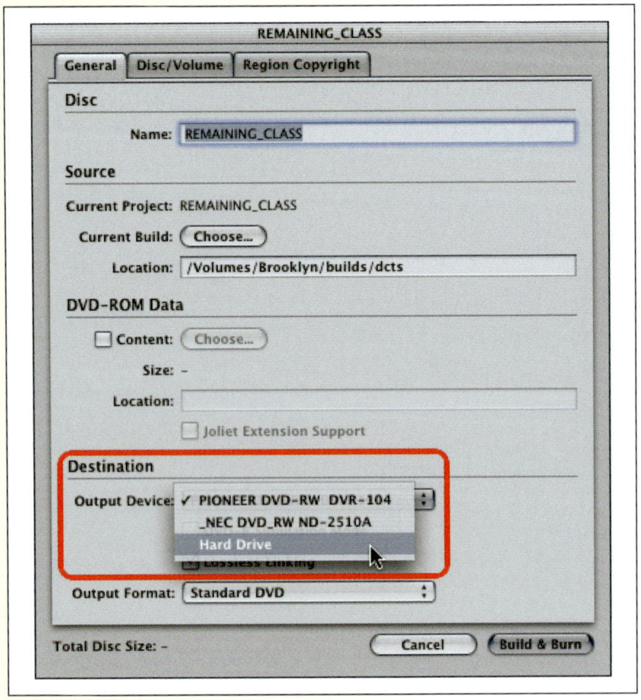

Figure C-7. Writing from DVDSP to a hard-disk image

We want to use OTP for our layout and set a specific break point. Figure C-6 highlights these choices in the Disc Inspector window.

Notice in Figure C-6 that we did not check the Seamless checkbox. The DVDSP manual warns against using this option. Not all players handle it properly, so we left it off to maximize compatibility.

Burning a Disc Image

As of DVDSP 3.0.1 (running on Mac OS X 10.3.5), DVDSP did not support dual-layer media. But it does support burning a dual-layer project to disc. That's exactly the route we'll take. Figure C-7 shows the Build & Format window. Notice that the destination is simply the hard drive and not the dual-layer burner.

The default output format of *.img* is exactly what we want. That's it; go ahead and burn the disc. You'll be prompted to name the file. We chose *rc.img*. We'll use that image in Toast 6.

Toast 6 to the rescue

We used a LaCie d2 dual-layer external FireWire drive. It shipped (quite proudly, at that) bundled with Toast 6. You can pick up Toast 6 separately as well. We had to install LaCie's drivers to help OS X recognize the drive for use as a burner with the various iLife products. For the disc itself, we used Verbatim 8.54 dual-layer +R media.

Toast 6 knows all about disc images. To get started we simply dragged our *rc.img* file onto the Toast application window. Figure C-8 shows the results.

After that, we clicked the Burn button and waited about an hour. Out popped a complete disc. It played fine in all the consumer set-top players and computers we had available.

You might not have a single movie that merits 8GB of space, but you also can use the dual-layer discs to hold more individual clips. You'll probably want to rely on the automatic break point feature for that scenario, though.

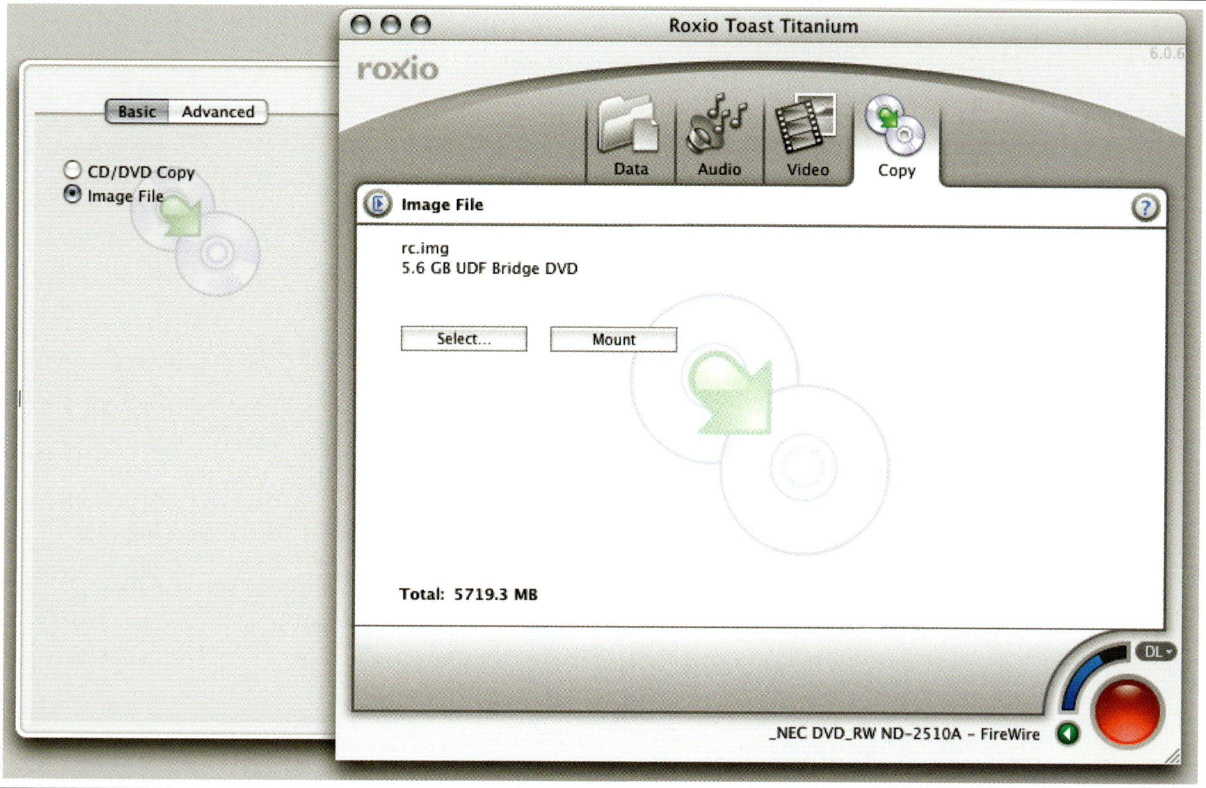

Figure C-8. Burning a dual-layer image from Toast 6

It's worth pointing out that DVDSP finally started supporting +R media (and +R drives, obviously) in version 3.0.1. We fully expect support of dual-layer media will show up directly in DVDSP eventually. In the meantime, Toast 6 can get you through the process with only minimal extra effort. And besides, as long as the cost of dual-layer media remains so high, you'll want to burn your DVD to disk anyway to make sure the burn is clean. Dual-layer coasters just aren't as much fun to make!

Index

Symbols

A

B

C

D

L

M

N

T

U

V

W

Z

About the Author

Marc Loy is a trainer and media specialist in Madison, Wisconsin. He has worked on or near computers for the past two decades. While he had fun programming web sites with Java, digital media represents his future. He can still be found teaching the occasional Java or Perl class out in the wild, but rest assured that he'll have his PowerBook and a video or animation project with him even when he's on the road.

Colophon

Our look is the result of reader comments, our own experimentation, and feedback from distribution channels. Distinctive covers complement our distinctive approach to technical topics, breathing personality and life into potentially dry subjects.

Genevieve d'Entremont was the production editor for *DVD Studio Pro 3: In the Studio*. Audrey Doyle was the copyeditor, and Leanne Soylemez was the proofreader. Rick Schlott, David Futato, and Genevieve d'Entremont did the typesetting and page makeup. Claire Cloutier and Emily Quill provided quality control. Caitrin McCullough and Sanders Kleinfeld provided production assistance. Reg Aubry wrote the index.

Emma Colby designed the cover of this book using Adobe Photoshop CS and Adobe InDesign CS, and produced the cover layout with Adobe InDesign CS using Linotype Birka and Adobe Syntax fonts.

Melanie Wang designed the interior layout, based on a series design by David Futato. This book was converted from Microsoft Word to Adobe InDesign CS by Julie Hawks. The text and heading fonts are Linotype Birka and Adobe Myriad Condensed; the sidebar font is Adobe Syntax; and the code font is TheSans Mono Condensed from LucasFont. The illustrations and screenshots that appear in the book were produced by Robert Romano and Jessamyn Read using Macromedia Freehand MX and Adobe Photoshop 7.

Related Titles Available from O'Reilly

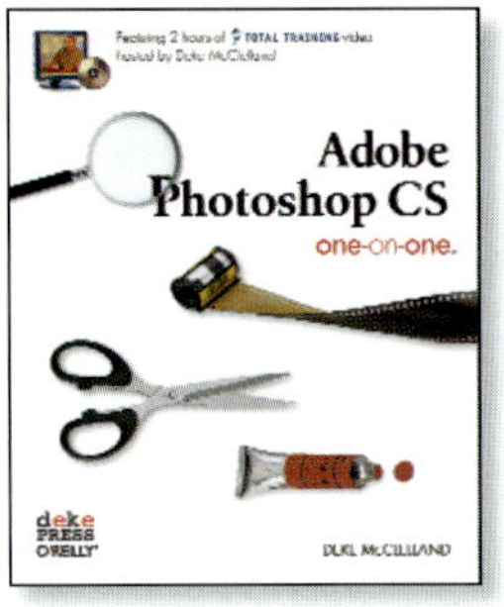

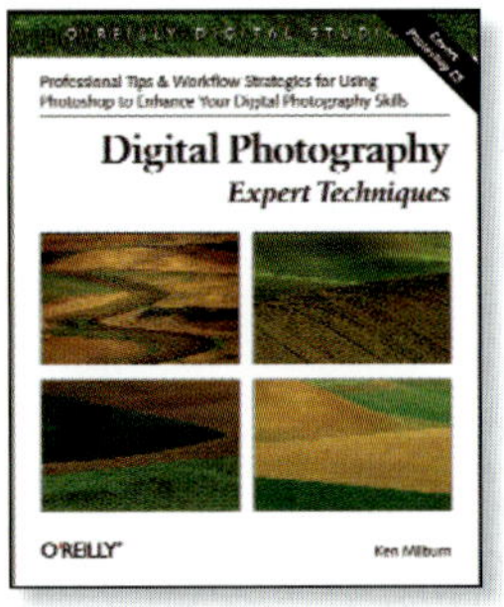

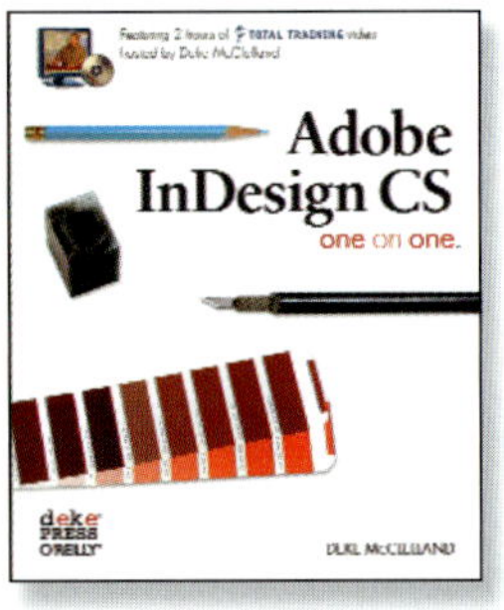

Digital Media

Adobe InDesign CS One-on-One
Adobe Encore DVD: In the Studio
Adobe Photoshop CS One-on-One
Creating Photomontages with Photoshop: A Designer's Notebook
Digital Photography: Expert Techniques
Digital Photography Hacks
Digital Photography Pocket Guide, *2nd Edition*
Digital Video Pocket Guide
DVD Studio Pro 3: In the Studio
Illustrations with Photoshop: A Designer's Notebook
In the Loop with Soundtrack
iPod & iTunes: The Missing Manual, *2nd Edition*
Photo Retouching with Photoshop: A Designer's Notebook
Windows Media Hacks

Keep in touch with O'Reilly

1. Download examples from our books

To find example files for a book, go to:

www.oreilly.com/catalog

select the book, and follow the "Examples" link.

2. Register your O'Reilly books

Register your book at *register.oreilly.com*

Why register your books?
Once you've registered your O'Reilly books you can:

- Win O'Reilly books, T-shirts or discount coupons in our monthly drawing.
- Get special offers available only to registered O'Reilly customers.
- Get catalogs announcing new books (US and UK only).
- Get email notification of new editions of the O'Reilly books you own.

3. Join our email lists

Sign up to get topic-specific email announcements of new books and conferences, special offers, and O'Reilly Network technology newsletters at:

elists.oreilly.com

It's easy to customize your free elists subscription so you'll get exactly the O'Reilly news you want.

4. Get the latest news, tips, and tools

www.oreilly.com

- "Top 100 Sites on the Web"—PC Magazine
- CIO Magazine's Web Business 50 Awards

Our web site contains a library of comprehensive product information (including book excerpts and tables of contents), downloadable software, background articles, interviews with technology leaders, links to relevant sites, book cover art, and more.

5. Work for O'Reilly

Check out our web site for current employment opportunities:

jobs.oreilly.com

6. Contact us

O'Reilly & Associates
1005 Gravenstein Hwy North
Sebastopol, CA 95472 USA

TEL: 707-827-7000 or 800-998-9938
(6am to 5pm PST)

FAX: 707-829-0104

order@oreilly.com
For answers to problems regarding your order or our products. To place a book order online, visit:

www.oreilly.com/order_new

catalog@oreilly.com
To request a copy of our latest catalog.

booktech@oreilly.com
For book content technical questions or corrections.

corporate@oreilly.com
For educational, library, government, and corporate sales.

proposals@oreilly.com
To submit new book proposals to our editors and product managers.

international@oreilly.com
For information about our international distributors or translation queries. For a list of our distributors outside of North America check out:

international.oreilly.com/distributors.html

adoption@oreilly.com
For information about academic use of O'Reilly books, visit:

academic.oreilly.com